The Urban Book Series

The Urban Book Series is a resource for urban studies and geography research worldwide. It provides a unique and innovative resource for the latest developments in the field, nurturing a comprehensive and encompassing publication venue for urban studies, urban geography, planning and regional development.

The series publishes peer-reviewed volumes related to urbanization, sustainability, urban environments, sustainable urbanism, governance, globalization, urban and sustainable development, spatial and area studies, urban management, transport systems, urban infrastructure, urban dynamics, green cities and urban landscapes. It also invites research which documents urbanization processes and urban dynamics on a national, regional and local level, welcoming case studies, as well as comparative and applied research.

The series will appeal to urbanists, geographers, planners, engineers, architects, policy makers, and to all of those interested in a wide-ranging overview of contemporary urban studies and innovations in the field. It accepts monographs, edited volumes and textbooks.

Indexed by Scopus.

Fernando Carrión Mena ·
Sebastián Rodríguez Alvarez
Editors

From the City as a Project to the City Project

Volume I: History

Springer

Editors
Fernando Carrión Mena
Inter-University Alliance URBS.TIC
Quito, Ecuador

Sebastián Rodríguez Alvarez
Inter-University Alliance URBS.TIC
Quito, Ecuador

ISSN 2365-757X ISSN 2365-7588 (electronic)
The Urban Book Series
ISBN 978-3-031-94225-9 ISBN 978-3-031-94226-6 (eBook)
https://doi.org/10.1007/978-3-031-94226-6

This Springer imprint is published by the registered company Springer Nature Switzerland AG
The registered company address is: Gewerbestrasse 11, 6330 Cham, Switzerland

Contents

Editors and Contributors

About the Editors

Fernando Carrión Mena is a Research Professor at the Interuniversity Alliance URBS.TIC. His focus of study are the topics of: housing, urbanization process, city, historical centers, cultural heritage, violence, security and drug trafficking, borders, decentralization and sociology of soccer, among others. He created 8 thematic magazines (political science, security, city, historical centers, borders), wrote more than 1000 journalistic articles and 306 academic articles, published 74 books (editor of 52 and author of 22) and edited 12 book collections (97 volumes). He produced 4 film documentaries. He has worked as a consultant for multilateral organizations and as a university professor. For his career he received 9 awards, 6 decorations and 5 distinctions of honored citizen. He was director of CIUDAD, FLACSO, Municipal Planning, as well as former Councilman and former Advisor of the Municipality of Quito. e-mail: fcarrion@flacso.edu.ec; URL: https://works.bepress.com/fernando_carrion/.

Sebastián Rodríguez Alvarez is a Researcher at the Interuniversity Alliance URBS.TIC. Master in Urban Studies with specialization in Geography and Territorial Processes from FLACSO Ecuador and Geographer from the Pontificia Universidad Católica del Ecuador. Consultant in diverse projects of land use planning, environmental management and urban mobility at local and national levels. His academic experience is related to the development of research projects on urban ecology and modeling of territorial processes. His research interests include Geographic Information Science, Data Science, Quality of Urban Life, Socio-Spatial Segregation and Urban Violence. Part of the POLIS.TIC networks. e-mail: msrodriguezfl@flacso.edu.ec.

Contributors

Paola Alfaro d'Alençon Frankfurt University of Applied Sciences, Frankfurt, Germany

Daniel Astorga-Poblete University of La Serena, La Serena, Chile

Alex Ivan Usca Baca Machu Picchu National Archeological Park, Cusco, Peru

Víctor Jiménez Barrado Departamento de Geografía. Universidad de Las Palmas de Gran Canaria, Las Palmas de Gran Canaria, Spain

Júlia Bianchi University of Brasilia, Brasilia, Brazil

Ana Elisa Carnaúba University of Brasilia, Brasilia, Brazil

Fernando Carrión Mena Inter-University Alliance URBS.TIC, Quito, Ecuador

Jorge González Aragón Castellanos Autonomous Metropolitan University - Xochimilco, Mexico City, Mexico

Francisco Colom-González Spanish National Research Council (CSIC), Madrid, Spain

Marcelo Corti National University of Córdoba, Córdoba, Argentina

Inés del Pino Martínez Pontifical Catholic University of Ecuador, Quito, Ecuador

Maribel Aliaga Fuentes University of Brasilia, Brasilia, Brazil

Jesús M. González-Pérez University of the Balearic Islands, Palma, Spain

Žaklina Grgić University of Georgia, Athens, GA, USA

Sonia Hirt University of Georgia, Athens, GA, USA

Mustafa Kahraman Istanbul University, İstanbul, Türkiye

Eduardo Kingman-Garcés Department of Anthropology, History and Humanities of FLACSO Ecuador, Quito, Ecuador

Fernando Luiz Lara University of Pennsylvania, Philadelphia, PA, USA

Rosa Cristina Laureiro Quito, Ecuador

Rubén C. Lois-González University of Santiago de Compostela, Santiago de Compostela, Spain

Lola Martínez-Fons Universidad de Sevilla, Sevilla, Spain

Francisco Javier Haro Navejas University of Colima, Colima, Mexico

Rodrigo Castro Orellana Complutense University of Madrid, Madrid, Spain

Salima Ouchen Rabat, Morocco

Antoine Paccoud LISER, Esch-Sur-Alzette, Luxembourg

Juan Manuel Parreño-Castellano Departamento de Geografía. Universidad de Las Palmas de Gran Canaria, Las Palmas de Gran Canaria, Spain

Helihanta Rajaonarison University of Antananarivo, Antananarivo, Madagascar

Amit Ranjan Research Fellow, Institute of South Asian Studies, National University of Singapore, Singapore, Singapore

Gina Rey Havana, Cuba

Albert Roca University of Lleida, Lleida, Spain

Sebastián Rodríguez Alvarez Inter-University Alliance URBS.TIC, Quito, Ecuador

Mireya Salgado-Gómez Department of Anthropology, History and Humanities of FLACSO Ecuador, Quito, Ecuador

Benny Schvarsberg University of Brasilia, Brasilia, Brazil

Yulia Shaffer University of Georgia, Athens, GA, USA

Part I
Introduction

Chapter 1
The City as a Project: A Historical Overview

Fernando Carrión Mena and Sebastián Rodríguez Alvarez

> Since it has taken more than five thousand years to arrive at even a partial understanding of the city's nature and drama, it may require an even longer period to exhaust the city's still unrealized potentialities.
> Lewis Mumford (1961, pp. 3–4).

Abstract This introductory chapter outlines the book's objectives and presents the underlying argumentative premise that guides the editorial production: the city as a project. This implies the systematization of the various city projects developed on a planetary scale throughout history, with the objective of establishing a basis for the delineation of new perspectives and alternatives for contemporary urban development. The structural axes of the thematic to be developed are formulated, starting from the affirmation that the city has been reflected from a sense of anticipation, of what it is aspired to be. This is an important interest of the text, as it aims to highlight the significance of the city's anticipation dynamics.

Keywords City · Project · Urban history · Urban theory

1.1 Introduction

The city has experienced a multitude of transformations from its earliest days to the present, a phenomenon that can be attributed to its status as a historically determined product. In its genesis and subsequent evolution, the city has been shaped by a multitude of factors, including the actions of diverse individuals and groups and

F. Carrión Mena · S. Rodríguez Alvarez (✉)
Inter-University Alliance URBS.TIC, Quito, Ecuador
e-mail: msrodriguezfl@flacso.edu.ec

F. Carrión Mena
e-mail: fcarrion@flacso.edu.ec

F. Carrión Mena and S. Rodríguez Alvarez (eds.), *From the City as a Project to the City Project*, The Urban Book Series, https://doi.org/10.1007/978-3-031-94226-6_1

the spatial organization of its urban fabric. These elements contribute to the city's intricate and multifaceted character (Mumford 1961).

While the earliest cities exhibit considerable diversity, this is also true of those that emerged subsequently. Nevertheless, despite the diversity observed in urban development, a common starting point can be identified: cities have evolved in accordance with a sense of desire or a desired future that has not been homogeneous.

The city has been conceptualized in terms of a sense of "duty to be," as well as in terms of a sense of project. In other words, the prefiguration of a future reality that should emerge from the negation of the existing urban context. Consequently, the city has been the subject of reflection from a prospective standpoint, with the concept of the project at its core. The term "project" has its etymological roots in Latin. It is composed of two elements: the prefix "pro," meaning "forward," and the verb "iacere," meaning "to throw" or "to pull." Thus, the word "proiectum" literally means "something thrown forward." It is the prefiguration of a future reality from the negation of a present one; that is, from the transformation and overcoming, in this case, of the Urbicide of the present city (Carrión 2023).

It is a matter of anticipating or prefiguring a desire, emerging from an idealized perspective that tends to overcome situations perceived as anomalous and thus facilitate the understanding of the city, based on perception or diagnosis, with the city of the present as a point of departure. In some contexts, the proposals were implemented, whereas in others, they were not, remaining as mere ideas. Nevertheless, they all served to foster social mobilization, facilitate consensus-building, promote knowledge exchange, anticipate outcomes, and ultimately contribute to urban development. It can be argued that this represents the historical trajectory of urban thinking on a planetary scale. It can thus be asserted that the prospective has been a dominant dynamic in urban thought throughout history, with urban planning emerging as a central concept (Hall 2014).

Similarly, it is challenging to assert that all cities have formal proposals. Nevertheless, it can be stated that each city has been putting forth solutions to its particular or structural issues within the context of its historical development. Furthermore, the influence of key actors is a significant factor in this process of change, as they are themselves undergoing a transformation.

In examining the historical development of urban settlements, it becomes evident that two fundamental characteristics have persisted throughout the centuries, continuing to influence the formation and evolution of cities in the present era:

- The initial transition from a nomadic to a sedentary lifestyle was made possible by two key developments: the domestication of animals and the advent of agriculture. This represents a historical turning point, namely the permanent settlement of the population in a specific territory, which was made possible by the overcoming of the need to move in order to obtain the food necessary for their survival. The establishment of a fixed population in a particular area led to the emergence of a distinction between the agricultural and pastoral zones, which came to be known as the countryside, and the settled residential areas of the city. From this point onwards, the city became the area with the highest population density, and the rural/urban relationship became a differentiating factor that gave rise to a sense of opposition between the city and the countryside, from its earliest stages of development (Capel 1975).

- The second factor was the quality of sedentarism, which was concomitant with specialization. This resulted in a new division of labor (Silveira 2009), which was linked to trades in the areas of production, commerce, and services. As a result, social and urban differences began to emerge in the nascent city, as each of the trades possessed distinctive qualities and yielded disparate economic returns. Consequently, the city became a conglomeration of individuals engaged in diverse pursuits, thereby establishing itself as a hub of considerable heterogeneity (Wirth 1938).

The city became the most densely populated and heterogeneous space on the planet, representing the most complex social product produced by humanity. Furthermore, it is a historical product that has contributed to historical processes, to such an extent that it constituted the fundamental basis for the emergence of the nation-state and currently serves as the foundation for globalization.

This entire process occurs within the context of the accumulation of prospective knowledge among urban actors, such that city projects have been a constant presence since their foundation. However, given the diversity of activities that were promoted and the population that was concentrated in that space, it was inevitable that their visions of the future would also be diverse. It can therefore be stated with certainty that from the outset, heterogeneity has been a defining feature of the city, giving rise to a multitude of visions for its future.

In developing this concept, the expository structure of this book does not adhere to the linear logic of the city's evolutionary timeline. Instead, it employs a method of articulating historical entries (periods), locations (continents), and themes to reinforce the central thesis of the duty to be of each city.

From the confluence of time and space, city projects have emerged, elucidated through a synthesis of pivotal theoretical and methodological advancements. For example, there are cases of the Classical City, the Modern City, and cities in Asia and Africa. The cities in question were selected on the basis of their prior designation as analytical typologies. Based on their description, the reflections and proposals were organized into a systematic framework to identify the emblematic projects of the city throughout history, thereby advancing from a conceptualization of the city as a project to a conceptualization of the city project.

This process is documented in the two volumes of this collection, entitled *From the City as a Project to the City Project*, each of which is inscribed in two different historical moments. The initial volume, entitled *The City as a Project: History*, is the result of the systematic aggregation of proposals that have been conceived throughout history. The second volume, entitled *The City Project: XXI Century*, presents the proposals currently being implemented. The two volumes include a compilation of cases from Latin America, Europe, Asia, Africa, and the United States. In essence, it is a compilation of research contributions made by a group of highly esteemed academics.

From this perspective, the universe of the selected cities has become a living catalog of city projects that are worthy of systematic analysis in order to identify more effective directions. This vision extends beyond the mere description of each city case, as the value of this editorial undertaking lies in the comparative compilation of all of them.

This Vol. 1 represents a preliminary attempt to categorize and organize the most significant city projects throughout history. The second volume not only corroborates the preceding analysis but also reaffirms the centrality of project dynamics. Since the fin-de-siècle period, there has been a proliferation of city projects seeking to address the significant challenges they are confronted with. Such projects include platform, democratic, feminist, or intercultural cities, which are solutions to problems arising from historical factors, the emergence of new actors, the development of technology, and other issues. These projects are driven by a range of social actors, including social movements, local governments, academics, and politicians, who play varying roles in the process.

As postulated by Eduardo Galeano (2001) about Utopia:

> She is on the horizon (…) I get closer, she moves away. I walk ten steps and the horizon runs ten steps further. No matter how far I walk, I will never reach her. What is utopia for? To walk.

1.2 The City as a Project Throughout History

As Mumford (1961) observed, the city is not merely a consequence of history; it is also a historical product. This line of thinking allows for the city to be understood and transformed in a rational manner, based on the fact that each moment of its evolution has resulted in a proposal for its future. In other words, the history of the city has seen the formulation of explicit solutions to urban challenges or proposals for urban transformation.

An interdisciplinary approach is essential to understanding the urban question, given its historical and distinctive nature. If this is the case, we can identify at least the following major historical periods and their respective proposals for the future: the ancient city, with its projects based on specific religious and cultural hierarchical components; the classical city, with utopias as a project; the industrial revolution, with urban planning as a discipline; modernity, with urban planning as a proposal; and globalization, with urban development linked to technology. This process can be delineated from the following historical typology of cities.

The ancient city was constructed according to a specific set of religious and cultural principles. These structuring axes are based on the movements of the sun and moon, as well as on the cyclical patterns of nature, and are oriented in accordance with the four cardinal points. In general, these cities are hierarchical in structure, exemplified by the paradigmatic cases of Mesopotamian and Egyptian cities.

The classical city represents a significant starting point of classical political thought (polis), where the theses of Aristotle (384–322 BC) are particularly notable in his work Politics (1984). In this text, the philosophical foundations of the city are elucidated. The following trio of definitions is structured there: *polis* (politics), *urbs* (space), and *civitas* (citizenship), articulated from the perspective of the good life. The classical thinkers not only marked the urban development of their time, but their ideas remain relevant today, at least in the cities of the West.

The thesis of the ideal as set forth by Thomas More (1478–1535) is presented in his book Utopia (2012), which was written in 1516. In this work, More delineates the concept of the ideal place or the place that does not exist (island) as a means of exploring the nature of a perfect society. Another key figure in this utopian tradition is Tommaso Campanella (2007), whose ideas are synthesized in his book The City of the Sun, written while imprisoned in 1602. These reflections emerged within the context of the major historical processes of urban consolidation, with the cities of Rome, Athens, and Florence serving as key reference points.

The modern city, with the advent of the Industrial Revolution, gave rise to Urbanism as a discipline that embodies and implements the vision of a city project. In this case, the impetus for urban planning was driven by the necessity to mitigate the accelerated growth of cities, attenuate the rural/urban contradiction, and reduce the environmental pollution that emerged from factory production. The initial expression of this concept can be seen in the differentiation of the workplace (factory) from the place of residence (housing), which gave rise to the concepts of zoning and land use. These were to become central elements of what was later designated as Urban Planning. In consequence, proposals were put forth by Haussmann in Paris, Ring in Vienna, and Cerdà in Barcelona, among others (Hein 2017).

In the late twentieth and early twenty-first centuries, Urban Planning emerged as a formal discipline, initially consolidated by Le Corbusier (1887–1965) within the framework of the International Congresses of Modern Architecture (CIAM). This initiative facilitated the global expansion of urban planning (Massey 2017). In the course of this evolution, the remnants of classical, walled, and industrial cities were discarded, and from that point forward, the functionalist concept of the city came to the fore.

The global city, typical of the globalization process, refers to current or contemporary city projects. In this sense, the emergence of the global city can be attributed to a number of factors, including the advancement of technology, which has driven the transformation of the material base of the city; the emergence of a new pattern of economic accumulation, which has reconfigured the trends of socio-spatial (re)production; and, most recently, the impact of the global pandemic of 2020, which has exacerbated urban contradictions and highlighted the importance of the socio-temporal dimensions of cities, leading to the formation of the Global Urban Network (Carrión 2024).

In the context of the historical process of cities, the meaning of urban development, which has historically been associated with the spatial dimension (*urbs*), is being reframed to encompass the social (*civitas*) and political (*polis*) dimensions. It is also crucial to recognize that urbanism and urban planning must be understood in a pluralistic manner, as there is not a singular urbanism or urban planning approach, but rather a multitude of them. The process of urbanization in Europe, Asia, Latin

America, and Africa is distinct, necessitating the formulation of specific proposals tailored to the unique circumstances of each region.

It is beyond question that technology has played a pivotal role in the historical evolution of urban settlements. The advent of the wheel, the steam engine, electricity, the internet, and artificial intelligence has had a direct impact on the functioning and development of cities throughout the world, although the impact has been observed to vary in different regions.

1.3 The Structure of the Book

The book, entitled The City as a Project: History, is structured around ten sections: an introduction, a section on theoretical perspectives, seven sections presenting various cases, and an epilogue. This results in a total of 26 articles, written by 34 authors from 4 continents.

In the second section of the book, entitled ***Central Ideas of the City as a Project***, a comprehensive theoretical, methodological, and critical reflection is presented. This serves as a framework for the subsequent analysis of the selected city cases. In order to achieve this, a systematic analysis is conducted of the principal concepts that have been employed to conceptualize the city from a variety of perspectives, including the prospective, decolonial, utopian, and historical. These fundamental lines of argument enable us to interpret the era and its expression in the selected emblematic cases. The main thinkers who marked the period are present, as are the axes that induced the proposals. This section is structured on the basis of five contributions.

The initial text, entitled *The City From Duty to Be*, is authored by Fernando Carrión Mena and Sebastián Rodríguez Alvarez. It examines and critiques the concepts of *civitas*, *urbs*, and *polis*, while also presenting a series of illustrative case studies exploring various city projects. In this manner, the text reflects on the hypothesis that the city has historically been conceptualized from a sense of project or anticipation, that is, from a sense of duty to be.

Fernando Luiz Lara offers a critical reinterpretation of the concept of the city as a project in his text *Decolonial Perspectives in the idea of city as a project*. Using a decolonial framework, he argues that the "project" of European colonization was a significant driving force behind the production of Latin American cities. This is in contrast to the position that the cities are the result of predetermined geometric projects.

In *City as Utopia*, Francisco Colom González examines the various connotations of Western character associated with the concept of the city, both in its broader urban context and in relation to the ideal of utopia. From classical antiquity to modern urbanism, the author traces the evolution of these associations, analyzing how the idea of utopia has served as a foundational element in the representation of urban ideals, from the Renaissance ideal city to the contemporary intelligent and sustainable city.

The authors of *Urban History and the Urban* are Eduardo Kingman Garcés and Mireya Salgado-Gómez. The text aims to illustrate how the city is a historical product and to underscore the central role that historical research, particularly urban history, plays in the contemporary understanding of cities. In this regard, the organization of social life, power relations, the production of subjectivities, and the socio-spatial configuration are examined from a historical perspective, encompassing strategies, ruptures, and overlaps.

The rural–urban relationship is a key factor in understanding the future of the city, as well as its origins and subsequent evolution. In their contribution, *The Evolution of Urban over Rural Areas*, Juan Manuel Parreño-Castellano and Víctor Jiménez Barrado examine the evolution of the relationship between urban and rural areas. They argue that from the origin of urbanization to the end of the twentieth century, this relationship shifted from an initial dependence on the rural to urban domination.

Sections three to nine constitute the core of the book. It presents a catalog of the most significant city projects, previously selected for examination. It offers a systematic account of the principal ideas and fundamental components of city projects developed throughout history on various continents.

A critical examination of the city projects is undertaken, with a particular focus on the historical context and the significance of the locations in question. It is based on a synthesis of diverse theoretical and methodological perspectives, as well as their empirical application in concrete cases. This study makes reference to the conceptualizations that have been employed throughout history in the prefiguration of cities, which have been previously defined by analytical typologies. It is enlightening to examine the most notable types of cities in order to gain insight into their conceptualization and prefiguration. The cases have been divided into seven sections according to historical, continental, and thematic criteria, as detailed in the following sections.

The third section describes ***Asian cities*** in accordance with the fundamental principles that underpin their development. To illustrate these principles, specific case studies are presented, including examples from Beijing, Mumbai, and Istanbul. Francisco Javier Haro Navejas's *Beijing: Ruptures and Endurances in an Asian intersocietal system* offers an analysis of Beijing's historical construction as a political capital. It does so by examining the complex historical process of Beijing's development, which has involved a series of overlaps, conflicts, and identity negotiations. In contrast, Amit Ranjan's *Postcolonial Mumbai: The Paradoxical City* examines the contradictory socioeconomic character of India's financial capital. Additionally, it assesses the dissolution of religious pluralism and the profound political influence and function that Hinduism has played in the city's historical trajectory. Finally, Mustafa Kahraman's *Istanbul: Constantinople* explores the evolution of Constantinople between the forth and fifteenth centuries, emphasizing its urban, commercial, military, and religious development following its designation as capital.

The section on ***African cities*** is presented from a central angle that defines them and discusses their characteristics in detail. It also critically examines the continent's city projects and presents the case of two cities: Antananarivo (the capital of Madagascar) and Nador (an intermediate Moroccan city). In her work, *Africa's cities as projects?* Lola Martinez-Fons challenges the applicability of the Western concept of the "city as project" to the African urban context. She argues that this abstract approach fails to account for the intricate and dynamic realities of urban living, which often challenge dominant Western rationalist paradigms. The case study of Antananarivo, Madagascar's capital city, illustrates the complexities of urban landscapes and the challenges of interpreting them. In their contribution, *Antananarivo, capital status and readings of the urban landscape Concepts of competing powers since the eighteenth century*, Albert Roca and Helihanta Rajaonarison present a historical overview of how the diversity of populations and socio-cultural contexts allow us to understand the multiplicity of city projects present in the capital of Madagascar. The second case study, presented by Salima Ouchen, is an analysis of the historical evolution of Nador (Morocco) based on the main changes in the city's architectural and urban planning trends. The case study is entitled *Nador: From Military Barracks to Intermediate City*.

Latin American cities illustrate an initial convergence between pre-Hispanic and European urban rationality. Furthermore, the urban developments of the Inca (Machu Picchu) and Aztec (Tenochtitlan) civilizations are discussed. In his work, *Colonial Cities: Vulnerability and Contingency*, Rodrigo Castro Orellana puts forth the hypothesis that colonial cities are not the direct result of classical-Cartesian-modern reason. Instead, he suggests that such urban dynamics were shaped by practical and conjunctural criteria that were responsive to a context of vulnerability. Alex Ivan Usca Baca addresses the case of the pre-Hispanic Inca cities. In *Historical context of the Q'ente and Santa Rita de Q'ente properties (Historic Sanctuary of Machupicchu) and its implications for the Machupicchu property*, the author provides a comprehensive account of the evolution of the Historic Sanctuary of Machupicchu from its origins in the colonial context of the seventeenth century to the agrarian reform processes of the last century and the management challenges currently faced. Tenochtitlan has a twofold entry. On the one hand, *Tenochtitlan: Agriculture, Urbanism, and Aztec Houses in the sixteenth Century* by Jorge González Aragón Castellanos provides a chronological account of the settlement process of the island of Tenochtitlán, as well as the settlement pattern recorded in the indigenous cartography of Aztec tradition. On the other hand, Daniel Astorga-Poblete's *The Plan of Tenochtitlan and the Symbolic Dominance of the Holy Roman Empire during the Diet of Nuremberg (1524)* challenges the publication of the Plan of Tenochtitlán in Nuremberg. The article demonstrates how the plan effectively conveyed the Holy Roman Empire's perceived dominance over the Aztecs.

The Classic City considers the implications of political community and utopian thought in order to elucidate the underlying rationale of its functioning. Rubén C. Lois-González's *The political community in classical urban thought (Utopia)* provides a historical analysis of the classical Mediterranean city, examining the idealization of fifth-century Athens as a model of civic urbanism and suggesting the expansion of urban research towards Asia and America. This would allow for a revaluation of their contributions to the history of urban development.

The Industrial City project resulted in the establishment of Urbanism as a distinct academic discipline, which subsequently underwent a process of global institutionalization. In this regard, Marcelo Corti's *The dialectics of industrial-immigratory city* endeavors to comprehend the industrial city as a technical-political project, predicated on three foundational elements: the substantial influx of immigrants that accompanied urban industrialization, the genesis of urban planning as a distinct scientific discipline, and the establishment of urban infrastructure, social housing, parks, boulevards, and other key contributions of this era. In *Haussmann and the disruption of the apparatus of capture in mid-nineteenth-century Paris*, Antoine Paccoud employs a case study approach to examine one of the core conflicts in urban development: the privatization of housing and land. The process of Haussmannization in Paris during the mid-nineteenth century is analyzed within this context.

The Modern City emerged in the early twentieth century as a result of the postulates promoted by CIAM, which sought to establish functionalist urbanism. The city of Brasília is presented as a paradigmatic case study. In accordance with these premises, Inés del Pino Martínez presents her argument in *Modern City. A Breaking Point in the History of Cities*, it is argued that the modern city project, as consolidated by CIAM, represented a significant turning point in the way of life in cities, as well as in their future growth. Despite the fact that these projects were not fully realized, they did, however, develop perspectives and debates that influenced modern urbanism. In a similar vein, Maribel Aliaga Fuentes, Benny Schvarsberg, Ana Elisa Carnaúba, and Júlia Bianchi examine the historical and documentary evidence in *From IAPI to Ceilândia: Occupation and Dislodgement in the Construction of Brasília*. They illuminate the tensions between the modern city project, which involved the construction of Brasília, and the daily experiences of the need for housing and the right to land.

Ultimately, ***The Socialist City*** serves as a significant theoretical point of reference, with notable examples including the cities of Berlin in East Germany, Moscow in the Soviet Union, and Havana in Cuba. In this way, the *Soviet and East European City during State Socialism: An Overview* by Zaklina Grgic, Sonia Hirt, and Yulia Fadina Schafer offers a comprehensive examination of urbanism and urban planning under state socialism in the Soviet Union and Eastern Europe. The authors present the socialist city as a unique project specific to this regime. In her analysis *Moscow, The Soviet City. The impact of the Soviet urban plan in the city of Moscow today* by Rosa Cristina Laureiro reviews, from architectural and urbanist perspectives, the metamorphoses that the Russian capital has undergone throughout the twentieth century, addressing the main challenges in terms of socioeconomic, political, and environmental inequalities that the city faces. As another case study, Paola Alfaro-d'Alençon examines the transition from *From the Industrial to the Post-industrial City. The Reunification of Berlin: Searching for the 'Centre' the New City Project*. This new city project analyzes the contrasting urbanistic and socio-political trajectories involved in the division and subsequent reunification project of the city of Berlin during and after the Cold War. Finally, Gina Rey's *Havana: Past, Present and Future* explores the genesis and evolution of the Cuban capital, mapping the diverse transformations that have shaped it since the city's embrace of socialism.

The concluding section comprises the ***Epilogue***, which synthesizes the findings obtained throughout the process and provides a summary of the journey undertaken. In his contribution, *The importance of the city as anticipation*, Jesús M. González-Pérez encourages readers to consider the historical development of urban planning and the emergence of urbanism. He argues that these processes have shaped a predominantly urban society, which faces significant challenges within its own urban context. Instead of initiating new debates, this section aims to present a series of reflections that facilitate a deeper understanding of the book's content.

References

Aristotle (1984) Politics. Chicago University Press, Chicago
Campanella T (2007) The city of the sun. Cosimo, Inc
Capel H (1975) La definición de lo urbano. Estudios Geográficos 36(138):265–302
Carrión F (2023) Urbicide. The liturgical murder of the city. In: Carrión F, Cepeda Pico P (eds) Urbicide. The urban book series. Springer, New York. https://doi.org/10.1007/978-3-031-25304-1_2
Carrión F (2024) La red urbana global: los ensambles que la constituyen. In: Vera F, Toro F, Mashini D (eds) Lo Metropolitano: escala, complejidad y gobernanza. Banco Interamericano de Desarrollo (BID)
Galeano E (2001) Las palabras andantes. Catálogos
Hall P (2014) Cities of tomorrow: an intellectual history of urban planning and design since 1880. Wiley, Amsterdam
Hein C (2017) The what, why, and how of planning history. In: The Routledge handbook of planning history. Routledge, New York
Massey D (2017) Key planning histories of the developing western tradition from the mid-19th century to the early 20th century. In: Hein C (ed) The Routledge handbook of planning history. Routledge, New York. https://doi.org/10.4324/9781315718996-13
More T (2012) Utopia. Penguin Books
Mumford L (1961) The city in history: its origins, its transformations, and its prospects. Harcourt Brace Jovanovich
Silveira ML (2009) Región y división territorial del trabajo: desafíos en el período de la globalización. Investigación y Desarrollo 17(2):434–455
Wirth L (1938) Urbanism as a way of life. Am J Sociol 44(1):1–24

Part II
Central Ideas of the City as a Project

Chapter 2
The City from the Duty to Be

Fernando Carrión Mena and Sebastián Rodríguez Alvarez

> But a city is more than a place in space, it is a drama in time.
> Patrick Geddes (2004, p. 108).

Abstract This chapter examines the notions of *civitas*, *urbs*, and *polis* as the "essence" of the city and the urban, while providing a brief overview of these concepts in the context of contemporary urban processes. The presentation of significant city projects throughout history is intended to illustrate the principles and proposals implicit in cases of cities belonging to different temporal coordinates. This approach supports the hypothesis that the city has been prefigured from a sense of project, that is to say, from the 'should be'. In this way, urban planning emerges as a technical-political tool for anticipating and overcoming structural and particular urban challenges of contemporary societies.

Keywords Civitas · Urbs · Polis · City project · Urban history · Urban planning

2.1 Introduction

It is estimated that by the middle of the twentieth century, <30% of the global population was residing in urban areas. By 2025, three-quarters of a century later, this figure is projected to reach 58.3% (UNDESA 2018). It can be stated, therefore, that this is an urban era.[1] This demographic transition, which saw the shift from an agricultural society to one primarily characterized by population agglomeration in urban areas, occurred during the initial decade of the twenty-first century. It was described by Davis (2004):

[1] For a critique of the Urban Age thesis see Brenner and Schmid (2014).

F. Carrión Mena · S. Rodríguez Alvarez (✉)
Inter-University Alliance URBS.TIC, Quito, Ecuador
e-mail: msrodriguezfl@flacso.edu.ec

F. Carrión Mena
e-mail: fcarrion@flacso.edu.ec

F. Carrión Mena and S. Rodríguez Alvarez (eds.), *From the City as a Project to the City Project*, The Urban Book Series, https://doi.org/10.1007/978-3-031-94226-6_2

> Sometime in the next year, a woman will give birth in the Lagos slum of Ajegunle, a young man will flee his village in west Java for the bright lights of Jakarta, or a farmer will move his impoverished family into one of Lima's innumerable *pueblos jóvenes*. The exact event is unimportant and it will pass entirely unnoticed. Nonetheless it will constitute a watershed in human history. For the first time the urban population of the earth will outnumber the rural. Indeed, given the imprecisions of Third World censuses, this epochal transition may already have occurred.

It is imperative to pause and contemplate the urban environment, as cities represent the primary hub of human activity and the primary conduit through which most individuals interact with nature (Pickett et al. 2013). One might reasonably inquire as to why anyone should concern themselves with the meaning of the term "urban." For the majority of everyday activities, this concept does not present any significant issues. Nevertheless, as soon as an attempt is made to elucidate the processes, objects, and imaginaries to which they are presumed to refer, such terms become polysemous and, consequently, almost enigmatic (Sayer 1984).

While it is relatively straightforward to identify an urban environment when it is observed, defining it is not always a simple process. It can be agreed that the term "urban" is a place-based characteristic, defined by the spatial agglomeration of people whose lives revolve around non-agricultural activities (Weeks 2010). The urban/rural binary opposition[2] that underlies this definition is of central importance. It can be stated that urban is equal to non-rural and rural is equal to non-urban. Accordingly, a settlement with an agricultural vocation and a population exceeding 5000 should not be classified as urban, whereas a tourist town with a population of <2000 may be designated as such.

An alternative approach to defining the concept is to consider an equation in which four elements are interrelated (Weeks 2010). The four elements that comprise this expanded definition are as follows: (1) population size/threshold; (2) spatial characteristics (fiscal-environmental); (3) the relationship between population and space (density); and (4) social, economic, and political structure. The accelerated pace of urban processes presents a challenge to this definition, simultaneously undermining the urban/rural opposition in the context of an increasingly unequal, fragmented, and digitalized planetary urbanization.

If the focus is not on the defining characteristics of a particular place, but rather on the place itself—the city—the conceptual delineation becomes equally intricate. Isin (2008) identifies three reasons for this difficulty: historical, geographical, and theoretical. With regard to the first one, it can be observed that each interpreter of the city is immersed in a historical context that provides and demarcates a specific perspective of the city. An Athenian in 594 BC would undoubtedly have observed the city from the tensions and disputes between the aristocratic and peasant classes. In contrast, a "European" of the fourteenth century would have interpreted his walled city from a perspective shaped by specific institutional, corporative, and territorial processes. Such variations are not only historical but also geographical. In the case of Lübeck (Germany) and Padua (Italy), for example, inhabitants of these cities

[2] For an examination of contradictions, tensions and implications of this opposition see Angelo (2017) and Baird (2022).

would have been aware of the significant differences between them in the fourteenth century. Similarly, an inhabitant of Tikal (Guatemala) would recognize Teotihuacan (Mexico) as a wholly distinct entity.

In addition to the historical and geographical difficulties, there are also those of a theoretical nature (Isin 2008). It has been assumed that in each particular spatio-temporal context, there existed a unified understanding of the city. The assumed unity is not borne out empirically, as each context presents a unique combination of actors, processes, and disputes that collectively constitute the city in question. This renders any theoretical perspective from which the city is interpreted necessarily partial and invariably fragmented.

This article aims to examine the historical origins of the "city," tracing its development from the concepts of Civitas, Urbs, and Polis in the West to its evolution as a concept in urban praxis. It then proposes that cities have been conceptualized as a "should be" in a sense of project, influencing the trajectory of urban thought.

The paper is divided into four sections. Following this introduction, it presents a discussion of the conceptualizations of *Civitas*, *Urbs*, and *Polis* as the "essence" of the urban, as well as a contemporary interpretation of these concepts. It then develops the thesis of the "city as a project" through an examination of the categories of ancient and modern planning ideologies, with a focus on several illustrative examples. Finally, some concluding thoughts are offered.

2.2 Civitas, Urbs and Polis

2.2.1 Three Approaches

The concept of the city has at least two distinct connotations when we examine its etymological roots in Greek and Latin. In classical Greek, the terms are *polis* and *asty*; in Latin, their equivalents are *civitas* and *urbs* (Jennings 2001). As posited by Isin (2008) and Grbac (2013), a significant aspect of the contemporary discourse on urbanism is shaped by the insights of Fustel de Coulanges (2003), Weber (1964), and Mumford (1961), particularly with regard to the analytical distinction between the concept of the city as *civitas* and the concept of the city as *urbs*.

Fustel de Coulanges (2003 [1864]) postulates that this distinction can be traced back to Greek and Roman philosophical traditions. While the term "civitas" was used to describe the religious and political association of family or tribal units, "urbs" was used to refer to the dwelling place and, most importantly, the sanctuary of this association. It is posited that the "essence" of the city was religious, whereby the unification of disparate tribes and the establishment of a shared cultus resulted in the founding of the city as a sanctuary for the practice of this cult. In light of this reasoning, it can be argued that the establishment of a city was inherently a religious act.

For this French historian, Greek and Roman thought deliberately sought to differentiate between *civitas* and urbs by maintaining that the city is first and foremost an association, which may (or may not) have a concrete spatial correlate. Consequently, Coulanges' most significant contribution to the urban debate was his assertion that the city cannot be reduced to its spatial attributes of concentration, distribution, and construction (*urbs*) (Isin 2008).

In his work, Weber (1964 [1921]) presents an "ideal type" of city, which he defines as a synthesis of the European medieval city and its Greek and Roman antecedents. Weber dismisses the notion of the city as merely a function of population size and instead delineates its essence through an examination of its economic, political, administrative, and military characteristics. In other words, he characterizes the city as a free and autonomous "urban community" composed of citizens who participate in decision-making through electoral means (Prato 2016).

While this approach enabled Weber to develop typologies such as the "producing city" or the "consuming city," his conceptualization is ultimately anchored in Coulanges' differentiation between civitas and urbs. For Weber, the city is not only a social association but also a space of collective identity and belonging, embodied by the condition of citizenship. In the view of Grbac (2013), this represents a significant contribution to the urban debate since German sociologist's urban debate, as it demonstrates that "only in the city is both the individual capable of developing associative dispositions of solidarity and the government capable of transforming the city into an association" (p. 23). In short, the city is a key site for the production and promotion of citizenship.

Mumford (1961) addresses two distinct conceptualizations of the city: the analytical distinction between *civitas* and *urbs* proposed by Coulanges and the Weberian understanding of the essence of the city based on the condition of citizenship. In regard to the latter, the American thinker argues that the modern Western city, by reducing citizens to a kind of subject, resurrects the despotic city. Conversely, his portrayal of the metropolis and megalopolis enabled him to elucidate their inherent tensions and limitations due to his assertion that the "essence" of the city is more closely aligned with *civitas* (association) than with *urbs* (space), despite the postulates of the Chicago School of Sociology that advocated otherwise (Wirth 1988 [1938]).

The approaches of these three authors collectively indicate an impetus to identify the "essence" of the city throughout the history of humanity. The concepts of "civitas" and "urbs" are beneficial in this endeavor, provided that they are interpreted as mutually reinforcing. The notion of "civitas" cannot be fully grasped without considering the concept of "urbs," and vice versa. In other words, when we use the term "cities," we are referring to actual geographical locations, whereas when we employ the term "the city," we are incorporating elements of those cities. However, the concept of "the city" is not limited to the characteristics of specific urban settlements. The city is a complex entity that cannot be reduced to the mere sum of its parts (Isin 2008).

The concept of *urbs* has undergone a significant transformation from its original function as a site for mass gatherings to its current role as a hub for economic activity and financial transactions. The loss of its religious association has contributed to this shift. In this way, the *urbs* has become a space where everything is a commodity and everyone is driven by the pursuit of material self-interest and instant gratification of desire. This is a space of anonymity, individualization, and perpetual reinvention of the self. In contrast, *civitas*, which Aristotle understands as *polis*, is associated with the political community established to pursue the good life and virtue. The existence of a city is contingent upon the presence of a shared moral and political identity among its citizens. Aristotle's concept of the *polis* is a civic structure ordered by principles of reciprocity, equity, and justice (Jennings 2001). The concepts of *civitas*, *urbs*, and *polis* enable us to comprehend the city as a complex entity, not solely as a mere aggregation of individuals in a geographical space. Instead, they allow us to perceive the city as an ethical, aesthetic, and political sphere of human existence.

In accordance with this line of reasoning, the city can be seen as the setting in which the rights and duties of citizenship are established on the basis of three fundamental elements. Firstly, the *polis*, that is to say, the space of participation and democracy, strives to represent the social collective in order to satisfy the political, social, cultural, and economic rights of citizenship. Subsequently, the *urbs* represents the insertion of the citizen in the city and is the space in which social rights are constructed. Finally, the *civitas* represents the space in which a community of meaning is constructed, symbolizing the citizen's right to identity and representation. In essence, the city is the setting in which the rights and duties of citizenship are established and exercised, and in which the citizen is constituted as the indispensable element of social life (Carrión 2010).

2.2.2 A Contemporary Perspective

If we conceptualize the city as a political community (*polis*), dialectically (re)produced in relation to its physical-material (*urbs*) and civic (*civitas*) context, it is possible to establish that when one or more of the structuring components of the system is altered, the system itself is also subject to change. In the context of neoliberal urban governance, the model of local government is transformed (*polis*), within the context of a deregulated market and the refunctionalization of the state. Concurrently, the urban structure mutates (*urbs*), within the context of the social function of space, public space, and nature is diminished. And society undergoes a transformation (civitas) in the context of a shift from a sense of citizenship to an emphasis on individual consumption (Carrión 2022).

With respect to the transformations of the contemporary *polis*, there is evidence of an important shift in both the institutional framework of the state and the design/implementation of urban policies beginning in the 1980s. This shift coincides with the reform of the state and the implementation of the neoliberal model. This resulted in the juxtaposition of various levels of public administration within an institutional complex, which, through the implementation of urban policies at multiple scales, led to the emergence of a multilevel government structure comprising local, intermediate, and national levels. In this sense, municipal decentralization processes have a distinctive impact on city governance in three key ways: they represent society directly; they establish a socially and politically legitimized institutional framework; and they define the municipal authority as the primary governing body for urban regulation.

Broadly speaking, at least two models of urban governance can be identified: one of a progressive kind and the other of a neoliberal nature. The one defined as progressive gives greater weight to the condition of citizenship (right to the city) and to the public sphere (institutional strengthening) (Carrión 2015). This leads to a local government that expands its field of urban intervention from a democratic and equitable perspective. In contrast, the refunctionalization of the state apparatus, in which the commodification of the city is privileged, constitutes the essence of the neoliberal model of urban governance. This model is framed within the so-called state reform, which has directly influenced local government by shifting the centrality of the sense of the public and the collective towards the strengthening of the private and the individual. This is within the framework of the principle of consumer sovereignty, which sees the public as a brake on development (Carrión 2022).

The reform of the local government structure resulting from this scenario tends to move towards privatization and free market. Municipalities facilitate the ascendance of corporations, companies, and foundations, fortifying sectoral policies with insular institutional frameworks and eschewing integration with comprehensive urban policies. This engenders fragmentation in municipal governance and obstructs the conceptualization and implementation of city projects. In this context, urban planning is concomitantly transformed through strategic planning (the ascendance of business agents in urban development) and project urbanism (large urban projects) (Carrión 2020).

Some of the most significant shifts in urban development can be traced back to at least the first third of the twentieth century. In Latin America, the modern and functionalist planning models that originated in Europe gave rise to the phenomenon of the "dual city," as described by Hardoy and Satterthwaite (1987). This urban formation resulted from a conflict between formal and informal approaches to urban development. In this context, popular sectors, which were predominantly impoverished and racialized, were not considered subjects of urban rights due to their non-compliance with the prevailing building regulations. This is evidenced by the forced displacement of the population and the lack of public space, housing, infrastructure, and other urban services that are essential for social reproduction.

The advent of neoliberalism marked a shift in urban determination, moving away from a normative framework based on legal and illegal criteria towards an economic one based on supply and demand. This shift occurred within the context of a deregulated market. This has resulted in the emergence of a city model characterized by a differential articulation of territorial divisions within the urban structure. The Industrial Revolution gave rise to urban segregation, exemplified by zoning based on land use. In contrast, the spaces of neoliberalism reinforce an insular and discontinuous fragmentation of territory (Duhau and Giglia 2012; Castells 2008).

In consequence, the new urban inequalities manifest in three distinct forms of social differentiation. Firstly, there is a notable segregation within urban environments with regard to access to markets and the primary social circuits of society. Secondly, the subsumption of common goods (air, water, soil) into the category of capital serves to exacerbate the consequences of climate change. A third consequence is the liquefaction of public space and the concomitant retreat into the private sphere. This is evident in the profound material and symbolic transformations of neighborhoods, including the placement of globalized artifacts, the construction of urban platforms, and the commodification of public space.

Civitas, or the condition of citizenship, has also been subject to significant transformation. This shift is a response to the growing prominence of individual freedoms, which are now more prevalent than collective rights. This results in a shift in the nature of urban conflict. During the latter half of the twentieth century, the primary demands centered on access to housing and infrastructure. In contrast, contemporary demands are more diverse and are situated within identity policies (gender, ethnicity, migratory status, or age), environmental concerns (climate change, disasters), urban violence (common, discriminatory, illegal markets), and other domains.

From a governance perspective, this occurs in a context in which local governments assume responsibility for urban issues as a result of the decentralization of the conflict, the popular election of their authorities, and the political constitution of their territories (Carrión 2015). This trajectory ultimately gives rise to the structural tensions inherent to the contemporary city. It is a dynamic wherein the processes of democratization of local governments and those of mercantilization/financialization of the city are in constant confrontation. This confrontation is played out against the backdrop of demands from citizens, who are pitted against the interests of capital and its repertoires of territorialization.

The democratic electoral route has been regarded as a prerequisite for citizenship, given its role as a means of resistance against the implementation of neoliberal policies (Carrión 2015). Elections have resulted not only in a change of authority but also in the victory of one project over another. This is because the population exercises its right to self-representation through the act of voting, thereby embodying an active form of citizenship. In this sense, the prescribed right also implies a formal component of citizenship, whereby the rights of peoples and nationalities are recognized and barriers to this condition are eliminated (Carrión et al. 2023).

It is evident that the historical connotations associated with the terms "civitas," "urbs," and "polis" have diverged from their original meanings. The processes of unequal and exclusionary urban development since the middle of the twentieth century have eroded this triadic relationship. This is evidenced by the tendencies observed in contemporary cities. The recovery of the meaning of these three "essential" components of the city, especially the meaning of *civitas* and *polis*, must be informed by an understanding of their evolution, particularly in relation to the constitution of the city. This understanding should be socially constructed and democratically negotiated. In this context, planning finds its rationale as a praxis and instrument that enables the prefiguration of the future and the formation of consensus. It is of critical importance to elucidate the concept of a project, the principles that inform it, and to provide illustrative examples from historical contexts.

2.3 The City as a Project

From its foundation, the city has been conceptualized as a prospective entity, a project yet to be realized. This is manifested in an aspiration that emerges from the negation of the preexisting socio-spatial order, whether in its genesis (the countryside) or in its contemporary manifestation (the existing city). It is the anticipation of a desired future, which may be conceived as a means of overcoming general or particular problems. Such proposals were frequently implemented, yet on occasion they remained merely as conceptual frameworks. Nevertheless, these proposals all contributed to the transformation of urban environments, whether through alterations in their operational dynamics, architectural configurations, or the conceptualization of the challenges they faced. It can therefore be stated with certainty that this constitutes an integral element of the history of the city.

The idea of a project can be defined as a set of strategies and practices that are oriented in a specific direction and exhibit internal coherence. These processes are carried out by different actors (Sevilla-Buitrago 2023). The practical application of this concept entails an intellectual and professional endeavor that is reflected in the traditions that view planning as a project. In a similar vein, Healey (2010) develops his notion of the project, defining it as the collective effort aimed at improving the qualities of places through the use of various methods and paradigms throughout history.

This vision of the project engenders a certain aura of neutrality that is, in fact, an illusion, given that the "ideology of planning" primarily responds to the aspirations of professionals in the discipline, to the detriment of the actual functioning of urbanistic praxis in specific historical contexts (Harvey 1985). We argue that city projects throughout history were not merely well-intentioned attempts to improve the qualities of places. Rather, they were ideological exercises in the creation and transformation of the built environment, responding to diverse motives ranging from military, religious, or communal to aesthetic, nationalist, or even global inspiration.

2.3.1 Planning Ideologies

In order to situate city projects in history, we employ the categories of ancient and modern planning ideologies as defined by Smith and Hein (2017). In addition, we undertake a brief revisitation of some paradigmatic cases of ancient and modern project systematized by Smith (2007) and Smith and Hein (2017), respectively.

In general, the focal point of ancient planning ideologies can be found in the government, politics, and divine destiny of their respective societies. This is evidenced by the numerous urban developments from Mesopotamia to the Aztecs. In other words, the blueprint of ancient planning ideologies was formulated by religious and/or economic elites with the intention of promoting specific political agendas. Contemporary planning ideologies are centered on the concepts of modernity and its representation. These ideologies were a significant influence on European practices from the nineteenth century onwards. Consequently, in contrast to the legitimization of monarchical rule through religious glorification, modern planning practitioners employed historical references to advance modernist and nationalist agendas, constructing narratives that would justify the establishment of new structures.

2.3.1.1 Ancient Planning

Shangdu City-Palace

Originally known as *Kaiping Fu*, the history of the city dates back to 1256, with its foundation ordered by the Mongol Grand Khan, *Mongke Qaghan*. This city served as the imperial capital until the construction of *Dadu* in 1267 (Steinhardt 1988). *Shangdu* was a three-walled city, comprising a palace-city, an imperial-city, and an outer-city, successively contained by walls (Fig. 2.1). The principal architectural complex of *Shangdu*, known as *Da'an Ge*, was situated in the northern part of the palace-city and resembled an inverted U (Steinhardt 1999).

The project behind Shangdu, a Chinese-influenced city in Mongolia, was the implementation of coordination and orientation attributes between built spaces (Smith 2007). These attributes imply a certain common distribution and orientation in the construction of individual buildings from urban sites, such as walls, avenues, or squares. The palace-city of Shangdu exemplifies this principle by presenting a common orientation with respect to the rectangular walls that contain it. In contrast to other ancient cities such as Becan in Guatemala, the palace-city of Shangdu demonstrates a high level of planning and belongs to a standardized Chinese imperial tradition of urban design.

Angkor Wat

It is widely believed that Angkor Wat was constructed by the Great King Suryavarman II in the twelfth century. Since its construction, the temple has served as an important religious center, and it has remained a symbol of Angkor identity and world architectural history to the present day (Carter et al. 2019). By the sixteenth century, the temple had become the primary destination for Buddhist pilgrims. Approximately a century later, a Japanese pilgrim created the first map of Angkor Wat (Chemburkar 2021).

It can be argued that the project underlying Angkor Wat is the notion of formality (Fig. 2.2). This refers to the fact that the organizational principles of an architectural group (axiality, walling, squares, etc.) remain clear to observers. In addition to their status as capital cities, ancient states such as Angkor Wat are also distinguished by their monumental character. This is evidenced by the fact that the dimensions of

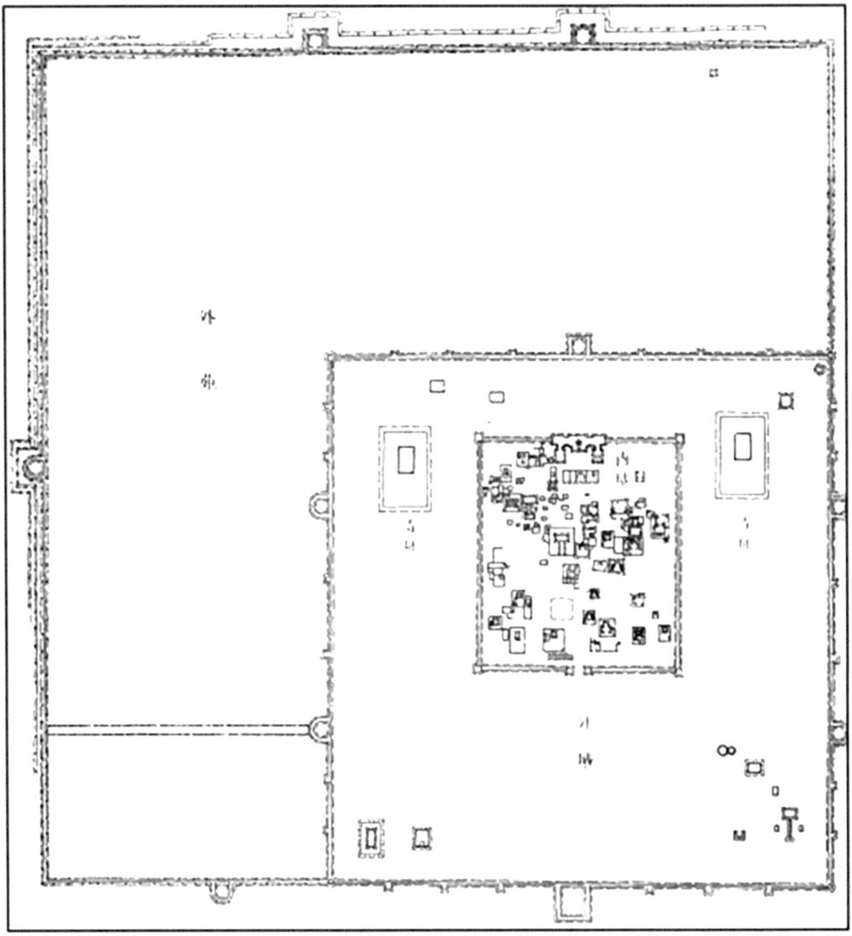

Fig. 2.1 Plan of Shangdu. *Source* Steinhardt (1988)

construction of buildings exceed the threshold of functionality, thereby promoting a variety of ideological agendas. The temple of Angkor Wat provides an exemplar of the integration of the projects of formality and monumentality in the ancient world (Smith 2007).

Teotihuacan

Situated in the northeastern basin of Mexico, Teotihuacan was a city-state of considerable influence from its establishment around 100 BC until it became the largest city in Mesoamerica. In the view of Nichols (2016), immigration constituted a crucial element in the formation of the multiethnic composition of the city, while also underlining the impact of neighborhoods, irrigation, markets, and warfare on the structure and governance of Teotihuacan. In this context, as Teotihuacan's regional influence

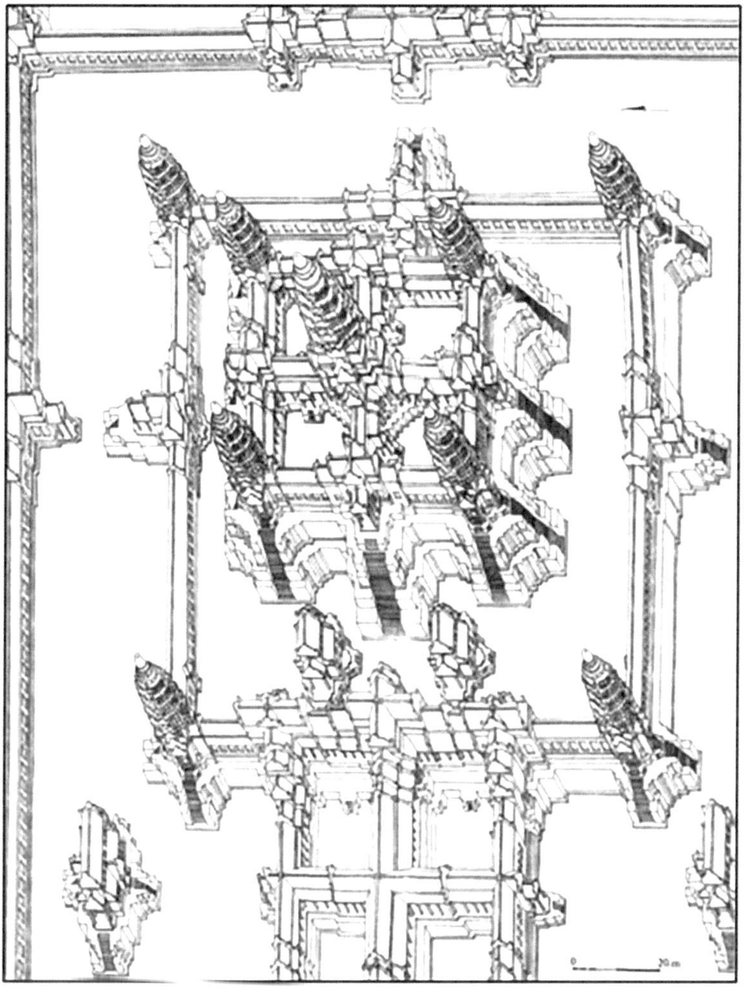

Fig. 2.2 Drawing of the Angkor Wat temple. *Source* Dumarçay y Royère (2001)

diminished, the urban center encountered internal challenges that ultimately resulted in its destruction and desecration.

The orthogonality or grid pattern Project is an integration of the principles of coordination between spaces and formality. This project employs right angles in urban design, manifested in the construction of buildings that are aligned orthogonally with respect to various large-scale features (Smith 2007). The case of Teotihuacán is illustrative (Fig. 2.3). The Street of the Dead and the East–West Avenue structure the entire city. Consequently, the majority of buildings in the city are aligned with the Street of the Dead, which in turn integrates the Pyramid of the Sun with the Pyramid of the Moon.

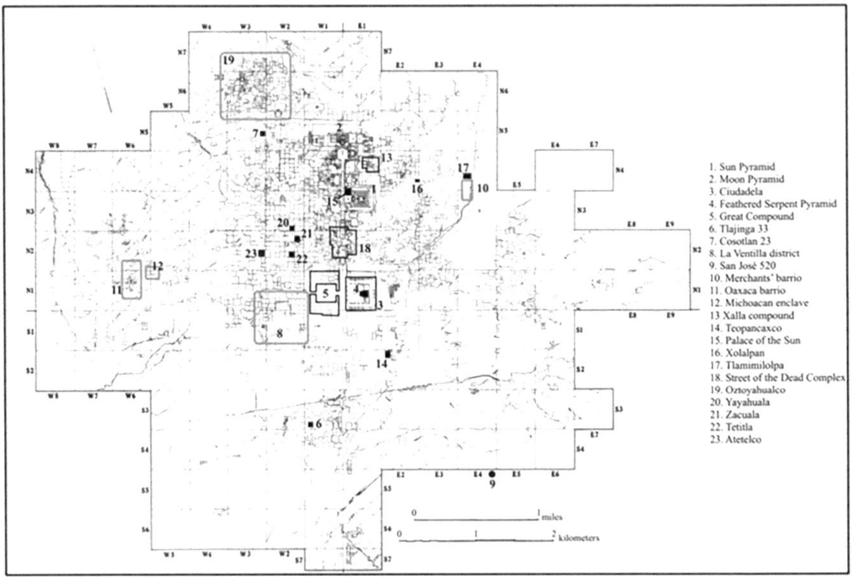

Fig. 2.3 Map of Teotihuacan. *Source* Nichols (2016)

2.3.1.2 Modern Planning

The Colonial City

A significant influence on the development of modern urban design can be attributed to the intellectual traditions that emerged in ancient Greece, Rome, and during the Renaissance. This has served as a source of inspiration for the prefiguration of new cities based on a range of criteria, including communal, military, and political considerations. In this regard, numerous European colonial powers pursued the dissemination of these urban planning ideologies to their colonies (MacCormack 2007).

A case in point is the design of orthogonal colonial cities in Latin America. The colonial "planners" of the sixteenth century turned to idealized concepts of Roman urbanism and architecture. The influence of Roman urbanistic practice can be traced from the writings and proposals of Vitruvius in Rome, through Alberti in the Renaissance, to the Laws of the Indies in the New World (Fig. 2.4). These laws, as outlined by Smith and Hein (2017), contained specific principles, such as orthogonality, for the design and construction of cities throughout the imperial territory.

The Modern City

Le Corbusier (1887–1965) was a renowned and seminal figure in the field of architectural and urban design, widely recognized for his contributions to modern urban

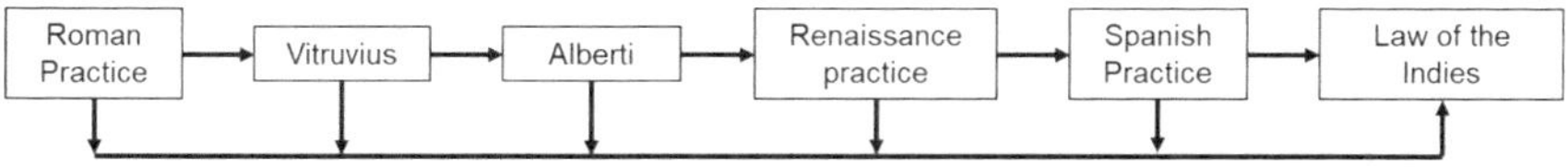

Fig. 2.4 The influence of Roman urban practice. *Source* modified from Smith and Hein (2017)

development in the twentieth century. His most significant contributions included the assertion that rectangular planning was indicative of a superior form of civilization, namely, modernity (Smith and Hein 2017). In his foundational work, Urbanisme (Le Corbusier 1985 [1924]), he provides a rationale for rectangular and gridded planning by revisiting various plans originating from Egypt, Babylon, and Peking. This served to reinforce his vision of a functional and orthogonal urbanism, which was subsequently reflected in the proposals of the Plan of Voisin (Fig. 2.5) and the various CIAMs (Congrès Internationaux d'Architecture Moderne) that emerged in subsequent years.

The modern city project aimed to rethink the existing urban landscape through functional planning, introducing the concept of zoning. This entailed the division of human activity into distinct zones, including residential, circulation, work, and recreational areas. Consequently, urban land values diverged according to their location within the city. Despite the theoretical potential of this approach, it rarely materialized in practice. Instead, it was championed by enlightened political elites with significant influence over urban governance.

The Fragmented City

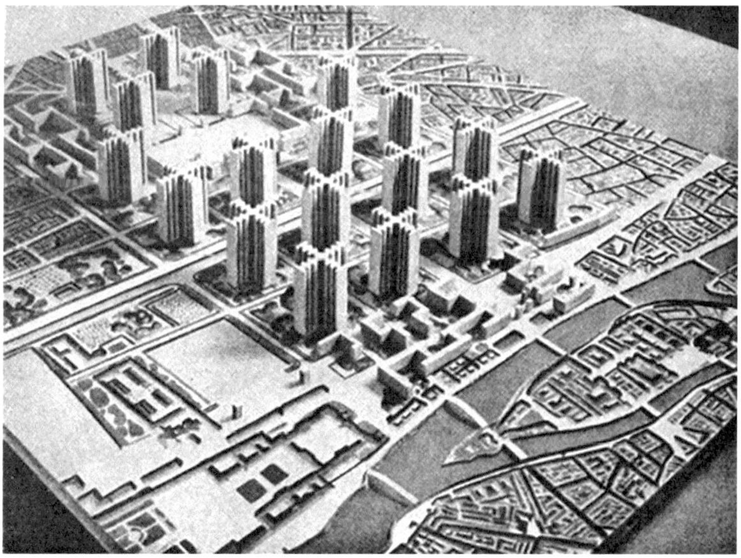

Fig. 2.5 Plan of Voisin. *Source* Le Corbusier (1985 [1924])

If we take Latin America as an example, rather than a project in itself, the fragmented city can be considered a product of a particular set of circumstances. The processes of neoliberalization, which entail the withdrawal and refunctionalization of the state, the deregulation of markets, and the privatization of urban space, transformed the structure of the Latin American city starting in the last decade of the twentieth century. Bähr and Borsdorf (2009) identify four distinct phases in the transformation of the Latin American city: the compact (colonial) city, the sectoral city, the polarized city, and finally, the fragmented city (Fig. 2.6).

The fragmented city is characterized by two structural principles inherited from past urban models: sector-linear tendencies and the cellular growth pattern. The construction of extensive intra-urban highways has been a significant factor in the attraction of the upper-middle classes to peripheral and peri-urban areas. In this context, the fragmented city is understood through the lens of new functions and socio-spatial forms, including the proliferation of gated communities and adjacent marginal neighborhoods separated by walls, fences, and other material and symbolic barriers.

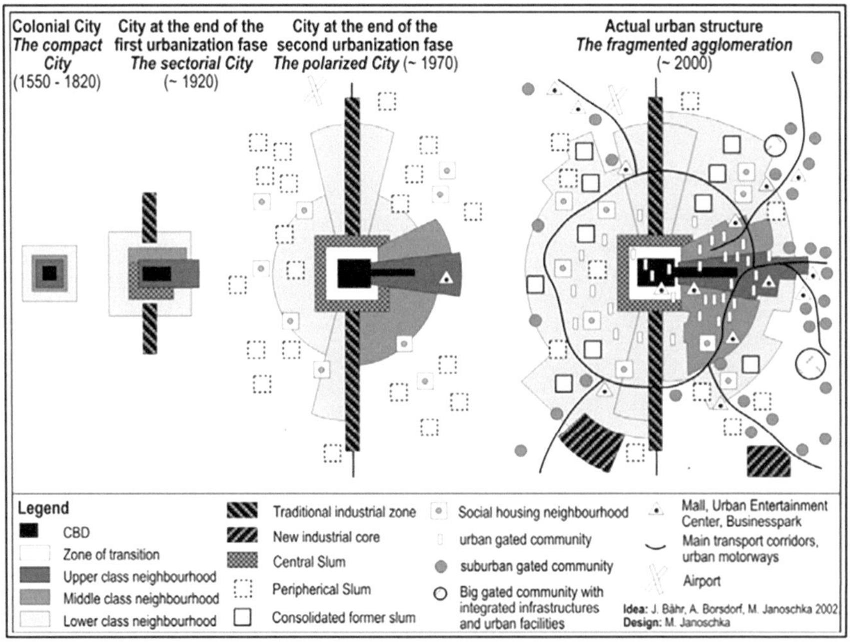

Fig. 2.6 Latin American city model. *Source* Janoschka and Borsdorf (2004)

2.4 Final Thoughts

This essay presents the argument that the existence of a city project has been a constant throughout history, influencing the development of cities on a planetary scale. It is crucial to explicitly acknowledge this premise, as it underscores the necessity to rethink theoretical frameworks and reformulate practices in order to effectively address the multifaceted nature of contemporary city projects. These responses must be guided by a dual temporal perspective. The first entails the formulation of a socially negotiated and updated diagnosis of the city in its present state, while the second pertains to the consolidation of technical and analytical devices of a prospective nature. This is with the intention of solving the main urban challenges within the framework of a sense of anticipation of the city.

By examining the various interpretations of *civitas*, *urbs*, and *polis* throughout the history of historical-philosophical thought, as well as a reading of these notions in the light of the contemporary city, it becomes evident that urban planning plays a significant role as a technical and political device for the prefiguration of a future city, or, in other words, for the conception of a city project. However, the ascendance of neoliberalism in urban governance has constrained the scope of traditional planning, limiting its capacity to exert influence at the local, intermediate, and national levels. This crisis is not only a consequence of the inability of conventional planning instruments to address the accelerated changes occurring in urban areas. It is also a result of the failure of these instruments to align with the envisioned city project, either due to their absence or their inadequacy.

A review of the historical development of city projects reveals that, in contrast to a linear historical reading, the planning history under consideration is one that reflects the deployment of ideologically directed strategies for the production and transformation of the built environment. In this sense, ancient planning projects must be understood in terms of the principles of coordination, formality, and orthogonality within the context of historical, religious, political, and military circumstances. Similarly, colonial, modern, and neoliberal city projects cannot be fully comprehended without reference to the Industrial Revolutions, human mobility, technological innovations, and other significant historical developments.

If we understand the notion of a project to imply the launching of something forward, we may define it as an invention or innovation of an object that changes over time, thanks to actions carried out in an arranged and methodical way: a plan. It follows that project and planning maintain an indissoluble and sequential relationship. In this context, the object of study is the city, which, depending on the context, can acquire different characteristics. It is therefore essential to establish a correspondence between the city project and urban planning.

References

Angelo H (2017) From the city lens toward urbanisation as a way of seeing: country/city binaries on an urbanising planet. Urban Stud 54(1):158–178. https://doi.org/10.1177/0042098016629312
Bähr J, Borsdorf A (2009) La ciudad latinoamericana. La construcción de un modelo. Vigencia y perspectivas. In: Centeno P (ed) Lima: diversidad y fragmentación de una metrópoli emergente. OLACCHI
Baird IG (2022) The problem of the urban–rural binary in geography and political ecology. Dialog Hum Geogr 12(2):218–222. https://doi.org/10.1177/20438206221102953
Brenner N, Schmid C (2014) The 'urban age' in question. Int J Urban Reg Res 38:731–755. https://doi.org/10.1111/1468-2427.12115
Carrión F, Silva E, Rodríguez A, Sugranyes A (2023) Constitucionalismo Urbano. La ciudad en los procesos constituyentes en América Latina. FLACSO Ecuador—Sur
Carrión F (2010) Ciudad, memoria y proyecto. FLACSO Ecuador—OLACCHI
Carrión F (2015) El giro a la izquierda en los gobiernos locales en América Latina. Café de las ciudades
Carrión F (2020) La red urbana global: los ensambles que la constituyen. The Metropolitan Unit: Intersectoral and Regional Integration. BID
Carrión F (2022) La ciudad neoliberal en América Latina. In: Carrión F, Corti M, Ramírez P, Abramo P, Cepeda P (eds) El futuro de las ciudades. FLACSO Ecuador
Carter AK, Stark MT, Quintus S, Zhuang Y, Wang H, Heng P, Chhay R (2019) Temple occupation and the tempo of collapse at Angkor Wat. Cambodia. Proceed Natl Acad Sci 116(25):12226–12231. https://doi.org/10.1073/pnas.1821879116
Castells M (2008) La cuestión urbana. Siglo Veintiuno
Chemburkar S (2021) Heritage, history and heterotopia at Angkor Wat. J Art Historiogr 25:1–21
Davis M (2004) Planet of slums. New Left Rev 26:5
Duhau E, Giglia A (2012) From the modern city to the micro-orders of the insular city. Espaces Et Sociétés 3:15–30
Dumarçay J, Royère P (2001) Cambodian architecture, eighth to thirteenth centuries, handbook of oriental studies. Brill
Fustel de Coulanges ND (2003) [1864] La Ciudad Antigua. Estudio sobre el Culto, el Derecho y las instituciones de Grecia y Roma. Editorial Porrúa
Geddes P (2004) Civics: as applied sociology. Sociol Rev 1:100–118. https://doi.org/10.1177/0038026104SP100110
Grbac P (2013) Civitas, polis, and urbs: reimagining the refugee camp as the city. Refugee Studies Centre. University of Oxford
Hardoy J, Satterthwaite D (1987) La Ciudad Legal y la Ciudad Ilegal. Grupo Editor Latinoamericano
Harvey D (1985) On planning the ideology of planning. In: The urbanization of capital. Blackwell
Healey P (2010) Making better places: the planning project in the twenty-first century. Palgrave Macmillan
Isin EF (2008) Recasting the social in citizenship. University of Toronto Press
Janoschka M, Borsdorf A (2004) Condomínios fechados and Barrios privados: the rise of private residencial neighbourhoods in Latin America. In: Glasze G, Webster C, Frantz K (eds) Private neighbourhoods. Global and local perspectives. Routledge, New York
Jennings B (2001) From the urban to the civic: the moral possibilities of the city. J Urban Health 78:88–103
Le Corbusier (1985) La Ciudad del Futuro. Ediciones Infinito
MacCormack S (2007) On the wings of time: Rome, the Incas, Spain, and Peru. Princeton University Press
Mumford L (1961) The city in history: its origins, its transformations, and its prospects. Harcourt Brace Jovanovich
Nichols DL (2016) Teotihuacan. J Archaeol Res 24(1):1–74. https://doi.org/10.1007/s10814-015-9085-0

Pickett STA, Cadenasso ML, McGrath B (2013) Introduction. developing a metalogue: ecology, society, and design. In: Pickett STA, Cadenasso ML, McGrath B (eds) Resilience in ecology and urban design. Linking theory and practice for sustainable cities. Springer, Dordrecht. https://doi.org/10.1007/978-94-007-5341-9
Prato GB (2016) European urban traditions: an anthropologist's view on polis, urbs, and civitas. Diogenes 63(3–4):9–19
Sayer A (1984) Defining the urban. GeoJournal 9:279–284. https://doi.org/10.1007/BF00149040
Sevilla-Buitrago A (2023) Contra la común. Una historia radical del urbanismo. Alianza Editorial
Smith ME (2007) Form and meaning in the earliest cities: a new approach to ancient urban planning. J Plann Hist 6(1):3–47
Smith ME, Hein C (2017) The ancient past in the urban present. The use of early models in urban design. In: Hein C (ed) The Routledge handbook of planning history. Routledge, New York
Steinhardt NS (1988) Imperial architecture along the Mongolian road to Dadu. Ars Orientalis 18:59–93
Steinhardt NS (1999) Chinese imperial city planning. University of Hawaii Press
UNDESA (2018) World urbanization prospects: the 2018 revision. https://population.un.org/wup/Download/
Weber M (1964) Economía y Sociedad. Fondo de Cultura Económica
Weeks JR (2010) Defining urban areas. In: Rashed T, Jürgens C (eds) Remote sensing of urban and suburban areas. https://doi.org/10.1007/978-1-4020-4385-7_3
Wirth L (1988) El urbanismo como modo de vida. In: Bassols M, Donoso R, Massolo A, Méndez A (eds) Antología de sociología urbana. UNAM

Chapter 3
Decolonial Perspectives in the Idea of City as a Project

Fernando Luiz Lara

Abstract When we think of cities as a result of "The Project" of European colonization, we immediately think of the grid, the damero. In its usual sense, the word "grid" suggests something of a geometric nature to select, relate, fix, or otherwise order a set of particulars or possibilities. The issue at stake in this essay is not that every city is the result of a geometric pre-determined project, but that most of our cities are the result of one specific project: European colonization. As I will explain later in this text, modernization and colonization are one and the same, indistinguishable facets of the same project. Let's call it "The Project". The questions that remain from this understanding are twofold: in a historical perspective, it is important to elaborate and dissect the past in order to understand how we get here. And from a present and/or future perspective, we need to ask what must we change in our design approach once we understand the colonizing roots of the Eurocentric project.

Keywords Decolonial theory · Urban space in the Americas · Abstraction · Project

> Perinthia's astronomers are faced with a difficult choice.
>
> Either they must admit that all their calculations were wrong and their figures are unable to describe the heavens, or else they must reveal that the order of the gods is reflected exactly in the city of monsters.
>
> Italo Calvino, Invisible Cities.

F. L. Lara (✉)
University of Pennsylvania, Philadelphia, PA, USA
e-mail: flara@upenn.edu

F. Carrión Mena and S. Rodríguez Alvarez (eds.), *From the City as a Project to the City Project*, The Urban Book Series, https://doi.org/10.1007/978-3-031-94226-6_3

3.1 Introduction

Every city has a project. As Italo Calvino elaborated so poetically in his Invisible Cities, every city is the reflection of its society, and every society is guided by some idea of a project, a future goal. I start this essay by reminding our English native-speaker colleagues that in the Latin languages we use Project as a synonym of Design. As our Latin concept of a design project comes from the idea of a future. Project are means to through something forward, to launch towards the future. In English, we were left with projectile (bullet) as the most common expression of this idea.

The issue at stake in this essay is not that every city is the result of a project, but that most of our cities are the result of one specific project: European colonization. As I will explain later in this text, modernization and colonization are one and the same, indistinguishable facets of the same project. Let's call it "The Project".

So what is "The Project"? A large body of scholarship commonly called "Decolonial Theory" has entertained this question for over 30 years now,[1] and their contributions are the basis of my assumptions, as I will elaborate further. Basically speaking, after 1492 the European powers projected themselves all over the world while at the same time erasing such extra-territorial experiences from the narrative of spatial rationality. The cities we have today are the result of this strategy, this "Project". The questions that remain from this understanding are twofold: In a historical perspective, it is important to elaborate and dissect the past in order to understand how we get here. And from a present and/or future perspective, we need to ask what must we change in our design approach once we understand the colonizing roots of the Eurocentric project.

3.2 The Project at the Center of Modernization/ Colonization

Peruvian sociologist Anibal Quijano is credited with advancing those ideas when he moved away from economic-centric Dependency Theory in the 1990s to propose a World System that explains the intricate relationship between modernization and colonization.[2] Before Quijano, Leopoldo Zea[3] and Edmundo O'Gorman,[4] both in

[1] Quijano, Aníbal, and Immanuel Wallerstein. "Americanity as a concept, or the Americas in the modern world." International social science journal 44, no. 4 (1992): 549–557; Escobar, Arturo. Encountering Development: The Making and Unmaking of the Third World. Princeton, N.J: Princeton University Press, 1996; Mignolo, Walter D. The Darker Side of Western Modernity: Global Futures, Decolonial Options. Durham: Duke University Press Books, 2011.

[2] Quijano, Aníbal, Op.Cot, 1992.

[3] Zea, Leopoldo *América como conciencia*, México: Cuadernos Americanos, 1953.

[4] O'Gorman, Edmundo. *La invencion de America: Investigacion acerca de la estructura historica del Nuevo Mundo y del sentido de su devenir*. (México: UNAM 1958).

Mexico, had already started to question the way we narrate the history of the Americas. Zea wrote in 1953 that the Americas were not discovered but "covered" with a Eurocentric identity.[5] O'Gorman elaborated in 1958 that the encounter with the Americas triggered European modernization and not the other way around,[6] an idea that has been corroborated by innumerable scholars, from James Blaut in 1989[7] to David Wengrow and David Graebner in 2021.[8]

Also in Mexico, Pablo Casanovas coined in 1962 the term "internal colonialism" to explain persistent inequalities.[9] It is unclear if Zea, O'Gorman, and Casanova knew about Franz Fanon's Black Skin White Masks of 1952, but their arguments are very much aligned and they precede Edward Said[10] by two decades, and Homi Bhabha[11] by three. In 1992, Quijano and Immanuel Wallenberg coined a sentence that has now become a classic. They argue that "it is not that the Americas had a significant role in the development of capitalism, there would be no capitalism if not for the occupation of the Americas".[12] As I explained in previous publications,[13] my main point, paraphrasing Quijano, is that it is not that the Americas had a significant role in the development of architecture as we know it, there would be no modern architecture if not for the occupation of the Americas. A few years later, Arturo Escobar published Encountering Development: The Making and Unmaking of the Third Wolrd, in which he elaborates the thesis that there is no modernization without colonization, both being different sides of the same coin.[14] In the words of Escobar, "some of the key notions that make up the conceptual corpus of this research program are thus: the modern colonial world system as the ensemble of processes and social formations that encompass modern colonialism and colonial modernities; although it is structurally heterogeneous, it articulates the main forms of power into a system". Walter Mignolo went further by demonstrating that "the emergence of the Atlantic circuit during

---

[5] Zea, Leopoldo *América como conciencia*, México: Cuadernos Americanos, 1953.

[6] O'Gorman, Edmundo. *La invencion de America: Investigacion acerca de la estructura historica del Nuevo Mundo y del sentido de su devenir*. (México: UNAM 1958).

[7] Blaut, James M. "Colonialism and the Rise of Capitalism." *Science and Society* 53, no. 3 (1989): 260–296.

[8] Graeber, David, and David Wengrow. *The Dawn of Everything: A New History of Humanity*. New York: Farrar, Straus and Giroux, 2021.

[9] Casanova, Pablo Gonzalez. "Internal colonialism and national development." *Studies in comparative international development* 1, no. 4 (1965): 27–37.

[10] Said, Edward. Orientalism, Harmondsworth: Penguin, 1978.

[11] Bhabha, Homi. "Of mimicry and man: The ambivalence of colonial discourse." *October* 28 (1984): 125–133.

[12] Quijano, Aníbal, and Immanuel Wallerstein. "Americanity as a concept, or the Americas in the modern world." *International social science journal* 44, no. 4 (1992): 549–557.

[13] Lara, Fernando Luiz, "American Mirror: the occupation of the "New World" and the Rise of Architecture as We Know it", *The Plan Journal*, 5:1, 2020; "Spatial Abstraction as Colonizing Tool", in *Routledge Companion to Decolonizing Art History*, (Flores, San Martin and Villaseñor, editors) New York: Routledge, 2023, 331–341.

[14] Escobar, Arturo. *Encountering Development: The Making and Unmaking of the Third World*. Princeton, N.J: Princeton University Press, 1996.

the sixteenth century made coloniality constitutive of modernity"[15] something that Quijano also explores in his discussions of the "coloniality of power".[16]

3.3 Grids and Axis as the Physical Manifestation of "the Project"

When we think of cities as a result of "The Project" of European colonization, we immediately think of the grid, the damero. In a now classic paper on The Grid, Barry Maitland explained that the grid is anything which exists less for itself than as a discipline or frame of reference for something else.[17] In its usual sense the word "grid" suggests something of a geometric nature (…) to select, relate, fix, or otherwise order a set of particulars or possibilities. Richard Sennet, in The Conscience of the Eye discusses how the grid subdues those that must live there, disorienting their ability to evaluate relationships. The planning of neutral space is an act of dominating and subduing others.[18] However, when discussing the history of the grid, Sennet jumped from the Roman Empire to the seventeenth century, deliberately ignoring the hegemony of the grid in the process of occupation of the Americas in the sixteenth century.

In a small book published in the Dominican Republic in 1998—based on a seminar held in Caracas three decades before (!), scholars Gabriel Guardia, Leonardo Benevolo, Paolo Portoghesi, Erwin Walter Palm, and Graziano Gasparini discuss the legacy of the colonial cities of the Americas. Arguing for the centrality of the Spanish experience, Guardia defends that we cannot call the American cities "daughters of the Renaissance", nor can we attribute them to Italian treatises either, since the Renaissance ideal cities were circular or radiocentric polygons, not abstract orthogonal grids prone to expansion.[19] Besides, the gridded city is not at all a Renaissance invention despite its roots in Roman military strategy. Fray Francesc Eiximeniç had published his Quinna forma deu haber Ciutat bella e be edificada in 1386 and Villareal, a few miles north of Valencia, was founded in 1274 on a rectangular city with orthogonal streets. Guardia also reminds us that gridded cities existed in Mexico before the arrival of Europeans, Tenochtitlan and Cholula had orthogonal plans from around the time of Eiximeniç and so did Teotihuacan 1000 years before that.

---

[15] Mignolo, Walter. "The geopolitics of knowledge and the colonial difference." *The South Atlantic Quarterly* (vol 101.1, 2002): 60.

[16] Quijano, Anibal. "Coloniality of power and Eurocentrism in Latin America." *International Sociology* 15, no. 2 (2000): 215–232.

[17] Maitland, Barry. "The grid." Oppositions 15/16, 1979: 91.

[18] Sennett, Richard. The Conscience of the Eye: The Design and Social Life of Cities. 1st ed. New York: Knopf: Distributed by Random House, 1990: 60.

[19] Guarda, Gabriel, "Santo Tomas de Aquino y las fuentes del urbanismo indiano", in La Ciudad Colonial del Nuevo Mundo: formas y sentidos (edited by Miguel D. Mena) Santo Domingo: Jardin de las Delicias, 1998: 26.

Guardia therefore prepared the terrain for the coup the grace by Leonardo Benevolo. The Italian scholar starts by arguing that the Renaissance changed architecture but did not change urbanism. "The examples cited in history books—Pienza, Urbino, Ferrara, Mantova, August, Anversa, Lisbon, Rouen, Vitry, Nancy, and Palmanova are in some degree exceptions, built by unique circumstances that do not conform to a general rule."[20] The Renaissance, continued Benevolo, did not create a new city typology, it just intervened in existing medieval cities.

Benevolo is in full agreement with Lewis Mumford and his concept of medieval organic planning.[21] According to Mumford, there is no Renaissance city. European urban concepts evolved from medieval to baroque. Mumford wrote that "practically all of the 'renascence' building took place on medieval street plans, within the walls of essentially medieval cities, erected by crafts and guild still organized on medieval lines".

The colonial territory was where the Europeans could operate to create the largest number and in certain aspects, the most relevant works of the sixteenth century. Moving closer to my argument that modernity has a significant American component since the sixteenth century, Benevolo wrote that the grid adopted in America comes from an operative tradition but also from a European cultural ideal applied only partially in the old continent but is present in everything considered modern. It is indeed an index of modernism (Benevolo, 1998: 86). More recently, Clare Cardinal-Pett wrote that the idea that the Mesoamerican city itself could have provided some precedent for Spanish colonial city planning gains much traction with the fact that many of the first "new" towns were in fact built on top of existing indigenous urban places.[22]

Could we say then that modernity started in the Americas? Benevolo did not pose such question but ended his essay by stating that "the urban design developed in the Americas in the sixteenth century, consolidated by the law of 1573 (Law of the Indies), is the only model of new cities produced by the Renaissance. This model continues to work for four centuries after being generalized by the neo-classical culture and served as a basis for the most impressive territorial transformation of the modern era: the colonization and urbanization of the United States".[23]

Back to the words that Fernandez uses to explain the Americas, "no territory as close to a pure modernity, from the Renaissance utopias to the energic application of the ideas of counterreformation (1600 s), illuminist (1700s) and positivism (1800s)". Lewis Mumford understood very well the relationship between modernity and abstract territorial control. In his words,

---

[20] Benevolo, Leonardo, "Las nuevas ciudades fundadas en el Siglo XVI en America Latina. Una experiencia decisiva para la historia de la cultura arquitetonica de "Cinquecento", in La Ciudad Colonial del Nuevo Mundo: formas y sentidos (edited by Miguel D. Mena) Santo Domingo: Jardin de las Delicias, 1998: 81.

[21] Mumford, Lewis. The Culture of Cities. New York: Harcourt Brace and Co., 1938: 303.

[22] Cardinal-Pett, Clare. *A History of Architecture and Urbanism in the Americas*, New York: Routledge, 2016: 146.

[23] Benevolo, L. OpCit, 1998: 84.

> capitalism became militaristic [in the sixteenth century] and a change in the entire conceptual framework took place with its abstract love of money and power. And first; new conception of space. It was one of the great triumphs of the baroque mind to organize space, make it continuous, reduce it to measure and order, and to extend the limits of magnitude, embracing the extremely distant and the extremely minute; finally, to associate space with motion and time. Those changes were first formulated by the painters and architects, beginning with Alberti, Bruneleschi, Uccello, and Serlio.[24]

Such power of abstraction is clearly visible in the Mexican cartography of the sixteenth century. In 1578, King Felipe II sent a questionnaire to all Spanish towns in the Americas, a kind of census aimed at organizing all this information on the other side of the Atlantic. The so-called Relaciones Geográficas are a priceless picture of the first decades of European occupation, and many of the reports include maps. Such maps display a variety of conceptions of space, from the early relational depictions of the local Nahuatl and Mexica to the gradual introduction of abstract lines, which were instrumental in assuring Spanish control. In the map of Teozacoalco (Fig. 3.1), we read the relational perception of space with its geographic accidents connected to local rulers. Here, the new churches built by the Spanish are already visible but are engulfed by a circular view of the world punctuated by the sacred mountains of the Nahua.

In the map of Ixcatlan (Fig. 3.2), the territory is already organized in European (birds eye) perspective, and the sacred Nahua mountains are all matched by a chapel or church. The pairing of mountains/sacred beings to churches/sacred beings was one of the most effective tools of colonization in sixteenth and seventeenth century Americas. Our great cities to this day are called São Paulo, Asuncion, Rosário, Santiago, Santo Domingo, San Antonio, San Francisco, and Los Angeles. In the map of Zempoala (Fig. 3.3) we see the abstract lines imposed over the territory and the geographical features reduced to markers in an overextended and overpowering grid. In Spacious World, Cartography, Literature and Empire in Early Modern Spain, Ricardo Padron shows that before the sixteenth century a map was basically an itinerary with men located on the ground and travelling along the landscape. After Colón, Vespucci, and Descartes, a map became an abstract tool of territorial description with men removed from it. The transformation from a relational map that places mankind in the middle and in direct contact with the landscape to an abstract map that places [European] men above and removed from the earth is all but complete in the Zampoala map of 1580.

The change in representation, from relational to abstract drawings, is another instance in which design became a tool for implementing processes of modernization. Recently, Catherine Walsh and Walter Mignolo wrote that:

> the rhetoric of modernity aims to persuade you through promises of progress, growth, development, and newness if objects, composed of two interrelated domains: first a field of representation, which grounds its power in the very idea that signs represent something existing;

[24] Munford, L. OpCit, 1938: 364.

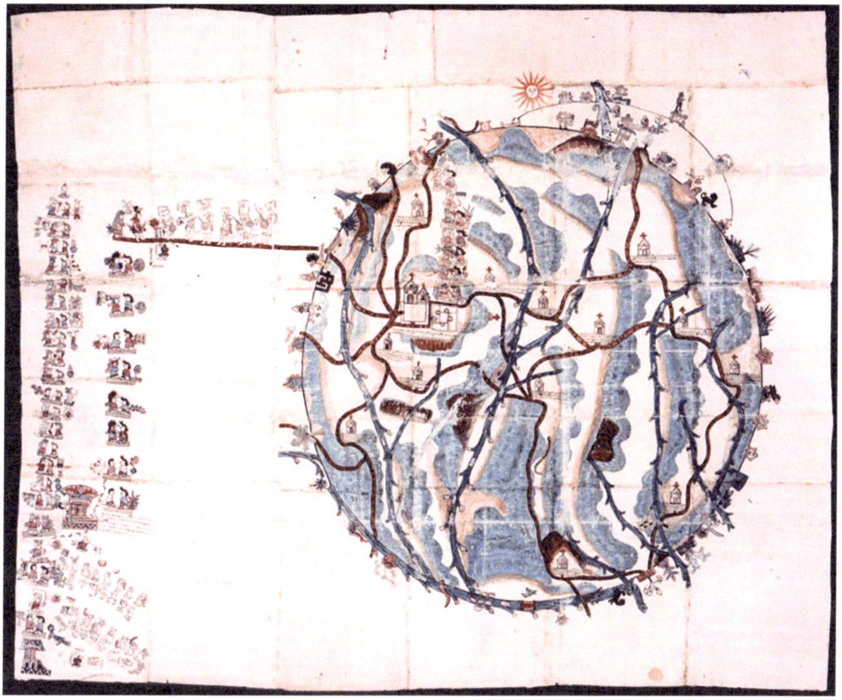

Fig. 3.1 Map of Teozacoalco, Oaxaca, 1580

and second, a set of rhetorical discourses aimed at persuading you that the world is as the field of representation tell you it is.[25]

The grid that Sennet overlooked and Benevolo acknowledged as the main urban form of the Renaisssance was a powerful tool of colonization and control. Padron continues his analysis of sixteenth century maps, matching the occupation of the Americas with the birth of cartography as we know it. The author elaborates that "the Iberian cultures that first planted European banners on non-European shores were—like the French, the English, and the others—cultures only beginning to think cartographically. They were only beginning to learn how to imagine their world, relate to it, and transform it in ways that depended upon abstract conceptualizations of space."[26]

Moving the issue of abstraction from instrumental maps to a whole epistemology Walsh and Mignolo explain that:

the course of action leading to the historical foundations of the modern/colonial project in the sixteenth century was not merely a question of physical actions (occupying lands,

[25] Walsh Catherine and Mignolo, Walter. On Decoloniality, Concepts, analytics, praxis. Duke University Press, 2018: 139.

[26] Padron, Ricardo. The Spacious Word: Cartography, Literature, and Empire in Early Modern Spain. 1 edition. Chicago: University Of Chicago Press, 2004: 12.

Fig. 3.2 Map of Ixcatlan, Oaxaca, 1579

> exploiting labor, etc.). It was above all a massive conceptual (epistemical) machine building and managing knowledge that the European institutions believed were superior and truer than others.[27]

Architecture was, and still is, an important component of such conceptual machinery. One question then remains unstudied: what was the impact of the encounter and the consequent rise in abstract thought on the architectural theories after 1492?

[27] Walsh and Mignolo, OpCit, 2018: 172.

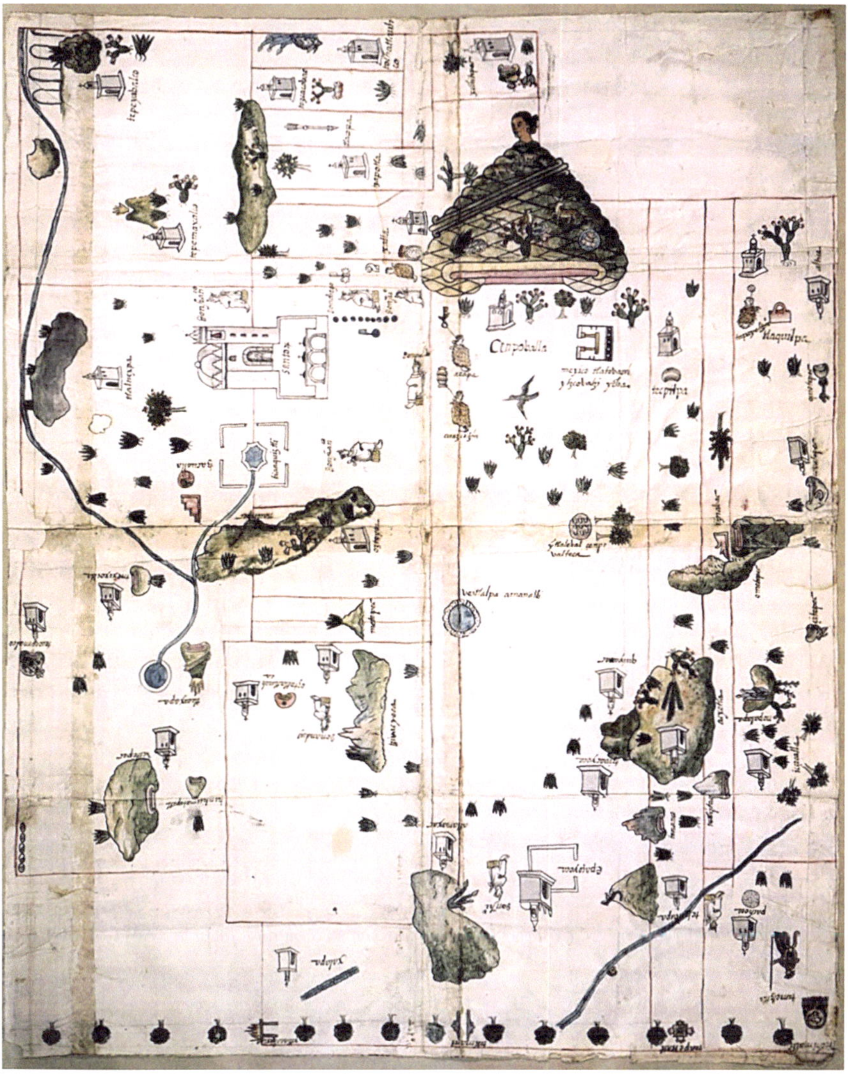

Fig. 3.3 Map of Zempoala, Hidalgo, 1580

3.4 Cities to Forget and Cities to Erase at the Core of "the Project"

Twelve years ago, I was standing on the steps of the cathedral of Lima, Peru, when a group of elementary school students walked by, and I could not help but hear their teacher explaining that they were going inside to visit the tomb of Francisco Pizarro who had founded that city in 1535. I immediately thought about the excavation of ancient settlements that are visible all over Lima and kept that moment in my memory because it summarizes a peculiar and quite damaging way in which we tell the history of the Americas. Growing up in Brazil, I was taught in the exact same way. The history of Minas Gerais, where I was born, started in 1693 when Portuguese explorers found gold pebbles at the foot of the Itacolomi mountain. Erasure and exclusion are operating in that very sentence, for Itacolomi is a Tupi word that means stone shaped like a boy, from itá (stone) e kunumĩ (boy).

One of the great joys of my academic career happened every August in Austin. Because classes always started on a Wednesday and I taught my undergraduate Signature Course on the history of urbanization in the Americas on Mondays and Wednesday mornings, I had the honor of delivering the very first college class for a group of 18 freshmen. In that first meeting, I asked where they grew up, the majority being from the suburbs of Houston and Dallas, some from the Rio Grande Valley, and a few from out of state. Then I told them that their first assignment was to write a fictional piece about a regular day of a person exactly their age and gender, who lived exactly where they grew up, but 500 years before. What they wrote is less important than the fact that they realized how little they knew about the long history of the land they call home.

In the Americas, we live in a land with a very short memory, a limitation that has been imposed on us and reinforced by centuries of Eurocentric narratives. Leopoldo Zea wrote in 1953 that the Americas were not discovered but covered with a narrative of European superiority and modernity that fogs our view to this day.[28] Until very recently, even intellectuals committed to understanding the identity of those lands suffered from and reproduced the idea of land without memory. Mexican poet and diplomat Octavio Paz, for instance, wrote:

> America was, if it was anything, geography, pure space, open to human action. Since it lacked historical substance—ancient social classes, established institutions, religions, and hereditary laws—reality presented no obstacles other than natural ones. Men struggled not against history but against nature. And wherever there was a historical obstacle—indigenous societies, say—it was erased from history, reduced to a mere natural fact, and dispensed with accordingly.[29]

Paz contradicted himself in the paragraph above, so I guess I am absolved of any heresy in disagreeing with the great Mexican scholar. The Americas were never lacking institutions, religions, laws, or social classes. They had it all, in a variety of forms. Our continent was home to over twenty-five million people in 1492: people who had lived here for thousands of years; people who built great cities such as Tenochtitlan and Cusco, to use the European concept of urbanism; people who also built thousands of other settlements, mostly low density and in deep synergy with nature. It was a nature transformed and modified by human hands but still very much revered and respected. Octavio Paz was wrong in the first sentence quoted above, and he knew it. The third sentence says it all: wherever there was a historical obstacle, it was erased.

[28] Zea, Leopoldo. *América Como Conciencia*. Universidad Nacional Autónoma de México, 1953.

[29] Octavio Paz, El Arco y la Lira , Ciudad de Mexico: Fondo De Cultura Economica, 1956: 279.

To lay out the basic tenet of this book, I will edit Octavio Paz's paragraph to show how it would read if the order of the sentences were reversed: America was the place where indigenous societies were erased from history, reduced to a mere natural fact, and dispensed with accordingly. In its quest to tame nature, men [of European descent] struggled against history. History was an inconvenient reality and an obstacle to the full domination of nature. To reduce America to a pure space open to [white men's] actions, was tantamount to erasing any historical substance: ancient social classes, established institutions, religions, and hereditary laws. How different is the meaning of Octavio Paz's sentences in reverse order? How striking is the violence of this erasure, a process of historical destruction that was indeed quite successful? Generations upon generations all over the Americas learned that our continent had no history prior to the arrival of the Spanish flotilla in October of 1492.

3.5 The Asymmetry of the Project in the Historiography

One of the most astounding features of the Eurocentric project is its hegemonic presence in the historiography, with the consequent erasure of any non-Eurocentric phenomena. This happened throughout the twentieth century without exception, our books on the history of architecture were overwhelmingly Eurocentric, and the language quite explicitly so. Even Spiro Kostof's A History of Architecture, Settings and Rituals published in 1985, which I consider the first successful attempt at an inclusive not-so-Eurocentric survey, could not escape the traditional narrative obliviating the Americas. Divided into 28 chapters, one of them is devoted to colonial Americas (Chap. 18) and one is devoted to the USA (Chap. 24). Kostof deserves praise for being the first to propose a survey that makes a legitimate and serious attempt to cover the built environment of the entire world. He also pushed the boundary of the disciplinary traditions by including many vernacular structures in his survey narrative and was the first major architectural historian to break away from the diffusionist argument around the urban orthogonal grid, explaining that its origins are multiple.

Despite all those advances, the overall tone of the book is still Eurocentric, starting with a Stone Age subchapter that only discusses Europe. A less informed reader would think that homo-sapiens evolved in Europe and lived only north of the Mediterranean Sea. The exuberance of urbanism in Mesopotamia and the Hindu Valley is discussed in the next three chapters until the chronology comes back to Greece, and Kostof labels "barbarian" all buildings that do not conform to classic Hellenism of the fifth century BC. In a subchapter labeled A Continent Alone, Kostof introduces the Americas, something that previous surveys failed to do. However, he insists on the idea that everything here was newer, claiming that the pyramids of La Venta in the Gulf of Mexico (which he wrongly dates as 500 BC) were among the oldest in the continent when information about the Olmec city of San Lorenzo (1000 BC), as well as about the Paracas (800 BC) and Chimu (900 BC) in Peru were easily available in English after WWII.

When presenting the high cultures of Aztecs, Mayas, and Incas, he placed them into the framework of Spanish expansion, as if "invented" or to use the now discredited term "discovered" by the conquistadores. The numbers say it all, Kostof devoted six pages to the presentation of the entire built environment of the Americas before 1492, followed by nine pages discussing Spanish colonial buildings. Before and after that, when discussing European architecture in the sixteenth and seventeenth centuries, Spain and its American colonies disappeared from the conversation.

Writing about the Renaissance, Kostof makes no mention of the Americas except to introduce a chapter about Spain and the New World where he states that "the rediscovery of the classical past was one of the two great adventures that informed the Renaissance. The other was the exploration and the conquest of America".[30] This statement alone is more than all previous authors ever said about the relationship between the European occupation of the Americas and the Renaissance, and consequently in the rise of architecture as an autonomous practice. All architectural history scholarship before the twenty-first century either ignored the Atlantic encounter or minimized its role in European developments. Kostof deserves the credit for being the first historian to break away from that ideology. Previous authors were deeply embedded in the traditional European narrative about the Americas acting as a secondary outcome of European modernity and therefore not worthy of much attention. The fact that Kostof did not ignore it was a significant advancement, but one paragraph down on the same page he held steady in his Eurocentrism by elaborating that:

> these newfound cultures should have proved that the worth of Western achievement was only relative, and forced it into fresh channels. They did not. The riches of the conquered New World added nothing to the enrichment of the Christian West except in the material sense.[31]

Much to the contrary, the recent works of Escobar, Mignolo, Dussel, Canizares-Esguerra, and Padron[32] demonstrate that the encounter of 1492 and the territorial occupation that followed played a central role in the development of Western culture in general, allowing me to extrapolate that to the architectural discipline in particular. Inverting Kostof's sentence, the main achievement of the sixteenth and seventeenth centuries was the erasure of those "newfound" cultures and the construction of a narrative in response to the American "indigenous critique".[33]

Facing the challenge of occupying and colonizing the vast territories of the Americas (soon to be extended to South and Southeast Asia), the European empires and

---

[30] Kostof, Spiro. *A History of Architecture: Settings and Rituals*. New York: Oxford University Press, 1985: 433.

[31] Kostof, Spiro. *A History of Architecture: Settings and Rituals*. Op Cit: 433.

[32] Escobar, A. "Worlds and knowledges otherwise: The Latin American modernity/coloniality research program." *Cultural studies* (vol 21.2–3, 2007) 179–210; Mignolo, W. *The Darker Side of Western Modernity: Global Futures, Decolonial Options*, Durham: Duke University Press, 2011.; Brading, D. *Orbe indiano: de la monarquía católica a la república criolla, 1492–1867* (México: Fondo de Cultura Económica, 2003); and Padron, R. *The Spacious Word: Cartography, Literature, and Empire in Early Modern Spain* (Chicago: University of Chicago Press, 2004).

[33] For the "indigenous critique" see Graeber and Wengrow, *The Dawn of Everything*, 2021.

their mercantile elites used architecture as a tool for controlling spaces, narratives, peoples, and goods, optimizing the binomial modernization/colonization as framed by Arturo Escobar. The City as a Project is an Eurocentric colonized city. As Italo Calvino reminded us, there is a monstrous side to this project that we are just beginning to unveil. More difficult still is to craft a narrative of resistance to this Project, to decolonize it. We are starting to move in this direction, and hopefully this new understanding of our cities will help us build better ones in the future. We must shift from "The Eurocentric Project" to "our project", even if our calculations are wrong and we reveal other monsters.

Chapter 4
City as Utopia

Francisco Colom-González

Abstract The idea of a collective life that is morally and materially fulfilling, associated in Western culture with the notion of "utopia", has often taken the form of a well-ordered urban community. This idea can be traced back to ancient Greece (the "virtuous city"), the Middle Ages (the "city of God"), and the Italian Renaissance (as the 'ideal city'), but also to the modern search for solutions to the problems of industrial and post-industrial cities. Urban planning and theory are therefore linked to utopianism in many ways. This chapter explores the different normative connotations that have been attached to the idea of the 'good city' throughout history, from classical antiquity to modern urban planning.

Keywords Urban utopias · Uchronia · Ideal cities

4.1 Introduction

The history of mankind is closely related to the history of urban spaces. In 300,000 years, *homo sapiens* has evolved from a small species of hunter-gatherers roaming the East African rift valleys to a true *homo urbanus* whose main habitat is the city. This change has resulted not only in a dramatic social transformation, but also in a process of anthropisation of the planet, posing social, environmental, and political challenges of unprecedented proportions. Throughout this process, the importance of cities has varied greatly, with their cultural contribution to the history of civilization being as significant as their political and economic role. While the earliest urban settlements were established in the Fertile Crescent region of the Middle East approximately 6000 years ago, it was not until the advent of Greek philosophy that we find the first systematic reflections on the meaning of the city itself. Since that time, the Western narratives of the city have been largely normative in nature: a reflection on the possibility of the *good city*. Its representation as an idealized or utopian reference for a morally and materially fulfilled life has taken many forms:

F. Colom-González (✉)
Spanish National Research Council (CSIC), Madrid, Spain
e-mail: f.colom@csic.es

F. Carrión Mena and S. Rodríguez Alvarez (eds.), *From the City as a Project to the City Project*, The Urban Book Series, https://doi.org/10.1007/978-3-031-94226-6_4

the *virtuous city* of ancient Greece, the *city of God* of early Christianity, the *ideal city* of the Renaissance, and the *smart*, healthy, or sustainable city from the Industrial Revolution to the present day. This is not to say that there are not more shadowy elements to such a discussion, as in the biblical tradition, where the first city was created by Cain, the primordial fratricide. Modern urbanism also contains a strain of scepticism about the capacities of the city: it is, in effect, projected against something that is thought not to work well and that can be improved. This chapter seeks to explore the idea of the city as a utopian reference and the different connotations that have been attached to it over time.

In a broad sense, the concept of utopia encompasses a wide range of topics, from accounts of journeys to exotic lands and descriptions of lost worlds to ideal formulas of government and prophecies of a happier time to come (Mumford 2011). What they all have in common is a reference to an optimal state of human affairs in the context of a particular organization of collective life. The idea of a perfect society, in another realm of existence, is predicated on the assumption that human beings are capable of acting on the existing social conditions in order to create a society that reflects the ideal. It can be argued that, even before the term was coined, utopianism was connected to the imagination of society and the justification of an alternative social order. In the elaboration of this universe, political essays and religious beliefs shared ground with narrative fiction. A conventional approach to defining utopia is to associate it with secularization and the abandonment of the redemptive image of paradise (Touraine 2000). This is why some authors have described the eschatologies that envisioned a Kingdom of God on Earth as social representations that did not rely on human purpose and resolve, which would distinguish them from the genre that emerged in the Renaissance (Sargent 2000). Thus, utopian thought is linked to the beginnings of modernity and the awareness that the *raison d'être* of society must be sought in itself, in the common good, to which all individuals must submit in order to attain happiness and self-fulfillment. In other cases, it was not so much a quest for an ideal society as a humanist aspiration guided by the Greek idea of *paideia*: the teaching of virtue and the arts that foster it.

Whatever the case, the role of the city as a utopian space has been crucial, both as an expression of a civic ideal and as a hope for earthly redemption. The notion of an idealized social bond in the form of a well-organized urban community was imagined simultaneously as a built environment, a political body, and an associative plexus. It is in this sense that this chapter refers to the city as utopia: as the urban representation of an emancipatory bond, a *social imaginary* that nestles in people's minds as an unrealized reference of collective desire or ambition. As Marx wrote in *Capital* (1990: 284), a spider carries out operations reminiscent of those of a weaver, and a bee would put more than one architect to shame with the construction of its honeycomb cells. But what distinguishes the worst architect from the best is that the former has already designed the structure in his mind before he builds it in reality. Ultimately, the outcome of the process is one that was already present in the worker's imagination prior to the commencement of his work. Humans do not merely transform the materials of nature; they fulfil their own purpose in these materials, creating a new environment that is imagined before it is realized. Utopias belong to this world of envisioned possibilities and potentials. In this particular sense, as we shall see, they are by no means limited to classical antiquity and the Renaissance. Modernity has also been strewn with multiple visions of urban utopias.

The history of urbanism can be interpreted as an ongoing pursuit of coherence between the physical dimensions of the city and the nature of human interactions facilitated within it. With the advent of industrial society, the classical concern with the aesthetic and political qualities of the city was superseded by the study of the urban impact of modernity. The pressing necessity to address the deplorable social and sanitary conditions of the urban slums where the new industrial proletariat was concentrated led to the search for technical solutions that would make modern cities physically habitable and socially dignified spaces. Nevertheless, modern urbanism did not entirely eschew the humanist tradition that had accompanied it in previous centuries. Le Corbusier's concept of the *radiant city*, a symbol of rationalist urbanism, was intended to be a habitation machine, but it also embodied a practical humanism at its core. His ultimate aim was to respond to the pressing social needs created by industrialism and wartime destruction by increasing urban density and efficiency. He offered "the city of our modern times, full of happiness, radiant with essential pleasures".[1] This movement subsequently gave rise in the 1960s to a counter-movement against the perceived *death* of modern cities, as Jane Jacobs famously denounced. This is the alternative, modern sense of urban utopia: the reverse of the industrial city, which was presented as a dystopian space conducive to human and social alienation. All these connotations reflect the normative continuity that exists between the search for the good life in the ancient *polis* and the projects designed to ensure the habitability of our contemporary metropolis.

4.2 Utopia and Uchronia

Traditional societies did not conceive of an alternative reality. The possibility of a better world lay in another time to come, at the end of days, or in another life, not in another place. Collective redemption could be sought, sensed, or hastened, but it could not be brought into existence in this present world. The idea of a *promised land*, which can be found in different forms in a variety of cultures, appears to contravene this principle. However, it is typically the fulfilment of a prophecy that is not the result of human agency but rather the expression of a supranatural will. This pattern would change with modernity, which, according to Bauman (2000), consists of the awareness of time from the moment it is endowed with *history*. The modern perception of temporality represents a departure from the messianic time of medieval Christianity, which fused past and future in a linear path of salvation from the first coming of the Messiah to the Day of Judgement. With the Renaissance and the Protestant Reformation, the Western concept of time recovered two emancipatory

[1] Le Corbusier delivered these words in a lecture at the Museum of Modern Art in New York on 24 October 1935.

references taken from its own cultural origins. On the one hand, the Judeo-Christian notion of Eden or primordial paradise, which also took on the millenarist format of a *new world to come*. On the other hand, the ancient Hellenic ideal of cultivating a virtuous life through the civil organization of human coexistence. As some historians have noted, "Utopia is a hybrid, born of the crossing of a paradisiacal, otherworldly belief of Judeo-Christian religion with the Hellenic myth of an ideal city on Earth. The naming took place in an enclave of sixteenth-century scholars excited by the prospect of a Hellenised Christianity" (Manuel and Manuel 1979: 15). This confluence gave rise to a utopian current that combined the two traditions, the time-bound and the space-based, and identified the ideal form of life with its urban materialization.

In the Abrahamic religions, human existence is viewed as a pilgrimage on Earth, with the ultimate goal of final redemption. Accordingly, a primordial state of felicity in human relations was proposed to have existed at the dawn of time, as a primordial paradise, or is expected at the end of days, as a heavenly kingdom. The utopian ideal is therefore constructed as a projection into the future, or conversely, as a *retrotopia* (Bauman 2013), that is, the recovery of a lost or degraded harmonious past. In the Hebrew version, the earthly pilgrimage is linked to a promised land, but salvation is ultimately a supranatural event and is associated with a holy city: Jerusalem. According to the Old Testament, at the end of time, God will erect a new city for his people, built on an imperishable foundation. This city is the New Jerusalem mentioned in the Book of Ezekiel. The prophetic texts provide a comprehensive overview of the dimensions and materials to be used in the construction of its temple. However, the most significant aspect of the rebuilding of Jerusalem is its symbolism as the advent of the Kingdom of God. This is the same magnificent city that, according to St. John, will be revealed at the Last Judgment. This is the reason why the city is named *Jehovah Shammah*, which translates as "the Lord is present."

> I saw the Holy City, the New Jerusalem, coming down out of heaven from God, prepared as a bride beautifully dressed for her husband. And I heard a loud voice from the throne saying, "Look! God's dwelling place is now among the people, and he will dwell with them. They will be his people, and God himself will be with them and be their God. He will wipe every tear from their eyes. There will be no more death or mourning or crying or pain, for the old order of things has passed away". He who was seated on the throne said, "I am making everything new!" (Revelation 21: 2–5)

This eschatological archetype of utopia was inherited by Enlightenment rationalism, transformed into a narrative of human liberation by Romantic philosophy, and finally presented as a model of social progress by various forms of positivism, including Marxism. In the twilight of the Middle Ages, however, it was still possible to imagine places endowed with mythical qualities and riches, such as the Fountain of Youth or the Seven Cities of Cíbola, sought in America by Spanish explorers like Juan Ponce de León and Francisco Vázquez de Coronado. In the biblical tradition, the exact location of Eden, the divine garden where our first ancestors lived, was never revealed. The Book of Genesis only alludes to it being in the East, but at no point does it say that *Paradise* (the Persian term used in the Old Testament to refer to this garden) had ceased to exist. The transoceanic expeditions that commenced in the Renaissance were inspired by medieval legends such as those of St Brendan's Island

and Antillia, but they still hoped to locate the Eden by travelling across the Atlantic to Asia. Columbus thought he had done so on his third voyage, when he reached the mouths of the Orinoco, which he identified in his report to the Catholic Monarchs of Spain as the Four Rivers of Paradise, whose sources flowed from a distant promontory in the interior of what he called *Tierra de Gracia* (the Land of Grace), on the coast of present-day Venezuela. The Earth, the admiral concluded, was not round, as the ancient Greeks believed, but pear-shaped "like a woman's breast", with Paradise at its highest point, inaccessible except by divine will.

The natural beatitude of the Garden of Eden stands in contrast to the Greek world. The Hellenic tradition favored a territorial representation of the ideal social bond and regarded civil institutions as indispensable for the proper organization of community life. However, we can also find in it the idea of a primeval era of peace, prior to a fall from grace. In *Works and Days*, Hesiod alludes to a golden lineage of humans who lived a life without hardship or suffering in the mythical time of the Titans. This mythological story was taken up by Plato in the *Laws* and in *Critias* (2008). In the latter, he describes the civil peace of primitive Athens, where its inhabitants, divided according to caste and social function, cultivated wisdom and the arts under the protection of the gods, without the need for any political order, until a natural disaster devastated the city, driving its inhabitants into the mountains and reducing them to savagery. The loss of this primitive order marked the end of natural harmony and the beginning of Athenian political history. This mythological tale was later adapted for Latin culture by Virgil and Ovid into a *golden age* of peace and justice that was to return in the eternal cycle of time. These two spatial and temporal references have since been combined in very different ways, but they share the idea that human potential can be fulfilled if the causes of social division and conflict are at some point removed. Since the sixteenth century, this kind of social imaginary has had a name and a literary genre of its own: *utopia*, the non-place. Social utopias have repeatedly upheld the city as the spatial framework for an ideal life, although the dominant version has followed the Judeo-Christian archetype and subsequently shifted from the notion of a collective life realized in an imaginary place to *uchronia*, the search for such emancipation in *another time*.

It may seem an oxymoron to wonder about the space of a social configuration characterized by the very absence of it. In Foucault's (1986) words, utopia is a place without a locus whose relation to the real space of society is either a direct or an inverse analogy: it represents the same society in a perfected version, or its reverse, as dystopia. In contrast to utopia as an unreal space, Foucault also conceived of places that are absolutely different from all the spaces they reflect and speak of: what he called *heterotopies*. Located outside all other places, heterotopies are capable of juxtaposing, making visible, inverting, and questioning spaces that are inherently incompatible. They function simultaneously as a mythic and real contestation of the spaces in which we live. He divided them into three types. In primitive societies, we find *crisis heterotopies*: privileged, sacred, or forbidden places where individuals in a state of crisis or transition, such as adolescents, the elderly, menstruating or pregnant women, the dead, etc. are located. This type of heterotopia would be increasingly disappearing in our modern societies, to be replaced by *heterotopies of deviance*, where people whose behavior deviates from the average or the required norms are confined: prisons, psychiatric hospitals, old people's homes, and the like. Finally, there are *heterotopies of compensation*, whose function is to create an illusory space that denounces the features of real space (for example, the uninhibited privacy of the brothel in the face of public morality) or, on the contrary, to produce an alternative space so perfect that it makes the original appear coarse and degraded. The latter would also be the case of certain forms of colonization encouraged by religious zeal (such as the *city on a hill*, an example to the world of a new alliance between God and his people, to which John Winthrop compared the Puritan settlement of Massachusetts Bay) or the meticulous planning of collective life carried out by the Jesuits in the missions of Paraguay. This compensatory element is what brings heterotopies closer to the universe of urban imaginaries.

In his phenomenological study of the concept of utopia, Bloch (1980) distinguished between the internal *topos* of utopian consciousness, guided by the search and the representations that prefigure the desired object as a yet-to-be, and an external *topos* constituted by the place of the longed-for object, the outline of which we can already glimpse, even if it is not yet complete. Roughly speaking, one could say that the first of these corresponds to the Judeo-Christian utopian imagination, committed to a *future to come*, and the second to the Hellenic one, focused on the adequacy of the existing world to a *telos* or ultimate end. The term utopia, however, lends itself to an amphibology. As is well known, More (2012), its inventor, used the transliteration of the Greek adverb *oὐ* (not) with the Latin character *u* to form the neologism *Utopia*: the imaginary island where a perfect society existed. In keeping with the satirical tone of his text, this orthographic twist also allowed him to play with the prefix *εὖ*, which has a positive connotation in Greek, so that in the poem that precedes the story, Anemolius, the local bard laureate, claims that his island should properly be called Eutopia, a *good place* where every wise man would want to go, since it surpassed Plato's Republic in terms of people, resources and good laws. Since then, utopias have been characterized as representing the optimal state of human sociability. Accordingly, the idealized cities imagined by philosophers and urban planners have focused on three aspects of them: their spatial structure, their political constitution, and their social organization. The emphasis placed on each of these dimensions implies significant differences.

4.3 The Classical City and the Good Life

The Greek *polis* and the Roman *civitas* were, above all, juridical-political entities. In contrast to this, the physical extent of their territories and the number of their inhabitants constituted a subsidiary dimension of this condition. The *polis* referred to both the core of the city (*asty*) and its periphery (*chora*), as well as to the social body that constituted it. In the Roman world, the term *urbs* referred to the physical dimension of the city, its buildings and temples, while *civitas* denoted its inhabitants as a civil community. This fusion of the built city with the living city distinguished the urban features of the classical world from the older civilizations that had emerged in the Fertile Crescent, where the city was a mere excrescence of the political and religious functions of a palatial or temple nucleus. This also explains why in the Hellenic tradition we find a political discourse on the city, but no definition of it or of the urban phenomenon as such. It is therefore difficult to distinguish between the original process of *synoikismos* (the clustering of primitive villages) and the emergence of the *polis* proper (Lafon et al. 2003). For the ancient Greeks, civilized life was synonymous with life in the city. Plato, in *The Republic* (2000), saw the origins of the *polis* in the search for collective subsistence, the specialization of crafts, and the complementarity of human needs met through exchange. The same idea is found in Aristotle's *Politics* (1984), for whom the proper functioning of the *polis* depended on its adequate location, size, and population, so that it could achieve *autarkhíā*, i.e., material self-sufficiency and political self-rule. But the city had a higher purpose: to enable the pursuit of the *good life* (*eudemonía*), which consisted in the full realization of human nature through the practical cultivation of virtue and reason. The Greeks believed in the influence of the environment on human behavior. Aristotle, for example, noted the affinity between the physical location of a *polis* and a given political regime.[2] For him, the ideal size of a city should be such that it could be traversed on foot in a single day and allow for civil life, i.e., personal interaction between all citizens, in order to form an opinion of the civic virtue of public officials. Larger states were typical of barbarians who, like the Persians, were ruled by despots. However, the relationship between the built environment and political power was not obvious. As Thucydides (2004) observed, Sparta, which had come to dominate the Peloponnese, had a relatively modest architectural legacy in comparison to the grandeur of Athens, which had been vanquished by the former.

The Hellenes' attachment to their cities was not driven by military glory or urban splendor, but rather by an institutional system that shaped the individual character and collective identity of their inhabitants. An ideal *polis* should therefore correspond to an optimal political condition. The utopian component of the Greek city was not dependent on its physical design; rather, it consisted of a moral-political imaginary generated by patterns of urban sociability. The cultivation of the good life was guided by *nomoi* (laws, customs, or conventions) rather than *physei* (natural qualities). To be a citizen of the *polis* entailed the sharing of a moral and political commitment. Such was the esteem in which Socrates held his civic dignity that he was unwilling to flee for safety after his sentencing. In *Crito*, Plato reiterates his master's conviction that the quality of citizenship is revealed in respect for the laws and acceptance of their corrective function. These are the same civic virtues that Pericles extolled in his funeral oration for the Athenian soldiers killed in the war against Sparta. Outside the laws of the *polis*, the individual is stripped of his dignity and even his natural condition, for man, according to Aristotle's celebrated definition, is an animal that lives in cities (*zoon politikón*). His rationality, moral judgement, and communicative competence are inextricably linked to his urban condition. The *polis* thus embodies

[2] "A citadel hill is suitable for oligarchy and monarchy, and a flat site for democracy; neither is favorable to an aristocracy, but various strong positions" (Aristotle 1984: 1330b17).

the highest good, both in terms of its intrinsic purpose (collective autonomy) and its superiority to lower-level communities such as the *oikos* (household) or the *phyle* (clan). The city is the primary means by which the individual fulfils his needs and satiates his appetites. Consequently, Aristotle concluded that an individual who lacks needs or is incapable of sharing is not part of the city: he is either a god or a beast (Aristotle 1984: 1253a).

In Greek philosophy, we can also discern reflections on the ideal urban form, yet these are always subordinated to its political functionality. A number of Plato's dialogues make reference to *kallipolis*, the just, beautiful, or ideal city. In *The Laws*, for instance, Clinias proposes to "build up by arguments the framework of a city, as if we were erecting it from the foundations" (Plato 1961: 702d), with the intention of imagining the optimal way to inhabit and live well in it. Plato's *Republic* is typically regarded as the epitome of utopian literature in the Greek tradition. In fact, *The Republic* is primarily a treatise on justice and the optimal configuration of a political system in accordance with the nature of the people and the social groups that comprise it. The Platonic idealization is thus concerned with the *politeia* (the political regime of the city) and not so much with its architectural order or spatial materialization. Plato's only allusions to urban morphology appear in the *Timaeus* and in *Critias*, where the island city of Atlantis, which is said to have been at war with Athens thousands of years earlier, is mentioned. In *The Laws*, the figure of the philosopher-king is advocated as the best form of government for the imaginary city of Magnesia. In both cases, the radiocentric structure is presented as the most perfect for a city, a morphology that would be taken up in later urban utopias, such as Campanella's *City of the Sun*. In contrast, Aristotle considered the orthogonal layout of the city to be the most beneficial and aesthetically pleasing, although less effective in terms of defending against enemy attacks. He proposed a combination of the grid layout devised by Hippodamus for the reconstruction of Miletus following its destruction by the Persians and the organic pattern observed in ancient Greek cities. In any case, such considerations had little bearing on the actual occupation of urban spaces, which were generally lacking any kind of prior planning. In Greek cities, the private sphere of domestic life was of secondary importance to the public sphere of community activities. These revolved around the core buildings and spaces of the polis (the walls, the *acropolis*, the *agora*, the *théatron*, the *gymnásion*, and the *stàdion*), whose location was determined by considerations of physical convenience or the sanctity of the site (Martin 1987). The residential quarters were simply located in the remaining spaces. The *agora* constituted the central hub of the *polis*, serving as a meeting place for the daily events of social and political life. In contrast, the *acropolis* was the elevated part of the city, a defensive enclosure that gradually assumed religious functions in Athens. From the sixth century BC onwards, with the spread of democracy, the Athenians' demands for protection extended to the whole city. Consequently, the limitation of the fortifications to the *acropolis* was interpreted, as Aristotle did, as the defence of oligarchic interests. Similarly, the rebuilding of Athens following the devastation of the Greco-Persian Wars did not adhere to a pre-established plan. Instead, it restored the traditional places of worship and assembly, while allowing the remainder to develop in an organic, disorderly manner. It may therefore be concluded that the Greek approach to urban planning was eminently pragmatic and did not appear to be based on any discernible theoretical framework (Morris 1994).

As with the case of Greece, the domination of the Mediterranean world by Rome was founded upon the establishment of urban colonies and an ethnocentric view of its relations with other peoples. The *inauguratio*, the practices associated with the founding of Roman cities, was a complex ritual involving auspices, prayers, and the symbolic organization of the territory. The objective was to create a specific spatial order. This entailed establishing a focal point on Earth from which the cosmogony could be replicated, thereby evoking the original act of creation (Rykwert 1976). Consequently, each urban foundation was imbued with a religious meaning and associated with a founding hero who became its protector. As Plutarch (2001) recounts in his *Life of Romulus*, the founders of a city first dug out the *sulcus primigenius*, the inviolable furrow that defined the city limits or *pomoerium*, with a plough drawn by oxen. At the center of the designated area, a circular pit, the *mundus*, was excavated. This was used to bury the sacrificial offerings and the earth brought by the founders from their places of origin. From this point onwards, the plan of the city and its *territorium* (jurisdiction) were established. However, the Romans held a different understanding of the superiority of urban life to that of the Greeks. The Roman ecumene encompassed the entire territory subjected to the process of civilization that emanated from its centre: *Roma caput mundi* (Rome, the capital of the world). The *barbaricum*, the peoples residing beyond the *limes* (the frontiers of the Empire) were excluded from the *humanitas*, the elevated culture of which Rome was the exemplar and disseminator. In contrast to the *ferocitas* of the barbarians, *humanitas* comprised qualities such as kindness, compassion, intellectual curiosity, education, and cultural refinement.

St. Augustine, a fourth-century Roman imbued with Christian values, reformulated the classical morals of the political community by describing human history as an earthly journey in which two *cities* or societies into which humanity is divided coexist in perpetual conflict: those who live according to man, corrupted by sin, and those who live in a state of holiness (Augustine 1984). The first one is the successor to the city originally founded by Cain and is destined to endure an endless cycle of conflict and torment at the hands of demons. The other one, the city of the saints, is a free and heavenly society that lives according to the grace of God and in the love of the common good. The Augustinian reflection occurred within the context of a profound social crisis. Following the establishment of Christianity as the religion of the Empire, Rome was subsequently sacked by Alaric in 410 AD. The dismayed citizens began to question what was to come and the relationship between spiritual and temporal power. The bishop of Hippo adopted Cicero's view that the *populus* was not merely a collection of individuals, but rather a collective agreement to adhere to the same laws and a shared community of interests, a *res publica*. He concluded that Rome had long ceased to be such due to the degeneration of its customs and the absence of true justice. Civil happiness was to be found in the city of God, where there would be no inferior who would envy a superior, and where it was impossible to sin or die. Earthly politics, in which the righteous and the sinners clash, as the flesh fights against the spirit, was only a transition to a perfected life, that of the City of God. With the advent of Christianity, the civic utopia of the classical world underwent a significant shift in its center of gravity, moving towards a transcendent dimension. However, the Kingdom of God continued to be represented as a heavenly republic.

4.4 In Search of the Ideal City: From the Renaissance to the New World

For over a thousand years after Augustine, the civil imaginary of utopia disappeared, replaced by an otherworldly ideal. In Western Europe, the Early Middle Ages saw a decline and abandonment of many Roman cities. Urban life fared better in the Oriental part of the Empire, where cities remained important centres of culture, commerce, and political power throughout the Byzantine period. By the end of the millennium, the population of Rome had decreased to approximately 30,000 from its peak of one million, reached under the rule of Emperor Augustus (27 BC - AD 14). In the Western cities that survived, their traditional spaces fell into disuse. Their buildings were often repurposed for new constructions, such as cathedrals and city walls, but the lack of a strong central power in feudalism worked against the realization of large public projects. In many instances, the vacant spaces within the walls of receding urban areas were utilized for agricultural purposes, including the grazing of animals. The squares and open spaces of medieval cities retained their commercial functions but lost their former political role (Colom González 2022). Consequently, the cathedrals, whose atriums were sometimes used for the administration of justice, and the town halls, the seat of civil power, became the main public buildings. They rivalled in importance with the palaces and towers of the nobility, as can still be seen in many historic Italian cities. Deprived of its ancient civic institutions, the medieval city became little more than an interstice, surviving under the authority of the bishops in a sparse network of territorial lordships. But wherever the freemen who inhabited these enclaves managed to assert their rights against the Holy Roman Empire, the papacy and the nobility by means of municipal charters (especially along the urban belt that began in northern Italy, crossed the Alps and the Low Countries and reached the Baltic Sea), seigniorial rule failed to take hold and the formation of dynastic states proceeded at a slower pace.

The new millennium marked a change in this trend of urban decadence. The restoration of trade routes and the generation of economic surpluses resulted in the petrification of public and private wealth in the form of new religious and secular buildings. The Renaissance saw a resurgence of urban life, which coincided with a renewed interest in the idea of the city as a virtuous space. This change was accompanied by a shift in the focus of attention from the previous emphasis on the governance of the city to a greater interest in the relationship between its physical form and the quality of the society it contains. This is reflected in the pictorial images of the *città ideale*, but also in the design of newly founded Italian towns such as Pienza, Giulianova, Cortemaggiore, Acaya, Sabbioneta, and others. Nevertheless, despite the fact that these cities were influenced by the cult of classical antiquity and constructed in accordance with geometrical and mathematical principles, their primary objective was to fulfil military or representational requirements, rather than to reorganize society. The pursuit of the ideal city was driven by an ethical imperative to foster *virtù* and human improvement through the perfection of the built environment. Theorists such as Leon Battista Alberti, Antonio Averlino (known as Filarete, "the friend of virtue"), or Francesco di Giorgio Martini represent a small number of cases in which the idea of beauty dominates the utopian imagination and links it to its physical materialization. This feature distinguishes them from the standard utopians of the sixteenth century. For these humanists of the Quattrocento, the search for *bellezza* in the city, through harmonious design and the perfection of architectural form, was a means of reflecting and cultivating the dignity of its inhabitants. The dimensions of the human body were to serve as the *misura*, or just measure, within which classical morality was interpreted. This interest in classical culture led to the rediscovery of the work of Vitruvius, with his theories on winds and their relationship to urban morphology (Lagopoulos 2009). The radial design of Vitruvius, adapted to the defensive needs of Renaissance cities, was employed by Filarete in his project of Sforzinda, an ideal city never built in homage to Francesco Sforza, the first Duke of Milan. It was also used in numerous Venetian cities and fortifications, such as Nicosia and Palmanova. The concentric model, adopted centuries later by the panopticon system in modern prisons, sought to imbue the city with a sense of harmony and a recognizable centre from which to control it. However, like other Renaissance urban designs, it ultimately reflected a geometric ideal animated by ethical and aesthetic motives with no direct connection to urban habitability (D'Autilia and Spada 2019).

The urban utopias that emerged from the literary imagination of sixteenth-century authors permitted social criticism to be expressed behind the façade of literary fiction. Tracts such as Thomas More's *Utopia*, Francis Bacon's *New Atlantis*, Tommaso Campanella's *City of the Sun*, James Harrington's *Republic of Oceana*, Johannes Andreae's *Christianopolis*, and numerous others contemplated the ideas of individual freedom, common property, the dissolution of private life and the family, eugenics, and the education of children by the state. Some of these ideals survived into the Industrial Revolution and were tested on a small scale in the Fourierist and Owenist communes. However, there was another kind of utopian transplant that travelled from the Old World to the New. As in Europe, the influence of Vitruvius can be observed in the decrees issued by Philip II of Spain with the intention of rationalizing the process of urbanization in the Americas. For the Spanish mentality of the Golden Age, imbued as it was with Aristotelian and Ciceronian values, urban life and its architectural correlate were the epitome of a civilized form of existence. The urban mentality of the Creoles and Crown officials contrasted with the naturalistic primitivism that distorted the interpretation of indigenous societies by some of the early conquistadors and missionaries, who often saw the natives as pagan barbarians or as naive souls untouched by the evils of the Old World. However, as Juan de Solórzano Pereira, one of the most renowned jurists of the Indies, recognized, the conversion of the natives required their reduction to Indian townships (*pueblos de indios*) in order to instruct them in the faith and force them to pay tribute. The foundation of cities, therefore, represented the materialization of the territorial rights granted by the Crown to the conquistadors, but it also expressed a belief in the malleability of social relations and the possibility of adapting them to a predetermined design (Romero 1999).

Just as the news of a New World can be recognized behind the utopian genre inaugurated by Thomas More, the urban foundations in the Americas sometimes revealed magical and religious elements. In the initial decades following their arrival, the conquistadors sought out mythical cities, drawn from medieval legends that fueled their feverish imagination. The millennialist doctrines propagated by the first mendicant orders to arrive in the New World imbued missionary urbanism with a utopian spirit, which saw the evangelization of the indigenous peoples as the culmination of human salvation and the prelude to the second coming of Christ. This apocalyptic vision was foreseen by Francesc Eiximenis, an influential fourteenth-century Catalan Franciscan, in *Lo Crestià*, his unfinished Christian encyclopaedia. His ideal city was a rough outline of the structure of the New Jerusalem described in the Book of Ezekiel, and was remarkably similar to that adopted by numerous missionary and secular foundations in Mexico. Such is the extent of this prophetic text's influence that it has been described as the most accessible treatise on urban planning in the New World, given that it was carried by the friars in their illustrated Bibles (Lara 2004; Navarro Segura 2006). Furthermore, there is evidence that an annotated copy of More's *Utopia* was in the possession of Vasco de Quiroga, the first bishop of Michoacán in the viceroyalty of New Spain. Indeed, there is a striking parallel between the ideas presented in the book and some of the practices tested with the Purepecha Indians in the mission town of Santa Fe de la Laguna (Zavala 1937; Florescano 1997). Additionally, from the seventeenth century onwards, the Jesuits, Franciscans, and other Catholic orders initiated the establishment of missions for the indigenous population, which were based on principles that differed significantly from those of the colonial towns and *encomiendas*. In the eyes of the Spanish Crown, the missions served to populate and secure border areas in dispute with other colonial powers or inhabited by hostile natives. Such areas included the present-day territories of the Californias, Paraguay, the Argentine province of Misiones, and the Brazilian state of Rio Grande do Sul. The Guarani missions, in particular, constituted an experimental utopia that tested novel forms of acculturation through the collective organization of labour and the regulation of the daily lives of their inhabitants. Despite the absence of explicit references to the planimetric and architectural concepts employed in the construction of the Jesuit missions, subsequent apologists of the Society of Jesus in exile advanced the notion that their colonizing model was an exemplification of the Platonic ideal city (Peramás 2003; Goycoolea Prado 2016). Nevertheless, it should be noted that not all accounts of cities in the Americas can be attributed to colonial origins. It is also the case that urban societies and narratives about their origins were engendered by pre-Hispanic civilizations. Of these legends, Aztlan is perhaps the most closely associated with the concept of an urban utopia. Aztlan was believed to be the ancestral home of the Aztec people, who migrated from what is now the southwestern United States in search of a new land. They found it on the shores of Lake Texcoco, in the

Basin of Mexico, where the sight of an eagle devouring a snake on top of a cactus is said to have inspired them to found Tenochtitlan, the future capital of the Mexica empire. After colonial independence, urban areas continued to play a central role in the socio-economic landscape of the new countries. In fact, it can be argued that what became independent were a series of head cities that exercised their influence over vast regional hinterlands while competing with each other. The creation of modern territorial states and the possibility of governing them from their capital cities was a conflictive process that dominated much of Latin American political history in the nineteenth century. The paucity of financial resources precluded the implementation of ambitious urban projects until the region's integration into the global export capitalist economy. Nevertheless, if there was a political utopia linked to an urban project in this period, it was Argirópolis, the imagined island capital of the Confederated Provinces of the River Plate, proposed by Domingo Faustino Sarmiento as a solution to the troubled relations between the different states bordering the Paraná Delta.

In Europe, the interconnection between the ethical and aesthetic values of the city persisted throughout the Enlightenment. Indeed, the Renaissance's separation of urban form and habitability demonstrated a certain affinity with the so-called French visionary architects (Kaufmann 1953) of the late eighteenth century, whose geometricism would anticipate the modern architectural avant-garde. Giambattista Piranesi's engravings, characterized by their oversized ruins and whimsical imaginary prisons (*carceri d'invenzione*), evince an air of unreality that prefigures the designs of Claude-Nicolas Ledoux. One of Ledoux's most ambitious projects, the Royal Saltworks at Arc-et-Senans in Eastern France, anticipated later experiments in communal living such as phalansteries and *familistères*. Although only the industrial component of the project was ever constructed, Ledoux continued to work until his death on the ideal city that was to complete it. His comprehensive vision of communal life and his concept of *architecture parlante* (speaking architecture, which would express function through form) are reflected in the numerous buildings he designed for the project. These include temples dedicated to equality, happiness, or the glorification of women, as well as the *oikema* or house of pleasure, a building with a phallic layout in which men would learn to educate their sexual appetites. The designs of Etiénne-Louis Boullée and Jean-Jacques Lequeu also belong to this category of visionary architecture. Among his many unrealized (and unrealizable) projects, Boullée designed a cenotaph in honour of Isaac Newton, a library and a grand opera house for Paris, all on a colossal scale and aiming for the sublime. Lequeu's inventive and prolific graphic repertoire encompassed a diverse range of genres, from monumental exoticism to erotomania. His rediscovery in the twentieth century was largely due to his extravagant designs, which in some ways anticipated the themes of expressionism and surrealism. The visionary nature of these architectural projects, isolated from the society they were intended to urbanize and improve, brings them close to the record of utopian cities. Nevertheless, the advent of industrialization eventually led to a shift in focus from aesthetic and political considerations of urban form to an analysis of the social impact of modernity on cities.

4.5 City as a Right and as an Experiment

The growth of European and North American cities in the early twentieth century, a foretaste of what was to happen exponentially in the rest of the world, revealed not only the unhealthy and explosive nature of urban overcrowding but also the technical difficulties of dealing with such expansion. The alleviation of sanitary and social problems in congested urban areas could not be achieved by piecemeal measures. This resulted in the initial efforts to implement comprehensive urban planning on a large scale. The tension between vertical and horizontal solutions would shape the urban debate throughout the twentieth century. In a seminal essay published in 1903, Georg Simmel proposed that the urban condition was associated with the loss of a sense of community and the risk of dehumanization (Simmel 2002). In Simmel's view, the overcrowded life of the big city produced a type of subjectivity characterized by a constant need to adapt, which ultimately led to a withdrawal of personality or a blasé attitude and a certain form of alienation. This anti-urban stance reflected a concern about the ambivalent effects of modernity, a concern that was shared by many of the founders of urban sociology. With Robert E. Park, an American disciple of Simmel, and the school of sociology he initiated at the University of Chicago, reflection on the city took on a new perspective: that of human ecology (Park and Burgess 1925). This approach, which was informed by Darwinian assumptions, posited that urban life develops in habitats that are the result of the interaction of individuals with each other and with the built environment. The city was regarded as a form of association that facilitated a range of relationships different from those found in traditional communities. As Louis Wirth (1938), a disciple of Park, summarized, urbanism was a new way of life. In the heterogeneous fabric of the urban milieu, traditional ties of kinship and community are weakened and replaced by competition. Urban social interaction was typified by a combination of physical proximity and personal distance. In order to avoid the potential disorder and lack of responsibility that such anonymity creates, urban society must adhere to rigidly predictable routines. For Wirth, the clock, traffic lights, and street signs are examples of the kind of formal control to which hectic urban life is necessarily subjected. Ernest Burgess expanded upon this concept, proposing that the process of urbanization can be understood as a dynamic of inter-group competition for space. This competition is guided by certain patterns of repulsion and aggregation, such as employment opportunities or ethnic affinity. Drawing on Chicago, then one of the fastest growing American cities, he formulated a theory of urban clustering through concentric zones based on ethnic background and social class around the economic and commercial core of the city (Park and Burgess 1925: 55).

Criticism of the alienation caused by modern urban life found a response in ruralizing currents such as the Garden Cities movement that began in England at the turn of the twentieth century. Ebenezer Howard (1898), a social reformer and its principal advocate, sought to combine the best aspects of urban and rural life through a system of small towns radially linked by rail around a large park and surrounded by an agricultural belt. With no higher education or architectural training, Howard's vision was strongly influenced by the economic theories of Henry George and by hygienist and ecological ideas. His view of urban life was based on social co-operation, not in the sphere of labour, but in that of leisure and civic association. The elementary unit of his ideal city was the neighborhood and the detached house with garden. The surrounding agricultural belt would ensure proximity to the food chain and lower prices for residents/consumers. These towns, such as Letchworth and Welwyn in Hertfordshire, England, soon deviated from Howard's original guidelines and became inaccessible even to the workers who built them. Nevertheless, they managed to inspire the New Towns movement in Britain after the Second World War (Hall 1988).

The Broadacre City project, designed by the American architect Frank Lloyd Wright in 1932, describes a similar horizon of urban decentralization. It is often credited with anticipating the American suburb, with its single-family homes and dependence on the automobile. In contrast to Howard's cooperativism, Wright's vision was dominated by Jeffersonian democratic individualism, but it also echoed the philosophical concerns of Transcendentalist writers such as Waldo Emerson and Henry D. Thoreau, who extolled the independence and freedom afforded by America's natural spaces. Individuality, Wright argued, is primarily an internal quality of the mind. His idea of an *organic spirituality* is reflected in his concept of urbanism and freedom. The *raison d'être* of cities, he believed, was originally the need for protection. Since this is no longer necessary, large cities are now redundant. In practice, his proposal dissolved the city into a disaggregated landscape of individual houses, farms, and small businesses linked by a network of highways. His manifesto was appropriately entitled *The Disappearing City* (Wright 1932). This text is a vindication of open spaces as opposed to the *sterile verticality* of life in large urban concentrations typical of the industrial age. Cities, in their exaggerated modern derivative, served speculative interests and destroyed individual character. In response, Wright proposed using technological advances to break up urban concentration and restore freedom of movement. This decentralizing vision of the city has often been criticized for failing to acknowledge the intrinsic complexity of the urban phenomenon and for promoting designs that are in fact anti-urban utopias. Furthermore, it has been argued that this approach ignores environmental sustainability in terms of land use and infrastructure consumption.

At the opposite pole is the concept of the *ville radieuse* (radiant city). This was conceived in 1924 by Charles-Edouard Jeanneret, better known as Le Corbusier (1964), but only partially tested in the post-war period. His idea of the radiant city was to respond to the urgent need for urban housing with economy of means by increasing density and architectural efficiency through the use of skyscrapers and a centralized organization of urban space. Despite its technocratic character, Le Corbusier ultimately claimed a commitment to practical humanism and individual freedom. Its avowed aim was to end the overcrowding and unhealthiness of the old industrial cities and to free their inhabitants, especially women, from the heavy burden of housework. His approach found support in the International Congresses of Modern Architecture, a forum organized around his figure, and was embodied in the Athens Charter of 1933, an urban manifesto written by Le Corbusier himself and the Catalan architect Josep Lluís Sert. The Charter identified the four human functions to which modern urbanism had to respond: living, working, leisure, and mobility. Although his approach advocated a decided interventionism in the urban fabric, it also reaffirmed the natural dimensions of man as the measure of architecture. In fact, his *Modulor* is basically a modernist replica of Leonardo da Vinci's Vitruvian Man.[3] In all other respects, Le Corbusier's postulates were radically different from those of Howard and Wright. For him, the garden city led to the sterile isolation of the individual and the annihilation of the social spirit. By increasing travel time, Broadacre City model was also an attack on individual freedom. His own urban model, zoned and functionally differentiated, was based on large apartment blocks with extensive green spaces and underground highways leading to the city center. According to his peculiar definition, the radiant city was a *vertical garden city*. This structure would take cars off the streets, do away with the suburbs, which he described as a mistake in urban planning, and bring community services such as daycares, canteens, and gymnasiums into the buildings. Modern housing should be, in his famous definition, a *machine à habiter*, a machine for living in. The social effects of this model, he believed, would be egalitarian: a city without class-based spatial organization. Ultimately, the radiant city came to merge two utopian currents of modernity: it was the Fourierist phalanstery at the heart of the industrial society that Saint-Simon had envisioned (Fishman 1982). Underlying this urban design, however, was a broader vision of modernity as a process of centralized homogenization induced by the geometric planning of the built environment. Ironically, the remnants of this conception can still be seen today in the degraded *banlieues*, housing projects, and *dormitory neighborhoods* of many European and American cities, often transformed into spaces of social marginalization.

The assumptions of these urban theories persisted until the mid-twentieth century, when the social sciences took a more critical stance. A landmark in the questioning of modern urban planning was the work of Jane Jacobs (1961), who described urbanism as a *pseudoscience* based on simplifications, superstitions, and symbols that were completely divorced from the real world. Jacobs denounced the type of urban planning that had been carried out in the United States since the beginning of the century. She also criticized the simplistic assumptions of the Chicago School about the alienation and impoverishment that urban life supposedly produced. On the contrary, her detailed analysis of Boston's North End, disqualified as a ghetto by official urban planners, highlighted the extraordinary vitality of community ties fostered by social diversity and economic complementarity. Whereas Howard, Wright and Le Corbusier had each, in their own way, eliminated the street as a space for social interaction, Jacobs saw beneath the apparent disorder of the downtown a complex but coherent socio-urban dynamic, like a dance unfolding on the city's sidewalks: "An intricate ballet in which the individual dancers and ensembles all have distinctive parts which miraculously reinforce each other and compose an orderly whole" (Jacobs 1961: 50). For her, the task of urban planning was precisely to manage the organized complexity represented by actual cities, not to imagine their ideal condition.

[3] The *Modulor* is an anthropometric scale used by Le Corbusier in his buildings and later codified as an architectural standard.

Jacobs' work preceded the turn to Marxism that urban sociology underwent in the 1960s. The political context of this period pressed to abandon the Darwinian paradigm of social ecology and to recognize the new dynamics of urban transformation, which no longer referred to the classical rules of production and reproduction of capital, but to new forms of interaction between space and society. This theoretical shift placed power relations and conflicts of economic interest at the center of spatial analysis, forcing Marxism to acknowledge the specificity of the urban phenomenon as a force of social structuring, its peculiar function in the processes of capital valorization, and the diminished role of the working class as a driving force for political change (Katznelson 1993). It was a French philosopher, Henri Lefebvre, who most directly linked the meaning of urban space to social practices, thus promoting a *spatialization* of Marxism. He no longer interpreted the city as a passive environment subject to the natural limits of space. The city was now seen as the product of social forces driven by capitalist relations of production. More specifically, Lefebvre defined the space of modern cities as a social production linked to the conflicts of capital valorization. According to this, the market has become the main dynamiser of cities and not the other way round, as was the case in the past. Nowadays, the economic cycles of capitalism do not simply take place *in* cities, but *through* cities. For him, Marx did not realize that the industrialization of society entails its urbanization and that the control of its development requires a specific knowledge of the processes that take place in cities. The effects of industrialization eventually have become problems of urban development, something that goes beyond the provision of decent housing for the working classes, about which Engels wrote in his classic study of Manchester in the mid-nineteenth century. In accordance with a praxeological ontology, Lefebvre denounced the abstract character that space assumes when considered in isolation from the human activities that imbue it with meaning. In this sense, urbanism would be "an ideology that formulates all the problems of society in terms of space and transposes everything that comes from history and consciousness into spatial terms" (Lefebvre 1996: 99). To articulate this critique, he drew on the classic Marxian distinction between exchange value and use value. Use value is the city itself, urban life, and time; exchange value is the spaces bought and sold, the consumption of products, goods, places, and signs. The urban is based on the use value of the city. Exchange value and the generalization of commodities as a result of industrialization tend to destroy it. In other words, urban society is governed by a different logic from that of the commodity, since the use value of cities escapes the demands of commodification. The *right to the city* would thus consist in the democratization of the urban process, transforming it into something more than the right of private access to the resources embodied in the city.

Written in the immediate context of the May 1968 revolts in France, Lefebvre's text defended urban life as a full form of existence and a necessary condition for a renewed humanism. For him, city dwellers were the bearers of the right to enjoy the goods of urban life, in particular the right to live in its core and not be relegated to the *banlieues* and ghettos of the periphery. This process had begun in France after the Revolution of 1848, when Napoleon III commissioned Baron Hausmann to remodel Paris with grand boulevards that broke up the medieval layout of the working-class districts. But for Lefebvre, the right to the city was not merely the right to keep or return to the traditional city. It was the right to a transformed and renewed urban life. This required the elaboration of a comprehensive theory of the city and urban society, using the resources of science and art. This science of the city had to be oriented towards a *new humanism* that sought a different type of man and a different social practice: the practice of urban society. Thus, Lefebvre argued, urban life has not yet begun. Insofar as urban spaces concentrate the contradictions of capitalism, any future revolution will necessarily have to be an urban revolution. The old Marxist utopia becomes for him an urban utopia, but this new condition cannot be created by the decree of the sciences. Only social life, praxis in its global sense, would have this capacity.

Writers such as Jacobs and Lebefvre shared, from very different perspectives, a critique of the technocratic, functionalist, and Darwinist assumptions of modern urban theory and practice. Some of the problems generated by this type of planning are the neglect of the existing city in favor of an idealized one, and the gigantism of its scale, with the consequent loss of control over inhabited spaces, the privatization and erosion of public life, the fragmentation and dispersion of the city, the destruction of heritage, and the anonymous and anomic dislocation of urban dwellers. These tensions have resurfaced with the environmental crisis caused by industrialism and the pressures of a growing urban population. To some extent, the concept of urban utopias has been superseded by the depiction of urban dystopias in science fiction, such as in the film *Blade Runner*. This celebrated cinematic work offers a dismal portrayal of the urban environments that will emerge in a world where population density and environmental pollution have reached critical levels. Nevertheless, there has also been the emergence of a range of innovative approaches to urbanism, imbued with utopian spirit. In the 1960s, Lyndon Johnson's Great Society in the United States gave impetus to the so-called Model Cities Program, a primarily social-oriented effort to combat urban poverty. This was, in any case, markedly distinct from planned cities, which are typically designed from the ground up with a specific technical, political, or social objective in mind. In some cases, such as the Minnesota Experimental City, conceived as a technological utopia, the project never went beyond the imagination of its promoters, as depicted in a series of science fiction comics. The Woven City project, currently being developed by Toyota in Japan around hydrogen technology, and the Sidewalk Labs initiative, sponsored by Google in Toronto, can be viewed as updated, smaller-scale versions of similar experimental projects. But the idea refuses to die, and Telosa, a new technological urban utopia, is being proposed in the United States, in an as yet undisclosed location, by e-commerce billionaire Marc E. Lore. In other cases, the purpose of planned new cities has been political, to decongest or relocate the administrative center of a country, the most famous examples being Canberra and, more spectacularly, Brasilia. There are older antecedents, such as La Plata (Argentina), and modern sequels, such as the New Administrative Capital being built in Egypt. In the present era, a number of urban projects of enormous proportions are confronted by equally formidable challenges, like Neom (The Line), a 170 km linear city to be built in Saudi Arabia and powered by renewable energy sources. Emblematic and rather dystopian *cities for tourists* can be found in Spain (Marina d'Or) and Malaysia (Forest City), while a replica of this dubious speculative model is under construction in New Alamain, Egypt. On a different scale, there are also *intentional communities* created to promote a social, spiritual, or cultural ideal. Twin Oaks, a small commune in Virginia loosely inspired by the behavioral theories of the psychologist B. F. Skinner, is of the first type. Auroville, in India, is of the second kind, in this case based on the metaphysical teachings of Sri Aurobindo. Arcosanti, in Arizona, originally designed by Paolo Soleri in the 1970s as an environmentally friendly and energy self-supporting township, has evolved into a self-contained urban laboratory with an emphasis on the architectural conditions of communal living. In Orania (South Africa), on the other hand, material self-sufficiency and cultural self-determination have been put at the service of an ethnic goal: the preservation of the Afrikaans language through a *Volkstaat*, an all-white Afrikaner homeland.

In conclusion, cities are socially and culturally diverse yet holistically integrated entities. They present themselves to us simultaneously as human communities, as spaces for social and symbolic interaction, and as architectural environments constructed over generations. The challenge of conceptualizing the city as a right or a common good lies precisely in recognizing that some social resources are inherently urban. This requires consideration of the multiple dimensions that comprise the quality of urban life, such as habitability, the capacity to identify with the built environment, and the advancement of opportunities and community cohesion. This is why the utopian elements inherent to modern urbanism resonate with a perspective that strives to reconnect the city with its human dimension.

Acknowledgements This research was funded by grant PID2020-120021GB-100R from the Spanish Ministry of Science, Innovation and Universities MICIU/AEI/10.13039/501100011033.

References

Aristotle (1984) Politics. Chicago University Press
Augustine (1984) The city of God. Penguin Books
Bauman Z (2000) Liquid modernity. Polity Press
Bauman Z (2013) Retrotopia. Polity Press
Bloch E (1980) Abschied von der Utopie? Suhrkamp
Colom González F (2022) Los espacios urbanos de la democracia. Del Ágora a la Plaza. Arbor 198(803–804):a635
D'Autilia R, Spada M (2019) Shaping ideal cities: the graph representation of the Urban Utopia. Environ Plan B: Urban Anal City Sci 46(3):423–444
Fishman R (1982) Urban Utopias in the twentieth century: Ebenezer Howard, Frank Lloyd Wright, and Le Corbusier. MIT Press
Florescano E (1997) Tomás Moro, la Utopía y el experimento de Vasco de Quiroga. La Palabra y El Hombre 101:53–81
Foucault M (1986) Of other spaces. Diacritics 16(1):22–27
Goycoolea Prado R (2016) De pueblos indios a ciudades platónicas. El ideario utópico de las misiones jesuíticas de la Provincia del Paraguay. In: Colom González F (ed) Forma y política de lo urbano. Planeta Colombiana, pp 263–289

Lafon X et al (2003) La ville antique. Histoire de l'Europe urbaine. Seuil
Hall P (1988) Cities of tomorrow. Blackwell
Howard E (1989) Garden cities of tomorrow. Swan-Sonnenschein
Jacobs J (1961) The death and life of great American cities. Vintage Books
Katznelson I (1993) Marxism and the city. Clarendon Press
Kaufmann E (1953) Three revolutionary architects: Boullee, Ledoux and Lequeu. Trans Am Philos Soc XLII:436–473.
Lagopoulos A (2009) The semiotics of the Vitruvian city. Semiotica 151(1/4):193–251
Lara J (2004) City, temple, stage. Eschatological architecture and liturgical theatrics in New Spain. University of Notre Dame Press
Le Corbusier (1964) The radiant city. The Orion Press
Lefebvre H (1996) Writings on cities. Blackwell
Manuel FE, Manuel FP (1979) Utopian thought in the western world. Harvard University Press
Martin R (1987) Architecture et urbanisme: aspects historiques et fonctionnels. École Française de Rome
Marx K (1990) Capital. Penguin Books
More T (2012) Utopia. Penguin Books
Morris AEJ (1994) History of urban form: before the industrial revolutions. Longman-Wiley
Mumford L (2011) The story of Utopias. Barnes & Noble
Navarro Segura MI (2006) Las fundaciones de ciudades y el pensamiento urbanístico en la era del Descubrimiento. Scripta Nova 10(218) https://www.ub.edu/geocrit/sn/sn-218-43.htm
Park RE, Burgess E (1925) The city. Suggestions for the investigation of human behavior in the urban environment. The University of Chicago Press
Peramás JM (2003) Platón y los guaraníes. Parroquia de San Rafael
Plato (1961) Laws. Harvard University Press
Plato (2000) The Republic. Cambridge University Press
Plato (2008) Timaeus and Critias. Penguin Books
Plutarch (2001) Plutarch's lives. Modern Library
Romero JL (1999) Latinoamérica: las ciudades y las ideas. Editorial Universidad de Antioquia
Rykwert J (1976) The idea of a town. The anthropology of urban form in Rome, Italy and the ancient world. Faber & Faber
Sargent LT (2000) Utopian traditions: themes and variations. In Shaer R et al (eds) Utopia. The search for the ideal society in the western world. The New York Public Library, Oxford University Press, pp 8–17
Simmel G (2002) The metropolis and mental life. In: Bridge G, Watson S (eds) The Blackwell city reader. Wiley-Blackwell, pp 11–19
Thucydides (2004) History of the Peloponnesian war. Dover Publications
Touraine A (2000) Society as Utopia. In: Shaer R et al (eds) Utopia. The search for the ideal society in the western world. The New York Public Library, Oxford University Press
Wirth L (1938) Urbanism as a way of life. Am J Sociol 44(1):1–24
Wright FL (1932) The disappearing city. Payson
Zavala S (1937) La Utopía de Tomás Moro en la Nueva España, y otros estudios, Biblioteca Histórica Mexicana de Obras Inéditas

Francisco Colom-González is a Research Professor at the Institute of Philosophy of the Spanish National Research Council in Madrid (Spain). Over the years, he has led and participated in a number of projects on cultural and religious diversity, national identities, and political space. He has written extensively on the political philosophy of the city in several of his recent edited books: *El espacio político* (2015), *Forma y política de lo urbano* (2016), *Narrar las ciudades* (2020) and *Urban Justice* (Springer Nature, 2024). He is currently co-directing the research project funded by the Spanish Ministry for Science and Innovation titled *The Just City. Exclusion, belonging and the commons: an urban perspective on theories of justice.*

Chapter 5
Urban History and the Urban

Eduardo Kingman-Garcés and Mireya Salgado-Gómez

Abstract This essay focuses on Andean cities to examine the role of historical investigation in contemporary understandings of the urban phenomenon. We situate the urban at the core of this analysis from which we look at changes in the economy, the governance of populations, social relationships, urban modernity, structures of class and the construction of subjectivities. We propose a research driven by observations of the everyday, of concrete realities rather than generalizations, of ruptures and links through time instead of a linear evolutive history of cities. We focus on an urban history that distances itself from foundations or origins, choosing instead to search for points of connection between different times and places. From within urban history, we conceive the Andean as particular ways of being and doing that originate in the rural and have continued to be reproduced in urban contexts and their transformations. History as we conceive it allows us to get close to these changes, to understand them in their complexity as well as their specificity; and to discover the minimal stories that make the city using the archive and people's narrations.

Keywords Urban history · Andes · Andean cities · Modernity · Peripheric modernities · Rural–urban relations · "Trajines Callejeros" · Genealogy of history · History of everyday life

5.1 Introduction

The goal of this essay is to reflect on the role of historical investigation in contemporary understandings of the urban phenomenon. This conceptual reflection leans on Andean societies for historiographic support. Although, it's possible that this concern

This article was translated from Spanish by Emilia Gonzalez Salgado.

E. Kingman-Garcés · M. Salgado-Gómez (✉)
Department of Anthropology, History and Humanities of FLACSO Ecuador, Quito, Ecuador
e-mail: msalgado@flacso.edu.ec

E. Kingman-Garcés
e-mail: ekingman@flacso.edu.ec

F. Carrión Mena and S. Rodríguez Alvarez (eds.), *From the City as a Project to the City Project*, The Urban Book Series, https://doi.org/10.1007/978-3-031-94226-6_5

is closer to those of historians than urbanists or urban sociologists, it remains relevant in relation to what we could call a genealogy of the urban.

We are interested in cities as a resource for understanding the complex contemporary phenomena related to the accumulation of resources, biopolitics, and displacement of populations, exclusion, structures of class, and the construction of subjectivities. It is important to note that this is only possible to conceive in the long run based on specific studies located in time and place, in other words, based on concrete experiences rather than generalizations. However, despite history, as a discipline, being geared towards the study of concrete realities and urban history falling in line with this, we argue simultaneously in favour of a history based on paradigms, hypotheses and concepts. Given the breadth of the subject, we will limit ourselves to discussing two concrete phenomena from within the experience of Andean cities: changes in how populations are governed and urban modernity.

Lefevbre (1974) says every social space has a history. What history is it? and why is that history relevant when trying to understand the present? The city becomes a privileged place from which to observe the changes in economy, social relationships, and the everyday that accompany modernity; however, there is not only one city nor one kind of modernity. Modern society is, in first instance, established in cities; but we are interested in the case of Andean cities, strongly influenced by the rural, be it that of the hacienda or of the communities.

If we follow the long process of the formation of the modern city, we come across dynamics of renovation and with parallel production of ruins and debris. In reality, the new and the old form part of the same process. The majority of the transformations that have been generated in the city as a result of modernity have been guided by a logic of accumulation, as well as, by the development of different forms of power which are continuously confronted and negotiated. This, in turn, has provoked changes in everyday life and in the structures of sensibility which Benjamin (2007), Simmel (1988), Sennett (1997), Kracaucer (2018) have referred to. It was, and still is, about urbanization without limits, which has the power to produce and destruct, and which has not always counted with alternative urbanism or counter-urbanism as counterforces.

We are not referring to a lineal, evolutive history, but rather, one of complex connections between different time periods, as well as ruptures in time. The presence of people with varied origins, in place of the old corporations, the changes in ways of living in and relating to urban space, and the development of a new kind of infrastructure, are phenomena typical of modernity, initially generated as urban forms. All this became evident first in literature and later in Sociology and History into the nineteenth century in Europe, and much later in Latin América, particularly in the Andes.

A historiography that breaks with all continuums or teleological history; that is, with the idea of an evolutionary process with a point of origin and a predetermined future, is needed to understand the complexity of these processes. The great urban interventions, promoted by Haussmann, Cerdá, Vicuña Mackenna, for example, cannot be perceived as successive moments in an evolutive, lineal and inevitable process, but rather, as moments of rupture that contributed to shifts in forms of life, systems of power and the imaginaries. Cities have been sites of advancement for economy, culture and other forms of social incorporation; in addition to, with the same strength, being sites for survival of other forms of life that gravitate from the past over the present. We are referring to points of connection between different times and places which constitute memory.

5.2 Towards a Critical History

For many, the problem of the city has to do mostly with its meanings and with what the Greeks understood as *Polis*. The urban tissue—conceived as the heart of the city—is permanently at risk of disappearing, taking with it an entire system of meanings. When we speak of urban plot we refer to streets, ports, plazas and all that they mean; we also speak about the life that is carried on within it, the dealings and human relationships, conflicts and contradictions. To forget this is to put to the side a large part of social memory or "collective unconsciousness", of the meaning of the city.

When talking about cities we are not referring solely to the dynamic of change that is produced in its plot; but rather to the type of relationships that, in the midst of these changes, are established with the past. There is a cognitive problem in all of this, which is not less vital given that we "walk through cities". When Walter Benjamin studied in Paris in the nineteenth century, he was not interested in the past for the past's sake, like traditional historiography; instead, he was interested in finding keys to understanding the present within the past. It is a political question or one related to the way politics is defined. Hanna Arendt similarly takes the city, particularly its classical forms (the Polis) as a metaphor for thinking of politics.

What we refer to as the *urban condition* is a complex phenomenon in which forces relating to the present, past, and future, are at play. Many Latin American thinkers have developed a concern for the relationship between the city and social life, or between the city and politics; radically distinct from that of the urbanists and the hygienists of the nineteenth century and the first half of the twentieth century, or of the current supporters of the doctrine of security, real estate, and the entertainment culture.

These insights into the relationship of the city with its past are different to the tendencies, normalized in the context of urban planning, around the idea of heritage (monumental history, historicism, antiquarianism) that turns the past into a justification for the present. They also distance themselves from those who establish a direct relationship between the history of cities and the search for origins whether those are pre-Hispanic (Espinosa and Salgado Gomez 2024), colonial or national origins and regardless of if they are imaginary (or real). The history of cities has continued to give space for the development of conservative concerns with origins, as much as to critical positions that question all origins.

In their evaluation of urban historiography in Latin America, Martinez Delgado and Mejia Pavony (2021), differentiate the history of cities related to the search for origins, from urban history as an expression of a rupture in terms of critical history.

> La historia de la ciudad ya no era la memoria de un ethos nacional, sino la preocupación por la vivienda, la salud, la agregación, el abastecimiento, en fin el acceso desigual de bienes que no solo son de consumo sino esenciales: el agua, el aire, el cosmos (...) Y estas preguntas nacidas como todas del presente de quien las formula, se llevaron al ámbito del pasado (...) (p. 15).[1]

The history of cities, particularly those in Latin America and in the Andes, -the place from which we write this essay- was committed to a positivist claim to reconstruct the past "as it happened", and to the application of universal models: the city as a "trace of Europe", the city as "Hispanicity", "modernity", "mestizaje", but also, later on, as a "city of transition". Although this type of historiography occupies a marginal place within academic spaces today, it continues to feed public memory (Cabrera Hanna 2019; Bustos 2007) and the production of monographs related to projects around identity, such as patriotic (Pérez Vejo 2010) and decolonial celebrations; or related to a tendency for the spectacularization of projects more directly anchored to capital, like patrimony and tourism. The problem is that these are the visions that are more widely available, cementing within common sense sweetened and aestheticized versions of cities empties of conflict, tensions and exclusions. (Salgado Gómez 2008). The city as tradition, but also as an advancement or evolution in the line of modernity and progress as opposed to the delay/regress of the countryside and "other neighbourhoods"; or in relation to the "barbaric" ways of urbanization, which is to say, those not legitimized by social or racial reasoning. The cities as modern refuges in the midst of tradition. The cities as monuments, or as coherent versions of the Lettered City. The history of cities is conceived, in this sense, as another form of the "partition of the sensible" (Rancière 2000); or in different moments within an evolutionary line marked by the aesthetic as opposed to the non-aesthetic, the orderly as opposed to disorder, the cultured versus the popular, the discourse versus speech.

We are looking to differentiate between this "history of cities", as a specialized field close to that of architecture and urbanism, and doing urban history, conceived as a game of complex problematics, related with the organization of social life, power, and the construction of subjectivities, spatial configuration and practices. These problems must be thought of in their historicity, in their common ground, but also in their differences, overlaps and ruptures. It is not about generic or abstract processes, these take specific forms in relation to different social, symbolic and material configurations. Rather than alluding to time as a shield for a teleology, historicity refers to process, junctions and concrete historical dynamics, and, to an internal logic that must be accounted for. In this sense, it is necessary to incorporate a reflection of the urban, minimal, micro and everyday histories and memories that

[1] Our translation: The history of the city was no longer the memory of a national ethos, but instead the preoccupation with housing, health, aggregation, provisions; in short, unequal access to goods that are not only for consumption but are essential: water, air, the cosmos(…) And these questions born from the present of those who ask them, were taken to the realm of the past (…).

respond to, for example, an empirical economy (Polanyi 2017), in which decisions and practices are informed from a variety of places. It is about a history of encounters and approaches, as much as of conflicts and struggles. A history of strategies deployed by the institutions that make up the city; of the tactics of everyday life, the scams and skirmishes of those who act in the fleeting moment of conjuncture (de Certeau 1996).

We know that urbanization provoked by early modernity meant moments of rupture with old forms of perceiving and living these cities, both in the idea of the corporation and in that of the Lettered City (Rama 1984). The lettered, in particular, cannot be understood outside of the non-lettered and of the rich social framework that is characteristic of cities (Adorno 1987; Beverley 1999; Rappaport and Cummins 2016). Nevertheless, something similar happens to us in the current context of post-urbanization, when many of the paradigms of the modern city have been overcome without necessarily entering into crisis.

If urbanization and urban life have made us modern, we are interested in knowing not only what kind of relationship that modernity established with its past, but instead what is the breaking point between us and urban modernity. This is to say, a type of society guided by ideas of progress and accumulation without limits. The modifications in both the natural and built environments provoked by urban expansion; the changes in land use, the expulsion of populations, the architecture of waste; the loss of the experience and proportion of the street, the spread of malls, entertainment spaces, businesses; the increase of informality and precariousness, ruins, debris, the collapse of social ties have all provoked deep changes in the present as much as in the way in which we relate to the past. In this case, it is not about pretending we can go back to the past, but instead to put into question the foundations on which our present is built. As Bruno Latour puts it:

> Saying that we've gone back to being 'archaic' is an understatement, we've become totally unused to resorting to the axe of 'modernization' (Latour 2021, p. 77).

In times of urban and social crisis, increased immeasurably by everyday violence, war and the ghosts of pandemics, one tends to view the past nostalgically. In this context, the city "is no longer what it once was", we must return to its emblematic places, to "traditional treatments (behaviours)", to stability; we have to return to "it". But who can legitimately take back the city, inhabit it, reinvent its past, and who can't? What past are we referring to? What city do we refer to? It's hard to answer these questions because it's the bases themselves that are in crisis, those of urban life, its relationship with ecology, with the future of the planet, and with it, the forms in which we relate to the past.

To speak of the relationship between the present to its past from within the city means giving way to archeology and genealogy of the urban where History occupies a relevant place. In contrast to what is proposed by traditional historiography, the cities themselves and the social world that defines them are made up of the past, by several temporal and spatial layers or stratifications, much like a palimpsest. Despite urban dynamics leaning towards a break of communication between these layers, there are a multiplicity of, sometimes secret, connections. Minimal histories and memories bring us back to them.[2]

If the attempt to characterize cities according to models no longer makes sense, it is necessary to seek out new paradigms, as well as different scopes and fields of visibility. Given the complexity of the object of study, pretending to build a "general history" of cities, detached from concrete processes, has been questioned in the same way as any purely descriptive perspective has. If the paradigmatic use of these studies is important, the elements from which the comparisons are drawn are the result of an archeological labour, meaning, from a reflection closely tied to a "cross inquiry" of different archives, memories and geological strata. What is of interest, in reality, is not the reconstruction of urban pasts in an empirical manner, but rather the development of new interpretations in close proximity to the archive, oral history and ethnography: something which allows us to think on specific kinds of cities, or to rethink them.

Even when we write from within social and cultural urban history, rather than from a history of urban forms, we account for the value of studies centered around place, its social uses, and its relationship with production of circulation channels like we have called "trajines callejeros" or street tactics (Kingman Garcés and Muratorio 2011; Santillán et al. 2022). Each new investigation on the social streams in relation to urbanization, or to the role that space has in the exercise of power and relationships, can open new possibilities for us to understand the urban.

5.3 Capitalist and Urban Modernities

If we continue the line of critical thought that extends from the School of Frankfurt passing through Deleuze, feminist theories, or the representatives of critical historiography into the genealogical perspective developed by Michael Foucault, we conclude that there is not just one course in the analysis of urban modernity. When we speak of modernity, or more specifically of urban modernity, we refer to a complex constellation of economic, social, and political forces; taking into consideration biopolitics and the administration of populations (but also of resistance) as parts of these forces. It is not about a homogenous modernity with a set path, but instead about its multiple historical possibilities.

Walter Benjamin (2007) demonstrated how the reach of the world of commodities in Paris in the nineteenth century, provoked changes in the organization of space, in addition to changes in the structures of sensibility and the relationships of the public. We can see now how these changes happened in parallel to the development of capitalist modernity. While Benjamin's analysis is centered on the city, the passages,

[2] For example, cultural forms related to informal economies and the ways in which streets and plazas are conformed and used, cannot be understood, in cases such as Mesoamérica, the Andes, Mexico or Brazil, without reviewing Baroque culture and the ancient "trajines" of streets with roots that take us back to the XVII–XVIII centuries (Ciriza-Mendivil 2019, 2021; Minchom 2007; Burke 2021; Pérez Vejo 2018; Kingman Garcés et al. 2018).

the great avenues, the journey of the flaneur, a more recent perspective examines the global flow of images as much as commodities, and the place of the city within them. To speak of the urban is to think of the city as a concentration, as well as a part of networks and circulation.

If we think of the ways in which the economic world system operates, or the ways in which information is produced and circulated, we can see it's not about systems born from a single core, it is about the incorporation of a disparity of social, economic and environmental phenomena in different times, forms and conditions (Sassen 2009). Let's consider, for example, the integration within the chains of production, into top-of-the-line technological resources with rudimentary technologies, or populations subject to precariousness, overexploitation, even slavery, in the midst of the euphoria of progress, consumerism and modernity. One could say we are living in a process of integration—of what powers life as much as what destroys it—within a unique system, be it under privileged conditions, or on the contrary, subordinate and subaltern ones.

This dynamic of urbanization and urban modernization has made the way for new relationships between individuals and classes based on a logic of governability, but also on the absence of government and administration. Furthermore, we refer to the demands of entertainment industries, colonization and dependency, the "democratization of consumption", as well as the efforts of the population to reconstruct their economy, their social bonds, or to resist plagues or criminal bands through networks of care.

5.4 Modernity and City in the Andes

The lasting image of the cities of Hispanic America in the colonial period is an extension of the image of these societies as static and rigid, organized into watertight compartments that respond to estate, ethnic, racial, or socioeconomic divisions. A rich historiography has been dismantling these images and showing us the complexity and dynamism of American societies and cities of the Ancient Regime (Cañizares Esguerra 2018; Ciriza-Mendívil 2019; Gonzalbo & Alberro 2013; Jouve Martín 2005; Lane 2002, 2019; Ramos & Yannakakis 2014; Yannakakis 2008, 2023). The colonial city was organized under an imaginary order, with figures like the grid; beyond that real or imagined order was the disorder of the "Indian barrios" of the Lettered City (Exbalin & Godicheau 2021; Ciriza-Mendivil 2019).

Guamán Poma (in Deler's reading) attempted a different representation of the city from that of the Hispanists. For Guamán Poma the city wasn't solely influenced by its colonial present, it was also determined by an indigenous world that despite the conquest, still existed, as a reality -transformed by the new conditions- and as a power (Deler 1992; Mumford 2012). In reality, when we talk about a colonial city we refer to the representation of an order, as well as, a space of flux and exchanges, of trajines between the city and the country and within the city itself; of fairs and plazas, as much as culture and religiosity, the result of mixing and juxtaposing; those that could be considered popular. It dealt with a dynamic and volatile world, deeply imbued with the indigenous as much as the Hispanic, defining what has come to be understood as colonial interculture (Yannakakis 2008; Espinosa 2000, 2015).[3] There is substantial historiography that highlights the vitality of this daily exchange, and of the active participation of subaltern groups in the economy and the paths of colonial cities. (Ares 2010; Webster 2021; Yannakakis 2008; Ciriza-Mendívil 2019; Borchart 1991). As Jouve Martín demonstrates in the case of Lima, subaltern sectors had an active role, both socially and culturally, putting into question the limits of the Lettered City:

> En una ciudad en la que aproximadamente la mitad de sus habitantes eran clasificados como negros, mulatos o zambos y en la que los individuos de estas castas se encontraban en una relación continua con personas de orígen europeo, necesariamente tenían que multiplicarse las posibildades de entrar en contacto con la escritura y con los productos del universo cultural de la sociedad dominante, particularmente en comparación con lo que sería posible encontrar en ámbitos más reducidos social y espacialmente como eran las plantaciones y haciendas (Jouve Martín 2005, p. 50).[4]

The urban, as a means of organizing power allowed, from the beginning, certain social mobility:

> A pesar del estigma ligado al trabajo manual los oficios sirvieron como vehículo importante de movilidad social para muchas personas de baja condición social. En una sociedad con fuertes prejuicios raciales, los oficios ofrecian oportunidades económicas a indios, negros y castas. Como aprendices estas personas podían desarrollar habilidades como sastres y carpinteros, como herreros y plateros, como hojalateros y curtidores. Algunos eventualmente lograban abrir su propio taller. Los oficios otorgaban a los sectores bajos no europeos, cierto grado de movilidad dentro de los parámetros de una sociedad jerárquica (García-Bryce 2008, p. 52).[5]

---

[3] We wonder how much one can speak of Andean specificity as a layer of the historical palimpsest that makes up the territories which formed part of the Spanish empire in America. We refer to the strength of the urban ideal brought from the sixteenth century by the conquerors, which was reproduced in the villas and Spanish cities as much as in the reducciones, doctrinas, and indigenous towns. The urban order was the civitas ideal, the model for social, political and Christian order that materialized governance of the city republic and the principle of common good. However, this urban order was always accompanied by flux and multiple forms of relation between different social strata.

[4] Our translation: In a city where approximately half of its inhabitants were classified as black, mulatto or zambo, and where the individuals of these casts where continuously engaged in relationships with people of European origin, it was necessary to multiply the possibilities for contact with writing and with the products of the cultural universe of the dominant society; particularly in comparison with what was possible to find in reduced social and spatial scopes such as plantations and haciendas.

[5] Our translation: Despite the stigma tied to manual labour, trades functioned as an important means of social mobility for many people of lower social standing. In a society with strong racial prejudice, trades offered economic opportunities to Indians, blacks and casts. As apprentices, these people could develop skills as tailors, carpenters, blacksmiths, silversmiths, tinsmiths, and tanners. Some eventually opened their own workshops. These trades allowed non-European lower classes, a certain degree of mobility within the parameters of a hierarchical society.

During the period of government that began with the enforcement of Bourbon reforms in Hispanic America, demographic politics and those that organized the space from the top down began to partially take shape. For cities, we are talking about enlightened proposals[6] directed at achieving a certain political and economic rationality in their governance, when speaking about police (Kingman Garcés et al. 2018). This meant the relocation of cemeteries, abattoirs, hospitals, the organization of supplies; and new ways in which to govern populations directed at rationalizing charitable practices beyond the corporations, through institutions such as hospices, hospitals, montepíos and the clear differentiation between "authentic" and "false" poor (Lempérière 2008). It was about proposals to generate a certain disciplinary order in the context of non-disciplinary societies and in moments before urbanization. These actions, directed at the baroque order and that of corporations that characterized a composite and counter-reform monarchy of the Ancient Regime, were not successful; amongst other things, because the population did not passively accept them, turning them into areas of contention.

Urban dynamics suffer a transformation from the beginning of the Republic when we witness, in the Andean area, a dual process in the configuration of the nation's territory. On one side, the constitution of a centralized state is found within urban contexts, and on the other, the generation of decentralized forms of managing populations, not just in the countryside but also in cities. These allow the State to act from a distance or through a delegation (Guerrerro 2010; Das and Poole 2002; Krupa and Nugent 2015). It was about, weekly linked, predominantly agrarian societies.

It's only towards the end of the nineteenth century, beginning of the twentieth century that cities gained back force in the Andes, yet their importance, in many cases, was not due to their industrial character as much as because of their rentier base. Simultaneously, cities sheltered proposals for the renovation and reproduction of a traditional (oligarchic) order. The conservative president Gabriel García Moreno, took advantage of the destruction of Ibarra, in 1968 due to an earthquake, in order to design an urban model conceived in terms of urban planning and policing as much as moral control. We are referring to experimental actions to manage modernities and populations under catholic principles. They were attempts at constituting the State from within its center, as well as from the localities and local power networks, made up of haciendas, parish heads, clerical action and communities. These networks were mediators in the relationship with the State.

Another instance in the process of modernization, situated between the beginning of the twentieth century into the second half of the last century, is specially relevant for urban history. In this period there is an entire, little developed, field of reflection on the transit between ways to manage populations from the private to the public (Poole 2009; Guerrerro 2010; Kingman 2006). Cities are the stage for these changes, with their projects for modernization, industrialization, urban and architectural renewal. These are proposals for change in the management of cities in their entirety, both in urban planning and social terms. They sought to incorporate entire populations within

[6] We are speaking about an Iberian and American enlightenment, as opposed to the canonical one.

a project of modernity that was civilized and disciplined, as well as constructed the public. As Drinot says:

> La imagen del indio y el obrero representando el pasado/presente y el futuro, respectivamente, articula la creencia de la élite en el poder transformativo de la industrialización, pero claramente ubica este cambio no en la esfera de la economía sino en el de la raza/cultura. El poder (¿la magia?) de la industria, sugiere la imagen, reside en su capacidad de transformar a las atrasadas poblaciones indígenas en trabajadores industriales blanco/mestizos civilizados (Drinot 2016, p. 17).[7]

It was about the challenge of managing a population that migrated and populated cities, tending to lose its traditional ties to the hacienda, the communities and the small settlements. They were changes to the economy and as a result to the construction of classes and relationships, that were not necessarily accompanied by changes in the habits or ways of engaging the sensible (Rancière 2000). The problem, posed by the elites, was how to access modernity without, in doing so, giving up the privileges of their aristocratic condition (Kingman 2006, 2009a, b). It's true that in Lima, Quito, or La Paz they were opening up to an early modernity, but in places like Abancay, Latacunga, Riobamba, or Cajamarca they continued to function under a complex urban–rural apparatus with the hacienda as its axis. An urban dynamic did exist; however, society was in one way or another entwined with an agrarian economy and with everyday relationships framed by servitude and racism.

Despite the first decades of the twentieth century opening up the realms of urbanism, encompassing extensive territories beyond the cities, not all sectors included in this process followed the logic of capitalist modernity. Indigenous people in the hacienda (colonos, yanaperos, conciertos) occupied, up to the first half of the twentieth century, the last step in the social scale of the Andes; contact with cities opened the door (as Argueda tells) to new kinds of relationships. With agrarian reforms and the modernization of society as a whole, cities became magnets for populations who kept material and social ties with their origins as they urbanized.

José María Arguedas was in charge of registering changes to cities in Peru. Cities like Huanacayo grew spontaneously next to roads, unlike main cities which had a grid or a urbanistic plan. "La primera iglesia fue levantada evidentemente sin propósito de convertirla en núcleo de una población importante sino con el objeto de superponerla a una huaca que por estar cerca del camino imperial debía ser adoratorio obligatorio de los viajeros" (Arguedas 1983, p. 201)[8] They were regional markets that, thanks to the roads and railroad, were incorporated into the national markets. The Huancayo fair in particular, became the most important market in the Mantaro valley, around 1950.

---

[7] Our translation: The image of the Indian and the worker representing past/present and future respectively, articulated a belief of the elite in the transformative power of industrialization, and clearly situates this shift not in an economic sphere but rather a racial/cultural one. The image suggests that the power (magic?) of the industry lies in its capacity to transform pre-modern indigenous people into white/mestizo, civilized, industrial workers.

[8] Our translation: The first church was evidently raised without an intention to turn it into the nucleus of a main town, instead it was built with the goal to overlap it with a huaca that in being close to the imperial trails, was a mandatory place of worship for travellers.

> Por su propia condición de feria popular los vendedores estaban exentos de ciertas reglamentaciones y restricciones que rigen el comercio regular. Por otro lado, la feria constituye un atractivo poderoso de tipo social. No es igual negociar fríamente en una tienda o en una oficina que hacerlo al aire libre, en un ambiente de fiesta y en compañìa de múltiples amigos y compadres a quienes se encuentra en la feria por ser un lugar de cita obligado y grato al mismo tiempo para los productores y compradores de toda la regiòn" (Arguedas 1983, p. 212).[9]

Not long ago, there was a specificity in the Andes that didn't directly respond to what had happened in Europe or the Cono Sur. We are referring to a particular form of urban society of a merchant character, yet still closely linked to agrarian spaces, with mobility of population, trajines, and simultaneously capitalist and non-capitalist commodities. A large portion of the population with peasant origins was in the process of urbanization, and interacted (still somewhat to this day) with commerce and street trades. That is to say, with independent and semi-dependent occupations established between the city and the countryside, that we have characterized as "trajines callejeros" (Kingman Garcés and Muratorio 2011). Another portion conformed the servitude and urban peonage, or was gradually incorporated into the working class sector. The urban, in this case, doesn't necessarily coincide with industrialization or with a prescribed model of industrialization, but neither does it coincide with urbanization as it happened in Europe.

In this period, conceived by Beatriz Sarlo as peripheric modernity, the city became an object of concern due to different kinds of expertise related to state planning. The sanitation of port cities and the control of plagues, was pushed by the Pan American Health Organization, with the support of national states, as part of the requirements for a budding global economy (Cueto 1994). Epidemics and the measures needed to fight them together with the issues around safety, moved to take a central place in debates around urbanization. It meant trying to guarantee the flow of commodities and resources, as well as labour and energy, in uneven exchange conditions. All of this was accompanied by a growing concern for racial decline, for encouraging European over Asian migration, a separation and control of spaces, and an increase in birth rates. This meant the organization of an urbanist and architectural order. At the same time, the administration of populations would have been impossible without the participation of local actors of different kinds, be it in plantations, haciendas, missions, mining enclaves or Indian governance.

As demonstrated by Goetschel (2018), the disciplinary, civilization and control dispositives built from within the cities extended to the countryside, although only under specific forms. While cities demanded the organization of systems like policing, public charity, school systems, hospital services, customs, centers for planning and statistics; agrarian spaces were organized around local micro

[9] Our translation: Due to its own condition as a popular fair, the vendors were exempt from certain regulations and restrictions that rule regular commerce. On the other hand, the fair constitutes a powerful social attraction. It is not the same to coldly negotiate in a store or an office than to do so in open air, in a celebratory atmosphere, surrounded by friends and compadres that you run into at the fair, since its a mandatory and pleasant place of gathering for vendors and buyers throughout the whole region.

powers that only with time gained access to modern dispositives. In contrast to what happened in previous centuries, this was about secularised spaces aimed at generating governability, although we now know that most of their actions were doomed to failure.

5.5 The City and the Government of Populations

Amongst the issues that became the focus of concern in Europe and America towards the latter half of the eighteenth century, and which had gained strength in previous centuries, we can find those related to the management or administration of populations. They revolved around the establishment of dispositives of control that kept growing in complexity, targeting not only humans but also non-humans, at both macro and molecular levels (Foucault 2006; Agamben 2003). Many of the elements that constitute what we refer to as capitalism have been collaterals of this process, and have been generated both within and outside of the State. We know, thanks to Foucault and to a detailed reading of Marx from the Grundrisse, that capitalism in its many stages, generated diverse means of structuring space. To what extent was this model of understanding fulfilled in Mesoamerican and Andean cities, where modernization and the administration of populations maintained classification criteria based on old data and postmodern relationships? (Kingman 2006, 2009a, b).

From the days of primitive accumulation, entire populations have been sheltering in cities. It is not solely an economic phenomenon, it is directly related to the way in which classes are configured and to the politics of the everyday. Unlike what happened in the past, when corporations were the, relatively stable, bases for functioning societies, the cities have started filling up with "strangers" (radically "other"), who like commodities or machines, needed administration. With modernity it becomes necessary that a long (and unfinished) restructuring of the urban stage takes place; a social and spatial differentiation between people with rights and "bare life", citizens and non-citizens. It was about a highly complex exercise of power (which continues to play out nowadays in relation to immigrant and displaced populations) in front of which it became necessary to develop planning, control and intervention systems at a distance. It was an attempt at organizing the administration of entire populations found in cities and engaged in urban activities.

In addition to that which directly relates to the need for accumulation, the requirements of order in fields such as sexuality, health, childhood education, and entertainment were expanded upon and enhanced. If the shift from workforce to commodities was one of the necessary conditions for the construction of disciplinary societies, on its own, this was not enough given that more sophisticated disciplines extended far beyond the workplace. Not only factories contributed, but also school, health and public welfare systems, psychiatric centers, jails and correctional facilities, the control of women, and currently, artificial intelligence and digital media. It is, according to Debord (2003), a colonization of all of life, including free time. Every one of these interventions was accompanied by changes to the organization of space, that is to say, to the configuration of cities and architecture.

The city itself constitutes a powerful device that requires a specific function in terms of governability or governance of populations. As they grow and broaden their boundaries, cities develop new links to the outside and multiply their exchanges. They operate from a place of concentration and territorialization as much as from deterritorialized networks of resources, flows of population, information, and images that make up the urban. The organization of frontiers and vanishing points as a part of the organization of space, became one of the objectives of modern logic, in a similar way as the control of bodies did.

The urban sphere is being made and destroyed, modifying its reach and its boundaries in a molecular way. It actually has a much greater mobility than the State, needing to be more creative, active and unpredictable. We also know that life must be understood not only in relation to the human, and that a city, beyond being an urban phenomenon or one defined in urbanistic terms, constitutes a somewhat violent manner for intervention in the natural and geographic environment.

5.6 Space, Security and Biopolitics

In this essay, we have attempted to highlight certain traces in the construction of an urban history that allow us to understand the complexity of cities, in particular to those in the Andes. On the one hand, we have pointed out the need for understanding the urban space from within minimal histories. On the other, the need to look at the growing relationship between the changes in ways of managing populations and the changes in ways of organizing space. We are speaking, in this case, of a technocratic order as well as of a form of social order, of the unpredictable scope of what was gradually constituted as a capitalist modernity with the city as its base. It is, according to Benjamin in his comment to Kafka, about the attachment of individuals to a complex bureaucratic gear:

> Hablo del ciudadano moderno que se sabe a merced de un aparato de funcionarios de dimensiones incalculables, cuya función está dirigida por instancias que permanecen imprecisas para los órganos ejecutores mismos (Benjamin 2014, p. 112).[10]

Many years ago Michael Foucault presented the relationship between disciplinary and biopolitical dispositives and the organization of space, introducing into the field a genealogical perspective. For Foucault there is a difference between the administration of leprosy, based on the separation and expulsion from cities, and the administration of the plague geared, on the contrary, towards the inclusion of the plagued into control mechanisms within the neighbourhoods, the blocks, the homes, the bedroom. "No se trata de expulsar sino, al contrario, de establecer, fijar, dar lugar, asignar sitios,

[10] Our translation: I speak of the modern citizen who recognizes itself at the will of an apparatus of officials of unmeasurable dimensions, whose only function is determined by instances that remain imprecise to the executive organs themselves.

definir presencias y ausencias en una tabla de registros" (Foucault 2001, p. 51).[11] It was about specific ways of intervening over the life of populations in urban contexts, shifted towards control rather than expulsion. Deleuze (2015) perceived the political, social and technological reach of the phenomenon, focusing on a control-based society rather than a disciplinary one.

To function, the State must develop a degree of administrative centrality and simultaneously, the capacity to decentralize it. Ministries, civil, ecclesiastical, military, and lettered authorities function in the cities, and it is from here that a number of actions, capable of integrating (and subordinating) other areas, no matter how remote, are deployed. The city also constitutes the place in which the modern imaginary takes shape, at the expense of the real city, deeply traversed by anomie.

The growth of cities, the increase of flux, supposes not only an urbanistic but also a social problem that has heightened in recent years. We are referring to the big waves of migration from one border to the next, one continent to another, present for various decades. We are also referring to the different forms of aggregation of small and medium-sized populations, as well as of traditionally agrarian zones, and, more severely, of jungle areas like the Amazon into the framework of urban life.

The world system (or what can most precisely refer to, following Spivak, as imperialism) could be assumed in terms of accumulation, but also of generalized urbanization. There are a series of new phenomena, like climate change, that demand not only the development of new ways to administer resources like water, but also solutions to its consequences. Consequences like the displacement and the extermination of populations, that society globally has not been prepared for, or has chosen not to be. On top of this, we find a series of dispositives related to artificial intelligence or mass culture that seek to make possible the domain of the planet as a whole, going as far as the management of emotions.

5.7 Closing Ideas

In this essay, we have ventured to differentiate urban history, conceived in terms of flows that encompass both the city and the countryside, from classical ways to perceive cities. This distinction is particularly important in the case of the Andes given their relationship with different forms of participation in the sensible ranging from Hispanism to urban segregation, passing through the Lettered City and urban beautification.

As it is conceived here, urban history, far from referring us to foundations or origins, allows us to understand the points of rupture and unfolding that construct who we are. The urban does not constitute a purely economic or geographic phenomenon.

> Entiendo por ´método Haussmann´ la práctica generalizada de abrir brechas en los distritos obreros, especialmente en los que tienen una posición céntrica en nuestras grandes ciudades.

---

[11] Our translation: It's not about expulsion, on the contrary, it is about establishing, setting, making way, assigning space for, defining presence and absence in a registrar.

> El resultado es en todas partes el mismo…las calles más problemáticas desaparecen bajo la autoalabanza de la burguesía…pero vuelven a aparecer de nuevo en otro lugar y a menudo en la más cercana vecindad (Benjamin 2007, p. 169)[12]

Haussmann's strategy to renew the city was developed in the urbanistic as well as the economic and security realms. It consisted of a radically new form or style of confronting the city, typical of modernity.

This manner of adapting spaces gave way, contemporarily, to the "city of separation" and the "city of walls" that limit and interrupt exchanges and streams, but also to spaces of resistance where the *City of the Others* tends to reappear. We are referring to a universal phenomenon that is particularly sensible in the case of the Andes, given the weight of past over present.

The big conglomerates of populations coming from many different places, have given way to the development of strategies of separation, whose point of deployment takes us back to the last third of the nineteenth century and the first decades of the twentieth century, but also to flexible forms of mixing and integration. We are referring to forms, constantly in change, that don't respond to a single mechanical order, instead responding to the life of the city itself; to the agency of populations that occupy spaces assigned by the "police" as well as by conquerors of the New World (Carlos Iván Degregori's term). It is, in any case, about ways in which cities, that are constituted in the long term, function based on different forms of administration of populations, as much as on erratic flows. We conceive cities, particularly those of the Andes, as forms of organization of power, built through time, but also as networks of kinship, reciprocity, moral economy and substantive economy (Degregori 2013, p. 324).

From within urban history, we conceive the Andean as particular ways of being and doing that originate in the rural and have continued to be reproduced in urban contexts, even if sometimes only residually. Alternatively, the Andean as the survival of pacts of colonial origin, like those that made states protect the land of communities close to urban areas in exchange for enforced labour as "servants to the city" or as hands in public works during early modernity (Wilson 2014; Kingman and Goetschel 1989). The Andean as displacement and informality during the Fordist urbanization. Or else, as a rural–urban gear that with a modern dynamic tied to migrations, the formation of barriadas, the incorporation of the working class and of urban consumption, as well as social and cultural matrices originating in the hacienda and ways of communally being and doing. The cities of the Andes as cities with populations of rural origin engrossed in urban trajines, but also as racialized, violent and hierarchical spaces.

In the eighties, Matos Mar spoke of the weakening of traditional systems of power in the Andes (those described by Arguedas) as a result of an agrarian crisis, modernization, urbanization and the irruption of popular sectors into cities. It was related to the economy, but also to politics, to modernization or rather to modernity

---

[12] Our translation: I understand by 'Haussmann method' the generalized practice of creating breaches in working districts, especially in those with a central position in our great cities. The result is always the same…the most problematic streets disappear under the self-praise of the bourgeoisie…but they reappear in another place, often in close proximity.

and ways of being a part of it. History as we conceive it allows us to get close to these changes, to understand them in their complexity as well as their specificity; to discover the minimal stories that make the city using the archive and people's narrations.

References

Adorno R (1987) La ciudad letrada y los estudios coloniales. Hispamérica 48:3–24

Agamben G (2003) Lo abierto. El hombre y lo animal. Pre-textos

Ares B (2010) Lima colonial (1535–1635): crisol de gentes, ¿crisol de culturas? In: Bernabéu S, Varea C (eds) La ciudad Americana: mitos, espacios y control social. Doce Calles, pp 115–132

Arguedas JM (1983) Los ríos profundos. In: Obras Completas, tomo III. Editorial Horizonte

Benjamin W (2007) El libro de los pasajes. Akal

Benjamin W (2014) Sobre Kafka. Textos, discusiones, apuntes. Eterna Cadena

Beverley J (1999) Subalternity and representation: arguments in cultural theory. Duke University Press

Borchart C (1991) La imbecilidad y el coraje: la participación femenina en la economía colonial (Quito, 1780–1830). Revista Complutense De Historia De América 17:165–181

Burke JL (2021) Architecture and urbanism in viceregal mexico: puebla de los Ángeles, sixteenth to eighteenth centuries. Routledge

Bustos G (2007) La hispanización de la memoria pública en el cuarto centenario de fundación de Quito. In: Bustos G, Büschges C, Kaltmeier O (eds) Etnicidad y poder en los países andinos. Universidad Andina Simón Bolívar Sede Ecuador, Universidad de Bielefeld, Corporación Editora Nacional, p 113

Cabrera Hanna S (2019) Hispanismo, mestizaje y representaciones indígenas durante el quinto centenario en Ecuador: Vistazo y El Comercio. Anuario Colombiano De Historia Social Y De La Cultura 42(1):213–244

Cañizares Esguerra J (ed) (2018) Entangled empires. The Anglo-Iberian Atlantic, 1500–1830. University of Pennsylvania Press

Ciriza-Mendívil C (2019) Naturales de una ciudad multiétnica. Vidas y dinámicas sociales de los indígenas de Quito en el siglo XVII. Sílex

Ciriza-Mendívil C (2021) Los arrabales de la ciudad de Quito en el siglo XVII. Del ámbito rural al control de los espacios. In: Exbalin A, Godicheau F (eds) Los arrabales del imperio. Administrar los suburbios de las urbes en la monarquía católica (siglos XVI–XIX). Prohistoria ediciones, pp 43–60

Cueto M (1994) El regreso de las epidemias. Salud y sociedad en el Perú del siglo XX. Instituto de Estudios Peruanos

Das V, Poole D (2002) State and its margins: comparative ethnographies. In: Das V, Poole D (eds) In anthropology in the margins of state. School of American Research Press Press and James Currey, pp 3–33

Debord G (2003) La sociedad del espectáculo. Pre-textos

De Certeau M (1996) La invención de lo cotidiano. Artes de Hacer, vol 1. Universidad Iberoamericana

Degregori C (2013) Del mito del Inkarri al mito del progreso. Migración y cambios culturales. Instituto de Estudios Peruanos

Deler J-P (1992) Ciudades Andinas: viejos y nuevos modelos. In: Kingman Garcés E (ed) Ciudades de los Andes (1-). Institut français d'études andines, pp 351–374. https://doi.org/10.4000/books.ifea.2257

Deleuze G (1998) Conversaciones 1970-B. Pre-textos

Deleuze G (2015) Derrames II. Aparatos del Estado y axiomática capitalista. Cactus

Drinot P (2016) La seducción de la clase obrera. Trabajadores, raza y la formación del Estado peruano. Instituto de Estudios Peruanos
Espinosa C (2000) El retorno del Inca: los movimientos neoincas en el entorno de la intercultura barroca. Procesos 18:3–29
Espinosa C (2015) El Inca Barroco. Política y estética en la Real Audiencia de Quito, 1630–1680. FLACSO Ecuador
Espinosa C, Salgado Gómez M (2024) La producción del paisaje "periurbano" de Quito en el siglo XVI (1534–1575). Hist Crít 91:31–55. https://doi.org/10.7440/histcrit91.2024.02
Exbalin A, Godicheau F (eds) (2021) Los arrabales del imperio. Administrar los suburbios de las urbes en la Monarquía católica (Siglos XVI-XIX). Prohistoria ediciones
Foucault M (2001) Los Anormales. Akal
Foucault M (2006) Seguridad, territorio y población. Curso en el Collége de France (1977–1978). Fondo de Cultura Económico
García-Bryce W (2008) República con ciudadanos:los artesanos de Lima, 1821–1879. Instituto de Estudios Peruanos
Goetschel AM, Kingman Garcés E (1989) La participación de los indígenas en las obras públicas y los servicios de la ciudad de Quito, en el último tercio del siglo XIX. In: Kingman Garcés E (ed) Las ciudades en la historia. CONUEP, Instituto de investigaciones CIUDAD
Goetschel AM (2018) Moral y orden: la delincuencia y el castigo en los inicios de la modernidad en Ecuador. FLACSO Ecuador
Gonzalbo Aizpuru P, Alberro S (2013) La sociedad novohispana: estereotipos y realidades. Colegio de México
Guerrerro A (2010) Administración de poblaciones, ventriloquia y transescritura. Análisis históricos: estudios teóricos. Instituto de Estudios Peruanos, & FLACSO Ecuador
Jouve Martín JR (2005) Esclavos de la ciudad letrada. Esclavitud, escritura y colonialismo en Lima (1650–1700). Instituto de Estudios Peruanos
Kingman EY, Goetschel AM (1989) La participación de los indígenas en las obras públicas y los servicios de la ciudad de Quito en el último tercio del siglo XIX. In: Kingman Garcés E (coord). Ciudad/CONUEP, pp 397–404
Kingman Garcés E (2006) La ciudad y los otros. Higienismo, Ornato y policía. Quito, 1860–1940. FLACSO Ecuador
Kingman Garcés E (2009a) Cultura popular, vida cotidiana y modernidad periférica. Quaderns De L'institut Català D'antropologia 25:47–69
Kingman Garcés E (2009b) Lo urbano, lo social: la historia social urbana. FLACSO Ecuador, Ministerio de Cultura
Kingman Garcés E, Salgado Gómez M, Bedón E (2018) La ciudad a través de sus tratos: los abastos de carne en el Quito de fines del siglo XVIII y primeras décadas del siglo XIX. Confluenze. Rivista di Studi Iberoamericani 10(2):237–264. https://doi.org/10.6092/issn.2036-0967/8857
Kingman Garcés E, Muratorio B (2011) Los trajines callejeros. Memoria y vida cotidiana, Quito, siglos XIX y XX. FLACSO Ecuador
Kracaucer S (2018) Calles de Berlín y otras ciudades. Ed. Errate Naturae
Krupa C, Nugent D (2015) State theory and Andean politics: new approaches to the study of rule. University of Pennsylvania Press
Lane K (2019) Potosi: the silver city that changed the world. University of California Press
Lane KE (2002) Quito 1599: city and colony in transition. UNM Press
Latour B (2021) After lockdown: a metamorphosis. Polity
Lefevbre H (1974) La production de l´espace. Éditions Anthropos
Lempérière A (2008) Orden corporativo y orden social. La reforma de las cofradías en la ciudad de México, siglos XVIII–XIX. Revista Historia y Sociedad (14):9–21
Mar JM (1984) Desborde popular y crisis del Estado : el nuevo rostro del Perú en la década de 1980. Instituto de Estudios Peruanos

Martínez Delgado G, Mejía Pavony G (eds) (2021) Después de la heroica fase de exploración. Universidad de Guanajuato, Pontificia Universidad Javeriana and FLACSO Ecuador, La historiografía urbana en América Latina
Minchom M (2007) El pueblo de Quito 1690–1810: demografía, dinámica sociorracial y protesta popular. FONSAL
Mumford L (2012) La ciudad en la historia. Sus orígenes, transformaciones y perspectivas. Ed. Pepitas de Calabaza
Pérez Vejo T (2010) Los Centenarios en Hispanoamérica: la historia como representación. Hist Mexicana 237:17
Pérez Vejo T (2018) Repúblicas urbanas en monarquía imperial. Crítica/Universidad Nacional, Imágenes de ciudades y orden político en la América virreinal
Polanyi K (2017) La gran transformación: los orígenes políticos y economía de nuestro tiempo. Fondo de Cultura Económica
Poloni-Simard J (2006) El Mosaico indígena. Abya Yala
Poole D (2009) Justicia y comunidad en los márgenes del estado peruano. In: Sandoval P (ed) Historia subalterna y cultura en América Latina y los Andes. Instituto de Estudios Peruano, pp 600–638
Rama A (1984) La ciudad letrada. Ediciones del Norte
Ramos G, Yannakakis Y (2014) Indigenous intellectuals: knowledge, power, and colonial culture in Mexico and the Andes. Duke University Press Books
Rancière J (2000) Le partage du sensible. Esthétique et politique. La Fabrique Éditions
Rappaport J, Cummins T (2016) Más allá de la ciudad letrada: letramientos indígenas en los Andes. Universidad Nacional de Colombia y Universidad del Rosario
Salgado Gómez M (2008) El Patrimonio Cultural como narrativa totalizadora y técnica de gubernamentalidad. Centro-h No 1:13–25
Santillán A, Kingman Garcés E, Salgado Gómez M (2022) Genealogías de la ciudad andina: conversaciones entre antropología e historia. In: Campo Imbaquingo C et al (eds) Antropologías hechas en Ecuador. Estudios Históricos y Sociales. Tomo III. Asociación Latinoamericana de Antropología, Abya-Yala, Universidad Politécnica Salesiana, and FLACSO Ecuador, pp 150–164
Sarlo B (1988) Una modernidad periférica: Buenos Aires, 1920 y 1930. Nueva Visión
Sassen S (2009) Ciudad y globalización. Organización Latinoamericana y del Caribe de Centros Históricos
Sennett R (1997) Carne y piedra. El cuerpo y la ciudad en la civilización occidental. Alianza Editorial
Sennett R (2002) El Declive del hombre público. Ediciones Península
Simmel G (1988) La Metrópolis y la vida mental. In: Bassols M et al (eds) Antología de sociología urbana. UAM, pp 47–61
Webster SV (2021) Artistas letrados y las lenguas del imperio: pintores y la profesión en el Quito colonial. FLACSO Ecuador
Wilson F (2014) Ciudadanía y violencia política en el Perú: una ciudad andina, 1870–1980. Instituto de Estudios Peruanos
Yannakakis Y (2008) The art of being in-between: native intermediaries, Indian identity, and local rule in colonial oaxaca. Duke University Press
Yannakakis Y (2023) Since time immemorial: native custom and law in colonial Mexico. Duke University Press

Chapter 6
The Evolution of Urban Over Rural Areas

Juan Manuel Parreño-Castellano and Víctor Jiménez Barrado

Abstract This chapter reflects on the temporally evolving relationship between the urban and rural spheres, highlighting how both dimensions have been intrinsically linked and subjugated to the interests of dominant power groups throughout history, which has benefited one over the other. This unequal relationship is examined from the origin of the urban to the end of the twentieth century, leaving useful pointers for the present and the near future. A process in which the city has gone from emerging from and depending on the rural, to total domination of the rural. Cities are centers of territorial control, whose growth and demands drive the transformation of rural space. Through a methodology that integrates historical, geographical, economic and sociological approaches, the complexity of this process of domination and its material, socio-economic and cultural implications is understood. The present moment evidences a continued process of urbanization and de-ruralization under global urban hegemony and the growing submission of the rural to the dynamics of capitalist accumulation. In short, this study illustrates how the rural–urban dichotomy is spatially and sociologically broken, although it underlies this duality as a conceptual reference. The old categories have been joined by others, which warns us of the urgent need to reconsider, and eventually reverse, this situation in the face of current challenges such as the lack of sustainability and socio-spatial justice

Keywords Global urbanization · Urban history · Urbanization process · Rural · Urbanization · New ruralities · Rural–urban relations

J. M. Parreño-Castellano (✉) · V. J. Barrado
Departamento de Geografía. Universidad de Las Palmas de Gran Canaria, Las Palmas de Gran Canaria, Spain
e-mail: juan.parreno@ulpgc.es

V. J. Barrado
e-mail: victor.jimenez@ulpgc.es

F. Carrión Mena and S. Rodríguez Alvarez (eds.), *From the City as a Project to the City Project*, The Urban Book Series, https://doi.org/10.1007/978-3-031-94226-6_6

6.1 Introduction

6.1.1 Urban Dominance Based on Subordination to Power Groups

Rural and urban have been defined by opposition as two morphological, social, economic and cultural realities differentiated by most researchers (Capel 1975). However, the essential difference since ancient times has been the preferential location of the privileged and dominant groups in one or the other space.

Since the city became home to power groups, the urban became the structure of domination of the territory, organizing and controlling its rural environment. Possibly during the period in which Rome dominated the Mediterranean and a large part of its surroundings in Europe, Asia and Africa, the city, as a political-military structure that hosted the dominant groups, reached its maximum dimension, redefining the distinction between city and countryside (Lois González et al. 2013). The Roman city was assigned a rural area around it, in such a way that it was composed of two complementary parts: the urban nucleus or *urbs* was the administrative, commercial and religious center from which a much larger territory, called *ager* or *territorium*, was administered. Thus, the city was a complex territorial fact that included the countryside. In this context, cities overflowed their own walls and the boundary between *urbs* and *ager* became less evident (Sicca 1977).

During the Middle Ages in Europe, this model changed. The collapse of the Western Roman Empire created a new territorial organization of a markedly agrarian character in which a new dominant group, the nobility, civil or ecclesiastical, controlled the territory from rural areas. Only a small part of the population lived in the cities and people were mainly engaged in agrarian activities in feudal socio-economic and cultural structures (Siqueira and Osorio 2000). At that time, cities were, above all, points for the exchange of goods, although they were not the only ones.

It was during the transition between the feudal system and mercantile capitalism that the city once again became the center of power and territorial control. It was home to the kings, the high clergy, a growing part of the aristocracy and a new emerging social class, the commercial bourgeoisie. This new concentration of the ruling classes led to the subordination of the rural nobility and, therefore, of the rural world. This was the way, for example, of the cities with which Castile and Portugal consolidated their territorial expansion in America.

The development of capitalism allowed the emerging urban classes to gain access to land ownership from the 18th to the nineteenth century, a definitive step in the subordination of rural spaces to urban strategies and interests. An example of this were the processes of disentailment and secularization of ecclesiastical properties, which were relevant in Spain, France, Austria and Italy.

The growth of industry and services since the late eighteenth century in Britain and since the nineteenth century in much of the rest of Europe and North America further empowered urban power groups and increased the subordination of rural areas. This initially maintained a certain closeness, but with the strengthening of the colonial system since the nineteenth century, and more recently with economic globalization, the urban ruling classes also control the rural space of distant areas. Rural areas in developing countries have been attributed the role of food providers, exacerbating signs of an urban bias where the rural poor are dominated and exploited by urban interests (Lipton 1977), while the nearer rural space has gradually taken on new functions, including, in addition to purely agrarian, environmental and landscape functions. In the latter sense, since the second half of the nineteenth century, urban conservationist movements have emerged which have often failed to take into account the rural reality, so that protection has been transformed into an act of subordination and dependence on the rural. At the same time, this relationship has been reinforced by the maintenance of urban prejudices about the rural as synonymous with poverty, lower cultural and educational levels, little innovation and, in some cases, a threatening or unrational environment.

6.1.2 The Configuration of the Urban in Relation to the Rural

From the nineteenth century onward, the growth of the means of production and wealth in Western cities led to an increase in the urban population. Many cities tore down their already useless city walls and expanded into their immediate surroundings. Rural land, often in the hands of the wealthy urban classes, became an economic asset that cities used in their expansion process. Through their urbanization, one of the most important mechanisms of accumulation was set in motion, demonstrating that the rural is indeed subordinate to the generation of wealth in capitalism (Wallerstein 2002).

On the other hand, the introduction, by the bourgeoisie, of new technologies in agricultural production and better life expectations in the cities has favored the depopulation of the countryside. Rural communities have been reduced and destructured as a consequence, in part, of the exclusionary subordination between the peasantry and the agricultural industry, as has been studied especially in the case of Latin America (Pradilla 2002; Rubio 2002), where processes of desruralization without mobility can also be found, with the proliferation of domestic industrial units (Wallerstein 2002).

This growth of the city at the expense of the countryside has materialized through different processes since then. Compact growth, suburbanization, peri-urbanization or rururbanization (Berry 1996; Ruiz and Delgado 2008) make up long-standing urban dynamics that have been the subject of geographical theory, such as the Differential Urbanization Model (Geyer and Kontuly 1993); the theory of the reversal of polarization in developing countries (Richardson 1980) or the Model of Urban Development Stages (Berg et al. 1982), which includes the phases of urbanization, suburbanization, deurbanization and counter-urbanization to understand the evolution of the city and its relations with the rural environment. We will devote a large part of the following sections to these processes.

Parallel to this process of rural loss, there was also a certain recovery of the countryside from the end of the nineteenth century, but with a new approach. The excessive artificialization of the city and its environmental problems led to a revaluation of the rural as synonymous with nature, identity, authenticity and even community life and solidarity. Low-density suburban growths are, for example, an expression of the agrarian or bucolic myth applied to the process of urbanization. The result is the emergence of a space of mixed rural and urban, what Lacour and Puissant (2007) refer to as reurbanity. According to these authors, we are in a time marked by the urbanization of the rural and the ruralization of the urban. This is the case in exurban communities in North America where the urban is connected to rural values of tranquility, stability and timelessness (Larsen et al. 2007).

Another fact that cannot be ignored is the growing movement toward the naturalization of built space in the city, as the population's environmental awareness grows and the objectives of sustainable development have become an accepted social ideology. In this sense, we are witnessing a return, at times anecdotal, cosmetic and more symbolic than substantial, to agricultural uses in urban environments, with the proliferation of urban vegetable gardens and agricultural activities for local consumption (Linck 2001).

In the following sections, we will analyze from a spatio-temporal perspective the different processes of domination that characterize the relationship between urban and rural spaces, as well as the emergence of new ruralities (Arias 2002). Successively, the rural origin of the city, the process of urbanization and its interpretation as a struggle from a historical perspective will be addressed, before focusing on the more recent processes of suburbanization, peri-urbanization, rururbanization and naturalization of the city in the last three sections.

6.2 The City as a Rural Product

The first known city was Uruk, founded approximately 6000 years ago (Nissen 1988; Liverani 2006), and although there are researchers who disagree about the specific location of the first glimpse of urbanity (Oates et al. 2007), the fact is that its appearance symbolizes the transition from agricultural, nomadic and dispersed societies to sedentary, concentrated ones with more complex political, social and economic structures. This, along with other cities located in Mesopotamia, arose because of the particularly prolific benefits offered by the combination of location and situation. For Strabo, the former provides permanent natural advantages, which in perspective implies a constant flow of capital accumulation. The situation, on the other hand, implies a relationship with the surrounding area and an optimal position among equals, which allows for the weaving of networks and the consolidation of the urban category over the rural one. The level of efficiency of both aspects will define the prosperity of the urban core. On this basis, the first cities are constituted as a distinct political unit with an agricultural domain through which they become independent and self-sufficient. Among others, these could be the Sumerian city-states, the nomos of Ancient Egypt or, later, the altépetl-meh in Mesoamerica.

Wittfogel's well-known hydraulic theory probably lacks the capacity to explain the emergence of cities holistically, but it does bring to the table the importance of two interlinked facts. Firstly, the dominance of human beings over space and its resources thanks to infrastructure (irrigation strategies, tools for cultivation, etc.), which, in the fertile crescent, coincided with a high productive capacity of the land that led to the appearance of notable surpluses. In this respect, his explanation finds points of connection with Childe's theory of the urban revolution, as he considers that technical evolution played a crucial role in social evolution.

Secondly, although based on the above, the emergence of relations of domination between people, with an absolute and despotic character for Wittfogel and of a more stratified nature for Childe, who considered a previous stage of demographic increase based on an improvement in food. In any case, both allude to the same event: the capacity of a reduced group to subtract, privatize and accumulate primarily the surpluses of space captured by a majority. This implies the formation of a very broad-based social structure attached to the land, which supports the emergence of other new, smaller socio-productive classes. In this way, the surpluses produced are the foundations for the specialization and the progressive distancing of another -dependent- class from working on the land, which is personalized by artisans, merchants, etc. At the same time, there appear the administrators or managers of the surplus, who are quantitatively inferior. This class, for Childe, is also the fruit of the overflow of family structures and the need for new organizational forms imposed on these social units of production and consumption.

The task of surplus management is initially indistinguishable from the rest and does not form a hierarchy within a subsistence and later barter economy, i.e., non-monetary. However, a factor of scale and pragmatism also plays a role here, which is exploited by these small groups. There is a shift from a vast space of free will, independence between beings and subsistence, to a new concentrated space where human–human and human-space power relations implode. In other words, the saving of surplus occurred within human-scale and apprehensible spaces (Arpal 1983), as opposed to the immense rural productive space. Eventually, this led to an urban control of rural space, not only from an economic but also from a social point of view. Gradually, this function of storing and administering the surplus was socially elevated, to the point of constituting elites who appropriated the wealth. The progressiveness of this process is demonstrated by the urban morphology and the size of the housing, which did not show great signs of distinction at the beginning.

For Mumford (2012, p. 55), the origin of the city is inseparable from the emergence of power - often indistinguishable from divinity—which, under conditions of abundance, is capable of artificially creating scarcity. Thus, he defines the city as "a structure specially equipped for storing and transmitting the goods of civilization, sufficiently condensed to provide the maximum amount of facilities in the minimum space."

This power play, which has materialized over space, has led to a parricidal relationship. In the following sections we will see how, although the urban arises from the rural, in its development it has disrupted the material, economic and sociological supports of the rural.

6.3 Global Urbanization Coupled with Nuanced Rural Retrenchment

According to the World Bank, 56.9% of the world's population (~7.88 billion people) lived in urban settings in 2022. According to the same source, in 2007 the urban population outnumbered the rural population. The latter has decreased while the urban population has continued to grow strongly, thanks in large part to rural to urban migration (Fig. 6.1).

Since 1960, the urban population has multiplied by 4.4 times while the rural population has increased by 1.7 times. This inexorable move toward an urban world is not a constant in the historical development of humanity. Until the second half of the twentieth century, urbanization was particularly concentrated in Europe, North America and Latin America. If we go back to earlier periods, only the first two continents had significant rates of urbanization in the first half of the nineteenth century.

Therefore, in a long time perspective, although in some areas there have been prolonged stages of strong increase in urbanization followed by others of clear deurbanization (for example, Europe under Roman domination in antiquity and the High Middle Ages, respectively), urbanization is a phenomenon of recent importance, with very marked spatio-temporal differences. Today, it is the least urbanized countries, in general terms, where the urban population is growing the most, especially in China,

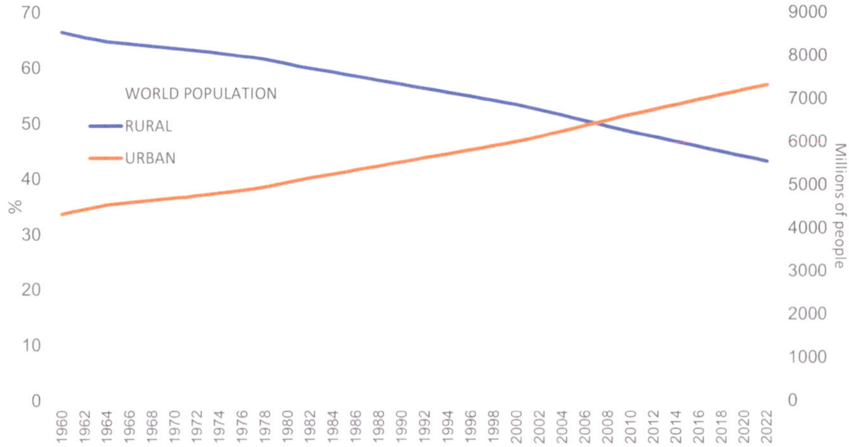

Fig. 6.1 Rural and urban demographic evolution (1960–2022). *Source* World Bank. Own elaboration

India and some African countries. On the other hand, it should not be forgotten that urban growth in Latin America has been highly concentrated. According to Gouëset et al. (2006), it was initially driven by rural–urban migration. However, since the 1970s, this growth became endogenous, driven by the high birth rate in cities, with a young and predominantly female demographic structure.

If urbanization rates are not homogeneous, neither are urbanization processes. While in some areas large cities are gaining importance, in many of the less developed countries the population lives in medium-sized or small cities (Robinson 2005).

This lack of uniformity in relation to the urban dimension on the spatio-temporal scale also determines a differentiated relationship with rural space. Nevertheless, the prevailing discourse is that we are witnessing a process of continuous occupation of rural space by urban growth or urbanization. The demolition of city walls in European cities during the nineteenth century is the best expression of this fact.

In the dynamics of urban expansion, rural space is conceived as a passive element, without its own capacity to make its own decisions and control its future. It seems to behave as an open or empty space that can be occupied for reasons of political-military or economic domination of the territory. In the latter sense, urbanization and the parallel desruralization is usually conceived as a consequence of commercial and industrial capitalist economic development at first, and as one of the main forms of accumulation of urban elites from the end of the nineteenth century onwards.

This is a basic outline that needs some nuances. Firstly, in the past, town and country maintained a more balanced relationship of interdependence. The city was functionally dependent on the rural, adopting the role of market for rural products. Even more recently, some cities are functional products of their rural environment. For example, Mediterranean agro-cities relate to the rural in a different way than European industrial cities do.

Secondly, urbanization is not only in the interests of the ruling class, although it can be interpreted as a consequence. Informal growths, especially in Latin American, African and Asian cities, seem to be on the margins, yet they have a major urbanizing dimension.

Thirdly, rural–urban migration also entails the transfer of rural practices and traditions to the urban world. Hybrid cultures are produced, especially nowadays in the countries of the Global South (Woods 2010), so that de-ruralization by urbanization is not complete.

Finally, it should not be overlooked that the rural population has also favored urbanization processes, as it participates in urban decision-making. Historically, rural elites have held property in urban spaces and have participated in urban power, as urban power has done in rural areas (Woods 2005). Even today, it is common for the rural middle class to own property in urban spaces to facilitate access to certain services.

6.4 The Battle for Space

The coordinates of power reveal the scoreboard of the contest between the rural and the urban, which roughly unfolds in three acts. The explanation for the origin of power is rooted in economic development. We have already mentioned that the city arose from the rural, as a primordial concentration of the surplus value generated by the land and the trade of its products. Metaphorically, these were tiny urban islands within vast rural lakes. This incipient spatial hierarchy is the basis of a social hierarchy. Thus, physical and forced labor takes surplus value from the land, and the lack of freedom of those who inhabit and work the rural space is used by the often-urban elites to capture it. The same happens in Mesopotamia or the Roman, Chinese, Inca empires and Medieval and Feudal Europe, where, respectively, kings and priests, aristocrats and patricians, emperors and bureaucrats, sapas and nobility and feudal lords exploit peasants, serfs and slaves and their living and working spaces.

For most of history, society has been agrarian-based and cities have had a symbolic rather than a productive function. This urban iconicity is associated with the representation and demonstration of elite power. While their commercial role cannot be underestimated, neither can their interrelationship and dependence on the rural space from which they sprang be denied.

There is a contradiction here, for while the city and the elites control and use the rural for their own benefit, they are also prisoners of what the rural offers them. The urban, therefore, is more a consequence than a cause. They were a landmark in space, whose main function, beyond representing power, was to entrench it in the territory and capture the surplus value of the space and the surrounding rural population. In other words, the cities (the border cities, more clearly) were the nodes of a network of containment of the real booty: the rural space. If the city and the elites wanted to prosper, there was no choice but to keep conquering more rural space. The slow pace of urbanization, weighed down by resistance and wars, is thus understandable.

Secondly, the Industrial Revolution in Europe and the United States, accompanied by the Transport Revolution, brought about a paradigm shift. In this second round, the city won on all fronts, supported by a new economic sector. The transformations gave both the urban and its new elites a quantitative and qualitative leap forward. Urbanity became associated with the concept of social progress, in the same way that industrialization meant economic progress (Baigorri 1995, p. 2). From all this positive charge emanated a vertiginous spatial advance supported by migrations, showing the relationship of communicating vessels between the rural and the urban (Fig. 6.2).

The advance of the cities and the retreat of the rural also meant an increase in the power of the eminently urban industrial and commercial bourgeoisie and a decline in that of the landowners. This pushed the coordinates of power further away from the rural space. At the same time, a significant fraction of the rural peasantry was transformed and fed into the urban proletariat. The Communist Manifesto records this subjugation of the rural (Marx and Engels 2000, p. 33):

Fig. 6.2 London going out of town or the March of bricks and mortar! *Source* The British Museum

> The bourgeoisie has subjugated the country to the city. It has created immense cities; it has prodigiously increased the population of the cities at the expense of that of the countryside, and has thus subjected a large part of the population to the idiocy of rural life. Just as it has subordinated the country to the city, so it has subordinated the barbarous or semi-barbarbarous countries to the civilized countries, the peasant peoples to the bourgeois peoples, the East to the West

But this view is spatially biased. This is because the development and role of the cities left out of industrialization (which were the majority) remained as they were before, although their backwardness deepened on a global scale.

The truth is that for this change to have a widespread impact, the development of transport was key, as it had an amplifying effect and further expanded the frontiers of the urban over the rural. However, the most important result was not the territorial transformation, but rather the annihilation of the territorial monopoly of cities and the emergence of a new competitive scenario. This means that many cities and their local elites lost their exclusive sphere of action and power, blurring their boundaries and exposing their content (the immediate rural space and society) to other more powerful cities and elites. Following the previous simile, the new scenario was made up of archipelagos dominated by large islands which, in turn, controlled entire seas.

The consequences can be read on two intersecting scales and approaches. On the one hand, this new and expanded hierarchy was a sign of globalization and the beginning of a geopolitics of remote territorial domains, which had its paradigmatic example in colonialism (resource depletion, slavery, etc.). The two revolutions combined expanded the role and size of large cities and turned colonial cities into secondary centers of power whose function was to suck from the coast all the resources of the rural and natural hinterland. On a different scale and to a different degree, the non-capital cities of the metropolis fulfilled the same function. That is to say, they were intermediaries between the big cities and the once rural environment they dominated. The role and organization of the states entrenched this political/economic/territorial structure. This was thus the beginning of a tele-directed and amplified domination of the urban over the rural, as well as the liberation of the big cities from dependence on the local scale.

Lefebvre (1978) warned that the urban tends to colonize the rural, taking advantage of a capitalist ideological-economic conjuncture that will result in the total urbanization of society. Fewer and fewer cities dominated a wider, more distant and spatially discontinuous rural space in relation to them and to each other, thanks to a hybrid control between the territorial and the economic. Cities had become what, in part, they still are: a sink for the resources, population and wealth of rural spaces, now forced to gravitate around global urban nodes and the specific interests of their elites.

In the third act, we see how power is installed in the urban space. There is a system of wealth accumulation that culminates in the emergence of large urban elites on a global scale. This small group conceptualizes the urban as a command center, from which it dominates with less and less physical necessity (accession of land, land ownership, legal dominion, etc.). Now, the conjuncture transforms our simile into a rural ocean in which large urban islands compete with each other and for global space.

Since the mid-twentieth century, the Industrial and Transport Revolution has been joined by the disruptive effect of the Technological Revolution, which has turned service providers into the new elites. This has not been problematic for the urban, on the contrary. The city continues to grow and accumulate population, power and wealth, thanks to its dual role as a well-connected production center and an accessible consumer center. For their part, the rural areas have recovered their productive weight, although based on a primary sector that is not an incentive for local development, without an incursion of the tertiary sector with real economic significance and with a depleted population. This combination of demographic scarcity and limited wealth generation gives them a notable irrelevance. For Segrelles (2015), "in this dialectic between the two spaces, it is the countryside that has lost out, being subjugated from the physical and socio-economic points of view to urban expansion, strength, influence and dynamism" (p. 7).

In this unequal contest, rural spaces and their inhabitants have functioned over the last decades as a puppet of the urban. This has given rise to political discontent that is expressed in democratic contexts. This role has been strengthened through legislation, enacted from cities and land and urban planning, which is sometimes scarce, due to a lack of profitability, in territories without real estate dynamics. Through it, and in the face of a world population that is increasingly urban and unfamiliar with the rural world, its image has gone from stigma and mistreatment to ignorance and subsequent rediscovery under urban concepts. An example of this is the residential movements from the city to the peri-urban and rururban space. One trigger for this centrifugal movement is the population's dissatisfaction with city life. The American Berry, in 1976, called it counter-urbanization. Rural space is conceived both as an escape route and as a space full of amenities that the city had made disappear. In Europe, Champion (1989) also describes this process.

In the following sections, we will examine the spatial and functional results of this incursion of the urban into the rural, but first we will briefly explore its significance for the urban/rural dichotomy. According to Lebrevre (1978), the extension of the urban fabric helped to break down the boundaries between the urban and the rural. Once the limits of the city have been exceeded, the expansion of the urban has given rise to peri-urban and rururban spaces. This means that, secondly, these confines of the city are "becoming more porous, hindering our ability to draw clear lines between what is inside the city as opposed to what is outside" (Soja 2008, p. 221).

This opens up a range of possible options. The result is functionally diverse and rich places, transiting between urban and rural nature. However, underlying the secular competition between the two is a tense and lively relationship. We are thus faced with fragmented and dispersed growth, identified in "a matter of degree, not in an absolute form" (Batty et al. 2002, p. 9). Borrowing the concept from other disciplines, it is an ecotone, i.e., a hybrid space. However, it should be borne in mind that these are born out of the dynamics of the city, the rural space being a passive entity. Therefore, it is one more moment in the urbanization process, generated by what the city expels, whether for environmental, economic or social and ideological reasons. The urban has won the battle.

6.5 Low-Density Suburbanization and the Myth of the Rural

Environmental degradation, poverty and the loss of a sense of community life in cities became evident during the nineteenth century, particularly in European industrial cities. In this context, the myth of the rural as an expression of a more socially and environmentally balanced way of life gained momentum.

What is certain is that during the Roman Empire, rural life was already assigned values in opposition to the degradation of the urban world. Theocritus (third century BC) was possibly the first author to promote the pastoral myth in Western literature, conferring on rural life values conducive to happiness.

With urban growth, Romanticism strongly revived the pastoral myth and gave special value to life in contact with nature, so that in the second half of the nineteenth century the idea of urbanization in imitation of the rural arose. Urbanization thus became a renewed idyllic representation of the rural myth, and the first low-density suburbanization formulas appeared, associated with wealthy social groups.

The development of the railway in the nineteenth century and the private car in the twentieth century were key, as without them this form of urbanization would not have been possible.

One of the earliest examples of ruralizing suburbanization is the city-park, such as Vésinet, on the outskirts of Paris. Designed by the landscape architect P. Lavenne de Choulot in 1856, it was planned to be integrated into the pre-existing natural landscape (Galantay 1977). The city center was planned in a checkerboard pattern and the rest of the urban space was organized on the basis of a network of roads and artificial rivers. On the different estates, the owners were obliged to build single-family dwellings with certain construction conditions, and the establishment of any industrial activity was expressly forbidden.

Bedfork Park (1877), west London, is considered the first garden or low-density middle-class suburb (Pacione 2009). Planned by the architect C.R.N. Shaw, it was envisioned as suburban growth 30 min by train from London. It included various typologies of single-family dwellings, many semi-detached, in an urban setting with a church, parsonage, club, shops, bars, restaurants and art school (Greeves 1975).

In Spain, the linear city, developed by Arturo Soria y Mata in 1882, was based on urban growth around a backbone street 50 m wide, which included a central corridor along which trains and trams ran. Around this corridor were the residential blocks occupied by single-family dwellings and, just behind, the countryside. Its materialization began in 1894 in a simplified and reduced form. The aim was to make the lifestyle as rural as possible, but without disregarding the comforts of the time (Terán 1999).

The suburban neighborhood gave way, at least in the theoretical field, to the suburban city with E. Howard's conceptualization of the garden city in 1898 (Howard 2010; Howard et al. 2013). Howard proposed a model of urbanization based on small, self-sufficient cities of no more than 32,000 inhabitants, combining the benefits of the countryside and the city and involving a new way of life. Cooperativism was advocated as a means of access to housing with the aim of lowering production costs, and the aim was to reduce distances in daily life, favor the decongestion of large cities through residential decentralization and improved connectivity between cities by rail. Howard sought both a new model of city and society based on cooperativism (Hall 1996). The urban structure of a garden city was based on a succession of concentric rings cut by radial axes. In the center was a park surrounded by a first ring of public buildings. The second ring consisted of a green strip for sports activities and the rest of the rings would be residential, articulated around ring avenues. Finally, the last ring would contain the agricultural area, open spaces and social and health services, where part of the population would also live. The dwellings were to be single-family homes, with private open space.

In 1904, Howard initiated his first garden city project: Letchworth, 55 km north of London, for a population of 33,000 inhabitants. A cooperative association was set up to buy land from 15 agricultural landowners. The project included a large central square surrounded by public buildings, from which radial avenues would lead off. The residential blocks were to be occupied by single-family dwellings surrounded by open spaces, following the theoretical principles announced. The fact that no industrial facilities were built meant that Letchworth did not become a self-sufficient nucleus and was reduced to a suburban neighborhood with a garden city morphology (O'Sullivan 2016). The first houses were occupied by artists, but later houses were built for blue-collar workers. However, the formal quality of the project made it of increasing interest to middle-class families, and larger houses were built. Over the years, after the Second World War, land speculation took place, transforming it into a middle- and upper-class neighborhood. In short, rather than a city for the non-wealthy, what was built was a middle-class suburban neighborhood with clear ruralizing characteristics.

Letchworth was followed by many other projects, but none fully reflected Howard's model. What did become consolidated in the first decades of the twentieth century was the low-density suburb as a housing model in which urban and rural values were fused. Welwyn (1920–26) was a good example, as were other utopian city projects, such as L.F. Wright's Broadacre City (1932).

Throughout the second decade of the twentieth century, low-density suburbia became established in much of Europe, North America and Australia, particularly as the private car became more widespread (Harris and Larkham 2003), and has become one of the dominant urban growth models today. After World War II, the first industrial production projects of suburban housing developments appeared with the construction of Levittown outside New York (1947–51). Standardization and chain production became a key element, at the cost of a gradual move away from the rural concept of the pastoral myth (Wagner and Wagner 2010).

We cannot forget that low-density suburbanization as an urban model was also assumed by public housing policies. For example, in Spain, the laws of Casas Baratas (Cheap Housing) (1911 and 1921) promoted these typologies and, even during Franco's autarkic regime, actions were carried out that recreated a rural way of life in city environments. The above shows that the relationship between the urban and the rural finds an apparently hybrid and, above all, commercial solution in suburban growth.

6.6 The Deployment of the Urban in the Rural Environment

The current incursion of urban elements into rural space is progressive, materially, economically, demographically and sociologically. This has led to notable changes such as the conversion of agricultural and natural lands into residential, industrial and commercial areas, the alteration of local ecosystems, the demand on resources, as well as other non-material changes.

From the first third of the twentieth century, Sorokin and Zimmerman (1929) perceived this, and a little later, Dewey (1960) and Pahl (1966) noticed the end of radical divergences by the emergence of the rural–urban continuum. Its emergence was motivated by changes in productive and spatial organization in the United States and showed that the *status quo* placed urban society in a predominant position. This was initially a reading for the Anglo-Saxon world, but American economic and cultural hegemony quickly made it valid for the rest of the world.

The variety of urban elements that colonize rural areas allows this heading to be organized in an infinite number of ways, but here we will take as a point of differentiation their contribution to the productive character of the rural space.

Starting with the non-productive ones, the most important colonization of rural space in terms of surface is that of housing. An invasion that for Lefebvre (2001) meant the desecration of rurality by the city and urban life. After the blurring of the boundaries between the urban and the rural, a process of creative destruction emerges in this way. The distinction between housing and residential function should be emphasized here. In fact, housing that is not used as a permanent residence has, as we shall see below, contributed in a differential way to the distortion of the rural. In any case, whether as a residential or recreational element, housing has allowed the infiltration of the urban into the rural (Naranjo 2009), and with it, sociological substitution. The protagonist of this transformation can be identified as a prototype: the *rur-alien* (Cabo 2008) is the person who, detached from the rural way of life, conquers or assaults the countryside, establishing uses for it that are neither proper nor traditional, including residential use without productive justification (residential, holiday, leisure, amenity). In short, it is a "being without belonging" in the countryside (Jiménez et al. 2020). The fact that this individual does not necessarily belong to the elite, but to more numerous groups such as the middle class or the vulnerable, multiplies the level of colonization and destruction of rural space.

The transformation and productive turn of spaces such as the Guadalupe Valley (Mexico) or the Alentejo region (Portugal) could be two concrete cases where the productive turns have meant an alteration of the previous space and a rapprochement between the concepts and materialities of the urban and the rural.

In this category, which promotes the emergence of an urban–rural interface, we can distinguish two processes that, in addition to other reasons, are identified according to their remoteness from the consolidated urban area.

First, peri-urbanization is discussed. This is "an often highly dynamic process in which rural areas located on the outskirts of established cities take on a more urban character" (Webster and Muller 2009, p. 280). Precisely because of this mobile and constantly reconfiguring nature, its definition is difficult (Salazar and Osses 2008). If we include it within the category of modern urban expansion processes, it would be an initial phase, both in terms of the space it occupies (normally tangential to the consolidated urban space) and the time at which it occurs (succeeding suburban growth and preceding rururban growth) (Fig. 6.3).

Due to its consolidation as a form of urban development, but not necessarily of city growth, it attains a conceptual dimension and its own entity (Quirós 1991, p. 396). However, its development tends to overcrowd and densify rural spaces, which sometimes results in their incorporation into the city and its growth (Martín 2016). Its success as a pattern of urban growth, together with the problems it generates, makes it one of the great challenges for spatial planning in the twenty-first century (Ravetz et al. 2013).

Fig. 6.3 Peri-urban development in Santiago (Chile). *Source* Maxtor. Own elaboration

In terms of their characteristics, there is a minimum common denominator based on three aspects. According to Andrés (1985), the first would be the constant relationship and proximity to the urban nuclei from which they originate; secondly, the high and diverse multifunctionality; and thirdly, their degree of renovation.

In morphological terms, the peri-urban space lacks the rational order of modern planned urban developments. This peculiar organicity is the result of, on the one hand, its apparently entropic rhythm and form of growth and, on the other hand, its location on a rural "template" crossed by roads.

As we mentioned, the fundamental vehicle of this process is housing and its residential use. Based on this constant use, other uses appear that diversify and complexify the space, such as services. In fact, rural space is the destination of not only demographic decentralization, but also the productive decentralization of the urban space (Delgado 1999). For both reasons, it follows that there is a tension between old and new uses, as well as between urban and rural society. For this to result in the advance of the urban over the rural, the collapse of agrarian resistance due to economic and profitability reasons is indispensable. This is currently an easy outcome, despite mainly governmental efforts to combat it. Public policies have been implemented along four main lines (investment in infrastructure and services, support for the establishment or maintenance of private enterprise, promotion of community participation, and tax incentives together with integrated spatial planning).

Although it is possible to think of these spaces as places of interconnection between the urban and the rural, especially in terms of mobility and transport, the use of resources and the location of waste, infrastructures and businesses and services (Tacoli 1998), the truth is that this penetration, due to the power of the urban and the capital gains that accompany it, leads to the total disappearance of any hint of rurality, with the exception of an inherited morphological structure, the semi-natural landscape and the appearance of small farms.

The second of the non-productive processes of urban encroachment on the rural is rururbanization. The conceptual glimpse is attributed to Galpin (1918), who in his work "Rural life" relates it to the growing influence of urban structures in the countryside. For Bauer and Roux (1976), creators of the term, the rururban space is a rural area close to urban centers, but characterized by the subsistence of a largely dominant non-urbanized space, which is subject to the residential input of a new population, usually coming from the city. Therefore, change begins with a sociological and cultural transformation (Precedo 2004) in which housing is the essential element. Through it, in this case with a greater prominence of the recreational function, the urban metamorphoses to the point of inoculating several of its characteristics into the rural environment, although this entails results such as habitat fragmentation, the alteration of ecological cycles, soil water contamination, loss of vegetation cover, displacement of local populations and changes in traditional community dynamics, among other issues.

Spatially, it occupies spaces much further away from the city, less dense, although with patterns of irregularity similar to those of peri-urban spaces. The persistence and survival of the rural is more evident, although the very appearance of the phenomenon is a potentially growing threat. The extension of transport networks or wireless communication and technology is helping to eliminate the handicaps that held back its development. In fact, the ways of inhabiting this space are no longer rural but predominantly urban (Claval 1980). In functional terms, urban integration in the rural environment takes place almost exclusively through housing. The low density prevents the achievement of a critical mass that would justify other uses. On the other hand, this preservation of the rural essence allows them to retain the attraction for those who evoke and idealize rural life. In fact, and although it may seem contradictory, the vocation of the rururban space and its inhabitants is not to serve urban expansion, although in fact they do so (more in function than in form), but to escape from it. Rururbanization does not create a city.

As for the productive processes that promote an advance of the urban over the rural, the fact is that urban agglomerations in the context of global capitalism have established behavioral patterns that distance production from consumption. At the same time, there is a separation of the urban population with basic self-provisioning skills, which is typical of rural societies. It is, in short, another capitalist strategy to keep the wheel turning and for every need there is a productive and commercial sector, controlled by mainly urban elites. This means that there are urban externalities that rural areas pay for without resistance.

If we separate these by sectors, we can observe that even in one traditionally linked to rural areas, there are urban traces. Without the hyper-demographic concentration in cities and their consumption patterns and fashions, one could not understand the appearance of huge deforestations, monocultures or macro-farms. In them, the sociological link between land and local community is broken, since the workers (increasingly reduced by technification) define themselves as proletarians, and in many cases, they are not even rural residents. On the other hand, they also contribute to the elimination of rural society, since the concentration of land and its massive occupation of space prevents other means of subsistence and expels population (Vilas Boas 2017). The rural landscape is also being swept away by a homogenizing veil with environmental and even detrimental effects on human health (Bikangui et al. 2023).

With regard to the secondary sector, the implantation of industries in rural areas has been the consequence of an improvement of the urban habitat that is not aware of the environment as an organic and interrelated whole. Under these strategies of new implantation, the magic solution was to remove and dissolve the problem in vast extensions of rural land. This declares the conception of this space as a storage room for cities and their inhabitants as people of lesser "status." Specific examples of this could be the "rural relocations" of the automotive industry in regions of Mexico, such as Guanajuato, or those of the textile industry in Bangladesh in rural spaces close to large cities.

Without leaving the secondary sector, at the crossroads of several contradictory aspects, the use of the rural environment by urban society continues to show it to be a provider of strategic resources. Far from being a qualitative improvement for the rural environment, the sustainable turn has further deepened its subordination to the urban environment. In the same way that monocultures provide food, there is an important global policy for large renewable energy plants located in rural areas to feed the productive processes of or for urban areas. These macro-energy plants condemn vast areas of land and even fragment agricultural developments. Moreover, the economic impact on local communities is miniscule, due to the effects of the technification of labor (Fig. 6.4).

On the other hand, the tertiary and quaternary sectors have been seen as a new hope for rural areas. Undoubtedly, the development of services and new economic activities such as tourism has revitalized areas in clear demographic and economic regression. Some see in this a rural renaissance, based on an advance in Rural Non-Farm Employment (RNFE), and the emergence of new ruralities. On the other hand, the facilities provided by technology to maintain an urban lifestyle within the rural environment have increased the contingent of neo-rurals, which is now also fueled by lifestyle migrants and digital nomads. The possibility of "being without being" (Hiernaux 2001, p. 32) leads to this. However, these phenomena are contributing once again to the conceptualization of the rural environment as a bucolic space, destined for leisure and relaxation for urbanites. The speed of change, as well as the ability of newcomers to adapt to the rural environment that receives them, will determine whether this new process is problematic or a solution.

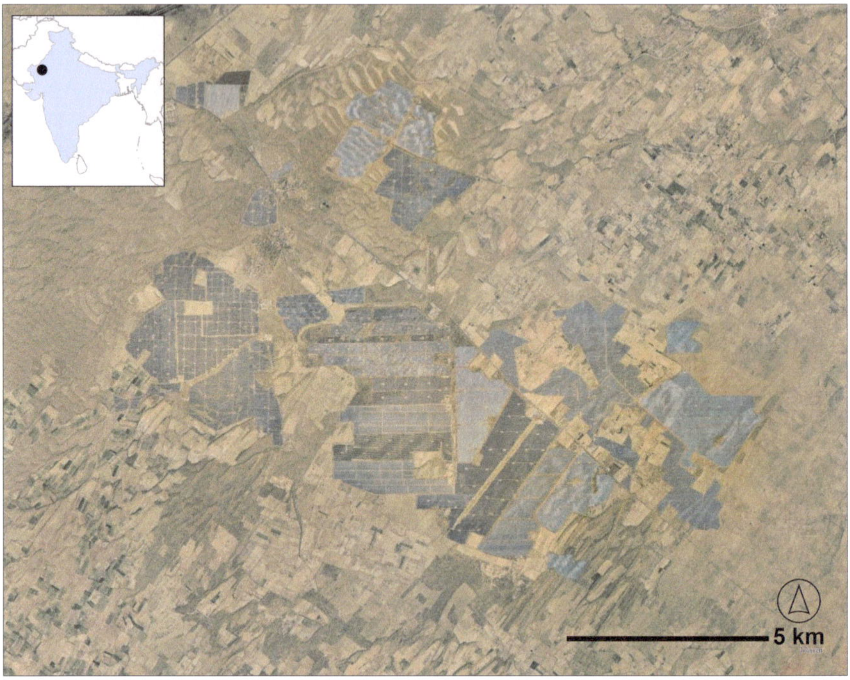

Fig. 6.4 Bhadla solar plant. *Source* Maxtor. Own elaboration

6.7 The Naturalization of the Urban

In recent decades, we have witnessed an attempt to naturalize or ruralize the city. This is not only a material or scenographic aspect, but also a new conceptual approach. The incorporation of rural landscape features into the city has its main antecedent in the appearance of gardens, parks and hunting grounds in the interior or surrounding areas of cities. At least until the sixteenth century, these open spaces were designed for the enjoyment of kings, nobles, the clergy or wealthy social groups and were usually part of palatial residences. The first open spaces with trees for public use—albeit for the ruling classes—were small in size, and generally corresponded to the typology of tree-lined boulevards, with the aim of allowing people to stroll in the shade (Jones and Wills 2005). Some of these avenues acquired great importance, such as the Alameda Central in Mexico City (1592) or the Champs Elysées in Paris (1640).

In the eighteenth century, the garden became an essential element of urban life. Hyde Park (1728) in London, the Prado and Recoletos Promenades in Madrid (1768) or the Luxembourg Gardens in Paris (1789) are some of the most outstanding examples. Many of them had been private recreational spaces that political changes opened up for the enjoyment of citizens, and the city had to order and conserve them from that moment onwards.

Finally, the development of the industrial city meant that the new middle social classes demanded these spaces, and the number of gardens and boulevards grew in the transition between the nineteenth and twentieth centuries. These spaces tended to include areas with an artificial structure but also other more natural-looking spaces. From the outset, these parks often included rural uses. Central Park (1858), for example, housed a flock of sheep and a pasture area for feeding them.

Architectural landscaping evolved throughout the twentieth century in such a way that, in the Athens Charter (1933), the need to promote green spaces in the city was defended. Open space was defended as an element regulating the environment, human well-being and social relations. Although *open planning* incorporated these principles, it was not until a few years later that the relationship between natural spaces in the city and sustainability became a matter of course (Beer 1994).

Thus, in recent decades, there has been a proliferation of projects for the natural recovery of open spaces located on the outskirts or in the interstices of cities. Actions for the recovery of degraded natural spaces and abandoned rural areas are multiplying, with the aim of converting them into spaces with educational, cultural or leisure value.

This process of naturalization of urban space would be incomplete if we did not consider the increasing presence of urban agriculture, in the form of vegetable gardens in public and private open spaces. The urban vegetable garden is that productive agricultural and livestock space located in and around cities in which the selection of plants and animals is made for productive rather than landscape reasons and which is dedicated to sale or self-sufficiency (Smit, Ratta and Nasr 1996). These gardens have appeared in recent decades in cities under different formulas such as community vegetable gardens, rooftop and balcony gardens, nurseries, greenhouses, public parks and gardens, vertical gardens (Borbón and De la Torre 2020).

Although the urban garden in developed countries has its earliest precedents in the educational farms of the 1970s in London, it is true that agricultural activity has remained in the city at least until the last century in developed countries. Urban dwellings have, since ancient times, had spaces set aside for agricultural production or animal husbandry to ensure food security, and this is even clearer today in developing countries. Thus, in countries such as Nigeria, the cultivation of unoccupied public land has been encouraged free of charge (Potter et al. 2008). In the same vein, the development of local farmers' markets in cities is also a further expression of the ruralization of the city (Jarosz 2008).

In any case, the development of these forms of agricultural production in recent decades has to be linked to the adoption of Sustainable Development Goal 11, "Sustainable cities and communities," which aims to make cities and human settlements inclusive, safe, resilient and sustainable. In this context, urban agriculture is a tool to improve air quality, reduce the impact of freight transport and support local economies. New forms of urban agriculture reflect current social changes, which is why it is increasingly present in developed countries.

In short, the excessive denaturalization of human settlements and the ecological crisis is affecting urbanization patterns and the ruralization of the city, blurring the artificial distinction between city and countryside and establishing a new point in rural–urban relations.

6.8 Brief Final Reflection

Throughout the chapter, we have presented the relationship between the rural and the urban as an unbalanced interaction of dependence and subordination, the result of the interests of power groups. With marked spatio-temporal differences, we have presented a process in which the city and its dominant groups have ended up controlling the rural space according to their interests and the mode of productive accumulation. In this process, the boundary between urban and rural has become blurred, with actions of occupation and appropriation of rural space such as suburbanization, peri-urbanization or rururbanization. More recently, there have been experiences of reconversion of the urban in pursuit of its naturalization.

For much of the historical process explained, relations have been neighborly, but with the development of colonial capitalism and communications and with today's globalization, relations have lost their proximity, while fewer and fewer power groups and cities organize rural space on a global scale.

Urban dynamics and the new ruralities themselves will condition rural–urban relations in the future, but possibly the current global change linked to the climate emergency will have a greater direct impact on the relationship between the rural and the urban. Beyond all these unknowns that currently exist, what we do know is that the changes that will occur in the future will be marked by the rhythms between spatial–temporal differentiation and hybridization, so that different models will coexist and appear.

Acknowledgements This work has been carried out within the framework of the R&D&I research project: "Cities in transition. Urban fragmentation and new socio-spatial patterns of inequality in the post-pandemic context" (PID2021-122410OB-C31), funded by MCIN/ AEI/https://doi.org/10.13039/501100011033/ and ERDF "A way of making Europe."

References

Andrés JL (1985) Las áreas periurbanas: espacios plurifuncionales en crisis. En *IX Coloquio de Geografía*. Universidad de Murcia-Asociación de Geógrafos Españoles, pp 11–22

Arias P (2002) Hacia el espacio rural urbano: Una revisión de la relación entre el campo y la ciudad en la antropología social mexicana. Estudios Demográficos y Urbanos 50(3):363–380

Arpal J (1983) Las ciudades: Visión histórica y sociológica. Editorial Montesinos

Baigorri A (1995) De lo rural a lo urbano: Hipótesis sobre las dificultades de mantener la separación epistemológica entre Sociología Rural y Sociología Urbana en el marco del actual proceso de urbanización global. En AA.VV. (Eds.), Actas del V Congreso Español de Sociología: La Sociología Rural en un contexto de incertidumbre [Detalles de la página faltantes], pp xx–xx

Batty M, Chin N, Besussi E (2002) State of the art review of urban sprawl: Impacts and measurement techniques. Recuperado de http://www.casa.ucl.ac.uk/scatter/download/Scatter_D1.pdf

Bauer G, Roux JM (1976) La rurbanization ou la ville éparpillée, 1ª ed. Seuil

Beer AR (1994) Urban greenspace and sustainability. In: Van der Vecht H (ed) Sustainable urban development: research and experiments. Delft University Press, pp 69–88

Berg L van den, Drewett R, Klaassen L, Rossi A, Vijverberg C (1982) Urban Europe. Volume 1: a study of growth and decline, 1st edn. Pergamon
Berry BJ (1976) Urbanization and counterurbanization. Arnold
Berry BJ (1996) The counterurbanization process: Urban America since 1970. In: Geyer HS, Kontuly M (eds) Differential urbanization: integrating spatial models, 1st edn. Arnold, pp 7–18
Bikangui R, Boussougou-Sambe ST, Saidou M, Ngossanga B, Ndalembouly AGD, Djida Y, More A, Beh Mba R, Abe H, Ushijima Y, Borrmann S, Adegnika AA (2023) Distribution of Aedes mosquito species along the rural–urban gradient in Lambaréné and its surroundings. Parasites Vectors 16(360). https://doi.org/10.1186/s13071-023-05901-2
Borbón DSU, De La Torre JMO (2020) Huertos urbanos como estrategia de resiliencia urbana en países en desarrollo. Vivienda y Comunidades Sustentables 8:81–102
Cabo Á (2008) Recuerdos y reflexiones en torno a la geografía agraria en España. Geographicalia 53:5–20
Capel H (1975) La definición de lo urbano. Estudios Geográficos 36(138):265–302
Champion AG (1989) *Counterurbanization: the changing pace and nature of population deconcentration*. Arnold
Claval P (1980) *Geografía Económica*. Oikos-Tau
Delgado C (1999) La urbanización del campo en Cantabria. En AA.VV. (eds) El Territorio y su Imagen. Departamento de Geografía de la Universidad de Málaga-Asociación de Geógrafos Españoles, pp 1011–1022
Dewey R (1960) The rural-urban continuum: real but relatively unimportant. Am J Sociol 66(1):60–66
Dureau F, Gouëset V, Mesclier É (2006) Géographies de l'Amérique latine. Presses Universitaires de Rennes
Galantay E (1977) Nuevas ciudades, de los orígenes a nuestros días. Gustavo Gili
Galpin CJ (1918) Rural life. The Century Co
Geyer H, Kontuly T (1993) A theoretical foundation for the concept of differential urbanization. Int Reg Sci Rev 15(12):157–177
Greeves TA (1975). Bedford Park: the first garden suburb: a pictorial survey. Peter Murray
Hall P (1996) Ciudades del mañana: Historia del urbanismo en el siglo XX. Ediciones del Serbal, Colección La Estrella Polar
Harris R, Larkham P (eds) (2003) Changing suburbs: Foundation, form and function. Routledge
Hiernaux D (2001) Las nuevas formas urbanas y reestructuración del mundo rural. In: Torres PA (Comp) Procesos metropolitanos y agricultura urbana. Universidad Autónoma Metropolitana, pp 31–42
Howard E (2010) Tomorrow: a peaceful path to real reform. Cambridge University Press
Howard E, Osborn FJ, Mumford L (2013) Garden cities of tomorrow. Routledge
Jarosz L (2008) The city in the country: growing alternative food networks in Metropolitan areas. J Rural Stud 24(3):231–244
Jiménez Barrado V, Larraín Suckel J, Trincado Olhabé B, Cabrera Cona F (2020). Promoted Urbanization of the countryside: the case of Santiago's Periphery Chile (1980–2017). Land 9(370). https://doi.org/10.3390/land9100370
Jones KR, Wills J (2005) The invention of the park: from the Garden of Eden to Disney's Magic Kingdom. Polity
Lacour C, Puissant S (2007) Re-urbanity: urbanising the rural and ruralising the urban. Environ Plan A 39(3):728–747
Larsen SC, Sorenson C, McDermott D, Long J, Post C (2007) Place perception and social interaction on an exurban landscape in central Colorado. Prof Geogr 59(4):421–433
Lefebvre H (2001) O direito à cidade. Centauro
Lefebvre H (1978) De lo rural a lo urbano. Editorial: Historia, Ciencia, Sociedad
Linck T (2001) El campo en la ciudad: Reflexiones en torno a las ruralidades emergentes. Relaciones 85(22):85–104
Lipton M (1977) Why poor people stay poor: urban bias in world development. Temple Smith

Liverani M (2006) Uruk, la primera ciudad. Bellaterra
Lois González RC, González Pérez JM, Escudero Gómez LA (2013) Los espacios urbanos. Biblioteca Nueva
Martín A (2016) Del rural agrario al rural periurbano: Instrumentos de desarrollo rural en la comarca de Torrijos (Toledo) [Tesis doctoral]. Universidad Politécnica de Madrid
Marx K, Engels F (2000) Manifiesto comunista. El Aleph
Mumford L (2012) La ciudad en la historia. Pepitas de Calabaza
Naranjo G (2009) El rol de la ciudad infiltrada en la reconfiguración de la periferia metropolitana de Santiago de Chile. Estudios Geográficos 70(266):205–229. https://doi.org/10.3989/estgeogr.0448
Nissen HJ (1988) The early history of the ancient near east. University of Chicago Press
Oates J, McMahon A, Karsgaard P, Al Quntar S, Ur J (2007) Early Mesopotamian urbanism: a new view from the north. Antiquity 81(313):585–600. https://doi.org/10.1017/S0003598X00095600
O'Sullivan K (2016) Letchworth: the first garden city's economic function transcribed from theory to practice. Berkeley Plan J 28(1)
Pacione M (2009) Urban geography: a global perspective. Routledge
Pahl RE (1966) The rural-urban continuum. Sociol Rural 6(3):299–329
Potter RB, Binns T, Elliott JA (2008) Geographies of development: an introduction to development studies. Pearson Education
Pradilla E (2002) Campo y ciudad en el capitalismo actual. Ciudades 54:3–8
Precedo A (2004) Nuevas realidades territoriales para el siglo XXI: Desarrollo local, identidad territorial y ciudad difusa. Editorial Síntesis
Quirós F (1991) Transformaciones urbanas recientes. En AA.VV. (eds) Sociedad y Territorio: XII Congreso Nacional de Geografía. Universitat de València-Asociación de Geógrafos Españoles, pp 393–398
Ravetz J, Fertner C, Nielsen TS (2013) The dynamics of peri-urbanization. In: Nilsson K et al (eds) Peri-urban futures: scenarios and models for land use change in Europe. Springer, Berlin, Heidelberg, pp 13–44
Richardson HW (1980) Polarization reversal in developing countries. Pap Reg Sci Assoc 45:67–85
Robinson J (2005) Urban geography: world cities, or a world of cities. Prog Hum Geogr 29(6):757–765
Rubio B (2002) La exclusión de los campesinos y las nuevas corrientes teóricas de interpretación. Nueva Sociedad 182:21–33
Ruiz N, Delgado J (2008) Territorio y nuevas ruralidades: Un recorrido teórico sobre las transformaciones de la relación campo-ciudad. Eure (Santiago) 34(102):77–95
Salazar A, Osses P (2008) La ruralidad en la Región Metropolitana de Santiago de Chile (2002): Determinación y relación con los grupos socio-profesionales. Scripta Nova. Revista Electrónica de Geografía y Ciencias Sociales, 12(270). Recuperado de http://www.ub.edu/geocrit/-xcol/435.htm
Segrelles JA (2015) Agricultura periurbana, parques naturales agrarios y mercados agropecuarios locales: Una respuesta territorial y productiva a la subordinación del campo a la ciudad. Scripta Nova. Revista Electrónica de Geografía y Ciencias Sociales 19(502). Recuperado de http://revistes.ub.edu/index.php/ScriptaNova/article/view/15107/18310
Sicca P (1977) La imagen de la ciudad. De Esparta a Las Vegas. Gustavo Gili
Siqueira D, Osorio R (2000) O conceito de rural. En: ¿Una nueva ruralidad en América Latina? (Colección Grupos de Trabajo). CLACSO
Smit J, Ratta A, Nasr J (1996) Urban agriculture: food, jobs and sustainable cities (publication series for Habitat II), vol 1. United Nations Development Program (UNDP)
Soja EW (2008) Postmetrópolis: Estudios críticos sobre las ciudades y las regiones. Traficantes de Sueños
Sorokin PA, Zimmerman CC (1929) Principles of rural-urban sociology. Henry Holt & Co.

Tacoli C (1998) Rural-urban interactions: a guide to the literature. Environ Urban 10(1):147–166. https://doi.org/10.1177/095624789801000105
Terán F de (1999) Historia del urbanismo en España III: Siglos XIX y XX. Cátedra
Vilas Boas LG (2017) Notas sobre a migração campo-cidade e a monocultura no Brasil. Ateliê Geográfico 11(1):189–209. Recuperado de http://www.revistas.ufg.br/index.php/atelie
Wallerstein I (2002) Conocer el mundo, saber el mundo: El fin de lo aprendido. Siglo XXI
Wagner RG, Wagner AD (2010) Levittown. Arcadia Publishing
Webster D, Muller L (2009) Peri-urbanization: zones of rural-urban transition. In: Sassen S (ed) Human settlement development. EOLSS Publishers Co Ltd., pp 280–309
Woods M (2005) Contesting rurality: politics in the British countryside. Routledge
Woods M (2010) Rural. Routledge

Part III
Asian Cities

Chapter 7
Beijing: Ruptures and Endurances in an Asian Intersocietal System

Francisco Javier Haro Navejas

Abstract I shall depict and analyze what elites built in and around the space of today's Beijing, where the ancient political hearts were, and where it lies today. I shall show how they combine the philosophical canons, their search for power, ideational aspirations, and architectural patterns of their cultural background to build a capital city. According to an official interpretation, history is a linear succession of facts that have shaped what is known as the Chinese civilization. Officially, thanks to the struggles of the masses and the Chinese Communist Party (CCP) guidance there is a so-called socialism with Chinese characteristics. Beijing, the capital of the People's Republic of China (PRC), would be a perfect example of the linear historical process that goes back to Western Zhou, 1045 B.C.E, to the present day. Here, in the framework of the volume *The City as a Project: History*, my focus shall be on three stages in the histories of the cities: Yuan dynasty, 1279–1368; Qing dynasty 1644–1911; and the first days of the PRC, 1949–1959. The methodological purpose of this decision rests on the fact that the three historical moments allow study ruptures and endurances. Those who challenged the dominant elites for power to some extent were natives and aliens at the same time to the dominant groups in spatial, ethnic, and cultural-ideological senses. Culturally and politically, the aspiring groups were part of overlapping and intertwined worlds of complex relationships, both of understanding and conflict. With a historical approach, the main goal is to portray the battle between continuities and discontinuities of (des)construction of Beijing as a northern political core in what is considered Chinese and non-Chinese worlds. There are common features in the *longue durée*, that could be conventionally named Chinese. It must be stressed that, on the one hand, concepts like China–Chinese–Chineseness belong to a group of cultural and political constructs that shape a contemporary meta-identity that cements the PRC under the CCP leadership. On the other, these constructs are components of the nonlinear historical process, negotiation, and imposition of identities, that materialize in cities built by wars and migration.

Keywords Beijing history · Intersocietal order · Urban identities · Ruptures · Endurances · Chinese architectural traditions

---

F. J. H. Navejas (✉)
University of Colima, Colima, Mexico
e-mail: fharo@ucol.mx

F. Carrión Mena and S. Rodríguez Alvarez (eds.), *From the City as a Project to the City Project*, The Urban Book Series, https://doi.org/10.1007/978-3-031-94226-6_7

7.1 Theoretical Approach to Chinese History

I shall depict and analyze what elites built in and around the space of today's Beijing, where the ancient political cores were, the Imperial Palace-*Gugong* (故宫), and where it lies nowadays, the headquarters of the Chinese Communist Party (CCP) leadership at *Zhongnanhai* (中南海). It will be shown how they combine the philosophical canons, search for power, ideational aspirations, and architectural patterns of their time.

According to the party interpretation, history is a linear succession of facts that have shaped what they consider the Chinese civilization. Because of the struggles of the masses and the CCP guidance, there is a so-called socialism with Chinese characteristics.

Beijing, the capital of the People's Republic of China (PRC), would be an impeccable example of the linear historical process that goes back to Western Zhou, 1045 B.C.E, to the present day. In the framework of the volume *The City as a Project: History*, the focus shall be on three stages in the histories of the cities: the Yuan dynasty, 1279–1368; Qing dynasty 1644–1911; and the first days of the PRC, 1949–1959.

The methodological purpose of this intellectual approach rests on the fact that the three historical moments allow study ruptures and endurances. Those who challenged the dominant elites for power were alien to the dominant status quo in spatial, ethnic, and cultural-ideological senses. Culturally and politically, the aspiring groups were part of overlapping and intertwined worlds of complex relations, both of understanding and conflict.

The main goal is to portray the battle between endurances and ruptures of a dual process destruction/construction of Beijing as a Northern political core in the Central and East Asian worlds.

I shall demonstrate that there are common features in the *longue durée*, that could be conventionally named Chinese. It must be stressed that, on the one hand, concepts like China–Chinese–Chineseness belong to a group of cultural and political constructs that shape a contemporary meta-identity that cements the PRC under the CCP leadership. On the other hand, these constructs are components of a non-linear historical process, negotiation, and identity imposition, which materialize in cities built by wars and migration.

Establishment academic constructs, successful all over the planet, have sinocentrist perspectives that go back to the late 60 s last century. Under the intellectual encouragement of John K. Fairbank, scholars began a research field to study the Chinese world based on a Sinocentric hierarchy. It had the following features: China was internal, large, and high; "barbarians" were external, small, and low (Yang, 1968).

Scholars studied this area as a space with one-way dominant interactions, where the Chinese civilized their neighbors that existed thanks to a superior culture. Some experts (Schuman, 1969) argued that throughout history there has been a fierce diversity that has been turned into unity many times. Chinese order, no matter what had the flexibility to endure the armies coming from the north. It could even tolerate historical defeats and even absorb foreign cultural components.

The historical perspective of this chapter considers there was an intersocietal regional order with diverse inhabitants and state structures. There was a wide spectrum of societies and governments. On one of the antipodes, there were uncountable (semi)nomadic groups near different mountain chains and deserts; on the other, the sedentary elites living around the large rivers, such as the Huang He (Yellow River), Chang Jiang (Yangzi River), or the Yue Jiang (Pearl River). Some societies were ruled by elites highly organized managing state structures capable of ruling over Han and non-Han populations. Exchanges were multi-tracks, not just relations between high and small. This order contained a wide range of institutional and kinship arrangements, state and non-state actors, This academic approach is "less political and more historical" not only were Mongol or Manchu "both separate (alien) and sinified" (Peterson, 2002), but components of a system where they Asianized each other, a mutual acculturation with created what is believed to be a Chinese civilization.

Dynastic elites have changed Beijing's face, even its name, many times. It has been a way to imprint how powerful they were. They try to leave a hallmark of their political and cultural achievements as a statement to venerate their ancestors, reliable proof of an amazing victory over the enemy, and testimony for future generations to emulate their greatness. In 1949 began a new phase of destruction-construction. Nowadays is still in motion with different intensities. The urban environment has been completely transformed: the Niujie Mosque, built in 996, survives with a dynamic social life; few reminders of the old city have been left, some temples, Tiantan, and gates, like Qianmen or the Gugong remain for the enjoyment of tourists.

In the modern Chinese language, officially named common language or *putonghua* (普通话), the word for city has two sinograms: *chengshi*-城市. From Western Zhou (1046–772 BCE), *guo* 郭 was used, and from the Warring States (476–221 BCE) cheng 城 appears, where the former refers to the exterior, the walls; and the latter, the interior, would be the city (Baidu, 2024a). Whereas today, *cheng* is preceded by *shi*-市 (Baidu, 2024c), appears in Shang (1766–1046 a.n.e) and it refers to the market, where economic transactions are carried out and is located the center of political power and bureaucracy.

Chapter XI of the *Daode Jing* (道德經), the Book of the Way and the Virtue (Lao, 1994), depicts the symbolism of a structure of walled cities. In my version:

> A wheel has 30 spokes around a hub; it is in the nothing that lies the usefulness of the cart. We transform clay into a vessel; it is in the nothing that lies the usefulness of the vessel. Chisel to make windows and doors to build a room; it is in the nothing that lies the usefulness. What exists makes things beneficial; therefore, nothing becomes useful.[1]

The city is inside a void protected by material boundaries and is filled with constructions that materialize power. By design, Chinese cities lack a spatial and symbolic center like Latin Americans have. In places like Mexico City, the space

---

[1] The text in Chinese: 三十當, 共一以, 當其無, 有車之用. 普道以為器, 當其無, 有器之用. 德戶城以為室, 當其無, 有室之用. 故有之以為利, 無之以為用.

of government, business, and religious power is visible to the naked eye in different material ways than in China.

There is a legitimate power in Beijing that comes from and with the Mandate of Heaven (*tianming* 天命), an emperor. This power could occasionally be in the hands of an empress or a regent. The person in charge had different levels of strength. Regardless of the force, negotiation and confrontation were a fundamental part of the life of a city. Inside, *nei* 内, there were powerful groups, like eunuchs, military, and bureaucrats; outside, *wai* 外, there were dangerous outsiders, such as regional powers, peasants, and (semi) nomadic groups. All of them belonged to the intersocietal and interstate above-mentioned.

Inside the framework of academics-theoretical-methodological Chinese approaches, it is hard to discern politics from science. The first dictates the conclusions, no matter what happens during the research process in archives or the field. The political debate revolves around where to establish the spatial origin of China and its civilization. The source provides the narrative elements to construct a story of continuity that flows to what today is the RPC: Archeology has proved a multi-spatial of both the origin and development of Chinese societies in embryonic stages once-called primitive cultures during the Neolithic period (Qizhi, 2015). Politics argues that no matter the various cultural regions, there is a common ancestor, a culturally dominant ethnic group, and a continuous civilization.

A sort of syllogism is constructed: if an early civilization arose in different spaces and similar times with shared specific traits in a linear time, today there is a multiethnic and unitary state with a Han dominant and a civilizational majority.

From the perspective of an encyclopedia, today *city-chengshi,* a disyllabic word for city, is considered the main component of civilization, at least at a similar level as writing.

Coinciding with the time-line and space of this chapter, semi-official arguments state that "[S]ince Yuan [, Ming and Qing dynasties], China has never been divided, it has been a unified country", a condition that "created the most majestic and glorious capital city in the history of Chinese urban development—Beijing" (Baidu, 2024b).

Chinese political and intellectual elites have created a discourse of linear historical process. Academics gathered information and analysis from archeology or history research and concluded that Chinese civilization is composed even by the contributions of politically branded pastoral national minorities without names. This interpretation stumbled upon two historical "revelations" (Yuan, 2010): the history of Chinese civilization lies, first, in the choice of peace, harmony, and tolerance; and second, in the choice of tolerance.

In the historical processes and in the discourse, there are a lot of tensions and paradoxes. Walls are to differentiate, horizontally, communities; and vertically, social strata. There are dichotomies between inner and outer, event and narrative, defense and control. They represent sociological differences related to wealth, subjecthood, and allegiance, among other aspects (Hay, 1994).

The main paradox is that the Chinese discourse to build a historical narrative is born from a European intellectual matrix. According to this perspective, the world civilizations are China, Egypt, Indus Valley, and Mesopotamia. While stressing continuity, unity, and basic atemporal cultural almost unchangeable traits, two aspects are avoided: one of them is that even when bureaucracies try to keep the status quo, there is no smooth succession and continuity of regimes; the other one is that before there was not total spatial domain and nowadays the controlled space is bigger than ever. More importantly, the historical rupture was deeper since 1949.

The current capital of China has been named successively, starting in 1270, Daidu-Dadu 大都, Yuan dynasty; Beiping 北平 since 1368, Ming; Beijing 北京 during Qing, 1644; and for a few years, 1945–1949, Beiping; named again Beijing since October of 1949. Few budlings have survived wars and revolutions, the city wall being one the main casualties of the CCP regime that began a feverish building process, especially around Tian'anmen. The dynastic architectonic is scarce and in Beijing "has stuck to the Moscow-style urban development pattern until today" (Wang, 2011). However, it tends to be a Westernized metropolis.

7.2 The Capital of the North: Outcome of Migration and Struggle for Power

The continental mass covering East and Central Asia has been inhabited by people with different approaches to their spatial context. The multitrack interaction between natural resources and geography shaped specific family-social structures, agriculture, forms of negotiation, and warfare.

Sedentary and nomadic interacted in many ways and for multiple reasons, mainly fighting for legitimacy-power, material resources, and space. In an intertwined social process, both exchanged and influenced each other with culture, religions, symbolic practices, wives, food, and other non-material goods. While the Chinese people, those who lived around the Huang He and Chang Jiang, thanks to powerful and efficient state structures made expansionist movements, mostly to keep at bay the strong groups that pushed to the South looking to build their state upon the existing status.

The most successful group was the Jurchen (*nüzhen* 女真 in Chinese) which built the Jin dynasty, 1115–1234; and self-labeled as Manchu (*manzu* 满族 in Chinese) during the Ming. It built the Qing dynasty, 1644–1911. Historically and because of spatial advantages, conducive to farming, Chinese dynasties had capitals not so far from the Yan Mountain and the Gobi desert. As a long-term capital, Beijing was a decision made by seminomadic elites. It offered them an advantageous spatial hinge to move to their cultural environment and to fight the displaced rulers. This choice pushed the ruling elites to the South where they protected material and cultural assets. Under new circumstances, they built alternative urban centers of power, mainly Nanjing.

Beijing has lived through deep changes from different sources. From a historical perspective, three are relevant and linked to each other: first, human flows, agriculture, and des/construction made by the elites; second, the ecological devastation, due to intensive farming and strong pressure coming from the Gobi desert; third, the devastation produced by several armies and the subsequent reconstructions practically from the foundations, especially of the spaces of the elites.

Drastic but relatively slow changes in fauna, such as the retreatment of the elephants from the North (Elvin, 2004), are a departure that shows that the transformations of Dadu-Beijing were part of a complex combination of struggles of elites between sedentary and semi-nomad elite negotiating-fighting for natural and person-made resources, and certainly for political and military power. Struggle for power had the destruction-construction of a capital as one of its main expressions with elements added like the exchange of goods and marriages.

The main political center of the current PRC has been demographically multiethnic, a magnet for poor farmers; provincial persons of different backgrounds, looking for power and fame, mainly in bureaucracy; and a seat of powerful and centralizers governments, Mongolian (Yuan dynasty, 1279–1368), and Manchu, (Qing dynasty 1644–1911), between these two arose an arguably Han dynasty (Great Ming, 1368–1644). Historically, the strongest force in the transformation process of the city has been the CCP, when it is possible to identify two main stages: the Stalinist blueprint during the first decade after 1949 and the actions to build a metropolis according to Euro-American patterns from the 90 s until today.

There is a classic model of how cities should be built in interaction with space composed of squares where there is a round Heaven (天) and a square Earth (地). Facing to the south, the direction of fire and summer should be the capital at the center protected from the northern enemies. The model had the philosophical goal of drawing the rules to reproduce a hypothetical order of the universe.

In Chinese cosmology (Feng, 1999), space was conceived as a group series of imbricated squares, at the center of which lay the capital of the empire strictly orientated to the points of the compass. The south is the direction of Red Phoenix, representing summer and fire. From the north came winter and destructive hordes. Its color was black. Therefore, the imperial palace and every important building faced southward, and the emperor faced away from the north when he addressed the gods or his ancestors. It is a convention that the *Rites of Zhou, Zhouli* 周禮, instructs how the center of power should be: three gates on every side, nine accesses-oriented north–south, nine-chariot wide, intersecting with similar ways east–west thoroughfares. The seat of power should have an Altar of Earth and Grain to the west and a market to the north. The south part aimed to house the court, on the opposite side of the market (Steinhardt, 1999).

It is impossible to find the ideal city portrayed in the Zhou Li. However, Beijing during Ming, 1368–1644, and Qing is close to the model. Therefore, it must be noted that, considering the situation of the quarters of the ruler, there is a type called Center Palace, the *Zhouli model; the North Palace, Chang'an, near Xi'an linked to the tradition of nomads from northern Mongolia. Dadu (大都), the capital of Yuan, represents a variation with bell and drum towers at the heart of city* (Funo, 2017). *There is not a single pattern, there have been three: Zhouli, Chang'an, and Da Du (Dai Du, Da Yuan, or Khanbalik). Beijing fits in the third model.*

7.3 Da Du: Born Out of War and Negotiated Identities

The Mongol conquest of what has been considered the Chinese cultural sphere during the thirteenth century was the product of a leadership capable of a highly organized management of natural and human resources. Leaders had the power and bureaucracy to mobilize large armies based on a decimal system and with strong ideological foundations around the concept of Eternal Heaven.

Mongol elites learned from their history, as well as from the Turkic and Chinese histories. Inside and outside factors paved the way to atomized Chinese and Jurchen elites who had to flee to the south and even fought each other allowing the northern enemies to become victorious. Mongols not only waged wars, but in their ranks, they had experts from different origins on different Chinese political and administrative matters (Allsen, 2006).

The future founder of Yuan, Khubilai, thanks to her mother, Sorkhakhtani Beki, learned from Muslims, Christians, Confucians, and Tibetans. She encouraged him to embrace government activities, and he understood the relevance of agriculture. Khubilai got involved in domestic fights and war against the Song in the south even beyond the Chang Jiang. After defeating the Chinese dynasty and his most fierce enemy, his brother, Ariq Böke, his rise to power was at hand. Even then, he had to face the rebellion of Li Tan, a powerful enemy against him being the emperor of a new dynasty (Rossabi, 2006).

In the seventies of the thirteenth century, the *Yeke Yuwan Ulus*-Great Mongol State or the Great Yuan dynasty was established with power over vast parts of today's China and other parcels of China under the fist of Shizu, temple name of Khubilai, Genghis' grandson and Tolui's son. It could be asserted that the similarities in the government were superficial, where Mongol elites administered appanages (Tanner, 2010).

His government had some ruptures with the Chinese systemic features, such as abandoning the examinations to enter the bureaucracy to empower himself and keep Confucians under his control leaving the selection of state officials to his personal decision. When there were examinations, 1315, specific amounts of places were decided in advance according to the state-designed ethnic identities.

The politics of identities is relevant because it was the foundation to build a capital and to decide public policies on administrative, fiscal, or judicial regulations. Discrimination was based on the Confucian rectification of the name operated by Shizu and his elite. Product of wars and negotiations, there were three pyramidal main identities: at the peak, there was the Mongol elite, which does not mean that every Mongol was part of it; in the middle, important to strengthen legitimacy, there were the Semuren (色目人), an array of various social groups from Central Asia that were already living in there or were with the troops in several capacities; finally, there were the Hanren (汉人), a general category of Chinese, that possibly included Jurchen, Khitan, Koreans. It meant subjects of Liao and Jin. It first referred to northerners and also the southerners (*nanren* 南人). Also, there were 32 more groups, such as Naimans, Russians, Turkish, Tibetans, Tuva (Gernet, 1996). This identity richness was key to ruling an enormous multicultural population with multilingual officials.

The new regime needed a magnificent urban framework to project, exercise, and legitimize its power. The new capital was thought to be there forever, like every powerful ruler of the past. Khanbalikh or Dadu was an outcome of cultural exchanges and negotiations. The construction of Dadu, in 1266, was a product of complex interactions between a society living a transition from nomadism to a sedentarism based on agriculture with a long urban tradition. Like their sedentary peers, Mongol rulers destroyed cities to their ashes. After the destruction, they built new cities. They learned from cultures in West Asia, but the main source of building was the Chinese ideal of erecting a *dingding* (定鼎*)* or *dingdu* (定都), a capital.

One of the main exchanges was that intellectual with the knowledge to choose the right place to build the urban space, perform divination rituals, draw a blueprint, decide on materials, consider military and economic aspects, manage thousands of workers, design a logistic control migration and to feed the population, among many other considerations.

The predecessors of Dadu were Qaraqorum, the capital and ceremonial center, and Shangdu (上都), the almost mythical Xanadu. There has been a debate where some experts (Rossabi, 2006) consider that the designer was a Muslim architect, Yeh-hei-tieh-erh. However, a different interpretation (Liu, 1992) explains that this version appears and disappears. Now experts consider that the designer and builder was Liu Bingzhong. Around 1253 he became advisor of Khubilai. A prototype of a Chinese Confucian scholar: astrologer, Buddhist, military expert, poet, politician, among other specialties. When the Mongol leader was commissioned to build Shangdu by his brother, he appointed Liu to divine the right place to build it. Because of the magnificence of this city, before establishing its power in the Chinese north, he was ordered to build Dadu, the core of the Yuan state.

Mongols had a central capital, Yuan Zhongdu, with three concentric walls, built between Shangddu and the magnificent Khanbalikh-Daidu created by Liu under the model portrayed in the *Rituals of Zhou*. It had a central axis running through the north and south.

In every direction, in the outer walls, there was a never-built-before stone (Steinhardt, 2019). These walls, more than 50 kms.^{2}, with a perimeter of 28.6 km., were a perfect rectangle. They were a heavy fortification of thick walls 22–24 m wide. The city, from the outer walls, was divided by boulevards into 54 wards and four bisecting streets. The former was 25 m wide in the center and 6–7 m on each side. Buildings of the elite and the flow of water broke the enormous avenues and streets. There were twelve gates, three at every cardinal point:

The construction had Mongol elements like curtains around the beds, wide avenues were thought to gallop and even the ruler preferred to sleep in the open air in tents located in ad hoc spaces. Rituals and sites of performance were political expressions to strengthen the legitimacy among elites from Central and East Asia and those of Jin and Song.

From a perspective of religion, Khubilai had a practical approach and even supported religious debates (Steinhardt, 2024). Tenggerism remained central and among some of the Mongolians, there was a fascination for esoteric Buddhism. There were many other expressions such as Christianity, Islam, Quanzhen Daoism, and many forms of Confucianism. Khubilai and his appanage, generally Mongols, worshiped mountains, and rivers. They combined their rituals with those of the conquered, like the Deity of the Ji River. In the Mongol cities, the ruler ordered build Daoist and Tibetan Buddhist monasteries, Khubilai always indicated to include Mongolian elements in every ritual with music played by Buddhists, Huihui, Tangut, among others (Ma, 2022).

7.4 Identity Flow and Spatial Expansion

The Great Qing Dynasty (Da Qing 大清) faced what could be considered traditional problems and successes. There was one that impacted almost every institution of the state and the fabric of the society. The impacts, combined, led to the end of the process of how Central and East Asian elites rose to state power. Crises were links that hooked together to an increasingly heavier chain that sank Qing: international monetary system-positive payment balance-silver, steam engine-invasions, Christianism-natural disasters-rebellions, opium trafficking-addictions-corruption-trade imbalances, European ideas-eagerness to change the political system, and many more.

I shall focus here on the transition and beginning of the regime and its capital, not on its evolution and end. Because of spatial and political-strategic reasons, Beijing remained the core of the elite's power. Mongol and Manchu rulers, and many years later Mao Zedong's regime, understood that the Northern Capital was the best point to interact with cultural identities from every cardinal cultural point to attempt to exercise power over them and negotiate several social actions. The intersocietal centennial order built by elites had many overlapped social spaces with extended families at the pinnacle of the structure of power that attempted to construct a center on that system

In the opposite direction of most dynasties, they did not destroy the sieged city. The Manchu (manzu 满族) elite, with its army and bureaucracy, after a long string of struggles, took possession of the Imperial Palace, Gugong 故宫, in 1644. Even when the control capacity and legitimacy were not even close to those of the future regime of the CCP, the new rulers conquered and colonized most of what the PRC is now. The emperor, Aisin Gioro Fulin, was a child, whose Chinese name was Shunzhi. Therefore, the defeat of the Ming and domestic rivals, the conquest of Beijing, and the seizure of power should be attributed to Aisin Gioro Dorgon, called Uncle and Prince Regent. He made Ming elites go to the far South and began the expansion to the West.

Although a new capital was not founded, significant changes were not made to the old one. Some popular readings (Clements, 2016) state that there were no changes and emphasize what fascinates tourists today: hutongs 胡同. These alleys, with a history dating back to the Yuan Dynasty, have been the space of several siheyuan 四合院, residences of Han origin with a central courtyard for extended or designated by the State families. If the Palace was home to the elite power, hutongs, the place where the barbarians are together, was home to the people, many of them immigrants at some point of their personal history.

Sometimes it was urbanization in the south, and in some places there were attempts to educate children, and one of the most important was related to the identity names imposed-negotiated. It is relevant, for two reasons. Firstly, different ethnicities clashed one against the other; sometimes, they acted together to face the State. Secondly, there were flexible state politics, mainly that pursued shape identities not linked to kingship or biology, but to social variables.

The complexity and ambiguity of the identity issues of Qing contrast with today's rigidity. Multiculturality was a structural feature of any society of this region and elites have had different ways to approach the social process. The transcendent were not in architecture, but those related to demographics, especially migration due to growing trade and lesser government controls. If the heart of Beijing did not suffer great changes, in its outskirts people sharing origins and religious beliefs helped to spur buildings to practice religion. Growing trade meant socially active merchants, who boosted cultural activities that led to new expressions such as jingju 京劇, known as the Beijing opera (Rowe, 2002).

Dorgon, more inclined to the field of battle than to that of politics, had to face many issues in Beijing, such as the coal supply, eunuchs' unrest, and changes in the bureaucracy to not to overstaff it, contain his fellow fighters to avoid the takeover of land belonging to the Ming elites, at the same time let bannermen to own land, and designing a tax system. Since Man and Han could share spaces, one of the main problems was the distribution of households according to ethnicity. In this process, old inhabitants lost their properties and fell into anomic activities becoming bandits or vagabonds (Wakeman 1985).

Even when the argument seems weak and obvious, it is considered (Hou, 2014) that a change was not necessary to build a new capital because the conditions of the city left by the Ming were excellent, even with an important inherence from Yuan and even before, like the Summer Palace or Imperial Garden (Yiheyuan 颐和园). On the other side, with European inspiration, Qing built the Yuanmingyuan 圆明园 with the ideas of two Jesuit priests Giuseppe Castiglione and Michel Benoist. Imperialist armies destroyed both places, but only the latter remains ruined as a testimony of foreign aggressive actions.

In a part of a classic study (Skinner, 1985), it is explained that the Qing capital is a model of an urban ecology found in late imperial China. It had two nuclei: on the one side, the center of commercial activities with two-floor buildings housing salesrooms and crafts business, some doubled as diners and palaces to spend the night. On the other side, a center destined for the gentry and officials' activities was located near the imperial school temple and the great examination hall.

7.5 Beijing Has Stood up!

There have been many political urban centers in what today is the PRC. Chang'an, near of today's Xi'an, has been the greatest and most cosmopolitan capital of several dynasties. Its golden age was during the Tang, 618–906. In the twentieth century, because of civil wars and invasions, there were at least five capitals: Guangzhou, 1925–1926; Nanjing, 1927–1937, 1946–1949; Wuhan, 1927, 1937–1938; Luoyang, 1932; and Chongqing, 1938–1946.

Since the Yuan Dynasty (1279–1368), with ruptures and endurances, sometimes with a different name, Beijing has been the magnet and political compass where politics are designed to control and manage massive amounts of natural resources and people. Its origin was part of the emergency of a Mongol state, the stage of the abdication of the last Qing emperor, and the scenery of almost any political expression struggling to build a State without recovering the dynasty cycle.

Beijing was built by warrior Mongol elites that belonged to a complex and diverse non-Sinocentric group of States that in the antipodes had several people with nomadic ways of life and sedentary societies. Purportedly Han-Chinese elites were cemented by a culture constructed by systematically trained and educated bureaucracy guardians of different philosophical and legal traditions materialized in institutions and regulations.

Mongols and Manchu elites, paradoxically, by marriage or alliances, belonged to that cultural-political universe, but at the same time, they were outsiders. However, they did not belong to the ruling bureaucracy or the powerful families down south the barren territories unsuitable for agriculture and sedentarism.

Displaced from the government decision-making process. The source of products that feed the elite, peasants were key social actors in countless rebellions. Even some of their leader became emperors, such as Liu Bang, Western Han Dynasty, 206 B.C.E.-6; Liu Xiu, Eastern Han Dynasty, 25–220; Shi Le, later Jie Zhao Dynasty, 319–351, and Zhu Yuanzhang, founder of Ming dynasty with the name Hongwu, 1368–1644. There are famous rebellions, like the one of An Lushan (755–763), a Sogdian beyond Tang cultural boundaries.

Following this historical pattern, Mao Zedong, in open opposition to the policies dictated by Moscow, designed a discourse to lead a peasant rebellion, with no intention to establish a dynasty. He had a minimalist goal: achieving the new democracy and, maybe, someday, instituting a local version of socialism. Mao was part of the East Asian cultural universe. As an outsider, even in a disordered society without a centralized state, he did not wholly break with his cultural tradition and environment but fought for the sinification of Marxism.

Beijing houses persons of every imaginable ethnicity and cultural background. Migrants take the capital by strategy to live, work, study, and rule a vast territory. They transformed the system and the city. In the same process, people are shaped by them. For instance, young Mao pretended to build a Republic of Hunan as part of a federation. Soon, in 1923, a member of the leadership of the young Communist Party, expressed that a federation would be a group of regional military separatist powers (Spence, 1999). Mao created a vertical political system with a military-party political machine controlling an unimaginable extension of territory.

Those who live there consider themselves lao Beijing (老北京), therefore sometimes they despise the waidi (外地), the outsiders, mainly peasants. However, beyond spatial divisive lines, almost everybody is laobaixing 老百姓, the people, components of a multi-identities society.

To prove its power, establish its path to the future, and achieve legitimization, Beijing has been partially destroyed and rebuilt many times by the bureaucracy under Mao's guidance. In the name of transformation and modernization the triumphant government since October 1949, the face of the capital changed to impose a new vision of the capital. Some spaces were barely touched, Gugong or Tiantan. The city wall and buildings around Tian'anmen were demolished, some *hutongs*-alleys disappeared and Chang'an Avenue was widened to increase its magnificence.

In the 50 s, not even one year after the beginning of the new government, the Gate of China, the south entrance to the Imperial City, was demolished by Soviet suggestion, and the Monument to the People's Heroes was raised there. Years later, in 1976, a mausoleum to venerate Mao's corpse was built around the same space.

To overshadow any square on the Planet, including the Red Square in Moscow, Tian'anmen was enlarged in 1959. Between 1958 and 1959, the tenth anniversary of the resolution, in the framework of the Great Leap Forward, was the reason to build the same number of buildings. The two most important are in the square, the Great Hall Of the People, the seat of the National People's Congress was erected on the western side of the square, and the Revolutionary History Museum (today called National Museum) was built on the east side. The structures have a strong modernist Soviet influence, symbols of the new power encircling the ancient dynastic buildings.

In a similar aesthetic fashion, the new buildings were the People's Revolutionary Military Museum, with a similar architecture to the Hall of the People; the Minzu (Nationalities) Hotel, designed to host foreign guests; the Workers Stadium, still functional and renovated several times, one to host events of 2008 Olympic Games; finally, the Overseas Chinese Hotel, now replaced by a similar business.

A mix of modern architecture and a Chinese flavor, especially on the top, other constructions were built: the Beijing railway station, not anymore the biggest in the country and situated inside of that was the walled quarters; the Diaoyutai State House, now a tourist attraction, once a royal spot for fishing, it was improved to guest foreign government guests, but also was a base to run meetings against Moscow and was the headquarter of the Central Cultural Revolution Group; the last one, a building pride of Zhou Enlai and designed by him, is the National Agriculture Exhibition Hall.

Every building aforenamed, by itself and as a group, has polysemic meanings. On the surface, they were an excuse to celebrate the first decade of the triumphant revolution. Secondly, they show, some more than others, the paradoxes of the inequal relation with the Soviet Union. Buildings are an expression of the eagerness to become a world power. Those constructions represent an interpretation of history, where past and present can be merged to boost progress. Finally, they picture the ethnic identity defended by the bureaucracy.

The Cultural Palace of Nationalities is not on that list. It has a strong political and "unambiguous" symbolism "of the government's ethnic policies in the early years" when the PRC and the CCP were in "a crucial period of consolidation" (Hung, 2012).

7.6 Beijing Capital of an Intersocietal Asian System: Final Considerations

The double-faced process of destruction-construction with Xi Jinping (2013–) has been the strongest in Chinese history. At first glance, Beijing has few physical vestiges of its dynastic past spatially dispersed: Gugong, Niujie mosque, Tiantan, Yiheyuan, Yonghe lamasery, Yuanmingyuan, Zhongnanhai, Chang'an Avenue, and a few *hutongs*. The Maoist buildings remain but some of them underwent considerable modifications. People from Central and East Asia, such as Koreans, Tibetans, Uyghurs, and many other migrants, both elites and *laobaixing,* are the main threads of historical processes.

In the academic world and politics, two intertwined sinocentrisms pretend to understand and transform societies: one rooted in American universities, and the other raised by Chinese elites. Both have subdivisions and have successfully spread throughout the globe. Their assumptions are simple: there has been a Chinese civilization culturally superior due to its writing system, Confucian bureaucracy, historical continuity, and the existence of a majority ethnic-biological group superior to the rest. They express weaker arguments such as a modern translation of Zhongguo 中國, center-state, originally guo 國 meant a walled city defended with weapons. Also, state that the once-called barbarians, now national minorities, were sinified, meaning assimilated and accultured.

Beijing has been and will be built by outsiders of the status quo of what traditionally has been labeled the Chinese civilization: Mongols, Jurchen/Manchu, and communists. These outsiders were actors of a complex system that belonged to communities with their history, language, culture, institutions, way of producing food, and practice beliefs. All of them have power and weaknesses. Even when every group had its own inner goals and rules, they had societal relations and shared values to engage in pacific and violent interactions of multiple tracks and mutual influence.

Their interactions produced a highly specialized and trained bureaucracy capable of governing and (re)producing ideas that materialized in new regimes that, among many aspects, built and rebuilt cities.

As a part of history and as a project, Beijing shows how the long-duration process works in two ways: ruptures and endurances. Even the most radical regimes, Mongols and Maoists, made public works to highlight the past. They constructed spaces to increase their legitimacy, strengthen their defense capabilities and face rebellions, deepen their relationships with other people of the intersocietal system, and confront enemies.

Symbolism in today's Beijing is polysemic: the building represents a fusion of architectural traditions, an officially sanctioned multiethnicity, with the Han identity superiority over the so-called national minorities. Its strongest significance is related to the party's power, the guide of the proletariat and the peasants.

References

Allsen T (2006) The rise of the Mongolian empire and Mongolian rule in north China. The Cambridge history of China, vol 6. Cambridge University Press, Cambridge, pp 321–413

Baidu (2024a) Cheng 城. [Consulta general]. 百度百科. https://baike.baidu.hk/item/%E5%9F%8E/2315874

Baidu (2024b) Chengshi 城市. 百度百科. https://baike.baidu.hk/item/%E5%9F%8E%E5%B8%82/33549

Baidu (2024c) 市. 百度百科. https://baike.baidu.hk/item/%E5%B8%82/22302

Clements J (2016) A Short History of Beijing. Armchair Traveller

Elvin M (2004) The retreat of the elephants. Yale University Press, An environmental history of China

Feng X (1999) Past and present: transformation of an ideal city—Beijing [Master, Virginia Polytechnic Institute and State University]. https://vtechworks.lib.vt.edu/server/api/core/bitstreams/89494a3c-ff2c-41ce-9e6e-f78f87e2ec34/content

Funo S (2017) Ancient Chinese capital models—measurement system in urban planning. Proceed Jpn Acad Ser B Phys Biol Sci 93(9):724–745. https://doi.org/10.2183/pjab.93.045

Gernet J (1996) A history of Chinese civilization. In: Foster JR, Hartman C (eds) Trans; 2nd edn. Cambridge University Press, Cambridge

Hay J (1994) Introduction. In: Hay J (ed) Boundaries in China. Reaktion Books, pp 1–55

Hou R (2014) An historical geography of peiping. Springer and Foreign Language Teaching and Research Press

Hung C (2012) The cultural palace of nationalities: ethnicities under one roof? J Contemp Hist 47(3):572–593

Lao T (1994) The way and its power: Lao Tzu's Tao Te Ching and Its Place in Chinese Thought (A. Waley, Trans.). Grove Press

Liu C (1992) The Yüan dynasty capital, Ta-Tu: imperial building program and bureaucracy. Toung Pao 4(78):264–301. https://doi.org/10.1163/156853292x00027

Ma X (2022) The mongol rule. In: May T, Hope M (eds). Routledge

Peterson WJ (2002) Introduction: new order for the old order. In: Twitchett D, Fairbank JK (eds) The Cambridge history of China, vol 9. Cambridge University Press, pp 1–8

Qizhi Z (2015) An introduction to Chinese history and culture. Springer and Foreign Language Teaching and Research Press

Rossabi M (2006) The reign of Khubilai khan. The Cambridge history of China, vol 6. Cambridge University Press, Cambridge, pp 414–489

Rowe WT (2002) Social stability and social change. In: Twitchett D, Fairbank JK (eds) The Cambridge history of China, vol 9. Cambridge University Press, pp 473–562

Schuman F (1969) In memory of a friend. In: Levenson JR, Schuman F (eds) China: an interpretative history. From the beginnings to the fall of han. University of California Press, pp v–vi

Skinner GW (1985) Introduction: urban social structure in Ching China. In: Skinner GW (ed) The city in late imperial China. SMC Publishing, pp 521–553

Spence J (1999) Mao. Weidenfeld and Nicolson

Steinhardt NS (1999) Chinese imperial city planning. University of Hawai'i Press

Steinhardt NS (2019).Chinese architecture. a history. Princeton University Press

Steinhardt NS (2024) Yuan. Chinese architecture in a mongol empire. Princeton University Press

Tanner HM (2010) China: a history from neolithic cultures through the great qing empire.10,000 BCE–1799 CE, vol 1. Hackett Publishing
Wakeman F (1985) The great enterprise. The manchu reconstruction of imperial order in seventeenth-century China, vol I. University of California Press
Wang J (2011) Beijing record. A physical and political history of planning modern Beijing. World Scientific
Yang L (1968) Historical notes on the Chinese world order. In: Fairbank JK (ed) pp 20–33). Harvard University Press
Yuan X (2010) The historical revelations from the Chinese civilization. Proced Soc Behav Sci 2:7006–7011. https://doi.org/10.1016/j.sbspro.2010.05.054

Chapter 8
Postcolonial Mumbai: The Paradoxical City

Amit Ranjan

Abstract This chapter examines the paradoxical character of Mumbai. The city has India's wealthy population and is also home to millions of poor people. The chapter analyzes how Mumbai's religious pluralism fell apart due to political rise of Hinduism and parochial demands. Due to constraint space, as the chapter shows, how socio-economic paradox intersects the everyday life of Mumbaikars (people living in Mumbai).

Keywords Business · Slums · Shiv Sena · Communalism · Communism

8.1 Introduction

Mumbai (known as Bombay until 1995) is India's financial capital. The city is the headquarter to the Indian banking system, houses India's prominent business persons and economic elites, the centre of India's soft power producing—Bollywood, as the Hindi film industry is prominently known. In terms of real estate value, in 2024, Mumbai ranked eighth in the Knight Frank Wealth Report. The report says that in Mumbai, one can purchase 103 m^2 of prime residential real estate in US$ 1 million (Ramani 2024). The estimated wealth of Mumbai is around US$ 950 billion, making it the 12th richest city in the world (Bendix 2018). Of the total 187 Indian billionaires, according to the 2023 M3M Hurun Global Rich List, a ranking of the US-Dollar billionaires in India, Mumbai is home to 66 individuals who fall into such category (Mint 2023).

Paradoxically, Mumbai is home to a large number of slum dwellers living in deplorable conditions. The poverty in Mumbai, in general, is attributed to increasing migration into the city. Demography-wise, between 1901 and 1941, the decadal growth rate in population in Greater Mumbai (Colaba to Dahisar in North to Mankhurd in East) was 18%. In the first four decades of post-independence years,

A. Ranjan (✉)
Research Fellow, Institute of South Asian Studies, National University of Singapore, Singapore, Singapore
e-mail: isasar@nus.edu.sg

F. Carrión Mena and S. Rodríguez Alvarez (eds.), *From the City as a Project to the City Project*, The Urban Book Series, https://doi.org/10.1007/978-3-031-94226-6_8

the population grew at decadal growth rates of over 35%. The decline in the growth rate began in 1991 and sharply reduced to 3.87% during 2001–2011. The Mumbai Metropolitan Region (MMR)[1] has continued to grow at higher rate. The population of Greater Mumbai has increased from 9.27 lakh (0.927 million) to 12.44 million in 2011, which has increased over 13 times (Municipal Corporate of Greater Mumbai Final Report 2016, pp. 2–1). According to the 2011 census, out of a total population of 9,356,962 in Greater Mumbai Municipal Corporation, 4,620,654 people live in various slums spread across the city. This is 49.38% of the entire population of Greater Mumbai (Census of India 2011). Another figure says that Greater Mumbai's population in 2011 was about 12,442,373 of which 41.84% lived in about 1, 135, 514 slums of the city (Mumbai (Greater Mumbai) City Census 2011). The total land occupied by slum dwellers in Mumbai is estimated to be about 43,000 hectares, which is 8% of the city's total land (Appadurai 2000). Figure 8.1 shows slums and high-rise buildings that define Mumbai's space, where people from two different classes live nearby despite different in their socio-economic positions.

Besides increasing migration into the city, historical discrimination favouring the business class and further sidelining the marginal section of the society in economic policies and planning has, over the years, widened inequality between rich and poor. According to a discussion paper to take a "progressive review" of the situation published by the National Institution for Transforming India (NITI Aayog), a government of India think tank, multidimensional poverty in Mumbai City in 2019–2021 sank to 1.21% from 3.59% during 2015–2016. During the same period, in the Mumbai suburban region, the percentage slid to 1. 15 in 2021 from 4.65 earlier (NITI Aayog 2023, p. 156). Across India, the discussion paper says that the incidence of poverty fell from 32.59 to 19.28% in rural areas and 8.65 to 5.27% in urban areas between 2015–2016 and 2019–2021 (NITI Aayog 2023, p. XV). According to some analysts such as Santosh Mehrotra, a well-known development economist, the basic flaw in the NITI Aayog's estimates is that it is using data from "the National Family Health Surveys or NFHS 3 for 2005–2006. It has claimed that the NFHS 4 of 2015–2016 applies from 2014 to 2016. Since there is no year-by-year data for the 12 indicators, just a straightforward compound annual growth rate projection is carried out by NITI Aayog, assuming that United Progressive Alliance (2004–2014) period improvement rates between 2005–2006 and 2015–2016 in the 12 indicators applied to 2 years (2015 and 2016) of the NDA government's term of office" (Mehrotra 2024).

This chapter attempts to answer two questions: How does city's history and economic development in Mumbai create a paradox? How is religious pluralism almost dying, segregating the people belonging to different confessions? This chapter chiefly argues that economic and social inequality in Mumbai has a history. The capitalist class flourished during the colonial period because of their close ties with

[1] The MMR is spread over 6328 km^2 of area comprising nine Municipal Corporations: Greater Mumbai, Thane, Kalyan-Dombivali, Navi Mumbai, Ulhasnagar, Bhiwandi-Nizamapur, Vasai-Virar, Mira-Bhayandar and Panvel. There are nine Municipal Councils Ambarnath, Kulgaon-Badalapur, Matheran, Karjat, Khopoli, Pen, Uran, Alibaug and Palghar, along with more than 1000 villages in Thane, Raigad and Palghar Districts (Mumbai Metropolitan Region Development Authority n.d.a).

Fig. 8.1 There is an intersection of slums and high-rise buildings in many parts of Mumbai. In this picture, the slums begin where the boundary wall of high-rise building lasts. However, the common thing in both areas is saffron Hindu flags, which were put up in January 2024 to mark the consecration of Hindu God Rama's temple by the Indian Prime Minister Narendra Modi in Ayodhya in the Indian State of Uttar Pradesh. *Source* Photo taken by the author

British officials. They were also connected with the Indian National Congress (INC) to secure their interests. On the other hand, many workers kept on landing in Mumbai in search of a job. Many of those labours spent their lives in one or the other slum of the city, spread across the city. The economic boom of Mumbai saw uneven benefits- a section of people earned a lot of money while many were pushed into poverty. After independence in 1947, although India adopted a socialist-welfarist economy model, Mumbai-based big business groups such as Tata, Birla, etc., remained important players in the Indian economy. After India adopted the new economic policy (NEP) in 1991, Mumbai stayed the centre of the country's economy and finance. Post-1991, Mumbai has seen an increase in the number of billionaires, and at the same time, there has been an increase in the number of poor people living in the city. Second, this paper argues that a shift in Indian politics towards the Hindu right-wing has changed the socio-political character of Mumbai. Unlike in many other parts of India, the Hindu right-wing in Mumbai mixes itself with "son of the soil" politics. This blend of politics often creates identity-based confusion for individuals but benefits the political class who shrewdly whip one or the other identity to reap political benefits. As identity-based politics strengthened, Mumbai's plural confessional space shrunk.

8.2 The Making of Mumbai

Bombay, as an urban centre, is a European creation and does not have much of a pre-colonial past (Farooqi 1996). Bombay was originally home to Marathi-speaking Koli people in pre-European days. Before the arrival of the Portuguese, the islands of Bombay and Mahim were under the Sultans of Gujarat. In 1534, to protect himself from the advancing Mughal power, Sultan Bahadur Shah established an alliance with the Portuguese, giving them seven islands of Bombay and Bassein. Portuguese gave names to all seven islands, called Bombaim (good bay) (Sequeira 2009). However, it was the British professionals who planned and developed Bombay. There was a conflict between the planners and administrators on the issue of city's development. Even the local inhabitants of the city resisted the British planning of Bombay (Dossal 1991). However, in the end, the imperialists managed to control the planning of Bombay.

8.2.1 Economic Rise of Mumbai

After they arrived in the western parts of India, the East India Company (EIC), since 1627, had a desire to possess Bombay due to its natural advantages over other places in the region such as, a natural harbour, and partly rocky and partly marshy terrain (Kosambi 1985, p. 33). EIC's wish was fulfilled after Bombay became a part of Catherine of Braganza's dower when she married to Charles II, King of England (1662–1685), in 1661 (Anderson 1854, p. 52). Physically, Bombay came under British control in February 1665. The city was later leased by Charles II to the EIC in 1668. On 1 September 1668, the ship *Constantinople*, arrived in Surat with a copy of the Royal Charter according to which the British king agreed to bestow Bombay upon the EIC "in free and common socage, as of the manor of East

Greenwich, on payment of the annual rent of £ 10 in gold on the 30th September of each year" (Cited in Ibid, p. 55).

In colonial years, Bombay highly benefitted from global events such as the American Civil War (1861–1864), which cut off the supply of American cotton for the British mills, the construction of railway lines connecting Bombay with the cotton-growing districts of India, and finally, the opening of Suez Canal in 1869 that shortened the distance between Bombay and London considerably (Kosambi 1985, p. 35). The cotton mills were set up in Bombay to cash the benefits from the increasing demands faced setbacks when the city was hit by severe bubonic plague in 1896. The bubonic plague virus was carried by ship merchants and sailors from plague-stricken Hongkong and lasted more than 20 years. The plague forced many migrant labourers to leave the city out of fear. In October 1896 an estimated 20,000 people fled the city. In November and December, 171,500 more migrants fled. It is estimated that by the end of January 1897, roughly around 400,000 people left Bombay (Sarkar 2014). The plague was so dangerous that by 1918 around 10 million people died due to it; of which, 1.6 lakhs (0.16 million) in Bombay alone (Fernando 2020). By the late 1920s, Bombay had begun to recover from pandemic-related economic severity.

The industrial class flourished in Mumbai by investing in small units during the colonial years. Simultaneously, a section of the business class also benefitted by indulging in the opium trade with China and other countries. Many Bombay-based traders and individuals were involved in opium smuggling with the help of the Portuguese, who had controlled Daman, a coastal town near Bombay (Farooqui 2013, p. 42). Many Indian capitalists, such as Jamsetjee Jejeebhoy Tata, were involved in the opium trade with China after 1842 (Markovits 2009, p. 115).

The end of the American Civil War terminated Bombay's fortune. In successive years, Bombay witnessed periodic booms during World War I (1914–1918), 1925, 1935, and World War II (1939–1945), alongside global aftershocks of the Great Depression (1929–1939). In these periods, the money that flooded in Bombay was spent on the nascent film industry. Three years before India's independence and the country's partition, in January 1944, "The Bombay Plan", a 15-year economic plan for India, was proposed by a group of industrialists and technocrats. It was signed by JRD Tata, G.D. Birla, Sir Ardeshir Dalal, Kasturibhai Lalbhai, D. Shroff, John Mathai, and Purushottamdas Thakurdas. The purpose of the Bombay Plan was to highlight that after the end of the war, the government may take populist measures if it faces organized political demands for the redistribution of income and wealth. As the plan underscored, such measures would harm India's economic development prospects in the long run. This could be avoided by taking caring steps. The plan proposed to initiate the process and make the government and the public aware of long-term issues of income distribution and development (Desai and Baru 2018).

8.2.2 Politics of Mumbai

Economic condition of Bombay facilitated the participation of many middle- and working-class people in political activities. Some educated rich and middle-class members from Bombay such as Pherozeshah Mehta, Kashinath Telang, and Badruddin Tyabji set up the Bombay Presidency Association in January 1885. On 28 December 1885, Mumbai hosted first meeting of the Indian National Congress (INC) at Gokuldas Tejpal Sanskrit College on the banks of Gowalia Tank in Tardeo. After the INC turned its character from indoor discussion and used constitutional methods to get civil and political rights for Indians to engage in mass activities to achieve political goals, the party received support from the industrial working class. For instance, on 22 July 1908, when Tilak was arrested for his "seditious" writings in his paper *Kesari* was convicted and externed to Mandalay in Burma for 6 years by the Bombay High Court around 400,000 textile mill workers of Bombay went on for a flash strike for six days. The army was deployed to control protests (Kulkarni 2022).

On 9 January 1915, Mohandas Karamchand Gandhi, popularly known as Mahatma Gandhi, landed in Bombay after spending 21 years in South Africa. On 6 April 1919, on Gandhi's call to demonstrate people's power against the Rowlatt Act, thousands of Bombay residents joined his twenty-four hour fast (Fernandes 2013). The city roared to protest when all seven members of all white Indian Statutory Commission, popularly called Simon Commission (known after its chairman Sir John Simon) landed in Bombay in February 1928. Years later, in August 1942, INC gave a call to last major mass movement called "Quit India" from Gawalia Tank Maidan, Bombay.

Many prominent businessmen from Bombay had close relations with the INC leadership (Kudaisya 2014). The businessmen opposed colonial government policies such as the Indo-British trade agreement which was concluded in March 1939 (Markovits 1985). At the Wardha session of the INC in 1937, national planning as a means of economic development was discussed and the session closed with a resolution favouring it. Subsequently, in the 1938 session of the INC at Haripura, the National Planning Committee was set up with Jawaharlal Nehru and K T Shah as chairman and secretary, respectively (Desai and Baru 2018). During the Second World War period, concerned about the prospects of allied forces in Europe, the Indian businessmen who earlier found Gandhi was "wandering in the wilderness" straddled between Congress and the British (Khan 2015, p. 86).

As an industrial city with millions of active workers and an educated class, Bombay could not evade the influence of communism and communist revolutions on them. Many of the workers were politically active and led by the Communist Party. The workers carried out strikes to meet their demands and ensure better wages and working conditions in the industrial units. For example, between 1928 and March 1929 there were 71 lightning strikes occurred (Chandavarkar 1998, p. 100). A round of wage cuts led to another strike wave in 1933–1934. Then, apart from several closures, eight general strikes occurred in the industry between 1919 and 1940, spreading to other cities such as Sholapur and Ahmedabad (Ibid 100). The communist Girni Kamgar Union (GKU) sustained the political momentum of the working-class neighbourhoods by holding regular processions and public meetings—at times these were an almost daily occurrence (Ibid 104). On 30 September 1929, the GKU had 334 members, which increased to 54,000 by the end of that year (Ibid 131). The labour movement was also marred with splits and victims of political suppressive methods used by the colonialists and other groups. Before the 1929 strike, communal disturbances in worker-dominated areas and in mills were introduced and communist and labour leaders were framed in the Meerut conspiracy case.[2] In Bombay, three forms of cleavage came at the forefront by the end of 1928, Brahmin-non-Brahmin; caste Hindu-outcaste; and; Hindu-Muslim. The main brain behind the anti-Brahmin agitation was B V Jadhav, the minister for agriculture in Bombay. Jadhav, however, could not succeed in splitting GKU on caste lines partly because by this time many Marathas, through the mill committees, had come up to positions in the top executive, and partly because the union members had gone through "a" year of intensive class schooling, both theoretically and practically (Lieten 1982, p. 698). Bombay Trade Disputes Conciliation Act of 1934 was replaced by another act in 1938 to deal with worker strikes. The workers opposed the new act. In 1942, many workers supported the "Quit India" call given by the INC. Many workers participated in the last days of the independence struggle, violating the official position of the Communist Party of India to not disturb the British in its fight against the Axis power.

Besides the INC and Communists, Mumbai was also place of Muslim nationalism in colonial India. Historically, by the mid-nineteenth century, Bombay was not only *urbs prima Indiss*, but also a primary city of Islam (Green 2011, p. 3). In the Persian *Jan-e Bomba' ı* (Bombay soul), written in 1816, the city's port is presented as the crossroads of the world where in addition to the English, Portuguese, Greeks, Dutch, Zoroastrians, Jews, Chinese and the many 'sects' (farqa) of Hindus, Muslim groups such as Arabs and Turks, Iranis and Turanis, Sindis and Hindis, Kabulis and Qandaharis, Punjabis and Lahoris, Kashmiris and Multanis, Madrasis and Malabaris, Gujaratis and Dakanis, Baghdadis and Basrawis, Muscatis and Konkanis lived in the city. In 1850, Bombay was already home to around 100,000 Muslims, which increased to around 137,000, about one-fifth of the city's overall population, and by 1901 over 155,000 Muslims were living there (Ibid 6). Furthermore, Mumbai is home to the founder of Pakistan-Mohammad Ali Jinnah. Jinnah was a prominent lawyer in Mumbai and began his political innings with the INC. He developed differences with the INC leaders and joined the Muslim League in 1913. In 1947, although Jinnah left for Pakistan with his sister Fathima, his only child, Dina, married to Neville Wadia, stayed back in Mumbai.

[2] A court case initiated in British Raj in March 1929 under which several trade unionists, including three Englishmen, were arrested for organizing an Indian railway strike. 27 trade union leaders were convicted and many arrested. The case closed in 1933.

8.3 The Postcolonial Mumbai

The postcolonial cities carry weight of Eurocentric culture, politics and space (Yeoh 2001). They provide scope to a few people to accumulate wealth and, at the same time, forcing many to struggle. Modernization-driven development activities in postcolonial cities have made many people suffer because of their consequences (Chandhoke 1991). The socio-economic inequities and identity constructed during colonialism continued in post-1947 Mumbai.

8.3.1 Mumbai's Economy

Between 1941 and 1951, Greater Bombay saw an increase of 5% in its population due to the influx of refugees from Pakistan and other parts of India searching for jobs. Since the 1960s, the city has undergone significant industrial restructuring. However, the mill owners refused to modernize their units, which caused frequent clashes between owners and the workers. Frequent strikes had caused many mill workers to close their units and invest in other industries. The strike of 1982–1983 led to the retrenchment of almost 100,000 workers. The number of city workers engaged in textile mills dropped from 27% in 1976 to 12.5% by 1991. Many workers joined the informal economy, where employment grew from 49% in 1971 to 66% in 1991 (Ghadge 2018, pp. 68–69). Statistically, the workforce participation rate in Greater Mumbai increased from 36 to 40% during 1961–2011. Work Force Participation Ratio as per the 2011 census is 40.34% (Municipal Corporate of Greater Mumbai Final Report 2016, pp. 2–1).

In 1991, India adopted a NEP based on liberalization, privatization and globalization (LPG) policies. The LPG replaced the welfare-based model of the Indian economy. Post-1991, India witnessed tremendous growth in its gross domestic product (GDP), became the fifth largest world economy, created many billionaires, and attracted several investors. Mumbai embraced the LPG model. According to the 2011 census, Mumbai City District's population was 3,085,000 (Economic Survey of Maharashtra 2022–2023, p. 27). Mumbai has 3,98,573 micro-enterprises, 10,007 small enterprises, and 931 medium enterprises (Ibid 145). With the expansion of economic activities, Mumbai has expanded. The Mumbai Metropolitan Region Development Authority (MMRDA) has undertaken various infrastructure projects in MMR (Ibid 179). According to media reports, the Government of India think tank NITI Aayog is drafting a plan to increase MMR'sGDP from US $140 billion to US $300 billion by 2030 (Gangan 2023).

Despite such planning and developments, as mentioned above, a big section of Mumbai's population lives in poor condition. Most of the poor lives in dilapidated buildings and slums spread across the city. To improve the life and life-style of slum dwellers, Slum Improvement Programme was started in 1971, in 1985 Slum Upgrading Programme and Prime Minister's Grant Project started, in 1991 Slum

Redevelopment Scheme launched, and in 1995 Slum Rehabilitation Scheme (SRS) was launched. The Slum Rehabilitation Authority implements the SRS. Under the scheme, tenements are being provided to families living in slums (Economic Survey of Maharashtra 2022–2023, p. 237). The SRS backfired, as a number of migrants tricked by the slumlords to buy the shanty homes believing they would get free homes at some point (Udas-Mankikar 2020). Few of the slums have been developed into buildings that give rooms to the slum dwellers, while a large number remain as they were earlier. The main reasons for it are misuse of funds, drying up of funds with the builder, delayed projects, and inability to shift slum dwellers to tenements have been the common complaints (Ibid). In 2018, the Devendra Fadnavis government came out with a new plan to free Mumbai slum. Under this new plan those living in slum tenements that came up between 2000 and 2011 are guaranteed a replacement house (Ashar 2018). In November 2022, the Adani Group, led by Indian businessman Gautam Adani, won secured the million dollars contract to overhaul Dharavi. The Adani Group, through its joint venture company Dharavi Redevelopment Project Private Ltd (DRPPL) with the Maharashtra government will build houses and commercial units and then hand them over to Dharavi Redevelopment Programme/Slum Rehabilitation Authority of the Maharashtra government for allocation to the people living there (Business Standard 2024).

8.3.2 Mumbai's Politics

A shift in the economic model has witnessed a change in the politics of Mumbai city. During the socialist economy days, the centrist, socialist and communist parties have their ground in Mumbai. After identity politics further strengthened, caste, religion and language became important markers in the state's politics. Despite the economic gain Mumbai made in post-1991, the city has seen a large-scale de-industrialization since the mid-1980s (Hatekar 2022, p. 27). Between 1961 and 2001, the employment structure in Mumbai underwent a significant change, with services together accounting for a little over 70% of employment in Mumbai in 2001. This trend has only continued in 2011 and beyond (Ibid 30).

After Maharashtra was divided and Gujarat was formed as a separate state, INC remained in power till 1995. A brake was put in by the Bhartiya Janata Party (BJP)-Shiv Sena (SS) government in 1995. INC returned to power in 2003 and remained in power till 2014, when BJP replaced the party. From 2019 to 2021, SS was in power with a main support from the Nationalist Congress Party (NCP) and the INC. The government fell when the current chief minister of the state, Eknath Shinde of the SS, rebelled against his party leader and many other legislature members, who formed a coalition government with the BJP. There was no particular reason for the break-up, but the Eknath Shinde faction maintains that the BJP is their natural partner and SS mistaken to join hands with rivals such as INC and NCP to form government in the state.

Unlike the visible politics, the change on the ground started in the 1960s when the issue of Marathi Manoos (Marathi people) was vociferously raised in the city's streets supporting Bal Thackrey. In 1966, SS was formed by Bal Thackrey. The SS attacked and protested against the presence of non-Marathi people in Mumbai. Bal Thackrey hurled abuses and made provocative speeches in Mumbai targeting the communists, south Indians and Muslims. The SS claimed that, unlike other political parties, it believes in action. Although Thackery calls communists anti-national because of ideological differences, the main reason for vitriol was to control the city, where many mill workers were supporters and sympathizers of communism (Prakash 2010, p. 235). Bal Thackeray was even accused of killing Krishna Desai, a popular trade union and political leader in 1970 (Ibid 204). Bal Thackrey's "son of soil" politics was supported by ripen material conditions in Bombay where many young Maharashtrian men with limited education faced unemployment. Between 1961 and 1971, the absolute number of unemployed persons educated up to primary and middle school increased by 109% (Hatekar 2022, p. 41). Many of those unemployed persons became foot soldiers of SS. Some of the unemployed recruits of SS became crucial players in establishing the party's dominance in street violence (Ibid).

After the rise of Hindutva politics in 1980s, the SS changed its political stance and turned mainly towards Hindutva politics. The SS workers participated in the *Ramjanambhoomi* movement and engaged in demolishing the tombs of the *Babari* mosque. Demolition of the mosque caused communal violence in different parts of India, including a serial blast in Mumbai carried out on instructions from gangster Dawood Ibrahim. Post-blast, Mumbai witnessed communal riots in which many members of SS were accused of carrying out attacks on the Muslims. Riding on big support, SS and BJP won the state assembly's elections and formed a coalition government in 1995. Manohar Joshi from SS was Chief Minister of the state. After coming into power, the SS-BJP government introduced a slum redevelopment scheme in 1995. The private builders were incentivized to provide free homes to 400,000 slum residents while simultaneously clearing the city of slums. In return, private builders were given transferable development rights through which developable floor area—Floor Space Index could be used in upmarket localities instead of slum redevelopment elsewhere (Ibid 45). Many ministers, party workers and relatives of the SS turned into builders and engage in building "social capital" instead of engaging in street activities. During this period, SS forged networks across wider sections of the population and carried out many outreach activities with the North Indians, South Indians, Christians, Muslims, and Dalits (Ibid 45). More than the outreach programmes, SS benefited from joining the BJP coalition and underlining the Hindutva causes. Shift in political gear made many north Indian Hindus living in Mumbai vote for the party. Figure 8.2 shows how Hindutva has gradually occupied Mumbai's plural socio-religious space.

Demographically, more than any other state of India, it is estimated that Mumbai has about 53% of the migrants from Maharashtra. Among the other states from where many migrants land in Mumbai is Uttar Pradesh (Chhapia 2019). On the religious count, since the British founded it, Mumbai developed into a plural city with people from different religious groups and sects embraced by it. Such plurality can be visible to anyone visiting the town or city. There are churches, mosques, temples, Agiary

Fig. 8.2 High-rise building, rented shop run by a poor and billboard calling people to participate in large numbers in a procession to support Rama Temple's consecration in Ayodhya. *Source* Photo taken by the author

etc., standing together or facing each other or at a stone throw distance. Agiary and churches are mainly located in the Mumbai region and not in many numbers in the suburban areas. Unlike Parsees and Christians, Muslims have spread themselves across the city and in the suburbs, they are in large numbers in Mira Road—a centre of communal tensions in the last 5 years. However, Mumbai remains a Hindu-dominant city. Of the total population, around 65.99% of the Mumbaikars follow Hinduism, 20.65% Islam, 3.27% Christianity, 4.10% Jainism, 0.49% Sikhism, 4.85% Buddhism about 0.40% "Other Religion", and 0.26% do not belong to any "particular religion" (Mumbai City Population 2023).

Among all other religious groups, majoritarian politics in India is targeted against the Muslims who are considered as "outsiders". In many gated societies, the Muslim family are not allowed to buy or rent a property. Not the ordinary but also Muslim celebrities have faced housing problem. In 2008, popular Hindi film actors Shabana Azmi and Saif Ali Khan stated that they could not buy or find it difficult to buy property in Mumbai because of their Muslim religion (The Times of India 2008). The confessional politics, religious differences, and confrontation have further escalated with the electoral rise of the BJP since 2014. Hindutva agendas, mainly to attack the Muslim economy, such as beef bans have been implemented in many Indian cities, including Mumbai. In 1995, a law was introduced in the Maharashtra assembly by the SS-BJP government. However, it was not passed until 2015 when Devendra Fadnavis was Maharashtra's Chief Minister. Under the Maharashtra Animal Preservation (Amendment) Act 1976 declared a law in 2015, the slaughter of bull and all varieties of buffalo, except the water buffalo, are crime and declared a non-bailable offence punishable by up to 5 years of jail (Mirza 2022, p. 91).

8.4 Conclusion

Due to its postcolonial character, Mumbai is a paradoxical city. Such paradox creates everyday tensions between Mumbaikars (people living in Mumbai). Some of those tensions are palpable in the form of "discrimination" one faces because of "difference". Often, the existing tensions create violence between the "different" groups living in the city. There is no single reason for the tensions and violence, as class, religion, and regional sentiments invariably intersect.

As s discussed in this chapter, Mumbai has a history of producing business class and poor people migrating to find jobs. It is a city full of billionaires who flourish and live with a number of poor who cannot afford proper housing and live in slums or roadside. In colonial years, the business class shifted their political support to suit their economic benefits. Some of the prominent businessmen of that period were also engaged in the illicit opium trade and made huge money out of trading with China. On the contrary, a large number of the working class lived in poor conditions and fought for proper wages and good working conditions. The swamping of migrants from different parts of India to get opportunities in Mumbai expanded the city's physical limit. Nothing much has changed after 1947: Mumbai is expanding itself to give space to more billionaires and, simultaneously, to poor people who come to the city with multiple dreams.

As observed in this chapter, Mumbai also has a Christian and Muslim past. Across the city one can see signs and buildings belonging to people from various confessions. The city has population belonging to different faiths. However, change in politics in India since 1980s has impacted its plural and religious accommodative character. The electoral rise of Hindu nationalist party such as BJP and SS has shrinked the plural socio-political space of the other religious group and increased the presence of Hindutva. Besides, Hindutva the regional groups with Marathi sentiments have also tried to dominate over the city. This regional sentiment is not pluralistic as it mixes Hindutva with Marathi sentiment.

References

Anderson P (1854) The English in Western India; being the early history of the factory at Surat , of Bombay, and the subordinate factories on the Western Coast: from the earliest period until commencement of the eighteenth century. Smith, Elder and Co., 65, Cornhill, London

Appadurai A (2000) Spectral housing and urban cleansing: notes on millennial Mumbai. In: Breckenridge C, Bhabha H, Chakrabarty D, Pollock S (eds) Public culture, special issue on cosmopolitanism, pp 627–651. http://www.arjunappadurai.org/articles/Appadurai_Spectral_Housing_and_Urban_Cleansing_Notes_on_Millennial_Mumbai.pdf. Accessed 20 May 2020

Ashar AS (2018) Maharashtra govt rolls out new plan for slum-free Mumbai. The Indian Express. https://indianexpress.com/article/cities/mumbai/maharashtra-rolls-out-new-plan-for-slum-free-mumbai-5180944/

Bendix A (2018) Drone photos of Mumbai reveal the places where extreme poverty meets extreme wealth. Business Insider. https://www.businessinsider.com/aerial-drone-photos-mumbai-extreme-wealth-slums-2018-9

Business Standard (2024) Dharavi project: no land to Adani group, transfer to Maharashtra govt dept. https://www.business-standard.com/companies/news/dharavi-project-no-land-to-adani-group-transfer-to-maharashtra-govt-dept-124061600167_1.html

Census of India (2011) Government of India "District Census handbook-Mumbai Suburban". https://censusindia.gov.in/2011census/dchb/DCHB_A/27/2722_PART_A_DCHB_MUMBAI%20SUBURBAN.pdf

Chandavarkar RN (1998) Imperial power and popular politics: class, resistance and the State in India, c. 1850–1950. Cambridge University Press, Cambridge

Chandhoke N (1991) The postcolonial city economic and political weekly, pp 2868–2873

Chhapia H (2019) 53 % of Mumbai's migrants from within Maharashtra. The Times of India. https://timesofindia.indiatimes.com/city/mumbai/53-of-citys-migrants-from-within-the-state/articleshow/70322261.cms

Desai M, Baru S (2018) The Bombay plan: blue print for economic resurgence. Rupa Publications, New Delhi

Dossal M (1991) Imperial designs and Indian realities: the planning of Bombay city. Oxford University Press, New Delhi, pp 1845–1875

Economic Survey of Maharashtra (2022–2023). http://mls.org.in/PDF2023/BUDGET/ESM_2022_23_Eng_Book.pdf. Accessed 12 Jan 2024

Farooqi A (1996) Urban development in a colonial situation early nineteenth century Bombay. Econ Polit Weekly 31(40):2746–2759

Farooqui A (2013) The decline of Mumbai's 'hidden' portuguese connection, c. 1831–1843. In: Kamat M (eds) Mumbai past and present: historical perspective and contemporary challenges. Indus Source Books. Mumbai, pp 41–54

Fernandes N (2013) City Adrift: a short biography of Bombay. Aleph Book Company, New Delhi

Fernando B (2020) How the 1896 plague epidemic shaped Mumbai. The Indian Express. https://indianexpress.com/article/express-sunday-eye/how-the-1896-plague-epidemic-shaped-mumbai-6434766/

Gangan SP (2023) NITI Aayog discusses plan to raise Mumbai's GDP to $300 billion by 2030. Hindustan Times. https://www.hindustantimes.com/cities/mumbai-news/niti-aayog-to-draft-economic-masterplan-for-mumbai-metropolitan-region-to-boost-gdp-to-300-billion-by-2030-101693392127891.html

Ghadge R (2018) Connections and disconnections: the making of Bombay/Mumbai as India's global city. J Glob Initiat 13(1):55–76

Green N (2011) Bombay Islam: the religious economy of the West Indian Ocean 184–1915. Cambridge University Press, Cambridge

Hatekar N (2022) Informality, missing markets, and political organisation: case study of Shiv Sena. In: Jose G, Parthasarathy D, Patel S (eds) Mumbai: majoritarian neoliberalism, informality, resistance and well being. Routledge, London, pp 27–48

Khan Y (2015) The Raj at war: a people's history of India's second world war I. Random House, Gurgaon
Kosambi M (1985) Commerce, conquest and the colonial city: role of locational factors in rise of Bombay. Econ Polit Weekly 20(1):32–37
Kudaisya M (2014) The promise of partnership: Indian business, the state, and the Bombay plan of 1944. Bus Hist Rev 88(1):97–131
Kulkarni D (2022) How Bal Gangadhar Tilak shaped the structure of Mumbai police. Hindustan Times. https://www.hindustantimes.com/cities/mumbai-news/how-bal-gangadhar-tilak-shaped-the-structure-of-mumbai-police-101659291304577.html
Lieten GK (1982) Strikers and strike-breakers: Bombay textile mills strike, 1929. Econ Polit Weekly 17(14/16):697–704
Markovits C (1985) Indian business and nationalist politics, 1931–1939: the indigenous capitalist class and the rise of the congress party. Cambridge University Press, Cambridge
Markovits C (2009) The political economy of opium smuggling in early nineteenth century India: leakage or resistance? Mod Asian Stud Expand Front South Asian World Hist 43(1):89–111
Mehrotra S (2024) Everything wrong with NITI Aayog's claim of 24.8 crore emerging out of poverty in 9 years. Wire. https://thewire.in/political-economy/niti-aayong-national-poverty-indicator
Mint (2023) Mumbai reigns supreme with 66 billionaires. Delhi. https://www.livemint.com/news/mumbai-reigns-supreme-with-66-billionaires-delhi-bengaluru-follow-11679476918509.html
Mirza S (2022) Neoliberalism and majoritarian politics: Hindutva and restructuring the meat business. In: Jose G, Parthsarthy D, Patel S (eds) Mumbai: majoritarian neoliberalism, informality, resistance and well being. Routledge, London, pp 84–95
Mumbai (Greater Mumbai) City Census 2011 data. http://www.census2011.co.in/census/city/365-mumbai.html
Mumbai City Population (2023) https://www.census2011.co.in/census/city/365-mumbai.html#:~:text=Mumbai%20Religion,%25%20and%20Buddhism%20by%204.85%20%25
Municipal corporate of greater Mumbai final report: executive summary comprehensive mobility plan (CMP) for greater Mumbai. https://portal.mcgm.gov.in/irj/go/km/docs/documents/MCGM%20Department%20List/Roads%20and%20Traffic/Docs/CMP%20for%20Greater%20Mumbai/Executive%20Summary/Executive%20Summary.pdf. Accessed April 2016
NITI Aayog (2023) Government of India. India: national multidimensional poverty index: a progress review. https://www.niti.gov.in/sites/default/files/2023-08/India-National-Multidimensional-Poverty-Index-2023.pdf
Prakash G (2010) Mumbai fables. Princeton University Press, Princeton
Ramani V (2024) Mumbai is now among the world's top 10 luxury real estate markets: knight frank wealth report 2024. Hindustan Times. https://www.hindustantimes.com/real-estate/mumbai-ranked-at-number-8-globally-in-terms-of-price-growth-in-luxury-housing-101709100088163.html
Sarkar A (2014) The tie that snapped: bubonic plague and mill labour in Bombay, 1896–1898. Int Soc Hist 59(2):181–214
Sequeira D (2009) Bombay before British. Deccan Herald. https://www.deccanherald.com/content/22328/bombay-british.html
The Times of India (2008) Indian democracy unfair to Muslims: Shabana Azmi. https://timesofindia.indiatimes.com/india/indian-democracy-unfair-to-muslims-shabana-azmi/articleshow/3371893.cms
Udas-Mankikar S (2020) Informal migrants in maximum city. Observer Research Foundation 21 April, 2020. https://www.orfonline.org/expert-speak/informal-migrants-maximum-city-64959/. Accessed on 12 June 2023
Yeoh BSA (2001) Postcolonial cities. Prog Hum Geogr 25(3):456–468

Chapter 9
Istanbul: Constantinople

Mustafa Kahraman

Abstract Throughout history, cities have grown and developed for various reasons. The city of Byzantion in the fourth century was a modest city. Constantine the Great made the city of Byzantion the capital of the Eastern Roman Empire and named it Constantinopolis. With the support of the administrative functions of Constantinople, it witnessed significant changes. In this section, the general situation of Constantinople between the fourth and fifteenth centuries will be discussed. The study first examines the city's overall development, followed by a brief evaluation of the reasons and consequences of its selection as the capital. Subsequently, the focus shifts to the general geographical layout of Constantinople as an administrative center. The other roles of the city, which was designated as the capital of the Eastern Roman Empire by Constantine the Great, are also explored. The city's trade, military, and religious functions were strengthened as a result of its administrative role. In addition, the city's administrative function contributed to a rise in its population and the advancement of its infrastructure systems. These urban functions and infrastructure indirectly influenced the city's cultural landscape.

Keywords Istanbul · Constantinople · Site characteristics · Administrative function · Cultural landscape

9.1 Introduction

The study covers a period of approximately a thousand years, from the founding of Constantinople in 395 to its conquest by the Turks in 1453. During this period, the city had a permanent and dense population (Ljungkvist et al. 2010a, b). Therefore, the area where the city was founded has a complex, multi-layered structure. Only a few important topics and anecdotes are highlighted to explain this complex structure in a limited page. The study first examines the location of the city and its development over time, and then investigates the reasons behind its selection as the capital, which

M. Kahraman (✉)
Istanbul University, İstanbul, Türkiye
e-mail: mustafa.kahraman@istanbul.eud.tr

F. Carrión Mena and S. Rodríguez Alvarez (eds.), *From the City as a Project to the City Project*, The Urban Book Series, https://doi.org/10.1007/978-3-031-94226-6_9

made it a "global" city. In the last section, important buildings that contribute to the cultural landscape of the city of Constantinople are examined according to their functions.

Before the city of Constantinople became the capital of the Eastern Roman Empire, it was a small city called Byzantium. The city of Byzantion was founded approximately where today's Topkapi Palace is located. In later periods, the city expanded towards the west. As the city expanded, new land walls were built (Fig. 9.1c). Therefore, determining the location of the walls will provide us with valuable information about the development of the city. Byzantine walls must have covered an average area of 3 km^2. The second wall, chronologically closer to the present day, is the Severus Wall. It is estimated that the walls, built by the Severus Dynasty in the third century, are approximately 700 m west of the Byzantine walls. The walls started from Eminönü New Mosque in the north and extended to the Hippodrome in the south. After passing the hippodrome, the road turned east and reached the Marmara Sea. According to this calculation, the city covers an area of approximately 4.8 km^2.

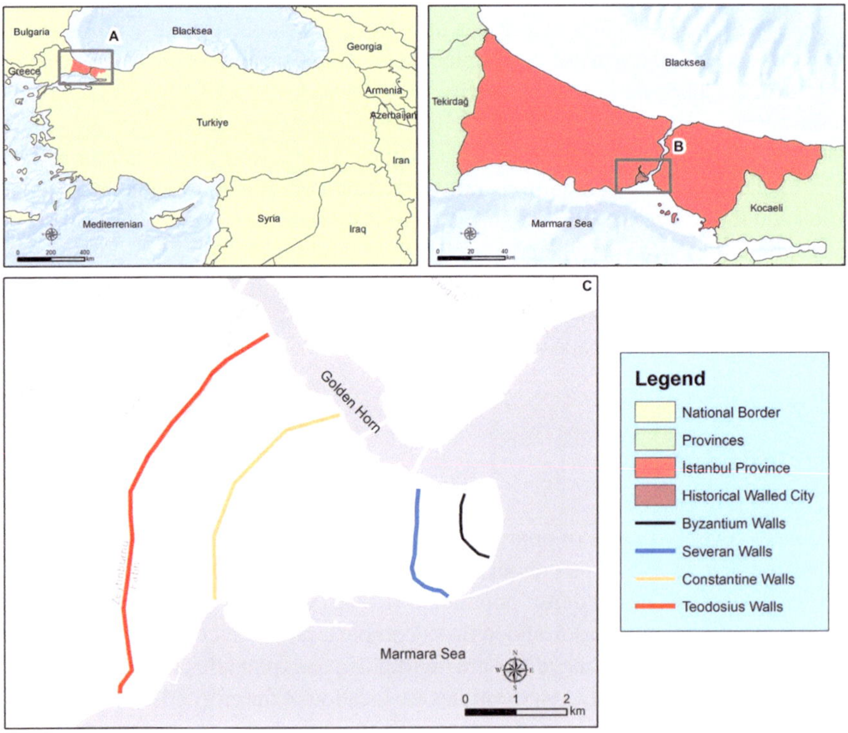

Fig. 9.1 Location and walls of Constantinople. *Source for Figure 1c* Prepared using Drakoulis (2012), Lordoğlu (2019), Safran (2007)

The fourth century was a turning point for the city of Byzantion. In 395, the city was declared the capital of the Eastern Roman Empire. After this, the city began to be called Constantinople. As a result of the city's administrative function and the implementation of policies to encourage population growth, the city expanded westward. After Constantine the Great founded the city, he determined its borders with a law called "limitatino". The western border of the city is surrounded by walls located approximately 2.7 km west of the Byzantine walls. With the construction of this wall, the area of the city within the city walls tripled (Fig. 9.1c) (Kahraman 2023). It is not possible to determine the exact location of the new walls, the construction of which started during the reign of Constantine. However, it is thought that the Constantine Wall started from today's Yenikapı, turned slightly northeast at Oruçgazi Park and extended directly to the Golden Horn (Fig. 9.1c).

In the fifth century, another city wall was built, which still exists today. Known as the 2nd Theodosian Walls, the construction of these walls expanded the area enclosed by the city walls from approximately 7–17 km^2. This area was not uniformly developed. The western parts of the city are less densely populated.

In the fourth and fifth centuries, Constantinople expanded beyond its walls. As a result, Galata was granted "police" status (Yılmaz 2015). The Galata area remained administratively connected to Constantinople until the early thirteenth century. Yet it is important to mention that the area commonly known as Constantinople is situated within the walls.

9.2 The Declaration of Constantinople as the Capital and Its Effects on the City

The location where the city of Constantinople was founded had many advantages. The historical peninsula was surrounded by sea on three sides, making it easy to defend. Additionally, the abundance of sea creatures in the estuary (the main reason why the estuary is called the Golden Horn), ease of transportation, and the presence of natural harbors provided significant benefits to the city. The historical peninsula also had some advantages in terms of its geographical characteristics. For example, between the fifth century BC and the fifteenth century AD, there was an intense relationship between the Black Sea and the Greek peninsula. The Black Sea coast was an area that Greek sailors visited annually for wheat. In addition to wheat, products such as leather, wool, and timber were also traded. All these goods passed through the Bosphorus. Byzantion/Constantinople, located at the entrance of the Bosphorus, served as both an equipment base and a resting station for sailors and merchants (Dominian 1917). Constantinople's strategic location persisted for many years, making it a significant cultural crossroads point. Constantinople is located on a road that connects Asia and Europe by land. By sea, it connects Russia and the Mediterranean Basin.

These positive features undoubtedly influenced the development, cultural landscape, and urban functions of the city of Constantinople. Additionally, Constantine the Great must have considered these advantages of the city when selecting it as the capital of the Eastern Roman Empire. However, there were also geopolitical reasons behind Constantine the Great's choice of Constantinople as his capital. In the fourth century, the Roman Empire's internal issues and economic problems had increased (Tahiroğlu 1981). On the other hand, one of the greatest dangers to the Roman Empire was the Sassanid threat on its eastern borders. A similar reason could be attributed to the movements of the Goths, who were on the move at the time (Şahin 2019). In order to respond more quickly to these threats, it was deemed more appropriate for the administrative center of the country to be located near the eastern borders.

The city became a "global" city after it acquired administrative functions. In fact, the city's assumption of administrative roles has propelled it to the forefront in various other functions. For instance, the trade functions evolved in response to the growing population and demands following the city's designation as the capital. The administrative function also bolstered the religious functions. It can be argued that the military function, along with other functions, has gained strength due to the administrative function.

From the time Constantinople was declared the capital in the late fourth century until the fifteenth century, the city physically expanded, and its population generally increased. The main reason for this increase was the prominence of the city's administrative function, which was accompanied by the strengthening of other urban functions. In addition, the city also implemented some policies to increase its population. For example, free land was distributed to senators who were already residing in Rome. The population was recruited from other settlements in the country. It is recorded that 8000 loaves of bread were distributed daily. It is known that products other than grain were also distributed. Although it is not known what these items were, it is estimated that they were probably products such as oil, wine, and lard (Türkoğlu 2010).

Although controversial, the estimated population of the city in 330 is 200,000 people. In the fifth century, it exceeded 300,000. This number exceeded the population of the city of Rome. It continued to increase until the plague epidemic in 541–542. Justinian I even made legal arrangements to control the flow of immigrants to the city (Magdalino 2015). In the sixth century, the city lost population due to the bubonic plague epidemic, wars with Arabs, Persians, and Avars. In the eighth century, the city's population was estimated to be between 30,000 and 40,000. From the eighth century to the eleventh century, the population is thought to have increased steadily. In the twelfth century, the population reached up to 400,000. However, in the middle of the fifteenth century, the population was estimated to be around 40,000 to 50,000 (Keçiş 2015).

9.3 General View of the City

In the fourth century, the city had the appearance of a classical Greek city with its acropolis, theater, and religious buildings. However, the overall appearance of the city started to change when it was chosen as the capital. Constantine the Great made plans inspired by the Roman city and initiated significant reconstruction efforts in the city. In this context, marble and bronze statues and works of art were brought from all over the empire. In addition, the construction of basic administrative and public buildings of Roman culture, such as a new senate, law schools, temples, and baths, began (Akalin 2015). These construction activities were actually aimed at transforming the city into a capital befitting Eastern Rome. These construction activities initiated by Constantine were continued by his son Constantius II (337–361). The city plan that emerged as a result of the activities carried out in the early Byzantine period became one of the determining factors of the city's image until the fifteenth century. It is possible to categorize the structures that influence the city image of Constantinople into six headings. These include administrative buildings, commercial buildings, roads and squares, religious buildings, and water structures.

9.3.1 *Administrative Structure*

The area where the administrative buildings in Constantinople are concentrated is the eastern part of the historical peninsula (Fig. 9.2). The Great Palace, built by Constantine the Great to the south of the Acropolis of Byzantion, is also located in this area. Built in 330, the Great Palace served as the central governing location for the city and the country until 1081. The palace was similar to the Roman Imperial Palace on the Plat with its guard barracks, schools, courtyards, living rooms, and reception halls (Müeller-Wiener 2016).

In fact, it is not a coincidence that the management structure is concentrated here. This field security perspective is provided in the most appropriate way on the peninsula. Byzantion City was founded on Sarayburnu hill, the first of the seven hills within the city walls (see Fig. 9.2). This area has a slight slope up to the first 14 m from the sea, and a high slope between 14 and 35 m. After 35 m, it partially flattens out on the hill (Fig. 9.2) (Kahraman 2023). The Grand Palace, which served as the imperial residence for a long time, is located in the southwest of the Byzantion acropolis. The geomorphological structure seen in Sarayburnu is similar to the structure in the area where the Grand Palace is located. The altitude of the place where the Great Palace is located is on average 10 m lower, and the slope is slightly less. On the other hand, the flat area required for the construction of a large palace complex is more.

The Great Palace's location high on a hill provides a security advantage. In the event of a naval attack, enemy ships can be spotted from afar. Moreover, a sea attack requires more effort, time, and resources compared to a land attack. Another security benefit of the eastern part of the historical peninsula is the strong Bosphorus current.

Fig. 9.2 Historical Peninsula. *Source* Kahraman (2023)

This area boasts the most robust current in the Bosphorus, making it a highly protected zone against sea attacks.

Between the seventh and eleventh centuries, it is estimated that the great imperial palace expanded to the Marmara Sea with additional constructions. (Magdalino 2015) (Fig. 9.3). Today, there are very few archaeological remains of the Great Palace.

Another administrative building is the Blachernai Palace. Blachernai Palace was built as a resting place for the emperor when he visited the nearby church. In the eleventh century, the Great Palace was abandoned, and the imperial residence was relocated to this site. During the reign of Manuel I (1143–1180), new buildings were added to the palace. Building elements were brought from the Great Palace to be used in the construction of these new buildings (Müller-Wiener 2016). The Blachernai Palace was used as the imperial palace until the early fifteenth century. In the early fifteenth century, Tekfur Palace was used as the imperial palace (Asutay-Effenberger 2024).

In addition to the Great Palace, Blachernai Palace, and Tekfur Palace, there were also palaces built at different times for important state officials. The Bukoleon Palace, situated on the shore of the Marmara Sea and considered part of the Great Palace, along with the Lausos and Antiochus Palaces located on the northwest corner of the Hippodrome, the Botaneiatas Palace (whose exact location is now unknown), and the 11th-century Mangana Palace are notable examples.

The hippodrome, located next to the Great Palace, is another remarkable structure in the city's skyline. Built at the end of the second century by Septimius Severus, the structure was expanded during the reign of Constantine, following the design of the Circus Maximus in Rome (Akyürek 2021). The Hippodrome is not only a place for entertainment and car races. For many years, it was the most important stage of city and imperial life. Some administrative offices were also located within the Hippodrome. It was also a place where general political discussions were held, punishments were executed, and some decisions were presented to the public (Ortaylı

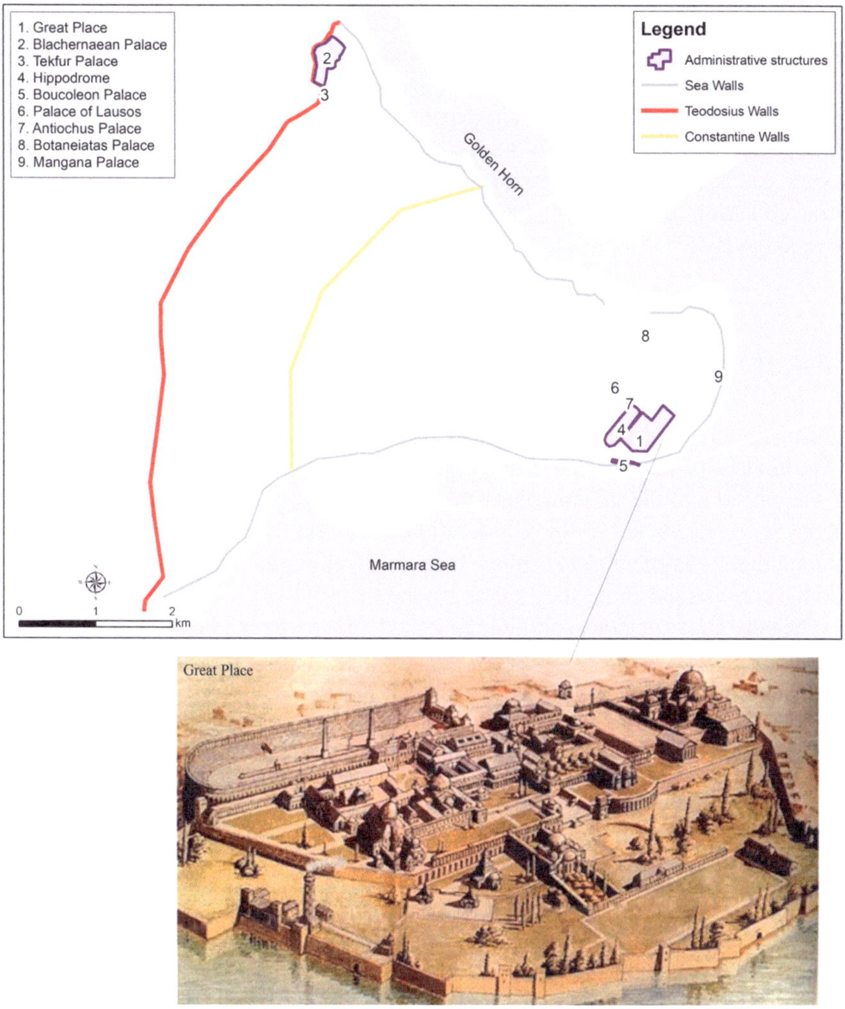

Fig. 9.3 Estimated reconstruction of major administrative buildings and the Great Palace in Constantinople. *Source for the picture* Kuban (1996) as cited by Kavrakoğlu (2016)

1977). It is no coincidence that the Nika rebellion, which emerged as a result of the activities of the Blue and Green factions, which held significant influence in the history of the Eastern Roman Empire, began in the Hippodrome.

On the other hand, the obelisks in the hippodrome are architectural elements that draw attention in the city and are prominently featured in medieval engravings. In fact, these obelisks were symbols of being at the top of the city hierarchy at the time and reinforced the idea of the capital (Bassett 2005). Among the figures depicted on the pedestals of these obelisks are representatives of foreign nations bringing taxes and gifts to the emperor (Akalin 2015).

The most powerful urban function of Constantinople was its administrative role. Being the capital also strengthened other functions, such as trade and transportation. For example, many wealthy members of the Senate owned houses in the capital and resided there for part of the year. The expenses of the senators and their entourage also stimulated a market for the city's artisans, craftsmen, and merchants who imported luxury goods. In addition, the games organized by the senators in Constantinople attracted various professions such as coachmen, grooms, actors, and singers to the city (Keçiş 2015).

9.3.2 Commercial Structure

Constantinople was not a major center of production. The growth of the city and its prominent position in the city hierarchy can be attributed to the establishment of the palace and the government here. The increase in the city's population and, therefore, its needs led to the proliferation of commercial districts and structures. Among these commercial structures, harbors stand out due to the area they occupy and the commercial zones they create around them.

The inhabitants of Constantinople benefited from many privileges, similar to those in the city of Rome. Rome's privilege of receiving free wheat from Egypt was also applied to the new capital (Asal 2010). In order to prevent product shortages in the city, only a minimal amount of goods were exported. While some measures were taken to restrict exports, imports were encouraged. For example, in the tenth century, a tax of two nomismata was levied on every Venetian ship arriving in the city and fifteen nomismata on every loaded ship departing (Demirkent 2005). Therefore, the ports of Istanbul were a city where ships arrived full and left empty, with large warehouses located next to the ports. For example, there were four warehouses (horrea) near the Port of Prosphorion (Portus Prosphorianus) and the port of Neorion (Fig. 9.4).

The first harbors of the city were the Neorion and Prosphorion harbors located in the present-day Sirkeci settlement (Fig. 9.4). These harbors have been in use since the Byzantine period. The Neorion harbor served as both a port and the location of the imperial navy and shipyard. The port of Prosphorion was a genuine trading port. Goods coming from Chalcedon were unloaded here (Türkoğlu 2010). In the warehouses near the harbors, products such as livestock and grain were stored and then distributed to the city. These two ports were the center of the city's maritime traffic until the fifth century.

These two harbors must have been insufficient due to the city's population growth and the fact that a significant portion of the city's needs were met from external sources. Large harbors were constructed on the southern shores of the city for this reason. From the fifth century onwards, it is thought that port traffic was shared between the north and south coasts (Magdalino 2000). There were two important harbors on the southern coast of the city, namely Sofia and Theodosius. The port of Theodosius experienced its most active period from the 6th to the eleventh centuries. Grains imported from Egypt held a significant position in the ports along the Sea

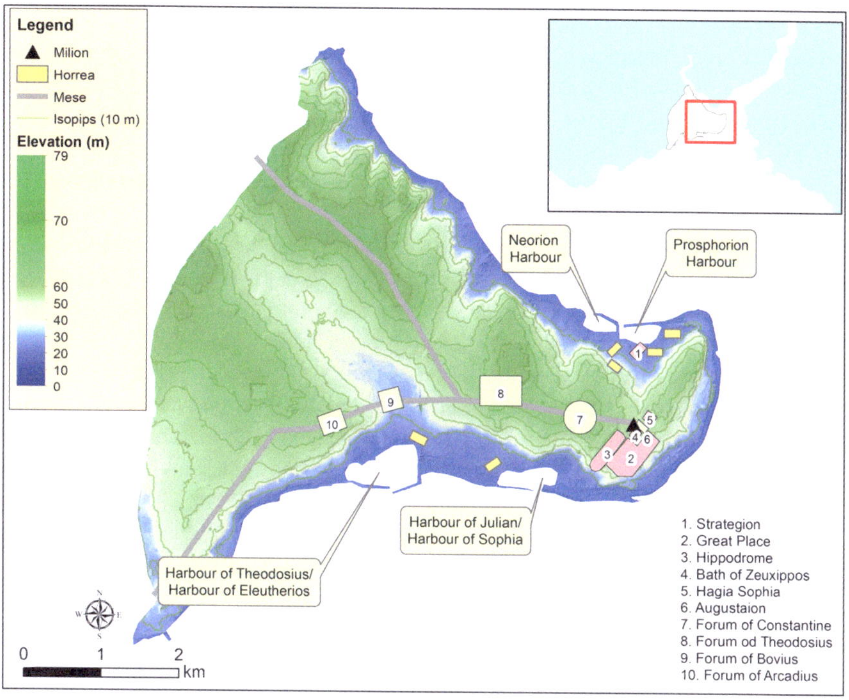

Fig. 9.4 Important ports of Constantinople. *Source* Nicolescu and Safran (2024) for the location of the Horreas

of Marmara. In addition, building materials such as timber, bricks, and Prokonessos marble were also imported from the port of Theodosius. These ports were utilized for significant construction projects, particularly in the fourth and fifth centuries (Türkoğlu 2010). To the west of the port of Theodosius was the port of Sofia, believed to have been constructed in the mid-fourth century (Mango 1986), and it remained operational until the eleventh century (Jones and Olgun 2024). In addition to these major ports, there were also smaller ports on the southern and northern coasts, such as the port of Bukeleon, which is part of the Bukeleon Palace, and the port of Kyanedion. It is also known that there were various piers both on the shores of the Golden Horn and on the Marmara coast.

In the ninth century, the commercial structure of the city was notably influenced by the Italian city-states, as well as Russian and Bulgarian merchants. Pisans, Genoese, and Venetian merchants settled in the Galata and Golden Horn regions with the privileges granted to them. Given this situation, the commercial center shifted to these regions. After the eleventh century, the Port of Theodosius lost its significance due to siltation caused by the Lycos Stream. Based on the findings, we can conclude that it was no longer a commercial port in the twelfth century (Asal 2010). On the other hand, the port of Prosphorion had become unusable by the end of the millennium due to siltation (Müller-Wiener 2016).

The locations of the ports also indirectly influenced the locations of the bazaars and market areas. Strategion is a good example of this. This area was probably the agora of the classical city. Strategion is considered a highly suitable location for trade within the city (Fig. 9.7). This was also the endpoint of Via Egnatia. It was possible to travel to Chalcedon from here. From Chalcedon, Nicomedia, and the Asian provinces were connected. On the other hand, the Strategion was located in an area with a gentle slope.

The surroundings of the great palace, the forums, and the main streets known as mese were the focal points for commercial establishments in the city. There were small perfume shops around the palace, among many monumental buildings such as the Hagia Sophia Church, Turkish bath, and hippodrome (Kuban 2000). There were many book sellers near Million. In the large streets where trade took place and in the streets adjacent to these streets, there were crowds of merchants coming from different places. This crowd included tradesmen such as Arab merchants, Byzantine craftsmen, money changers, candle makers, bakers, fishermen, butchers, carpenters and shovel makers (Keçiş 2015).

It is thought that the main streets of the city have porticoes (Şekil x). The porticoes in Istanbul have two floors. On the first floor, there were shops called Tabernae. In some of these shops, both production and retail sales were carried out. These structures were made of brick and had wooden roofs. In addition to these, there were also grocery stores. These could be found more or less throughout the city.

9.3.3 Military Structure

As the capital of the Eastern Roman Empire, Constantinople served as the administrative and military hub for all regions under the empire's control. Political stability and security in the city were crucial for the empire's well-being. Consequently, dense military infrastructure was present throughout and around the city. Despite this, only a few military structures in Constantinople remain today, with the walls being the most notable surviving example.

The walls built by Constantine the Great have not survived to the present day. Although the walls constructed in later periods have been demolished in places, they still stand. Today, the walls in the historical peninsula can be categorized into three groups based on their location: Land walls (Theodosian Walls), Marmara walls, and Golden Horn walls. The walls along the Golden Horn were comparatively weaker. This can be attributed the proximity of the historical peninsula to Galata. The distance between Galata and Sarayburnu Hill is approximately 700 m. While this distance was considerable in terms of the weapon technology of that era, it was close enough to allow for a chain to be drawn between Pera and Sarayburnu. This chain prevented enemy ships from entering the Golden Horn.

Today, the Maramara walls remain inland due to the filling of the coastline (Fig. 9.5b). However, in the Middle Ages, they were located directly on the coast. This situation makes a possible sea landing difficult. Additionally, the presence of currents in the Marmara Sea, in contrast to the calmness of the Golden Horn, facilitated the defense of the southern and eastern lines of the city. These natural factors enhanced the defensive capacity of the walls and significantly contributed to protecting the southern and eastern facades of the city.

The land walls in the western part of the city were more durable than the Marmara and estuary walls. The land walls have three levels. There is a moat in front of the walls and two separate wall systems behind them (Fig. 9.5c). One of the reasons why the land walls were built so strong was that there were no natural obstacles outside the walls that would slow down or stop attacks from the west. There is only one mountain range in Eastern Thrace, extending in an east–west direction. Until the first quarter of the sixth century, the Thracian agricultural areas and the Theodosian walls of Constantinople were vulnerable to enemy attacks. The Anastasian Walls were built in the first quarter of the sixth century to stop the raids from the west (Fig. 9.5a). However, a large number of soldiers were needed to protect this 65-km-long wall. On the other hand, the construction and repair of walls in the Middle Ages were quite costly.

As seen in Fig. 9.5b, there are towers on the land walls with a length of up to 19 m. Although there were exceptions, these towers were two-story, and there were vaults between the floors. Tower entrances are from the city side. It is thought that the lower floors were used as warehouses for military equipment or military outposts during wartime (Millingen 1899).

There were law enforcement forces with various names in the city of Constantinople. For example, there were the Varangian Guard (from the tenth century to the fourteenth century), the Themata army, the Navy, and garrison troops from various regions of the Empire, who were responsible for protecting the emperor in the city. The Varangian Guards were located in the Augustaion square near the Hippodrome (Fig. 9.5b). There were military units in strategic places such as the area where the Theodosius port was located and the Blachernae palace. The garrison was located in the western part of the city (Ortaylı).

9.3.4 Religious Structure

In Byzantion, which was a pagan city before it was declared the capital, there were various religious buildings built in accordance with pagan traditions. During the construction activities initiated by Constantine, old pagan temples were renovated, and buildings belonging to the Christian faith were built. After Theodosius I (379–395) made Christianity the official religion, the temples of polytheistic religions in the city were closed and repurposed as building materials (Çetinkaya 2015). The area where pagan temples were located was repurposed for other uses. For example, houses attached to Hagia Sophia were built instead of the Helios temple on the Acropolis (Türkoğlu 2010). Especially in the years 450–600, the city became a Christian city. (Magdalino 2015). Many churches and monasteries were built at various times on the hills of the city. Monasteries and churches are concentrated primarily in the

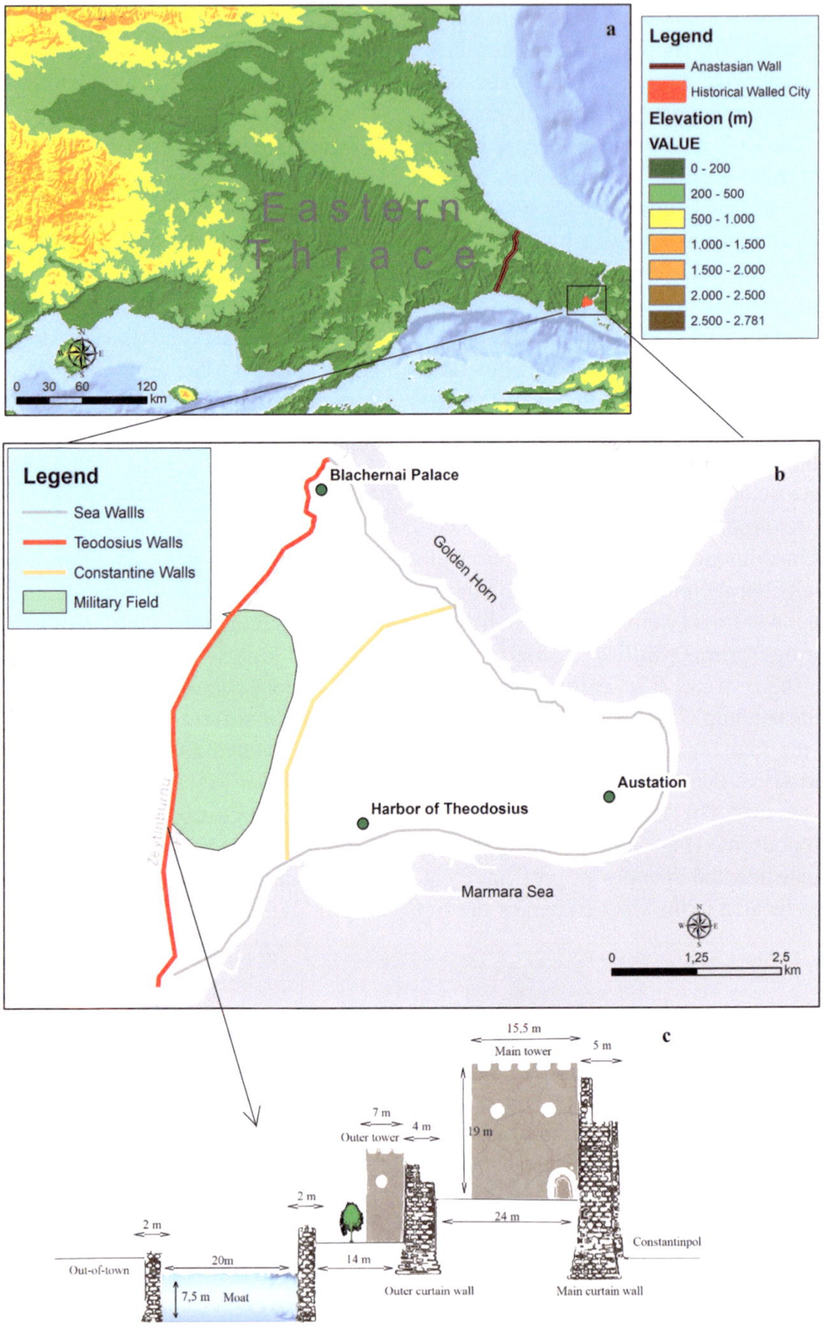

Fig. 9.5 Anastasian and Theodosian walls (land walls)

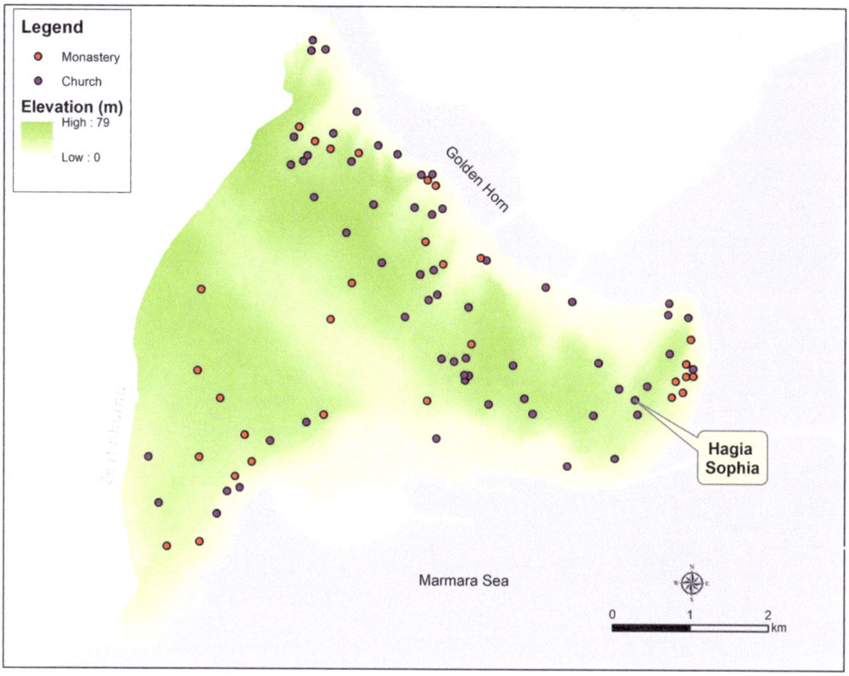

Fig. 9.6 Religious buildings in Constantinople. *Source* Redrawn from Öztepe (2001)

northeast of Hagia Sophia and on the slopes overlooking the Golden Horn (Fig. 9.6). Among these structures, Hagia Sophia, built in the sixth century, was the center of religious authority in Constantinople (Kuban 1996).

Hagia Sophia is a building that attracts architectural attention as well as religious significance. It was the largest cathedral in the world for a thousand years after its construction. Therefore, it is an extraordinary building for the city's image during that period.

As seen in Fig. 9.10, there are monasteries and churches in Constantinople. Initially, monasteries were established in remote provinces, far from settlements, with the purpose of escaping worldly affairs and focusing on worship. Due to changing conditions, monasteries were also established in settlements to ensure the well-being of the ruling class in their old age or following a political change that could potentially harm them (Çetinkaya 2015). The establishment of monasteries also formed the core of new settlements (Hatlie 2007).

Religious buildings in Constantinople were not just places of worship. In the sixth century, notary offices were found adjacent to churches. In the following years, schools and baths began to be built next to these churches (Magdalino 2015). For example, in the twelfth century, the Pantokrator Monastery had a hospital with 50 beds (Eyice 2011).

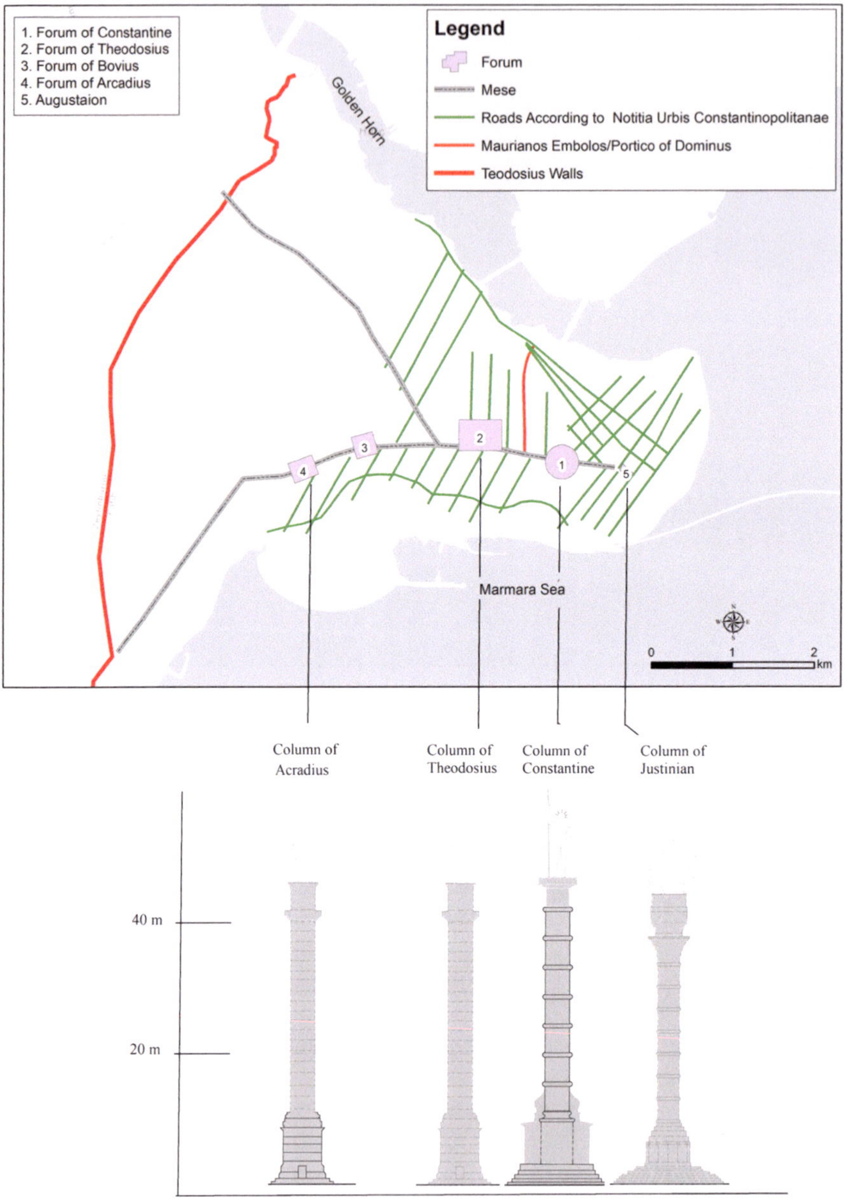

Fig. 9.7 Main streets and squares of Constantinople. *Source for the picture:* Yoncaci Arslan (2016)

9.3.5 Square and Roads

Another noteworthy aspect of the cityscape of Constantinople is the squares, which are typically designed in square and rectangular shapes. These forums were located on the road to the mese, and except for the Bovius forum, the others had an obelisk. These obelisks were erected by emperors and named after the respective emperor (Fig. 9.8). Augustan Square was built by Septimius Severus (193–211). The square in front of the walls of Byzantion was surrounded by porticoes (Müller-Wiener 2016). In the seventh century, the Hagia Sophia Church was known to be used for religious and imperial ceremonies. The forum located to the west of Augustaion Square, in today's Çemberlitaş, is the Forum of Constantine. Unlike other forums, it has a circular structure. 1.5 m of soil were filled for the forum built on the old necropolis. The entrance gates of the forum were adorned with numerous statues. To the west of the Forum of Constantine, there is the Forum of Theodosius, located near today's Beyazıt Square. Forum Bovius was located in present-day Aksaray. The Bovian forum was the location where Christians were executed during the reign of Julian (361–363). It had a similar function during the iconoclasm period after the seventh century (Müeller-Wiener 2016). The Forum of Arcadius, located near the Cerrahpaşa Mosque today, is the final forum on the southern branch of the mese.

Forums are not just places where trade is lively. Apart from this, these areas serve as venues for administrative and religious ceremonies, social gatherings, and the dissemination of news. In fact, the main gathering places of the people in the early period were baths and theaters, but later hippodromes were used for similar purposes. As the Hippodrome fell into disuse over time, the main gathering places for the city's people shifted to the main road and the squares along it. The statues in the squares were perceived as protective talismans by the public (Çetinkaya 2015).

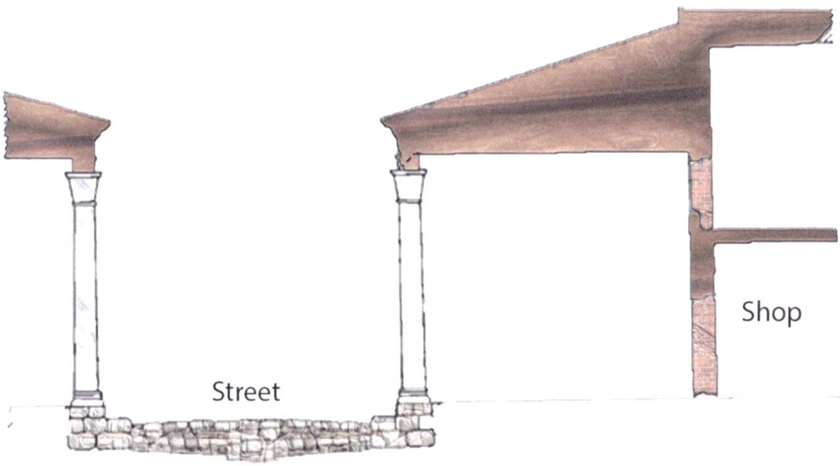

Fig. 9.8 Building with Portico. *Source* Redrawn from (Magdalino 2021)

Streets called "mese" are another striking element in the city's image. The concept of Mese, meaning "middle way," began to develop in the city of Byzantion and eventually became the main thoroughfare of Constantinople (Karakuyu 2010). Mese was the main thoroughfare of the city. It was also a socially and commercially busy route.

Although it is not possible to determine the exact route of the mesne today, an estimated route can be drawn. Accordingly, the starting point of the mesne is the Million structure, which is located in front of Hagia Sophia. It was considered the starting point of all roads in the Roman period, but today only remains as a stone. From the Million monument, he would pass the Konstantinos Forum (today's Çemberlitaş) and reach the Tauri Forum (today's Beyazıt). From Tauri Square, one of its branches curved northwards around present-day Koska-Vezneciler and followed Fevzi Pasha Street, as it is known today. Another branch passed through the Forum of Bovis and Arkadius and reached the Golden Gate in today's Yedikule Dungeons (Fig. 9.7).

Many roads cut through the Mese at right angles and formed secondary road routes (Fig. 9.7). Apart from this information, we do not have detailed information about minor roads. However, it does not seem possible to talk about a simple monolithic structure in the avenue and street system of Constantinople. The development of the city occurs in various stages. On the other hand, the topography of the inner city area is rugged and the coastline is uneven (Magdalino 2015). Moreover, this coastline was much more indented in the fourth century than it is today (Yoncaci Arslan 2016). Therefore, although the grid plan generally dominates the city, it is likely to be present in some regions rather than throughout the entire city.

The average width of the Mese is 25 m. Other secondary roads within the city had to be 2–3 m wide. However, there were streets in the city where houses were built very close to each other. This causes it to be thought that it was a model consisting of narrow streets, similar to those in early imperial Rome (Dark 2004).

Another difference between Mese and smaller roads is that Mese is built either on the tops of hills or on their sides, resulting in a relatively low slope rate. It was suitable for travel by passenger car. On the other hand, the streets constructed in the new residential areas of the city during later periods were situated on steep slopes. Due to this slope, roads with stepped were observed in some locations (Türkoğlu 2010).

Another striking feature in the streets and avenues of Constantinople was the embolos (a road with a portico). It is the name given to streets covered with porticoes, rows of columns, or porches. Such paths connected different forums. It is known that there were four porticoed roads in the city during the reign of Constantine. Two of them were located on the cape of the peninsula and on the coast of Marmara. Another road with a portico well-known road with a portico is the Severus Portico. The Severus Portico ended at Çemberlitaş. The fourth portico started from Forum Tauri and went to Aksaray, then to the location of the Forum of Arkadius, which is where Cerrahpaşa stands today (Kuban 1970). These porticoed roads are a distinctive feature of Roman cities in the Mediterranean basin, where summers are hot.

9.3.6 Water Structures and Baths

While the location of the city was discussed in the first section, many positive features were mentioned. Apart from this, the most significant drawback of the city was the scarcity of fresh water and the potential for a land attack from the west. Interventions have been implemented to address both deficiencies. These interventions had an impact on the city's image.

Comprehensive projects have been implemented in the city to address the lack of fresh water. Water was transported to the city from high-altitude areas in the northwest using slopes, transmission lines, and aqueducts (Fig. 9.9). During the reign of Hadrian (117–138), the pipeline, which was approximately 35 km long, brought water to the city from 20 km away. After the city was declared the capital, its population increased, leading to the expansion of the transmission lines to 172 km to meet the growing water demand. In the fifth century, the length of the transmission lines was increased to 258 km and extended to the Pazarlı settlement, which is 125 km away from Istanbul and is now located within the borders of Kırıkkale province. During the early years of the Ottoman period, the total length of active transmission lines was around 56 km.

Water from the north of the city was transported to the city through the transmission lines within the walls and the Valens aqueduct. The water brought to the city was stored in both open and closed cisterns. Some of these cisterns were built as cisterns. Some of them are the cellars of buildings and monasteries. Its walls were plastered with waterproof mortar and converted into a cistern (Eyice 2011).

There are 158 cisterns that can be identified today (Altuğ 2013). The distribution of these cisterns is similar to the distribution of religious buildings (Fig. 9.10). Among these cisterns, notable ones include: the Aetius cistern, currently serving as a stadium; the Aspar cistern, transformed into a green area; the Basilica (Yerebatan) cistern, utilized as a historical tourist attraction; and the Mocius cistern, repurposed as a green space. The Basilica Cistern is the largest enclosed cistern in Istanbul. The Aetius, Aspar, and Mocius cisterns are open cisterns.

There are basically two benefits of constructing numerous cisterns in Constantinople. These cisterns serve as a backup in case there is an interruption in the transmission lines outside the city walls during times of war. Apart from that, another issue is that there may be water scarcity during the summer months. For example, it is known that the Basilica Cistern was built in response to drought during the summer months.

Baths played a significant role as in the daily life of early Byzantine society. In Roman culture, baths were not used solely for cleaning purposes. At the same time, they were buildings that functioned as public spaces where people socialized and even made business deals from time to time. Some even had libraries, and sports facilities. According to statistics from the second quarter of the fifth century, there were a total of 161 baths in the capital, and 8 of them belonged to the public (Emecen 2015).

Although the names of many baths in Byzantine Istanbul are known, there is uncertain information about the locations of very few of them. It does not seem

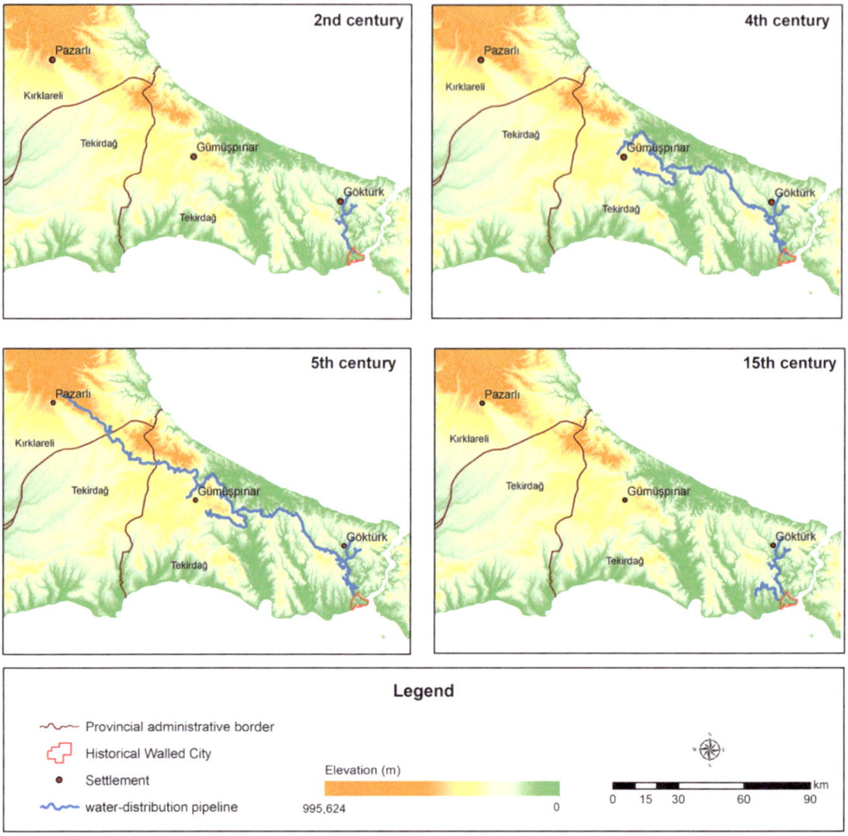

Fig. 9.9 Water channels of Constantinople according to periods. *Source* Drawn using Crow (2015)

possible to obtain detailed information about these structures from the data obtained as a result of excavations. However, as Müller-Wiener points out, these structures did not change significantly until the sixth–seventh centuries (Müller-Wiener 2016). It is thought to have similarities with Roman baths.

The area where baths are concentrated is also related to population density and water availability. During the Roman period, the absence of a hydraulic system made it easier to access the water required for the operation of the baths in areas near the transmission lines. For this reason, it is estimated that baths were established in locations near transmission lines.

On the other hand, it is known that the Byzantine Church had a negative attitude towards baths because it believed that they were luxurious and overly worldly (Ertuğrul 2015). Due to the pressures of the church, the decline in water resources, and economic factors, the construction of baths decreased, and the "thermae," known as public baths, lost their purpose. However, palace baths and baths belonging to private houses continued to be used (Magdalino 1990).

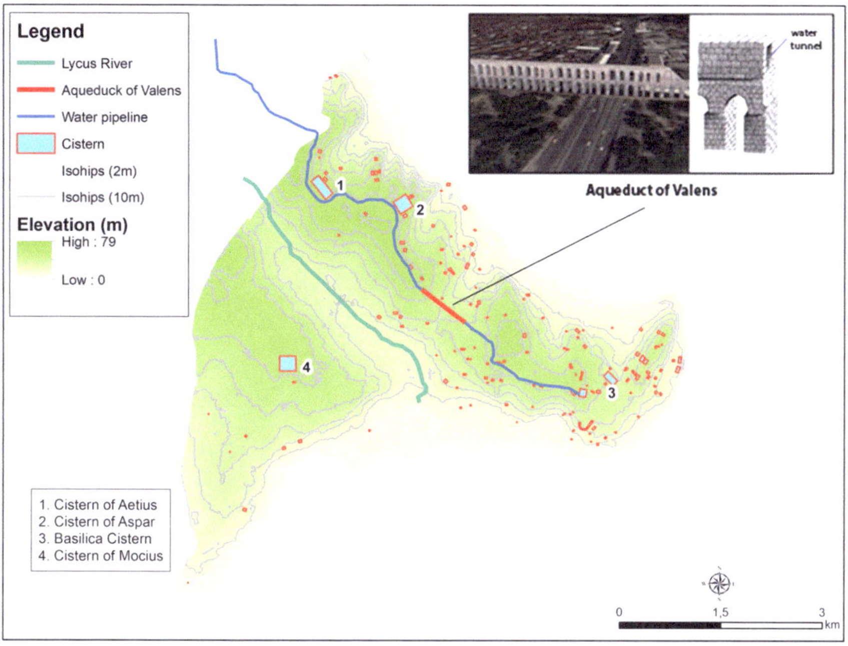

Fig. 9.10 Cisterns and Valence Water Line in Byzantine Istanbul. *Source* Drawn using Altuğ (2013), Crow (2015). *Source for picture* Adapted from http://www.brewminate.com (2018), Kahraman (2018)

9.4 Conclusion

The city of Constantinople played a crucial role as the capital of the Eastern Roman Empire for over ten centuries. The site and geographical characteristics of the city were significant factors in its selection and development as the capital. The historical peninsula where the city was established was easily defensible, except for the western part. Moreover, it possessed other favorable geographical features such as natural harbors, abundant sea life in the Golden Horn, and sea currents that facilitated transportation. Additionally, its strategic location at the crossroads between the Aegean and Black Seas by sea, and between Asia and Europe by land, further enhanced its appeal as the capital. These positive attributes greatly influenced the decision to designate it as the capital. However, it is important to note that the city was relatively modest before becoming the capital. Political events of the time played a crucial role in the city's development. Later, the city was established as the capital of the Eastern Roman Empire and its place in the city hierarchy gradually increased.

The city of Constantinople became the capital, significantly strengthening its commercial, religious, and military functions over time. On the other hand, the cultural landscape of the city began to change after it was declared the capital. The population growth in the city and the plans made to develop the city have also led to the development of infrastructures such as streets and water structures. When we look at the cultural landscape of Constantinople, it is seen that the above-mentioned functions and infrastructures are also effective on the cultural landscape of the city. Palaces and hippodromes with administrative functions, city walls with military functions, and ports with commercial functions, along with other churches and monasteries, especially Hagia Sophia, with religious functions, are important parts of the city's cultural landscape.

References

Akalin K (2015) Bir Dünya Şehri Olarak Konstantinopolis. In: Büyük İstanbul Tarihi (1. bs, C. 19, ss. 222–233). İSAM, Türkiye Diyanet Vakfı İslam Araştırmaları Merkezi

Akyürek E (2021) The hippodrome of Constantinople. Cambridge University Press

Altuğ K (2013) İstanbul'da Bizans Dönemi Sarnıçlarının mimari özellikleri ve kentin tarihsel topografyasındaki dağılımı [Doktora Tezi, İstanbul Teknik Üniversitesi]. https://polen.itu.edu.tr/items/22e121e2-6d8a-4e00-871c-2f2088f49dd2

Asal R (2010) Yenikapı Marmaray ve metro kurtarma kazıları buluntuları ışığında İstanbul'un Roma ve Bizans Dönemi Deniz Ticareti [Master's Thesis, Trakya Üniversitesi, Sosyal Bilimler Enstitüsü]. https://dspace.trakya.edu.tr/xmlui/handle/trakya/1281

Asutay-Effenberger N (2024) Blakhernai Sarayı ve Sözde Anemas Zindanı—İstanbul Surları. Blakhernai Sarayı ve Sözde Anemas Zindanı - İstanbul Surları. https://istanbulsurlari.ku.edu.tr/tr/essay/72/blakhernai-sarayi-ve-sozde-anemas-zindani

Bassett S (2005) The urban image of late Antique Constantinople (C. 49). University Press. Cambridge. https://muse.jhu.edu/pub/77/article/591383/summary

Crow J (2015) Bizans Konstantinopolis'inin Su Tedarik Sistemi. İn: Aydoğan A, Kaçar T (Çev). Büyük İstanbul Tarihi (C. 6, ss. 44–55). https://www.istanbultarihi.ist/assets/uploads/pdf/bizans-konstantinopolisinin-su-tedarik-sistemi-190.pdf

Çetinkaya H (2015) İstanbul'un Bizans Dönemi Mimarisi, Bizans Mimarisi. İn: Çağ'dan A (ed), vol XXI. Yüzyıla Büyük İstanbul Tarihi (C. 8, ss. 24–56)

Dark KR (2004) Houses, streets and shops in Byzantine Constantinople from the fifth to the twelfth centuries. J Mediev Hist 30(2):83–107. https://doi.org/10.1016/j.jmedhist.2004.03.001

Demirkent I (2005) 12. Yüzyıla Kadar Bizans' ta «Loncalar». In: Bizans Tarihi Yazıları, Makaleler, Bildiriler, İncelemeler, İstanbul: Dünya Kitapları: I (ss. 159–173). Dünya Yayıncılık

Drakoulis D (2012) The functional organization of early Byzantine Constantinople according to the Notitia Urbis Constantinopolitanae. In: Doukellis P, Korres TH, Sfetas SP. ve Toloudi F (eds) "Ανοιχτοσύνη"-"Openness".Vanias, Thessaloniki, pp 153–184

Dominian L (1917) The site of Constantinople: a factor of historical value. J Am Orient Soc 37:57–71. https://doi.org/10.2307/592905

Emecen F (2015) Emperyal Dönüşümlerin Muhteşem Kenti İstanbul. In: Antik Çağ'dan XXI. Yüzyıla Büyük İstanbul Tarihi (C. 1, ss. 118–181). İSAM, Türkiye Diyanet Vakfı İslam Araştırmaları Merkezi. https://istanbultarihi.ist/6-emperyal-donusumlerin-muhtesem-kenti-istanbul?q=domus

Ertuğrul AD (2015) İstanbul hamamları ve mimarisi. In: Antikçağdan XXI. Yüzyıla Büyük İstanbul Tarihi (C. 8, ss. 450–467). İSAM, Türkiye Diyanet Vakfı İslam Araştırmaları Merkezi. https://istanbultarihi.ist/assets/uploads/pdf/istanbul-hamamlari-ve-mimarisi-321.pdf

Eyice S (2011) Bizans İmparatorluğu'ndan Osmanlı İmparatorluğu'na İstanbul In: Hut D, Kurşun Z, Kavas A (eds). Şan Ofset Matbaacılık, ss. 15–24

Gökdaş R (2006) Haliç'te rehabilitasyon çalışmaları sonrası mevcut su ürünleri. Ph.D. thesis, Sakarya Universitesi (Turkey). https://search.proquest.com/openview/234b5f23a3f600ab1a84449812fd937e/1?pq-origsite=gscholar&cbl=2026366&diss=y&casa_token=6qqzKOckEC8AAAAA:qUpLRnjsvOZ53VtnsPinu8J2HbcLSeY-piyIsJSegr6TmVR921TwNHFkgQrn3tltwm6p4qY2u0I

Hatlie P (2007) The monks and monasteries of Constantinople, ca. 350–850. Cambridge University Press. https://ixtheo.de/Record/535339585

Jones M, Olgun Z (2024) İulianus veya Sofia Limanı—İstanbul Surları. İulianus veya Sofia Limanı - İstanbul Surları. https://istanbulsurlari.ku.edu.tr/tr/essay/35/iulianus-veya-sofia-limani

Kahraman M (2018) The causes of the formation of sebil and fountain in ıstanbul and the effects of these constructions on the city of Istanbul. Uluslararası Sosyal Araştırmalar Dergisi/ J Int Soc Res 11(59):368–383

Kahraman M (2023) İstanbul'un Tarihi Coğrafyası. In: İstanbul İlinin Coğrafyası. Türk Coğrafya Kurumu, ss. 353–374

Karakuyu M (2010) İstanbul'un tarihsel topoğrafyası ve literatür değerlendirmesi. Türkiye Araştırmaları Literatür Dergisi 16:33–60

Kavrakoğlu F (2016) Bizans İmparatorluğu 118| Bizans Sarayları 1 Büyük Saray 1. Füsun Kavrakoğlu. https://kavrakoglu.com/bizans-imparatorlugu-118-bizans-saraylari-1-buyuk-saray-1/

Keçiş M (2015) Bizans Döneminde İstanbul'un Kent Yönetimi. İn: Antik Çağ'dan XXI. Yüzyıla Büyük İstanbul Tarihi (C. 3, ss. 18–37). İSAM, Türkiye Diyanet Vakfı İslam Araştırmaları Merkezi

Kuban D (1970) İstanbul'un Tarihi Yapısı. Mimarlık Dergisi 5:25–48

Kuban D (1996) İstanbul, bir kent tarihi: Bizantion, Konstantinopolis, İstanbul. Türkiye Ekonomik ve Toplumsal Tarih Vakfı

Ljungkvist J, Barthel S, Finnveden G, Sörlin S (2010) The urban anthropocene: lessons for sustainability from the environmental history of Constantinople

Ljungkvist J, Stephan B, Finnveden G, Sverker S (2010) The Urban Anthropocene: lessons for sustainability from the environmental history of Constantinople. İn: Sinclair PJJ, Nordquist G, Herschend F, Isendahl C (ed) The urban mind cultural and environmental dynamics. Uppsala University, pp 367–390

Lordoğlu N (2019) Byzantion ve Kalkhedon'un şehirci lik açısından incelenmesi: kuruluşlarından Roma İmparatorluk Dönemi'ne kadar. Cedrus 7:169–194

Safran L (2007) http://individual.utoronto.ca/safran/Constantinople/Map.html

Magdalino P (1990) Church, bath and diakonia in medieval Constantinople. In: Church and people in Byzantium. University of Birmingham, pp 165–188

Magdalino P (2000) The maritime neighborhoods of Constantınople: commercial and residential functions, sixth to twelfth centuries. Dumbarton Oaks Pap 54:209–226. https://doi.org/10.2307/1291840

Magdalino P (2001) The Porticoed Street at Constantınople. In: Necipoğlu N (ed) Byzantine Constantinople: monuments, topography and everyday life (C. 33). Brill. https://books.google.com/books?hl=tr&lr=&id=tG0p_sZH-fEC&oi=fnd&pg=PR9&dq=Byzantine+Constantinople:+Monuments,+Topography+and+Everyday+Life,+2001&ots=V61ByBV7Oy&sig=d9kKnAk76LbFEpfcJP0FvWS72TE

Magdalino P (2015) Bizans İstanbul'u: Topoğrafya ve Yerleşim. İn Antik Çağdan XXI. Yüzyıla Büyük İstanbul Tarihi (ss. 410–427). İSAM, Türkiye Diyanet Vakfı İslam Araştırmaları Merkezi

Mango CA (1986) The art of the Byzantine Empire 312–1453: sources and documents (C. 16). University of Toronto Press. https://books.google.com/books?hl=tr&lr=&id=rSvf_KMYQiwC&oi=fnd&pg=PR11&dq=Cyril+Mango+&ots=iiLrYzMS42&sig=JCWvtqiyE_7KrNUcdbtLfFV5_MU

Müeller-Wiener W (2016) Istanbul'un Tarihi Topografyası. Yapı Kredi Yayınları, Istanbul

Necipoğlu N (2001) Byzantine Constantinople: monuments, topography and everyday life. BRILL

Nicolescu E, Safran L (2024) Interactive map of Constantinople. http://individual.utoronto.ca/safran/Constantinople/Map.html

Ortaylı İ (1977) İstanbul'un mekansal yapısının tarihsel evrimine bir bakış. Amme İdaresi Dergisi 10(2):77–97

Ozturk M, Sahin C, Yuksel Y (2017) Current power potential of a sea strait: the Bosphorus. Renewable Energy 114:191–203

Öztepe O (2001) İstanbul Suriçi'nde bulunan Bizans dönemine ait kiliselerin günümüzdeki durumları ve koruma sorunları/The conservation problems and the present status of the Byzantine churches which are situated at Istanbul, Suriçi. Ph.D. thesis. İstanbul Teknik Üniversitesi/Fen Bilimleri Enstitüsü/Mimarlık Ana Bilim Dalı

Saliou, C. (2007). The Byzantine House (400–912): rules and representations. Verlag der Österreichischen Akademie der Wissenschaften. https://www.austriaca.at/?arp=0x00178c57

Şahin M (2019) Roma kaynaklarında Gotlar ve Anadolu akdenizi [Yüksek Lisans Tezi, Akdeniz Üniversitesi Akdeniz Uygarlıkları Araştırma Enstitüsü]. http://acikerisim.akdeniz.edu.tr/handle/123456789/6720

Tahiroğlu B (1981) Roma Devleti'nin İktisadî Krizleri. J Istanbul Univ Law Fac 45(1–4):677–706

Türk Deniz Araştırmaları Vakfı. (2024) İstanbul ve Çanakkale Boğazı Akıntı Haritaları. TÜDAV. https://tudav.org/calismalar/deniz-alanlari/turk-bogazlari/istanbul-ve-canakkale-bogazi-akinti-haritalari/

Türkoğlu AY (2010) İstanbulun Roma Dönemi topografyası: Byzantion-Constantinople. Master's thesis. İstanbul Üniversitesi Sosyal Bilimler Enstitüsü Arkeoloji Anabilim Dalı. https://acikbilim.yok.gov.tr/handle/20.500.12812/182680

Van Millingen A (1899) Byzantine Constantinople: the walls of the city and adjoining historical sites. John Murray

Yılmaz EC (2015) Fetih ve Sonrasında İstanbul'da Barınma Kültürü. İn book, Antik Çağ'dan XXI. Yüzyıla Büyük İstanbul Tarihi (ss. 152–172). İSAM, Türkiye Diyanet Vakfı İslam Araştırmaları Merkezi

Yoncaci Arslan P (2016) Towards a new honorific column: the column of Constantine in early Byzantine urban landscape. Metu J Fac Archıtect 33(1):121–145. https://doi.org/10.4305/METU.JFA.2016.1.5

Zengin M (2008) Karadeniz'de Yunus Avcılığı. Aquaculture Stud 2008(1). https://dergipark.org.tr/en/download/article-file/204326

Part IV
African Cities

Chapter 10
Africa's Cities as Projects?

Lola Martínez-Fons

Abstract This introductory chapter raises the question from the outset of whether the North-centric approach to 'the city as a project'—an ideal anticipation of a frozen future based on abstract needs of the present—is relevant in African urban history. Understanding how Africans imagine, build, and inhabit their cities requires a journey through their ways of seeing and being in the world, which guarantees alternative cultural expressions. It also involves tracing their rich and heterogeneous urban heritage, the result of rhizomatic linkages of relationships and social dynamics over centuries of encounters and misencounters with other peoples and cultures—all of which are projected today in the Africanized city. A synoptic journey—and therefore not without generalizations and omissions— to encounter other forms of urban life beyond Western rationalism: heterogeneous urban forms that the Global North-West 'informalizes' (or exoticizes) because they disrupt its colonizing norms, yet Africa normalizes within its own systems of organization, exchange, control, regulation, and social reciprocity. A positive 'informalitude' that reaffirms and updates the original experience of city-making; that refuses to conceive of an alternative within monological spaces of containment and seeks its own space of enunciation, free from binary oppositions.

Keywords African urban knowledge · Urban creolization · Colonial urbanism · Informality · Conflicting rationalities

10.1 Who Are the Strangers to the City?

Reconstruire l'histoire des villes africaines,
c'est aussi, en ce sens, tenter de rendre
à cette partie du monde sa part

L. Martínez-Fons (✉)
Universidad de Sevilla, Sevilla, Spain
e-mail: lolamartinezfons@gmail.com

F. Carrión Mena and S. Rodríguez Alvarez (eds.), *From the City as a Project to the City Project*, The Urban Book Series, https://doi.org/10.1007/978-3-031-94226-6_10

d'universel qui lui longtemps été déniée[1]

(Coquery-Vidrovitch 1993, p. 10)

Understanding African urban thought, the way Africans imagine, build, and inhabit their cities, first requires a detachment (unlearning) from hegemonic Eurocentric theories and models in order to recognize (relearn) that, within the complex and diverse African urban landscape, there exist 'others' (in the plural) ways of conceptualizing the city, for which geographic location is only one variable among many. Thus, when we speak of African urban thought, we must not forget that it encompasses a heterogeneous spatial, temporal, cultural, political, and social variability, with its own cosmological, cognitive, semiotic, or psychological specificities.

The very idea of cataloging the most representative anticipatory and prefigured city projects in African urban history—'the city as a project in Africa,'—, due to its Eurocentric conceptual connotations, necessarily and primarily directs us to the colonial city projects developed by European powers on the continent from the late nineteenth century to the mid-twentieth century, as well as to some quasi-projects of port or military cities established along the Atlantic coast in the four centuries preceding colonization. Later, in the emancipatory adventure—'beyond colonialism'—of independence, some city projects will emerge as symbols of African nationalist and developmental ideals, though they will still be of Western origin both in authorship and theoretical foundations. Finally, in this cataloguing exercise, we encounter contemporary urbanism in Africa that oscillates between mega-projects replicating 'fabulous' images of 'Global Cities'—'urban fantasies'[2] exclusive and excluding, detached from the African reality, and aimed at (and only achievable by) a small segment of the population—and the autonomous construction and management of entire neighborhoods (from below) of greater relevance and significance for the daily lives of millions of African urban dwellers.

Traditional African forms of conceiving the city are difficult to trace in the urban history of a continent predominantly characterized by oral tradition[3]—whose historical (written) account was claimed by the West—where archaeological evidence is scarce, still unexplored, or presents technical and economic challenges for its exploration and excavation.[4] African urban thought, reflecting its cultures and its forms of social organization, will 'creolize'[5] diverse and heterogeneous external

[1] "Reconstructing the history of African cities is also, in this sense, an attempt to restore to this part of the world its share of universality that has long been denied to it." (English translation by the author of this article).

[2] South African planning theorist Vanessa Watson (2013) coined the term "African urban fantasies" to describe these decontextualized urban mega-developments in Africa.

[3] Some African historians, following the work of the Nigerian scholar Kenneth Dike, have reconstructed (rediscovered) African history through oral sources and evidence. (It is also important to remember that Ethiopia and North Africa had written records and used writing).

[4] An archaeological heritage that today faces multiple threats to its conservation, including agricultural intensification, uncontrolled urban sprawl, natural resource extraction, large-scale infrastructures projects, climate change, internal violence, terrorism…

[5] Creolization is the concept developed by the Martinican philosopher and writer Édouard Glissant to describe the process of constructing a culture (or a language) born from a melting pot of cultures,

influences throughout its history and geography. However, this urban evolution will be profoundly altered by the violent distortion of four hundred long years of the slave trade followed by the brief but intense mental colonization imposed by European modernity and its capitalist recipes. In addition to the difficulties of exploring pre-colonial urban concepts in Africa, there are the historiographical and conceptual consequences of the (arrogant) Eurocentric assumption that Africans were 'strangers to the city' (Plotnicov 1967, p. 45) and that the 'true city' was a civilizational contribution to the rural (and 'savage') Africa by Europeans—definitively the true 'strangers to the African city' (Coquery-Vidrovitch 2005a, p. xxxiii).

We must not forget that, although Africa is currently the least urbanized continent on the planet,[6] its heterogeneous process of urbanization precedes that of other regions by centuries. It is a largely rural continent that, nonetheless, throughout its history, has generated unique urban experiences. Places of cultural encounter and diffusion, cities that have been generators of social and political changes, and which have undergone very different trajectories throughout their history due to demographic, climatic, political, or economic accidents that led them to disappear, be refounded, relocated, or reconfigured as genuine urban palimpsests. Cities conceived, imagined, built, and inhabited according to the logics and languages of their inhabitants: cultural forms of traditional African thought.

10.2 African Urban Thought

Talking about urban thought in Africa entails going beyond Western critical and scientific rationalism to make room for the essential wisdom of traditional African societies. African urban knowledge is constructed from 'another' intellectual sensitivity, 'another' way of being sensitive, and 'another' way of seeing the world.

The conceptualization of urban ways of life in Africa is rooted in traditional African thought, which thinks, perceives, and intuits the world as an immense organic unity in which humans are vitally linked to 'nature conceived as continuity with the spirit that orders it and that it itself expresses through its harmonious and hierarchical diversity' (Ndaw 1983, as quoted in Iniesta 2010, p. 27): a Supreme Principle ordering chaos that is incarnated in successive descending levels, from divine

resulting from the contact and clash of diverse and heterogeneous cultures that mutually alter, interweave and mingle to "give birth to something absolutely unpredictable, absolutely new" (Glissant, 2002, p. 17). Creolization requires the heterogeneous elements brought into relation be intervalorized, that the resulting identity be rhizomatic, with multiple roots connecting and intersecting other roots (Boyer, 2009).

[6] Although it features, together with Asia, the highest annual urban growth rates (3.5%), it remains the least urbanized region of the planet. Urbanization on the continent accounts for approximately 50.4% (as of 2015), with an estimated population of 567 million urban dwellers (2015), and North Africa being its most urbanized region (78%). In 1950, the estimated urban population was 27 million inhabitants. (OECD/SWAC 2020). *Africa's Urbanisation Dynamics 2020. Africapolis, Mapping a New Urban Geography*. West African Studies).

royalty—catalyst of cosmic forces and backbone of social organization—to the ultimate concrete material diversification, even beyond of what is sensorially perceptible to human beings (Iniesta 2001, p. 82–3). A harmonious natural order of divine origin that pervades all existence—social disorder is an expression of cosmic imbalance (Iniesta 2010, p. 80)—, structuring social thought and political systems in Africa in a more or less similar manner, and which, consequently, is also expressed symbolically and spatially, over centuries, in its very diverse urban configurations. The notions of order, nature, community, and hierarchy will be present both in materials, techniques, and architectural forms as well as in the spatial situation of the buildings: They will be inherent principles of the African urbanization process (Coquery-Vidrovitch 1995).

In Africa, the place of origin is not just a physical attribute; it is a mark of community and a link to ancestors (both living and deceased). There is an intimate relationship between the group and the place that defines its geographical identity: The land is an essential part of what it means to 'be'. The communitarian and local dimensions prevail over the individual (Chabal 2011). There is a social sense of belonging nurtured by the dynamic networks of local and translocal kinship woven through systems of authority, respect, and participation, which translate into structures of reciprocity and impact daily life. These social dimensions are present in the urban dynamics of African cities and interconnect the rural and urban in complex systems of interdependence that remain vital for their contemporary cities.

Thus, traditional urban habitat in Africa is based on a solid social fabric with mechanisms for connectivity and communication personalized in social units of multiple affiliations (clans, tribes, ranks, fictive kinships, age groups, etc.) (Roca and Iniesta 2006). A logic of segmentary society with a strong communal component (lineage, clan) that is reflected in urban patterns. The spatial organization of the African urban environment pays more attention to human relationships than to geometric design (Hull 1976, p. 74), or seen from another perspective, it subordinates space to the multiple and multifaceted collective human relationships: groupings of dwellings based on African kinship ties (led by a lineage or clan chief) form communities or neighborhoods (under a representative authority) according to ethnic, military, occupational, social, or religious identities.[7] More or less hierarchical social structures that, in some cases, can lead to urban segregation—with differentiated areas for elites and commoners, free and unfree individuals, or immigrant or merchant communities—in which each group maintains firmly its identity or remakes it anew by adapting to the environment.

The forms of their cities will be the crystallization of their social order (Wynne-Jones and Fleisher 2014, as cited in Van Oostrum 2024, p. 3). The city is a macrocosm of these social microcosms, characterized by strongly cohesive and dynamic structures that promote stability, security, cooperation, and a strong sense of communitarian identity and responsibility. Urban spatial form and social practices are interdependent, from the micro-level of the design of the basic social unit to the macro-level of city design. A 'plurality of spatial structures' organically produced by urban

[7] The ancient city of Kano is organized into different ethnic neighborhoods, while Benin City, with only one ethnic group (the Edo people), is organized into artisan and religious neighborhoods.

society—'embodied planning' (Van Oostrum 2024, p. 3)—that promotes different modes of collective use of the city (Baumanova et al. 2019).

A built space in which intangible voids are key to the urban ways of life of its inhabitants: from the streets—essential material reflections of the daily life of the urban community, which allows for access, interactions, and links while also delimiting or containing parts of the city (Baumanova et al. 2019)—to small, open and dispersed public or semi-public (or semi-private) spaces, where wells, squares, and washing places are arranged; or to large public spaces either on the periphery, used for agriculture, or in the urban center, serving interaction, movement, public gatherings, or congregation (markets, festivals, public justice, processions, rituals, etc.) (Hull 1976; Coquery-Vidrovitch 1993). The urban morphologies that these macrocosms adopt are varied. We find urban forms that reflect authoritarian social systems: circular walled cities with radial structures and the royal palace at the center, characteristic of the Yoruba and Hausa; or grid-like structures typical of the Ashanti, Dahomean, or Beninese cities. And societies with more diffuse authority express themselves spatially in labyrinthine structures, such as the cities of the Ibo culture or Islamic cities (Hull 1976, p. 123).

These urban patterns and ways of life will evolve and exhibit variations as a result of the interrelationships of their inhabitants with the environment, and the combination of 'endogenous'[8] cultures with external cultural influences. A cultural creolization, primarily synthesized in the cities (though not exclusively), shapes diverse cultural and architectural characteristics that will be radiated to other enclaves. A mixture of memories that will enable the coexistence of apparently opposing urban conceptualizations within the same city, while maintaining the African-rooted material or symbolic expressions of attentiveness to human relationships, adaptability to nature, and complementarity with the surrounding landscape.

And if, as the anthropologist Ferrán Iniesta points out, the only guarantee for alternative cultural expressions are traditional thoughts, which, far from dying out, persist all over the planet (Iniesta 2010, p. 11), we must (re)know the urban cultural forms that, in their diverse manifestations—from urban planning and architecture to paintings or reliefs on walls and facades—distill traditional African thought: a tradition that is filtered, perceived, and experienced today in the structural, cultural, and social hybridity and plasticity of African cities.

[8] The Cameroonian anthropologist Francis B. Nyamnjoh (2012) proposes the term 'endogenous' instead of 'indigenous' to characterize African knowledges, thereby assigning them their rightful dynamism, and freeing them from pre-colonial boundaries: they "do not need colonial rationalism to come alive" (Nyamnjoh 2012, quoted in Winkler 2022, p. 64).

10.3 'Dis-Covering' African Urban History: Tracing Urban Footprints

Placing in context and valuing the diverse material and symbolic urban landscape of Africa requires 'dis-covering'[9] its rich and heterogeneous urban heritage. This heritage, accompanied by demographic dislocations, social upheavals, and rapid and/or dramatic reurbanizations (with the disappearance and abandonment of old cities), captures periods of endogenous splendor and fruitful cultural creolization. Briefly and succinctly traversing this rich urban history (composed of multiple histories), which has been shaped over centuries by both African-rooted urban civilizations and heterogeneous global influences—Greek, Roman, Phoenician, Byzantine, Arab, Persian, Mesopotamian, Ottoman, Indian, Chinese, Southeast Asian—that have molded the relationships and social dynamics of Africans, and consequently their spatial practices of urban production, is the objective of this synoptic urban journey we begin in the Nile Valley… A journey that, despite its inevitable level of generalization, reaffirms the inclusion of African cities in world history.

This journey tracing urban footprints takes its first route in the Nile Delta and from there heads toward the Mozambique Channel. The next route takes us after the Phoenician and Roman traces along the African Mediterranean coast, which become lost into the Berber states, from where they will leave already Islamized to intersect with the deep urban footprints of the Niger River Basin. In Central Africa region, the traces will disperse much later between the Zambezi and Limpopo Rivers or across the Congo River. Indeed, the urban history of Africa south of the Sahara is linked to megalithic cultures and can be traced back to the time of Pharaonic Egypt. It features autonomous urban developments—such as those in the Niger Bend since at least the second millennium Before Common Era (BCE)—and displays a diversity that challenges our understanding of evolution of cities. However, given the complexity of studying this period and the forward-looking nature of this text, the focus of this introductory chronology will be on those African urban societies whose memory can still be traced among contemporary communities.

The earliest traces of urbanization in Africa date back to the fourth millennium BCE and are located in the Nile Valley, one of the most urbanized regions in the world until the Common Era (CE). In the area of Ancient Egypt (Kemet), ruled by successive dynasties of god-pharaohs, the cities of Abydos (3200 BCE), Memphis[10] (3114 BCE), Thebes (1700 BCE), and Amarna (1370 BCE) (Chandler 1994a) are sited.

The northern part of Africa will experience the impact of fundamentally urban cultures such as Greek, Phoenician, and Roman, which will establish colonies and city-states along the entire African Mediterranean coastline and the Nile Delta, where they will encounter endogenous cultures. The highest exponent of urban cultural

---

[9] The concept of 'dis-covering'—observing without pre-judgments, as advocated by landscape architect Bernard Lassus (2015)—suggests adopting a "floating attitude" for an "inventive analysis" that allows for approaching the place in its uniqueness.

[10] The current city of Cairo stands on Menphis's original site.

creolization in the region will bethe cosmopolitan city of Alexandria[11] (fourth century BCE). Conceived by Alexander the Great as a Greek polis, its urban structure, buildings, ways of life, and cultural confluences from Classical Greece, Rome, and Ancient Egypt will make it an urban reference.

Following the urban footprint that starts its journey in Egypt, we cross the land of Kush, calling at Kerma and Meroë,[12] and navigate along the Red Sea coast toward the land of Punt. Upon reaching the highlands of Abyssinia, we find Axum,[13] the capital of the Aksumite Empire, from where urbanization will be directed toward the Horn of Africa and the eastern African coast along the Indian Ocean (Coquery-Vidrovitch 1993, p. 55). From Egypt to the 'far side ports' of the Indian Ocean, trade will be the main nexus. A pre-Islamic trade flow that will connect Europe and Asia, touching the ports of the eastern African coast from Somalia to the Mozambique Channel. Hubs for the exchange and integration of people, goods, and ideas from inland Africa with the Indian Ocean world, which will give rise to the Swahili city-states—Lamu, Pemba, Gedi, Zanzibar, Mombasa, Kilwa, Mogadishu, Paté, Malindi. Effervescent spaces of cultural production and social resilience that, while adopting some features of traditional Islamic cities with the Islamization of the region at the end of the eleventh century, will maintain unique traits of a common culture and language, reflected in their architecture and hybrid urban forms (Gensheimer 2004). Cities characterized by their multicentric spatial and social organization (Baumanova 2022 as cited in Van Oostrum 2024, p. 144), with narrow streets and alleys in apparent labyrinthine disorder, characteristic of Arab cities, organized into neighborhoods, *mtaa*,[14] which, unlike Arab neighborhoods, reflect the stratification of their society through residential hierarchy: permanent stone constructions of the ruling class and wealthy merchants, and adobe houses with thatched roofs of the lower class, primarily the indigenous African population (Gensheimer 2004).

When the Portuguese arrive on the eastern African coasts at the end of the fifteenth century, they will destroy within just a few years the prosperous medieval Swahili cities—populous, bustling cosmopolitan cities integrated into a dense intercontinental trade network that had been consolidated over centuries.

To the coasts of North Africa will arrive from Canaan (the Near East) and through the city-states founded by its sailors—Tangis, Mogador, Lixus, Utica, Tapso, Tripoli—the Phoenician urban culture. This culture will remain relatively isolated from the indigenous *Amazigh* (Berber) culture. The Tyrian city of Carthage (in present-day Tunisia), founded in the ninth century BCE and strategically positioned

[11] In the fourth century BCE, Alexandria is the most populous city in the world, with about 300,000 inhabitants (Chandler 1994a, p.13).

[12] Kerma (2500 BCE to 1500 BCE) is one of the earliest cities and centers of power in the ancient Kingdom of Kush, located in what is today Sudan, it is the capital of the Kerma civilization. Meroë (800 BCE to 350 CE)—situated further south than Kerma, also in the territory of present-day Sudan—will succeed Napata as the capital of the Kingdom of Kush and will become famous for its iron industry, unique architecture, and pyramids.

[13] The city of Axum is already described as a 'metropolis' in *Periplus of the Erythraean Sea*, which dates from the first century CE.

[14] The term for 'neighborhood' in Swahili language.

as the nerve center of the network of Phoenician satellite cities, will, after gaining independence from Tyre (sixth century BCE), establish its commercial hegemony in the Mediterranean and the northwest African coast—Mauretania and Numidia—where it will found new urban settlements until its fall to the Roman Empire in 146 BCE.[15]

In northwest Africa, the Roman Empire promotes a network of cities in its African provinces, with their forums, markets, aqueducts, temples, baths, libraries, stadiums, and cemeteries. This process of urbanization stagnates (with some cities even disappearing) as the Roman Empire declines and ultimately collapses in the fifth century CE (Freund 2006).

From the eleventh century onwards, Morocco, Algeria, and Tunisia, under Berber[16] political control, experience a period of three hundred years of economic development accompanied by a process of urbanization. Cities or small merchant states—Tunis, Algiers, Meknes, Fez, and Marrakesh[17]—are precursors to the later caravan trade across the Sahara (Iniesta 2001). Between the sixteenth and nineteenth centuries, the Ottoman Empire, which will dominate the northern coast of Africa—Egypt, Tunisia, Libya, and Algeria (as well as Eritrea and northern Ethiopia)—will leave its traces on cities such as Tunis, Tripoli, Cairo, and Algiers.

Archaeological remains in the Niger River Basin provide evidence of urban settlements dating back to the first millennium BCE. The remains of the walled city of Jenné-Jenno (250 BCE to the ninth century CE) offer proof of pre-Arab interregional riverine trade, a testimony to ancient and autochthonous urban life (Coquery-Vidrovitch 1995). In the West Africa region, significant trans-Saharan trade will influence the development of important urban centers in various kingdoms and empires, though they will not match the Maghreb cities in terms of population.

During the period of the Ghana Empire (fifth-thirteenth centuries CE), its capital, Kumbi Saleh, together with other cities in the Niger Bend region, such as Djenné and Gao, will be significant commercial centers of the time, to which trans-Saharan caravan routes will reach from Egypt, Algeria, or Morocco. It will be succeeded by the Mali Empire (thirteenth-sixteenth centuries), which will extend to the Atlantic, with Niani as its capital and Timbuktu (a city founded around the eleventh and twelfth centuries, and annexed to the Mali Empire in the fourteenth century) as an important commercial and intellectual city that will survive the collapse of the empire. Following this will be the Songhai Empire (fifteenth-sixteenth centuries), which will revive the city of Gao and briefly dominate some Hausa states in the sixteenth century (Chandler 1994b)—politically organized since the tenth century into city-states (such as Kano, Rano, Katina, Zaria, Zamfara, Daura, and Gober,

---

[15] As a Roman colony, the city of Carthage revives its splendor around the first century CE. In the fifth century CE, it will be conquered by the Vandals from the Iberian Peninsula and will become a province of the Byzantine Empire in the sixth century CE until it falls into the hands of the Muslim Arabs in 698 CE.

[16] The Almoravid Empire (eleventh century), led by Zenaga Berbers of Western Sahara, unifies the territory (which will later extend to al-Andalus) and implements Maliki Sunnism (a variant of Islam). They will eventually be overthrown by the Almohad Caliphate.

[17] The cities of Fez and Marrakesh each will have populations exceeding 100,000 inhabitants.

among others). The Songhai Empire will be invaded by Morocco at the end of the sixteenth century, which will move the capital to Timbuktu.

The Yoruba urban civilization (spanning parts of Nigeria, Benin, and Togo) secures its political and economic dominance in the region between the seventh and tenth centuries through the expansion of urban centers southward. The first Yoruba Empire will be the Ife Empire, which founds its mother city, protected within a circular wall, in Ile-Ife (between the tenth and eleventh centuries). The Benin Empire (thirteenth-nineteenth centuries) will have its capital in the walled city of Benin, a center of a commercial network monopolized by the king, which will come into contact with Portuguese trade in the fifteenth century. The Oyo Empire (seventeenth-nineteenth centuries) will turn its capital, Oyo-Ile, into an important cultural and commercial center until its abandonment in the 1930s due to the decline in the slave trade. The constant conflicts and civil wars at the end of the nineteenth century in the Yoruba region will lead to the abandonment and founding of new cities for displaced persons and refugees, which will lose 'the living community character' (Coquery-Vidrovitch 1995) distinctive of the Yoruba cities: Ibadan (1829) will become the most significant city of that period.[18]

The Asante Empire (seventeenth-twentieth centuries in Ghana and parts of Ivory Coast and Togo) will also use urban expansion to extend its dominance: its cities will become centers of power and trade. The capital city, Kumasi (seventeenth century), strategically located for trade with Europeans between inland trade routes and the Atlantic Ocean (Hull 1976), will become a dominant cultural center. After a fierce defense, the city will be destroyed by the British in 1874, and the Asante Empire will become part of the British colony of Gold Coast (1902).

The Arab conquest and the subsequent process of Islamization (both religious and cultural) in North Africa will begin in the seventh century in Egypt with the foundation of the Muslim capital Fustat (which will be integrated into the new capital, Cairo, three centuries later). Near Carthage, the Arab Muslims establish the city of Kairouan (Tunisia), the central seat for the governance of their African territories, which will become an important and prosperous cultural, religious, and intellectual center, attracting Africans, Europeans, Persians, Berbers, Jews, and Christians (Carreño and Ana 2019). Muslim expansion will extend across the Maghreb—with Fez as the most representative city—to the Atlantic coast of Morocco and the Senegal River Basin, and will advance southward, inland into the Sahara, hand in hand with trans-Saharan trade, transforming important cities in the Niger River Basin like Kano, Djenné, and Timbuktu under the rule of Islamic elites. These cities will adopt some characteristics of Islamic cities, but will retain elements of pre-Islamic urbanism (Freund 2006:26): *madinas*, mosques, *madrasas*, baths, narrow and labyrinthine quarters or *hara*—lively and colorful with hundreds of small shops and workshops organized by guilds—(Freund 2006) will coexist or merge with purely African urban elements.

[18] By the end of the nineteenth century, at least nine cities in the region will have populations of over 50,000 inhabitants (O'Connor [1983] 2012): Ibadan will reach 200,000 inhabitants.

Between the eleventh and fifteenth centuries, in Central Africa and within a mainly rural subsistence economy, the Kingdom of Zimbabwe represents the center of an urban civilization that emerges from the development and control of trade with the Swahili city-states, artisanal specialization, and mining monopolies (gold, copper, bronze). The architecture of large stone blocks arranged (without mortar) in circular walls enclosing the *kraal*,[19] where adobe dwellings or *daga* are distributed, characterizes Mapungubwe, Great Zimbabwe, and Khami cities. Numerous indications of such stone structures are found scattered from the Zambezi River to the Limpopo,[20] possible evidence of a stone-built urban civilization in the region.

In the Great Lakes region, inhabited by agro-pastoral societies dispersed across the territory, urban traces dating from the thirteenth century are found in the capitals of their small kingdoms, where the ruling aristocratic class will monopolize trade with the eastern coast. Extensive radio-concentric configurations around the royal residence or *ibwami*, generally situated on a hill, are radially connected to residential quarters (*insago*), where, replicating the rural model, clan or lineage chiefs and their subjects settle in their *kraal*. This circular spatial organization consciously represents the social order.

In the western region of Central Africa, the cities of the powerful Kingdom of Kongo (fourteenth–nineteenth centuries) and the Kingdom of Loango (sixteenth–nineteenth centuries)—and their respective capitals, Mbanza-Kongo[21] and Mbanza-Loango—share the same urban layout and political organization: The city (*mbanza*) centralizes power and controls the resources obtained from the surrounding villages (*mabata*), where slaves perform productive labor. The arrival of the Portuguese in 1491, the establishment of Christianity as the kingdom's religion, and the slave trade will leave their mark on the cities and alter the region's urban development.

10.4 Urbanization and Atlantic Slave Trade

The long period of the Atlantic slave trade, just as Islamization or the conquests of African territory by other civilizations did in their day, will act as a (sometimes violent) catalyst for processes of fusion, resilience, and cultural, social, economic, and political adaptation, leaving its mark on urban areas.

The new global economic order into which Africa became immersed starting in the sixteenth century will be reflected at the urban level. Between the fifteenth and seventeenth centuries, during their first contacts with Africa, the dominant European mercantile empires established a series of strategic points along the Atlantic coast of Africa for trade with Africans (manufactured goods in exchange for raw materials and

---

[19] Settlement of huts arranged in a circle with lifestock kept in the central space.

[20] The stone ruins of the town of Mapungubwe on the banks of the Limpopo, together with other archaeological remains along the river, provide evidence of the possible extent of Shona urban culture (Coquery-Vidrovitch, 1993:80).

[21] The Portuguese will later rename it São Salvador.

slaves), where they built fortresses—Gorée Island and Saint Louis (Senegal), Cape Coast (Ghana), Cape Town (South Africa), San Sebastián (Mozambique Island)—and small port cities like Elmina (Ghana) and São Paulo de Luanda (Freund 2006). These enclaves, protected by walls or natural geography, will develop into agglomerations of buildings without spatial regulations until the nineteenth century, when European authorities, faced with disorder and persistent sanitation issues, reorganize them with orthogonal streets and permanent and durable constructions that will contrast with the 'winding' streets and perishable native structures (Sinou 1981).

A warlike-commercial culture (Coquery-Vidrovitch 2005b) dominated by the slave trade—the cornerstone of the triangular trade—from the seventeenth century until the abolition of slavery in 1889, that will produce negative effects (if not 'catastrophic' (Mabogunje 1994, P. xxv) ones) by eroding traditional political institutions, instigating conflicts and local wars, and catalyzing significant structural changes in African societies (Frances Mobley 2015). Four hundred years of *Fanga*,[22] involving material and moral destruction, will leave a traumatic imprint on the African cultural world (Iniesta 2010).

New cities will emerge either as completely new constructions or built upon preexisting settlements and small towns, modeled after European cities, such as Elmina and Saint Louis, or established based on African urban models, like the Yoruba in Lagos (Nigeria). Some cities will adopt an eclectic urban style that combines European and African influences, along with Latin American and Indian elements brought from other European colonies, showcasing new architectural forms that reflect the new lifestyle of Europeans and the emerging African elites (Freund 2006, p. 55). Other cities will present particular idiosyncrasies and urban image—such as Saint Louis, Cape Coast, Elmina, Ouidah, Bonny, Old Calabar, Loango, Malemba, or Cabinda—while urban centers along the Zaire River Basin (now Congo River) will, in the nineteenth century, maintain their traditional layout, orientation, and social and political organization based on African traditional knowledge (Anderson and Rathbone 2000, p. 4).

Paradoxically, the end of the slave trade and the 'regulation' and reorientation of commerce in the mid-nineteenth century[23] will result in greater European influence on the layout of some Atlantic cities, such as Dakar, Freetown, Accra, Lagos,

---

[22] The *Fanga*, as anthropologist Ferrán Iniesta (2010) explains (p. 91), is a Bambara term that refers to unlimited force, without restrictions, without moral constraints, an attribute of the adult male, of the biological father (*Fa*) capable of both procreation and protection or repression. The unleashing of this force poses a danger to individuals and collectives: Controlling the *Fanga* is essential for a community to function without excessive shocks. The slave trade unleashes and enthrones the *Fanga* in large regions of the African continent.

[23] By the end of the nineteenth century, the native cities of Southern and Eastern Africa, although not comparable in scale to those developed in Western Africa, will become important centers in their respective regions: Ankober and Addis Ababa in Ethiopia; Mombasa and Lamu in Kenya, and Zanzibar, Lindi, and Bagamoyo in Tanzania, as centers of Indian Ocean trade, with both floating and permanent Arab population; or inland cities like Taveta and Baringo in Kenya, Kazembe in Zambia, and Sonjo in Tanzania, which will become strategic for supplying caravan routes. In the southernmost area of the continent, urbanization is largely considered a product of European colonization.

and Luanda. These cities, having become important urban commercial centers, will experience economic and population growth at the expense of smaller nearby towns.

10.5 Spatial Representations of European Colonialism

Over centuries, European travelers, explorers, missionaries, merchants, and conquerors—much like the Greeks, Romans, and Arabs did—will draw the maps of Africa, name its places, narrate its history, and describe its geography, its cities and its ways of life using European languages. By using sketches, names, images, and words that are not native to those lands and their people, they will differentiate, construct, and signify African nature and spaces. But assigning the names ‘city’, ‘town’, ‘village’, ‘square’, or ‘street’ to unfamiliar realities can be inappropriate when these names evoke images of other realities; when they serve as anchors, familiar words that turn the unknown into temporary places of habitation or orientation (Carter 1988, cited in Tuan [1991] 2018, p. 120) with a tendency toward dominant permanence.

Naming, mapping, or narrating (with words or images) is power: the creative power to bring something into existence, to make the invisible visible (Tuan [1991] 2018) , and vice versa; it is not a neutral act, as it is grounded in the subjectivity of perceived, felt, and conceptualized experience. Words and maps end up completing the landscape, becoming part of it, and in doing so, inevitably altering it; the interaction between environment and mind is transfigured into new patterns of action, perception, and thought. European urban colonization will come to superimpose (impose) its subjective imprint, its habitational anchoring, onto an African landscape that becomes physically and mentally ungraspable to them.

Thus, originating from its Western capitalist societies and globalized through its colonial expansion, Europe will put into practice in Africa a spatial production developed on the basis of nineteenth- and twentieth-century European urban contexts and its social and cultural models. An urban spatiality sustained by Western urban theory and praxis, and endowed with its corresponding institutional, legal, and regulatory frameworks produced and managed by European experts and professionals.[24] The city is pre-designed in the (European) planner's[25] mind as a whole; the future is anticipated, and an urban image to be achieved is established based on abstract (European) needs estimated in the (European) present. Africans will be arbitrarily assigned to live in predetermined spaces without the possibility of adopting other patterns or directions.

Urban planning will be the tool that, implemented in the nineteenth century in Europe to redesign cities like Paris or London on a large scale, will guide colonial

---

[24] The machine-like efficiency of industrialization transferred to the production of cities and the design of the social ‘parts’ determined by a pre-visualized urban whole (Sennett 1973, p. 79).

[25] The urban ‘messianic-scientist’ who ‘with limpid innocence and absolute knowledge orchestrates the human condition’ (Koetter and Rowe 1980, p. 111).

urban practices and policies. Urban and architectural forms and spaces that, driven by European spatialities and representations of reality, institutionalize the idea of the city ('the true city') and its conception ('the city as a project' of a frozen future). For European colonists, the cultural senses of spatial organization and everyday spatial practices of Africans denote a lower (non-urban) level of spatiality that ascribes a superior order to the European city. This self-deduced and self-attributed spatial superiority—which actually reveals a lack of willingness and capacity to understand other ways of seeing, feeling, imagining, thinking, living, and making cities—will intellectually (civilizationally) justify the geographic and racial segregation of the colonial urban space.

Thus, unlike other forms of urban colonization deployed in African history, European colonization is implemented through spatial representations with intellectually elaborate systems of signs[26] that, with the aim of 'ordering', controlling, and dominating the African territory, will reaffirm European intellectual (and racial) superiority and the dominance of geopolitical structures of knowledge production, granting Western logos and Cartesian postulates universal status. Western urban theories and models will hegemonically dominate (and colonize) the ways of making cities, eclipsing and discarding other cultural ways of conceiving, planning, and managing the urban environment.

10.6 Colonial Urban Disruption

The colonial partition of Africa at the Berlin Conference[27] (1884–85) marks the milestone from which the continent experiences the most significant development of newly established cities and the changes—radical in some cases and subtle in others—to some existing ones: cities serving a minority European population in both

[26] The geographical map of Africa with the new borders resulting from the division of the continent among European powers could be considered the largest-scale example of it. The colonial borders set during the Berlin Conference will regroupe or divide ethno-cultural groups without consideration, provoking the confrontation or separation of communities. The metropolises will use ethnicity—under the premise of 'divide and rule'—and clientelist ethno-networks to gain the support of local authorities (Ruiz-Giménez 2000).

[27] The partitioning and establishment of sovereignty rights over administratively and/or militarily occupied (and to be occupied) African territories by European powers—the Kingdoms of France, Great Britain and Ireland, Portugal, Belgium, Spain, Italy, and the German Empire—will be agreed upon and formalized at the Berlin Conference (November 15, 1884, to February 26, 1885). In addition to the seven interested European powers, the conference will be attended by a limited international representation including the United States, the Ottoman Empire, and five European countries with no direct interest in the African partition. African representation will be nonexistent, evidencing the absolute disregard of the convened countries for the perspective of Africans regarding who, when, and how (and why) their territories and peoples will be divided, occupied, and administered.

form and function (Hanna and Hanna 1981, p. 18) that will gradually form a colonial urban network organized hierarchically around macrocephalic cities or nodes—administrative, transportation, or social surplus centers (Potter and Lloyds-Evans 1998, p. 123)—located in strategic places near markets, railways, roads, resource extraction centers (plantations or mines), or on the coast. Gears for a perfect system of 'accumulation, legitimation, and domination' (Myers 2003, p. 21). A colonial network that shapes a new and varied urban landscape in which the colonized will form an African population of temporary migrant workers who will settle in the cities either permanently or semi-permanently as significant economic or political changes occur (Pons 1969).

Colonial cities are built from scratch or by consolidating native settlements, European military fortifications, or pre-colonial ports; following predefined site plans or, more frequently, through successive unplanned building additions (Nunes Silva 2015, p. 8). The initial layouts of colonial cities tend to adopt a grid configuration upon which designs and architectures prevalent in the respective metropolises are replicated. The early site plans, mainly drawn up by European military engineers stationed in the territory, will be succeeded, starting in the first decade of the twentieth century, by formal urban designs developed by architects and civil urban planners (almost always) from the metropolises.

The main references of this period will be the British Howardian garden city model of low-density and scattered green spaces, and the French hygienist Haussmannian model[28] of broad tree-lined avenues, squares, and parks. These urban models will be transferred into African contexts seeking to reproduce metropolises lifestyles and urban services for European colonists. They will later be replaced by the principles of modernist urbanism with 'universal applicability' formalized in the fourth *Congrès Internationaux d'Architecture Moderne* (CIAM 4) Athens Charter[29] of 1942: a functional city with linear and zoned organization. Africa becomes a vast 'virgin' territory for urban experimentation and the development of anticipatory and prefigured city projects. During the interwar period, the British, French, Italians, and Portuguese will introduce the long-term comprehensive territorial control paradigm of urban planning—the Master Plan—to design the cities of some of their colonies. The Master Plans for Casablanca, Algiers, Tunis, Tripoli, Nairobi, Accra, Lusaka, Dar es Salaam, and Luanda correspond to this period. In the years following colonization, master planning will proliferate, evolving into a tool for identity reinforcement and development.

Colonial planning practices, along with legislative and policy reforms in urbanism and housing, will not only be decisive for the development and physical formalization of their cities but also for the geopolitics of their colonial empires. Urbanism used as an instrument of power (Foucault 1977; Scott 1998; Yiftachel 1998; Dovey 1999;

[28] As in Paris Plan of 1853, this 'hygienistic' urban design—characterized by regular geometry, with wide, straight, and tree-lined boulevards and avenues converging in large squares—will intend to facilitate the movement of colonial troops within the city as well as the control and surveillance of the colonized population.

[29] Published in 1943, ten years after the CIAM 4, this urban manifesto, inspired by Le Corbusier's *Cité Radieuse*, will become central to the ideological foundation of the Modern Movement.

Njoh 2009) will facilitate control—both demographic and spatial—and surveillance of the racialized and Subdued African population through order and spatial zoning, ensuring a reserve of cheap (confined) labor at the service of the capitalist economy introduced by Europeans.[30] Thus, colonial urban projects will reflect the geopolitical strategies deployed by the different European powers: Germans and British will maintain the autonomy of the political, religious, and familial structures of the colonized by interposing an acculturated African social group (kings, clan chiefs, and bourgeois elite) that, as a hinge with the native societies, will introduce the objectives of the colonial machinery (Iniesta 2001, p. 108). The French and Portuguese, guided by a more centralist and assimilationist political culture, will suppress this mediation.

The French assimilation policy will aim to integrate, from its hegemonic particularism, the twenty million Africans from its colonies[31] (the *originaires*[32]) as citizens of the French Republic. However, these citizens will have their rights curtailed by legal and socio-cultural barriers (Bigon 2016): full assimilation will never be implemented in practice (Crowder [1967] 2023) . Direct administration will tend toward an associationist rhetoric and the disabling of native institutions deemed inefficient. In the built urban environment, colonial power and control will be reinforced: through urban policies and city regulations[33]; by erecting large infrastructures and architectures—"stones of imperial memory" (Demissie 2016); through the toponymic appropriation of places and memories; and, finally, (setting aside its rhetorical principle of integration and avoiding the 'dangerous proximity' of Africans (Attahi et al. 2009, cited in Nunes 2015, p. 17) by separating the population into neighborhoods for Europeans—*les villes nouvelles* or *villes blanches*—and *les villes des indigènes* or villages (Bigon and Njoh 2018). This urban duality will be justified as 'cultural segregation' between two different residential forms (Bigon 2016), and as a (pseudo-)sanitary measure against the 'diseases of Africans': The white neighborhoods will occupy, surrounded by a 'sanitary cordon'—green belts, vacant urban areas, militarized zones, or physical barriers such as rivers or railways—, the best locations in the city, generally in elevated and well-ventilated areas, as will be seen in the cities of Nouakchott or Dakar (Demissie 2016; Nunes Silva 2015).

---

[30] From the sixteenth century onward, European powers establish mercantile capitalism in Africa, fueled by slavery. By the nineteenth century, this will evolve into industrial capitalism, replacing slaves with the exploitation of cheap African labor and resources in the colonies.

[31] French colonies in Africa will include: Algeria, Tunisia, and Morocco in North Africa; Mauritania, Senegal, Guinea, French Sudan (now Mali), Upper Volta (now Burkina Faso), Niger, Togo, Côte d'Ivoire, and Dahomey (now Benin) in West Africa (AOF); Gabon, Middle Congo (now the People's Republic of the Congo), Ubangi-Shari (now the Central African Republic), and Chad in Central Equatorial Africa (AEF); French Somaliland in the Horn of Africa; and the non-federated territories (TNF) of Togoland, French Cameroon, and Madagascar.

[32] The *originaires* who attain higher education will grant the status of *évolués*.

[33] The prohibition of native cultural practices, the denigration of local construction materials, along with the adoption of European urban and building codes and regulations, curfews imposed on non-Europeans, European systems of land acquisition, parceling, and distribution, as well as socio-economic restrictions on urban land access…

The indirect administration of governmental structures parallel to those of the native population that the British Empire imposes in its colonies[34] as a perfect control system will be translated at the urban level—using the 'method of divide and contain by codifying a visible hierarchy of spatial order' (Mitchell 1988, p. 33)—into dual cities: a city built only for European colonists (with restricted access for Africans) physically separated from the 'native reserves' where Africans will be confined, again by means of 'building-free zones' (BFZ) or other kinds of sanitary cordons. While the former will be provided with infrastructure and basic services, the latter will consist of simple barrack settlements or endogenous settlements reconstructed by the inhabitants themselves, lacking infrastructural facilities. A socio-racial and geographic-spatial hierarchy (in the colonies of Eastern and Southern Africa and Nigeria, a third level of sectorization or hierarchization for intermediate racial groups not European will be included) driven by the 'sanitary syndrome' (Swanson 1977)—"racist in orientation and segregationist in practice" (Mabogunje 1981, p. 74)—will be established in Lagos, Accra, Nairobi, or Freetown.

The Portuguese,[35] who initially advocate a policy of assimilation and integration in their colonies, will not consider racial segregation as part of their urban policies. However, from 1945 onward, in an attempt to prevent independence movements, they will exacerbate racial segregation in Angola and Mozambique. During the interwar period, the Germans[36] and Italians[37] will deploy a colonial urbanism with strong intra-urban racial segregation policies: extreme in the Italian colonies of East Africa and Libya, and explicitly manifest in the urban plans of the German colonies. The smaller colonial powers, such as Belgium, Italy, Portugal, and Spain, although they will deploy their own urbanization mechanisms, will take as references the French and British colonial urban models.

And if race and urban zoning are the main determinants of social order and residential patterns in colonial cities, without a doubt, the most extreme case of these racial urban policies is found in the colonies of the Union of South Africa and Namibia during the *apartheid* regime[38] and its distinctive urban form that will confine non-whites (Blacks, coloureds or mestizos, and Indians) to the townships on the outskirts of the cities. The imprint of this urban apartheid—the apartheid city—remains visible today in South African cities such as Johannesburg, Pretoria, Cape Town, or Windhoek.

[34] British colonies in Africa will include: Nigeria, the Gold Coast (now Ghana), Sierra Leone, and Gambia in West Africa; Kenya, Uganda, British Somaliland, Zanzibar, Sudan, and Egypt in East and North Africa; as well as the Union of South Africa, Botswana, Malawi, Zambia, Zimbabwe, and Lesotho in Southern Africa.

[35] Portuguese colonies in Africa will include: Angola, Mozambique, Portuguese Guinea, Cape Verde, Cabinda, Sao Tome and Principe.

[36] German East Africa (1884–1919) encompasses present-day Burundi, Rwanda, and Tanganyika; German South West Africa, now Namibia; and German West Africa, which includes the territories of Kamerun and Togoland.

[37] Italian colonies in Africa will include Libya, Eritrea and the Italian Somalia.

[38] Established in 1948 by the National Party, apartheid will favor a white minority that represents only 21% of the population.

"In the late 1890s, segregation escaped from the laboratory (as it were [staid bourgeois professional's synonym for 'separation']) and transformed itself into an electric political rallying cry. You could hear it in cities almost everywhere" (Nightingale 2012, p. 164). Intra-urban social and racial segregation will be a highly complex and heterogeneous phenomenon in its implementation[39] (even under the same colonial power[40]), but it can be concluded that in all colonial cities, colonial urbanism will follow criteria of urban zoning (with restricted access for Africans), with their residential zones subdivided—following criteria of racial, 'cultural', or pseudo-sanitary segregation—into officially planned (and 'modern') areas for Europeans and 'spontaneous' development areas for natives, to be built and equipped in an autonomous and precarious manner[41] (O'Connor 2012), without order or plan from the European subjectivity. The degree and nature of interdependence, mobility, and physical separation between these subdivisions will vary from one city to another: in Kano, Khartoum, Bamako, or Brazzaville, the physical distinction will be clear, while Lagos, Kumasi, or Ouagadougou will be hybrid cities with more blurred boundaries.

In the transition from theoretical abstraction ("the city as a project") to execution (which does not always occur), colonial urban projects, with their forms of intra-urban racial segregation, will undergo unforeseen adjustments and alterations. The urban configuration may vary depending on cultural, social, or topographical contexts, the interests, divergences, or internal conflicts (administrative or bureaucratic) of local colonial governments, or the different degrees of resistance from the colonized Africans[42] (which, if present, will manifest in various forms: protests, strikes, disregard for eviction or displacement orders, appropriations, or changes in the use of planned spaces).

Furthermore, urban planning practices will not follow only the unidirectional routes of knowledge and communication from the metropolises (center) to the colonies (periphery). Reverse feedback will also emerge from these African territories of urban experimentation; they will not be limited to a simple 'copy and paste' but will involve both transnational (arriving through different *modi operandi* of dissemination) and local forms of 'editing' (Beeckmans 2013). Thus, the colonial city will

---

[39] Racial segregation is imposed in urban plans from the outset (as in Lusaka, Nouakchott, or Abidjan) or adopted —justified in various ways— on pre-existing racially mixed urban fabrics, by forcing the displacement of native populations to new, undeveloped sites to be developed 'in their own way' (such as Medina in Dakar, Pumwani in Nairobi, Bacongo and Poto-Poto in Brazzaville, or the extreme cases of townships on the outskirts of South African cities during apartheid, like Soweto in Johannesburg).

[40] In French Equatorial Africa, racial segregation in its cities will be more explicit than in West Africa.

[41] The stigma attached to peasants temporarily migrating to the 'strange' urban environment, and thus being perceived as transient, will be used to justify economically and demographically the exploitation of African labor and the lack of investment in native-populated areas of colonial cities (Coquery-Vidrovitch, 2005b). These 'measures,' intended to discourage rural exodus and control urbanization, ultimately will prove (with some exceptions) ineffective in the face of African cultural, social, and spatial dynamics.

[42] French assimilation policies will leave more room for active protests compared to British or Belgian policies, where passive resistance is more commonly observed (Beeckmans, 2013, p. 619).

become a privileged space for encounters and syntheses of traditional values reformulated under the action of Western incentives: colonization, which entails either forced or optional adaptation of the colonized to a new order (legible to the colonizer), gives rise to a constant and ever-dynamic process of popular mediation from which new ideas, techniques, and knowledge emerge (Coquery-Vidrovitch 2005b, p. 19–20).

The colonizer-colonized bidirectional relations and exchanges, with their complex processes of social and urban *mestizaje* and their multiple forms of resistance, negotiation and appropriation, will also elude the spatial planning (with all its systems of signs) deployed by the colonial powers as a tool for management, control and social reconfiguration (Nunes Silva 2015, p. 3). The colonial (social and urban) reconfiguration objective will be achieved mainly and mostly in the physical appearance of cities—architecture, landscape, and design. The different modes and degrees of interaction between colonial and endogenous urban planning practices will produce distinct and even divergent results (Van Oostrum 2024).

Thus, colonial urban hegemonic projects will be incomplete transformative processes due (in different conditions and degrees) to necessary internal contradictions and the elusive persistence of the urban majority: internally and necessarily, colonial governments will preserve social and cultural differences that justify their dominance over the Africans 'subaltern'; and externally and elusively, the colonized, as the only alternative to lack of governmental support and uneven distribution of resources, will maintain (evading the colonial order) the 'persistent disorder' (Myers 2003, p. 161) resulting from their local cultural senses of spatial organization. Neither Africans will be passive recipients of ideas, nor will colonial spatial practice be unidirectional: It will involve multiple and ambivalent spatial, top-down and bottom-up, processes and interactions (Bigon 2016), as well as linkages and flows of people, ideas, products, processes, and patterns operating over, through, between, and beyond political systems and societies.

If European colonizing urbanism subsumes the concept of urbanization under modernization, exacerbating the traditional-modern, indigenous-colonial, or native-civilized (or Western) dichotomies, its spatial order will collide with a pre-colonial urban reality based on African foundational reasons, with its own urban traditions and aesthetics, and its ways of understanding space and reading the landscape. Relegated to the realm of the 'traditional', Africans will build their own cities or neighborhoods from this standpoint, improvising and devising new survival strategies by combining the old and the new: construction techniques, systems of solidarity, work, and beliefs (Coquery-Vidrovitch 2005b). A 'collective effervescence' (Macamo 2013, p. 296) that displays its capacity for adaptation and creative potential to generate new social structures with their own particular urban spatial forms, which colonizers, unable to see and understand, will categorize as 'informal', that is, those that do not conform to or submit to Western rules. 'Informality' as an urban (sub)category emerges when Western spatial order is hegemonically imposed on urban Africa. But does what we call informal not enclose another cultural way of imagining and creating the city? (The very prefix 'in' negates or deprives); Does it not carry another notion of 'city as a project'?. Failing to at least pose this questions would contribute to the invisibilization of an urban reality that has not only endured right up to the present day in African cities but also represents and has represented the urban habitat of millions of Africans.

Africans will consequently have to face the challenging stage of independence and their aspirations for an Africanized and cosmopolitan modernity with their urban forms of social, economic, and kinship organization either weakened, distorted or stimulated by colonial rule and its urban planning systems (spatial, normative, institutional, and managerial).

10.7 City Projects for the 'Liberated People': Beyond Colonialism?

Designing a dream city is easy; rebuilding
A living one takes imagination
(Jacobs 1958, p. 131).

With African independences and their modern statist and developmentalist projects,[43] the city will become a key political stage—the stage of the 'liberated people'—for the transition from colonial to postcolonial Africa—to a world defined by themselves, free from external interference, where they will 'reappropriate their history' (Balandier [1954] 1985, vii) . Socially and economically, it will become a territory that generates opportunities, prosperity, and wealth, where they can 'manifest their capacity to achieve progress' (p. vii) and insert themselves of their own initiative into the international system. Thus, at the beginning of the 1960s,[44] an unprecedented urbanization takes off across the continent[45]: The pull effect of the

[43] Although Liberia's independence is recognized in 1847, South Africa's in 1910, and Egypt's in 1922, the 1950s mark the beginning of African independences: Libya in 1951, Morocco, Tunisia, and Sudan in 1956, and Ghana in 1957; from then until 1965, most other African countries achieve independence. By 1975, the last states gain independence: Angola, Mozambique, São Tomé and Príncipe, and Guinea-Bissau. Namibia, however, does not gain independence from South Africa until 1990, with some of its territories remaining under South African control until 1994.

[44] In the early 1960s, less than one-fifth of the African population (18.3%) lives in urban areas. However, this average does not reflect the regional exceptions of Southern Africa (41.9%) and North Africa (30%) compared to the much less urbanized East Africa (7.4%), though even here there are exceptions, such as Somalia and Zambia (UN 1993, p. 74).

[45] From 1960 to 1995, the average annual rate of urban growth remains around 4.61%, with no significant differences between regions. However, differences are notable between countries, with less urbanized nations like Mauritania and Rwanda experiencing high growth, while countries like Congo and Senegal, which already had large capitals—Brazzaville and Dakar—during the colonial period, see more moderate growth (Rakodi, 1997).

cities, together with the massive and multifactorial rural exodus,[46] will generate a 'perverse urban boom' (Davis 2007, p. 14) that the territorial administration systems inherited from the colonial era, in a dependent (neo-colonial) and weak economic situation—marked by failed industrialization[47] and a deficit in employment, housing, and services—, will not be able to manage. Reality will assert itself, and urban life will seek its ways to develop 'outside the plan'.

While the 'surplus population' (UN-Habitat 2003, p. 40)—labeled as social parasites and threatened with expulsion by governments and local and central authorities—self-constructs its urban forms of insertion into modernized and cosmopolitan African cities, national construction—National Development Plans (NDP)—is accompanied by large-scale urban projects aspiring to break with the colonial urban legacy (and its racial segregationism) and turning their cities into symbols of development, modernity, independence, and national identity. Paradoxically, instead of turning to their cultural ways of imagining and making cities, most of the new states will give continuity to the urban planning systems implemented by the colonial powers. Now managed by African bureaucrats trained in visions of the 'true and good city', they will remain dependent on international 'experts' and consultants disconnected from African urban contexts.

The construction of new capitals for the new nations will be an essential goal for the aspirational nationalist processes. Most of these capitals will be projected onto existing ones, perpetuating the macrocephalic and asymmetrical pattern of the colonial urban network, now reinforced by the centralist model of the new states—Addis Ababa (1959, 1965, Ethiopia), Abidjan (1962, Ivory Coast), Conakry (1963, Guinea), Kinshasa (1966, DRC), Dar es Salaam (1968, Tanzania), or Nairobi (1973, Kenya). Only a minority of the new capitals, due to historical, strategic, economic, or ethnic reasons, will be relocated to other existing cities or constructed *ex novo* at different sites[48]—Abuja (1979, Nigeria), Lilongwe (1965, Malawi), Dodoma (1976, Tanzania), Gaborone (1964, Botswana), Kigali (1970, Rwanda), or Nouakchott (1958, Mauritania). Visions of future will be projected in the form of master plans imbued with different ideologies: the expectant modernization (with the revival of development engineering) featured in master plans like Abuja's (1979), the new capital of Nigeria, or the return to the authenticity of pre-colonial ways of life (with the 'invention of tradition' (Hobsbawm 1983, cited in Beeckmans 2018, p. 2.)) seen

[46] The rural exodus is accelerated by de-agrarianization—understood as the gradual decline of agricultural activities' contribution to income generation in rural areas (Bryceson, 1996, p. 99)—primarily due to market deregulation, civil wars, the implementation of agrarian policies that prioritize intensive monoculture production for export over subsistence farming, reliance on fluctuating prices, lack of technological and infrastructure investments, protective subsidies, agricultural imports, droughts, and other environmental factors. Additionally, cities are perceived to offer better economic opportunities, as well as improved access to education, healthcare, and social services.

[47] With notable exceptions in the northern and southern regions.

[48] Abuja (Nigeria), Lilongwe (Malawi), and Dodoma (Tanzania) will become new capital cities, while Gaborone (Botswana), Kigali (Rwanda), and Nouakchott (Mauritania) will be entirely new city projects developed from scratch. The relocation of the Ivorian capital from Abidjan to Yamoussoukro will be also proposed. The considerable economic cost of these relocations proves detrimental to the development of other areas (Peil and Sada, 1984, p. 26).

in the master plan for Dodoma (1976). An 'ambivalent scenario' (Enwezor 2001, p. 12) where, regardless of the adopted ideological discourse, city projects will be carried out by international teams, with minimal local involvement and participation, and commissioned by central 'ad hoc governments' (Fox 2013, p. 2).

Speculative planning (due to financial, technical, and theoretical deficiencies along with inefficient urban governance) proved inadequate for anticipating and addressing the housing, services, infrastructure, and employment needs of a rapidly growing urbanization in a climate of economic and, in many cases, political instability. The elitist nature of building and land regulations (a colonial legacy) and master planning—'informalized'[49] to cater to the political and/or economic interests of a minority elite—will guide the transition 'beyond colonialism' for African cities: an 'oppressive force for the poor' (Berrisford 2011, p. 210) that drives a large majority of urban dwellers toward the social and spatial exclusion.

As a result, urban spatial segregation now becomes socio-economic: the formally and orderly planned (top-down) neighborhoods for the middle[50] and upper classes versus the 'spontaneous' and 'informally' self-constructed (from below) neighborhoods of the underclass. In the name of the public good, order, and modernity, urban policies in most of the new African states will tend to eradicate neighborhoods established 'out of plan', forcing the eviction of their residents, who, without housing solutions, will rebuild their neighborhoods in other locations.

Between 1972 and 1981, and mainly influenced by Turnerian thesis[51] of autonomous processual habitation, a paradigm shift takes place, and, driven by the World Bank, urban policies of laissez-faire and assisted self-help are institutionalized. Thus, in African countries receiving this conditional aid, the transition will be from (insufficient and inadequate) housing provision plans to ("prone to 'leakage'" Stren 1990, p. 49) site-and-service plans; and from the 'vicious elimination' of informal settlements to their (controversial) upgrading. These self-management and self-construction aid policies, aimed at addressing the problem of unregulated and

---

[49] Urban planner Ananya Roy (2009) uses this term in reference to urban systems inherited from colonial power in India, indicating their 'flexibility' based on the political and financial interests of those in power (cited in Watson, 2011, p. 205).

[50] The African Development Bank (AfDB) defines the 'African middle class' as those with a daily spending capacity between US$2 and US$20 (UN-Policy Brief, 2023, p. 5). According to the World Bank, extreme poverty is defined as having less than US$2.15 per day for spending. It is evident that the so-called 'African middle class' —subdivided into three categories: floating, lower-middle, and upper-living— lacks the purchasing power to buy or rent homes in the planned areas of most African cities. As a result, much of this population lives in informal neighborhoods.

[51] Architect and urban planner John F.C. Turner's thesis emphasizes the progressive development that self-building processes in informal settlements can lead to, involving mutual aid within the community without the need for external 'expert' assistance. The inhabitants take control of their own decisions in the design and management of their homes (Turner 1976, pp. 177-179). Turner's ideas are influenced, as he himself acknowledges, by the work of Lewis Mumford, Patrick Geddes, and Peter Kropotkin (Harris, 2003).

uncontrolled urban sprawl, will often prove not only inefficient or counterproductive but also come into conflict with informal real estate systems.[52]

In the 1980s, indebted African countries (strangled by neoliberal precepts and austerity packages) will enter a 'silent period of spatial planning' (Amedzro, cited in Harrison y Croese 2023, p. 32) to be reawakened in the late 1990s with the Habitat Agenda Declaration (1996) and initiatives, supported by UN-Habitat, the World Bank, and Cities Alliance in partnership with International Development Agencies, for a more strategic, participatory, and decentralized planning led by 'good governance'. An intricate (sometimes turbulent, in both senses: murky and agitated) relationship between politics (for 'development') and urban and architectural design will characterize post-independence urban Africa, with significant growth in transnationally assisted development aid projects and a renewed importation of foreign urban models.

Therefore, in contrast to the 'ideal' postcolonial city projects, reality will present a stubborn 'unregulated and marginal' urban overgrowth that rational analysts and quantifiers of poverty and precariousness will measure in order to label African cities—and, more broadly, those of the Global South—as 'urban failures'. But to what extent is this quantified 'failure' merely the result of a modernity both promised and simultaneously denied[53]? Have not African city dwellers navigated, and continue navigating, this ambivalence as they transition toward a world defined spatially and temporally by themselves and their daily lives?

10.8 The Boom of 'African Urban Fantasies'

With cities, it is as with dreams: everything imaginable
Can be dreamed, but even the most unexpected dream
Is a rebus that conceals a desire, or, its reverse, a fear
(Calvino 1977, p. 44).

At the beginning of the twenty-first century, a series of factors converge in most African countries—a continuous process of urbanization, sustained economic growth, increased foreign investment, the rise of the (upper)middle class, and a certain degree of political stability—transforming the continent (especially in the aftermath of 2008 global financial crisis) into the 'last frontier' of development. A development that will be mainly driven in cities—the hungry engines of the global economy—through an urban planning at the service of capital in the form of mega-developments

[52] A real estate market that, from a 'formal' perspective, is referred to as invisible or pirate: Baróss, 1990, Mohan 1994; Davis, 2007.

[53] Mozambican sociologist Elísio Macamo (2005) identifies the transition from colonialism to African postcolonialism as an ambivalent process of resistance and adaptation, stemming from a colonial project that promised modernity but simultaneously denied it, as the 'detribalization of Africans' (Cooper, 1996, pp. 168–170) could threaten colonial interests.

aimed at enhancing global positioning and urban competitiveness: what economist Paul Homer has termed 'charter cities'.[54]

Master plans revive. The 'wave' begins in North Africa with New Cairo City (1997), a new satellite city designed for a population of five million inhabitants in Egypt.[55] It is followed a few years later by Morocco with the *Villes Nouvelles* projects (2004), an initial proposal of fifteen new cities as part of its national development strategy. Countries in other regions will soon join the wave, either proposing brand-new cities—such as *Tatu City* and *Konza Techno City* in Kenya, satellite cities integrated into the Nairobi 2030 Metropolitan Plan; *Hope City* in Accra, designed as Ghana's new tech hub; *Kigamboni New City* in Dar es Salaam (Tanzania); or *Nova Cidade de Kilamba* in Luanda (Angola); comprehensive action plans for existing cities, which must deal with complex urban situations and the interests of numerous stakeholders—such as *Kigali Vision City* (2011) in Rwanda or *Addis Ababa's Master Plan* (2014); or 'edge cities' (De Satgé y Watson 2018), mixed-use mega-projects inserted into existing cities—such as *Eko Atlantic City* in Lagos (Nigeria) or *La Cité du Fleuve* in Kinshasa (Democratic Republic of Congo).

They are already academically known as 'African Urban Fantasies' (Watson 2013), developed by international corporations and teams based on iconic images of contemporary cities from other regions (Dubai, Shanghai, Singapore, Hong Kong, Shenzhen) and Western lifestyles. The boom of fantasies is relentless[56]: labeled with hyperbolic terminology (and even personalized[57]), largely financed with private capital, and accompanied by significant and speculative infrastructure investments. Many of these mega-projects are totally or partially abandoned, delayed, or put on hold, entangled in complex plots involving various actors, differing interests or visions, issues of corruption or funding, environmental controversies, political or social resistance... All of them are preceded by major marketing campaigns promoted by political or economic interests. When the campaign ends, most remain on paper or on websites and, undoubtedly, engraved in the colonized urban imaginations of aspirational and aesthetically exclusive representations of 'modernity'.

The clash of rationalities[58] (Watson 2003)—already evident in both colonial and postcolonial cities—intensifies: The fantastical visions of urban planning not

---

[54] Economist and 2018 Nobel Laureate in Economics, Paul Homer, unveils his 'charter city' concept in a 2009 TED Talk —'Why the World Needs Charter Cities?': innovative cosmopolitan cities focused on emerging markets, centers of economic growth, and connectors of the global economy.

[55] Currently, the population census reports 297,387 inhabitants (CAPMAS 2017). '2017 Census for Population and Housing Conditions,' CEDEJ-CAPMAS. Accessed 21 February 2023.

[56] Nkwashi City, a satellite city of Lusaka (Zambia); Itana, near Lagos (Nigeria); Enyimba Economic City (Nigeria); Morgan City in Khartoum (Sudan); Ribat Aal-bahar (Mauritania); Moderfontein City in Johannesburg (South Africa); King Mohamed IV Green City (Morocco); Machakos New City (Kenya); Tunisia Economic City (Tunisia); Huangjing City (Ethiopia); El-Menia (Algeria); and many others.

[57] One example is Akon New City, a new city proposed in Senegal by the Senegalese-American musician and rapper Akon.

[58] A clash of rationalities documented in the 1998 report by the Consultation Commission, commissioned by the municipality of Cape Town, on the resistance led by the Women's Power Group against the 'formalization' plan for the 'informal' Crossroads settlement. This conflict will later be

only conflict with the urban reality of African cities but also dramatically highlight social and spatial inequality, redounding to exclusion and marginalization. The dreams (of accumulation) of a few are fueled by the nightmares (of dispossession) of many others: evictions, relocation, demolition, expropriation, destruction of social networks, ruination of ways of life and work...

Decades go by, and we continue to lack effective answers: Is there operational space between normative theoretical processes (and production) and the 'unregulable' ones? Is it feasible to conceive the city as a project in between different forms and meanings of urban order? Do we have the intellectual and ethical capacity to value differences and plan (imagine, build, and manage) the urban by integrating diverse or even opposing frameworks?

10.9 Beyond Decolonization: The Africanized City

Urban territory in Africa develops within a context of conflicting rationalities, where the normative theoretical forms of capitalist urbanism compete with those 'other' urban forms that inhabitants autonomously construct from their own daily dynamics, encompassing their ways of being, perceiving, and imagining the world, in which they integrate the contributions—Africanized—from centuries of encounters (and misencounters) with other peoples and cultures. A concatenation of lifestyles projected into the present, interweaving with global (and globalized) forms and dynamics filtered and modulated by their African wisdom—holistic, amodern, and meridional—and their 'popular' memories (Roca and Iniesta 2006). A rhizomatic cultural singularity that catalyzes their urban singularity: the Africanized city. With its challenges, yes, but also complexly autonomous, dynamic, creative, generative, adaptive, and flexible. Heterogeneous urban forms that the Global North-West 'informalizes' (or exoticizes), as they disrupt its colonizing norms, yet Africa normalizes within its systems of organization, exchange, control, regulation, and social reciprocity. A positive 'informalitude'[59] that reaffirms and updates the original experience of city-making; that refuses to conceive of an alternative within monological spaces of containment and seeks its own space of enunciation, free from binary oppositions.

A place of cultural reconfiguration—about which there is a certain consensus rooted in the collective imagination of the Global South-East and promoted, over the past two decades, by the southern turn in urban studies (Simone 2004; Pieterse 2008; Parnell & Watson 2009; Myers 2011; Obrist et al. 2013; Parnell & Pieterse 2014; Simone & Pieterse 2017; Yiftachel & Mammon 2022)—for urban particularisms that, from 'other' relational geographies, propose ways of imagining and

---

analyzed by South African urban planner Vanessa Watson (2003) as a common struggle in African cities between state or municipal formalization ambitions and 'informal' structures.

[59] The term "informalitude", instead of "informality", is used deliberately to attempt to reclaim and celebrate, much like the "Negritude" cultural and literary movement once did, the cultural and social forms of urban practices by Africans that colonialism relegated to a marginal place.

making cities that are specific to their systems of thought, their social dynamics and practices, their ways of perceiving reality, and their means of making the world intelligible. A process that must avoid any ideologized decolonial approach predisposed to another 'walled segregation in the particular' (Césaire 2006[60]) and move beyond a decolonization dissolved in the 'universal': transcending the zero point perspective[61] by recontextualizing (spatially and temporally) and reconceptualizing (from different levels of rationality) universalist learnings, while collectively incorporating the knowledge, narratives, and experiences articulated by subjects with diverse collective or individual identities from those particular loci of enunciation.

Therefore, proposing an approach of 'the city as a project' implies a unique way of anticipating, representing, and analyzing the city that distances us from that place of cultural reconfiguration and connects us to the idea of dominant planning governed by walled structures that technocratically spatialize, regulate, and manage orderly modern cities interconnected with advanced infrastructures. In Africa, this North-centric approach to the city can conceal or segregate a deeply African urban reality that requires an alternative conceptualization (reconceptualization) constructed from the perspective (and the (re)context) of the Africanized city.

Turning this entrenched conceptualization requires an exploration 'from below' of those 'other' urban forms, intertwining and hybridizing not only academic knowledge[62] and technological and scientific advances, but also collective knowledges and the ways of synchronizing territory, body, mind, and affections that inhabitants deploy in their daily lives to build and transform their urban habitats. Because the way of imagining our cities cannot be universal, and it is always exciting to perceive other forms of (urban) life to 'dis-cover' the diverse ways in which men and women make their world intelligible.[63]

---

[60] "There are two ways of losing oneself: by walled segregation in the particular or by dissolution in the 'universal'", writes Aimé Césaire (2006), as quoted in Grosfoguel (2008, p. 209).

[61] In Santiago Castro-Gómez's (2005) philosophy of point zero, "the epistemic (neutral) subject has no sexuality, gender, ethnicity, race, class, spirituality, language, or epistemic location in any power relation, and produces the truth from an inner monologue with himself without relation to anyone outside" (Grosfoguel, 2008, p. 202). Therefore, transcending the zero point requires taking in account the social and cultural dimensions of the knowledge-producing subject.

[62] From various disciplines: urban planning, architecture, anthropology, sociology, geography, psychology, environmental neuroscience, and more.

[63] «Cities in Africa should be studied in their own right. [A group of scholars working in Dakar, Dar er Salaam or Johannesburg] encourage us to think out of the box, to feel the pulse of city life and to encounter diverse ways of living, not with Western gaze but with sharp, open and fresh minds» (Obrist, 2013:12).

References

Amedzro K (2021) History and current politics of master planning in Accra. https://www.ucl.ac.uk/urban-lab/news/2021/mar/history-and-current-politics-master-planning-accra. Accessed 22 Oct 2021

Anderson DM. y Rathbone R (ed) (2000) Africa's urban past. James Currey, Oxford

Attahi K, Hinin-Moustapha D, Appesika K (2009). Revisiting urban plannning in the Sub-Saharan Francophone Africa. United Nations Human Settlements Programme-Regional Report Francophone Africa, Nairobi

Balandier G ([1954] 1985) Sociologie de Brazzavilles noires. Armand Colin, Paris

Baróss P (1990) Sequencing land development: the price implications of legal and illegal settlements growth. In: Baróss P, van der Linden J (eds) The transformation of land supply systems in third world cities. Avebury, Aldershot

Baumanova M, Smejda L, Rüther H (2019) Pre-colonial origins of urban spaces in the west african sahel: street networks, trade, and spatial plurality. J Urban Hist 45(3):500–516

Baumanova M (ed) (2022) Urban public space in colonial transformation. Springer, Swistzerland

Beeckmans L (2013) Editing the African City: reading colonial planning in Africa from a comparative perspective. Plan Perspect 28(4):615–627

Beeckmans L (2018) The architecture of nation-building in Africa as a development aid project: designing the capital cities of Kinshasa (Congo) and Dodoma (Tanzania) in the post-independence years. Prog Plan 122:1–28

Berrisford S (2011) Why it is difficult to change urban planning laws in African countries. Urban Forum 22(3):209–228

Bigon L (2016) French colonial Dakar The morphogenesis of an African regional capital. Manchester University Press, Manchester

Bigon L, Njoh AJ (2018) Power and social control in settler and exploitation colonies: the experience of New France and French Colonial Africa. J Asian Afr Stud 53(6): 932–951.

Boyer A (2009) Archipelia. Lugar de la relación entre (geo)estética y poética. Nómadas 31:13–25

Bryceson D(1996) Desagrarianization and rural employment in Sub-Sharan Africa: a sectorial perspective. World Dev 24(1):97–111

Calvino I (1977) Invisible cities. Vintage, London

Carter P (1988) The road to Botany Bay: an exploration of landscapes and history. Knopf, New York

Carreño L, Ana M (2019) Africa's first city of Islam. AramcoWorld 26–33

Castro-Gómez S (2005) La hybris del punto cero: ciencia, raza e ilustración en la Nueva Granada (1750–1816). Editorial Pontificia Universidad Javeriana, Bogotá

Césaire A (2006) Discurso sobre el colonialismo. Akal, Madrid

Cooper F (1996) Decolonization and African society: the labour question in French and British Africa. Cambridge University Press, Cambridge

Coquery-Vidrovitch C (1993). Histoire de Villes d' Afrique Noir. Des Origins à la Colonisation. Ed. Albin Michel S.A., Paris.

Coquery-Vidrovitch C (1995) Histoire de l'urbaniztion africaine: la ville coloniale: «lieu de colonisation» et mestissage culturel. En Biget, J. y Herveé J (eds.) Panoramas Urbains: Situation de l'histoire des villes. ENS Editions, Lyon, pp. 13–31.

Coquery-Vidrovitch C. (2005a) African urban spaces. History and culture. En Salm SJ, y Falola T (eds) African urban spaces in historical perspective. University of Rochester Press, New York. pp Xv–xl

Coquery-Vidrovitch C (2005b) Urban culture: relevance and context. En Falola T, y Salm SJ (eds) Urbanization and African cultures. Ed. Carolina Academic Press, Durham, North Carolina, pp 17–22

Chabal P (2011) África. *La política de sufrir y reír*. Oozebap, Barcelona

Chandler T (1994a) Urbanization in ancient Africa. en Tarver JD (ed) Urbanization in Africa. A handbook. Greenwood Press, Connecticut/London, pp 3–14

Chandler T (1994b) Medieval and early Modern Africa. en Tarver JD (ed) Urbanization in Africa. A handbook. Greenwood Press, Connecticut/London, pp 15–32
Crowder M ([1967] 2023) Senegal. A study of French assimilation policy. Londres, Routledge, New York
Davis M (2007) Planet of slums. Verso Books, London y Brooklyn
Demissie F ([2012] 2016) Colonial architecture and urbanism in Africa: an introduction. En Demissie F (ed) Colonial architecture and urbanism in Africa. Intertwined and contested histories. Routledge, New York
De Satgé R, Watson V (2018) Urban planning in the Global South. Palgrave Macmillan, Switzerland
Dovey K (1999) Framing places mediating power in built form. Routledge, New York, London
Enwezor O (ed) (2001) The short century. Independence and liberation movements in Africa, 1945–1994. Prestel, Munich, London, New York
Foucault M (1977) Discipline and Punish: the birth of the prison. Phanteon Books, New York
Fox S (2013) The political economy of slums: theory and evidence from Sub-Saharan Africa. In: Working paper series 2013, No. 13–146. London School of Economics, London
Frances Mobley C (2015) The Kongolese Atlantic: Central African slavery and culture from Mayombe to Haiti. Dissertation submitted in the Department of History in the Graduate School of Duke University
Freund B (2006) The African city. A history. Cambridge University Press, New York
Gensheimer T (2004) Globalization and the medieval Swahili city. En Falola T, y Salm SJ (eds) Globalization and urbanization in Africa. African World Press, Trenton, Asmara, pp 171–185
Glissant E (2002 [1996]) Introducción a una Poética de lo Diverso. Ed. Planeta, Barcelona
Grosfoguel R (2008) Hacia un pluri-versalismo transmoderno decolonial. Tabula Rasa 9:199–215
Hanna WJ, Hanna JL ([1971] 1981) Urban dynamics in Black Africa. Aldine Publishing Company, New York
Harris R (2003) A double irony: The originality and influence of John F.C. Turner. Habitat International
Harrison P, Croese S (2023) The persistance and rise of master planning in urban Africa: transnational circuits and local ambitions. Plan Perspect 38(1):25–47
Hobsbawm E (1983) Introduction: inventing traditions. In: Hobsbawm E, Rangereds T (eds) The invention of tradition. University Press, Cambridge, pp.1–14
Hull RW (1976) African cities and towns before the European conquest. Norton & Company Inc., New York
Iniesta F (2001) El Planeta Negro. Aproximaciones Históricas a las Culturas Africanas. Los Libros de la Catarata, Madrid
Iniesta F (2010) El Pensamiento Tradicional Africano. Regreso al Planeta Negro. Los Libros de la Catarata, Madrid
Jacobs J (1958) Downton is for people. In: Whyte WH (ed) The exploding metropolis. University of California Press, Berkeley
Koetter F, Colin R (1980) The crisis of the object: the predicament of texture. Perspecta 16:109–141
Lassus B (2015) La transversalité, une démarche essentielle à la conception paysagère. Conferencia pronunciada. en 16ª Réunion du Conseil de l'Europe des Ateliers pour la mise en oeuvre de la Convention européenne du paysage. Andorra, 1 Octubre
Mabogunje AL (1981) The development process: a spatial perspective. Holmes and Meier, New York
Mabogunje AL (1994) Introduction: cities and Africa's economic recovery. En Tarver JD (ed) Urbanization in Africa. A handbook. Greenwood Press, Connecticut/London, pp xxi–xxxii
Macamo E (2005) Ambivalent transitions: colonialism, space and difference in Mozambique. In: Werbner P (ed) Postcolonial subjetivities. Zed Books, London, pp 85–104
Macamo E (2013) Living effervescence: the social in African urban settings. In: Obrist B et al (eds) Living the city in Africa. Processes of invention and intervention. Schweizerische Afrikastudien. Lit Verlag, Zürich

Mbembe A (1996) La naissance du maquis dans le Sud-Cameroun, 1920–1960, l'histoire des usages de la raison en colonie. Éditions Karthala, Paris
Mitchell T (1988) Colonizing Egypt. Cambridge University Press, Cambridge
Mohan R (1994) Understanding the developing metropolis: lessons from the city study of Bogotá and Cali. Oxford University Press, New York, Colombia
Myers G (2003) Verandahs of power. Colonialism and space in Urban Africa. Syracuse University Press, Syracuse, New York
Myers G (2011) African cities: alternative visions of urban theory and practice. Zeed Books, London/New York
Ndaw A (1983) La pensée africaine: recherches sur les fondements de la pensée négro-africaine. Nouvelles Editions Africaines, Dakar
Nightingale C (2012) Segregation: a global history of divided cities. University of Chicago Press, Chicago and London
Njoh AJ (2009) Urban planning as a tool of power and social control in colonial Africa. Plan Perspect 24(3):301–331
Nunes Silva C (2015) Urban planning in Sub-Saharan Africa. An overview. In: Nunes Silva C (ed) Urban planning in Sub-Saharan Africa. Routledge, New York, pp 8–40
Nyamnjoh F (2012) Potted plants in greenhouses: a critical reflection on the resilience of colonial education in Africa. J Asian Afr Stud 47(2):129–154
Obrist B (2013) Introduction to living the city in Africa. In: Obrist B et al. (eds) Living the city in Africa. Processes of invention and intervention. Schweizerische Afrikastudien, Lit Verlag, Zürich
Obrist B, Arlt V, Macamo E (eds) (2013) Living the city in Africa. Processes of invention and intervention. Schweizerische Afrikastudien, Lit Verlag, Zürich
O'Connor A ([1983] 2012) The African city. Hutchinson University Library for Africa, Johannesburg
OECD/SWAC (2020). Africa's urbanisation dynamics 2020. Africapolis, mapping a new urban geography. West African studies
Parnell S, Watson V (2009) Planning for cities in the global south: a research agenda for sustainable human settlements. Prog Plan 72(2):233–241
Parnell S, Pieterse E (eds) (2014) Africa's urban revolution. Zed Books, London/New York
Peil M, Sada PO (1984) African urban society. Wiley, Chichester
Pieterse E (2008) City futures: confronting the crisis of urban development. Zeed Books, London
Plotnicov L (1967) Strangers to the city. Urban man in Jos. University of Pittsburgh Press
Pons V (1969) Stanleyville: an African urban community under Belgian administration. Oxford University Press for the International African Institute, Oxford
Potter RB, Lloyd-Evans S (1998) The city in the developing world. Addison Wesley Longman Limited, Essex, UK
Rakodi C (1997) Global forces, urban change, and urban management in Africa. In: Rakodi C (ed) The urban challenge in Africa. United Nations University Press, Tokyo, pp 17–73
Roca A, Iniesta F (2006). Raíces: ¿Por qué la historia de África es un conocimiento vital en el África del siglo XXI?. En Santamaría A, y Echart E (coords) África en el hori zonte. Introducción a la realidad socioeconómica del África Subsahariana. Instituto Universitario de Desarrollo y Cooperación/Ediciones La Catarata, Madrid. pp 20–54
Roy A (2009) Why India cannot plan its cities: informality, insurgence and the idiom of urbanization. Plan Theory 8(1):76–87
Ruiz-Giménez Arrieta I (2000) El Colapso del Estado Postcolonial en la Década de los Noventa". en Peñas FJ (ed) África en el Sistema Internacional. Los Libros de La Catarata, Madrid, pp 165–207
Scott JC (1998) Seeing like a state: how certain schemes to improve the human condition have failed. Yale University Press, New Haven and London
Sennet R (1973 [1970]) The uses of disorder. Personal identity and city life. Pelican Book, Manchester

Simone A (2004) For the city yet to come: changing African life in four cities. Duke University Press, Durham

Simone A, Pieterse E (2017) New urban worlds. Inhabiting dissonant times. Polity Press, Cmabridge/Medford

Sinou A (1981) Les moments fondateurs de quelques villes coloniales. Cahiers d'Études Africaines 81–83, XX1-I-3, pp 375–388

Stren RE (1990) Urban housing in Africa: the changing role of government policy. en Amis P, Lloyd P (ed). Housing Africa's urban poor. Manchester University Press, Manchester y New York

Swanson MW (1977) The sanitation syndrome: bubonic plague and urban native policy in the Cape Colony, 1900–1909. J Afr History XVIII(3):387–410

Tuan YF ([1991] 2018) El Lenguaje y la Producción del Lugar: un Enfoque Descriptivo Narrativo. En Nogué J (ed) Yi-Fu Tuan. El Arte de la Geografía. Icaria, Barcelona. pp 111–142

Turner JFC (1976) Housing by people: towards autonomy in building environments. Marion Boyars, London

UN (1993) World urbanization prospects: the 1992 Revision. United Nation Department of Economic and Social Information and Policy Analysis, New York

UN-Habitat (2003) The challenge of slums: Global report on human settlemets, London

UN-Policy Brief (2023) Growing middle class and import substitution. Connecting the dots to unlock made in Africa. Africa dialogue series, 10 mayo 2023

Van Oostrum M (2024) Walkability and colonialism: the divergent impact of colonial planning practices on spatial segregation in East African Cities. Int J Urban Policy Plann 144 (Art. 104662)

Watson V (2003) Conflicting rationalities: implications for planning theory and ethics. Plan Theory Pract 4(4):395–407

Watson V (2011) Changing planning law in Africa: an introduction to the issue. Urban Forum 22:203–208

Watson V (2013) African urban fantasies: dreams or nightmares? Environ Urban 26(1):215–231

Winkler T (2022) Contemplating planning knowledges and ethics. En Yiftachel O, y Mammon N (eds) TheoriSE: debating the southeastern turn in urban theories. African Centre For Cities, Cape Town, pp, 60–66

Wynne-Jones S, Fleisher J (2014) Swahili urban spaces of the eastern African coast. Making ancient cities: Space and place in early urban societies. Cambridge University Press, Cambridge, pp 111–144

Yiftachel O (1998) Planning and social control: exploring the dark side. J Plan Lit 12(4):395–406

Yiftachel O, Mammon N (eds) (2022) TheoriSE: debating the southeastern turn in urban theories. African Centre For Cities, Cape Town

Chapter 11
Antananarivo, Capital Status and Readings of the Urban Landscape. Concepts of Competing Powers Since the Eighteenth Century

Albert Roca and Helihanta Rajaonarison

Abstract The authors show how the history, in the "longue durée", of the populations of Antananarivo (Madagascar), and the diversity of the socio-cultural contexts where they have been evolving, allows us to understand the current multiplicity of their "city plans" as an adaptive phenomenon, in which the influence of colonization and subsequent development is only one of the factors at stake. The authors propose that the study of this pluralism could contribute to shedding light on the complexity of many other urban dynamics and symbolisms south of the Sahara. Antananarivo therefore contains potentially useful African lessons for the management of diversity in global cities, lessons that require multidisciplinary approaches.

Keywords Urban space · Power relations · Urbanism · Madagascar · Precolonial Africa · Cultural pluralism

11.1 Introduction

Antananarivo has been postulated as the capital of Madagascar since the Merina expansion,[1] triggered at the end of the eighteenth century, although this status was not unequivocally established for the whole island until colonization, imposed by

[1] Merina is the designation of the largest ethnic group in Madagascar (nearly 30% of the current population). All groups on the island (variously called ethnicities, peoples, or more loosely, and euphemistically, "population groups ") speak variants of a linguistic complex that is referred to as the Malgache or Malagasy language, and all share cultural traits, despite their differences.

A. Roca (✉)
University of Lleida, Lleida, Spain
e-mail: roca.albert@gmail.com

H. Rajaonarison
University of Antananarivo, Antananarivo, Madagascar
e-mail: helihanta@gmail.com

F. Carrión Mena and S. Rodríguez Alvarez (eds.), *From the City as a Project to the City Project*, The Urban Book Series, https://doi.org/10.1007/978-3-031-94226-6_11

force at the same time as the rest of Africa south of the Sahara. This condition made Antananarivo the city par excellence in Madagascar, a city that would have adapted to the changes of the last two centuries and that would have served as a reference—despite its singularity—for urban development on the island.

This vocation as a capital was embodied, since 1787, in a certain "city plan". This plan responded to the urban symbolism of a new power, seeking to culturally visualize the adaptation of the habitat or defence functions of the original urban agglomeration in its transformation from a high place of a small warlord into the capital of a territorial state. The discourse of this plan, conceived and materialized in houses or streets before colonization, is still visible, even unavoidable. This is an exceptional fact in Africa, but the most striking thing is that this resilience reveals that the old plan of the kings has been accompanied, intersected and modified by a myriad of other so-called "plans" or visions, articulated or juxtaposed for two centuries, both top-down and bottom-up. A multiplicity that we believe to be the basis of its durability.

Apart from colonial action, the concept of city project has been generally ignored in the study of African cities, maybe especially south of the Sahara (see Martínez in this same volume; Biao 2019; Kamana et al. 2024; van Noorloos and Kloosterboer 2018), or even rejected: the more engaged analysis of the evolution of modern African urban planning focuses on the reappropriation—or even the "diversion"—by local populations of colonial city plans, always substantially failed, exclusionary and illegitimate (see Ouchen in this same volume). And yet, Antananarivo can provide an appropriate comparative scenario to complicate these approaches, by introducing the *longue durée* and cultural analysis,[2] considering, therefore, the agency of actors who are generally ignored or underestimated. It is in this sense that this chapter aims to be a contribution to the production of knowledge for the urban planning of capitals in Africa, knowing that everything indicates that they will be called upon to play an important role in the global city, in planetary urban planning (see Bearak et al. 2021).

Methodologically, the article relies on the systematic crossing of diverse perspectives —coming from the social sciences adscription, human experience or personal relationship with the object of study—this diversity, in approaches and epistemologies applied to the object of study, is necessarily embodied in the author-team composed of a historian and an anthropologist, woman and man, Malagasy and *vahiny* (native-foreign) (see Roca 2015). With this perspective, the argument is constructed diachronically. We first analyse the adoption of a cultural models of space distribution in the city of Antananarivo, to then examine the changes and adaptations

[2] The case of Antananarivo can be compared to those of cities with *medina* (ancient Islamic city) in the north and on the east coast of the continent, but knowing that it goes far beyond this comparison because the "old pattern" of Antananarivo was already partly an adaptation to contact with globalization and its European key (see architecture, for example), being able to incorporate a whole plurality of actors and cultural identities. The urban planning and architecture of Antananarivo have given rise to an abundant bibliography, collected in good measure by the PAUPA, Urban and Landscape Architectural Heritage of Antananarivo (see Department of History nd). Esoavelomandroso (1989), Guérin (2002) and Rajaonarison (2012) offer introductions and in-depth studies at the same time, relevant for the perspective of this chapter.

produced with its rise to the rank of capital (of a kingdom, a colony, a republic), accumulating distinct visions, of collectives associating differently distinct territories with this capital, to generate a multiplicity of "city plans", a true urban pluralism: pre-colonial Merina project, French colonial project with its cosmopolitan ambition, "decentralized" projects of post-colonial migrant populations...

11.2 Antananarivo, Capital of Imerina

The city of Antananarivo, which has never really left its original site, is today the greatly expanded version of what King Andrianampoinimerina had created during his reign between 1787 and 1810. This king is the re-founder of the city of Antananarivo, an urban site developed on the heights of the hill which was then called Analamanga ("In the blue forest", that is to say, beautiful, wild, even sacred), after having conquered it from their former sovereigns. Antananarivo then became the capital of all the powers that succeeded one another in the subsequent history. It was the capital of Imerina, of the Kingdom of Madagascar—which, despite the royal vision, did not encompass the entire island, but only 2/3 of the territory—then of the French colony of Madagascar and finally of the Republic of Madagascar since its independence in 1960.

The constant elements in the representations of Antananarivo over time include the Analamanga hill and the Betsimitatatra plain (in reference to the "large number of irrigation channels"). Thus, in a popular iconography of the city, emphasised by the Hollywoodian letters of Antananarivo, we can still see the Manjakamiadana palace and the Andohalo square, on high ground, with Lake Anosy and the Mahamasina stadium at the base. The heights were home to the royal city, the ancient power; at its foot, the stadium, once an enthronement field, connects the two cities, while Lake Anosy, once a marsh transformed into a pool for royal entertainment, today contains the memorial to the Malagasy soldiers who died in France during the two world wars.[3]The observer has the impression that attempts have been made, constantly, to harmonize the signs of Merina royalty with the emblems of the regimes that followed it up to the Republic of Madagascar (see Fig. 11.1).

The choice of the hillside site of Analamanga by the Vazimba,[4] who were the first occupants, had been dictated by the context of the most remote period, which was marked by the search for naturally protected places, namely perched sites, in the case of the Highlands. The eco-political pattern of the region can be summarized with two fundamental features: the development of intensive rice cultivation, using oxen force, on the lowlands and slopes converted into terraces; the rise of warrior principalities, installed on the height of the hills, disputing the control of these rice

[3] The authors invite readers to explore the urban landscape of Antananarivo with interactive maps, such as https://fr-fr.topographic-map.com/map-l3zcz/Antananarivo/?center

[4] Designation of the former occupants of the island, especially perhaps in the Highlands before the *andriana* and *hova* came to power (see the text above), but also in the West or elsewhere.

Fig. 11.1 General view of the heart of Antananarivo: Anosy, Mahamasina and Haute Ville. *Source* 1973 (Agence Nationale Taratra)

fields and of trade, local and with the coasts, including that of slaves (Urfer 2022; Bloch 1983).

11.2.1 Renivohitra: The "Mother Hill", the "Mother of Hills", a Local Concept

A socio-spatial logic has marked these settlements and has consistently influenced the Merina symbolism of power. It could be said that the distribution of the population in each village corresponds to a dichotomy: *ambonivohitra* (above, at the top of the hill) *vs. ambanivohitra* (below, at the bottom of the hill). The notion of *ambony* (high, above, perched) was reinforced in the eighteenth century due to the widespread insecurity throughout the island, caused by the search for slaves to sell to merchants waiting on the coast, a demand that forced people to settle on perched sites or to fortify and protect their habitat in various ways. *Vohitra*, on the other hand, means hill, but it also ended up designating the city or the town: *ambonivohitra* (high town) can be translated as urban site (a "civilized" place, seat of power as in the Greek *polis* or the Roman *civis*), while *ambanivohitra* does not mean lower town, but indicates the hinterland or the countryside, even the wilderness, the bush, a scenario of domination and struggles between the powerful. This is why everything that is rural and peasant is *ambanivohitra*. In principle, the occupants of hills are distributed in height according to hierarchical principles, according to their belonging to *stamenta* or social orders: the summit, the citadel often specifically fortified, was above all the space of the *andriana*, the "nobility", the lineages related to the local dynasty; lower down, in the village, were the *hova*, "free", and in the lower depths, even outside the enclosure, it was the place of the dependents, the *mainty* ("blacks"), the *ondevo* ("slaves") (see Mille 1970; Rafolo 1989).

It was when Andrianampoinimerina decided to make Antananarivo the capital of his kingdom, or *renivohitra*, "mother hill" or "mother of hills", that the city inexorably developed on the heights of Analamanga. The district of this ensemble is what is today called "the Upper Town" and the former rice-growing lowland is known by the name of "Lower Town". The notion of *renivohitra* or capital is much closer, therefore, to the value that is attributed to Antananarivo as a city par excellence, instead of *ambonivohitra*, which only indicates its city-ness (as matrix of citizenry of political affiliation), rather than its urbanity: even the fortified villages fell into the category of *ambonivohitra*, although today they are often considered to belong to the rural world.

Coveted by successive conquerors, the site of Analamanga was at the heart of several issues that led King Andrianjaka, in 1610, and his successors to settle on its heights. The presence, at the bottom of the city's hills, of the large Betsimitatatra marsh is significantly the major reason for this anchoring, because the domination of the plain transformed into rice fields will guarantee the economic system by making intensive agriculture the main means of subsistence for the whole of Tananarivian society. It is also this political-economic choice that will dictate the "development plan" of Antananarivo from the seventeenth century to the present day, as well as will explain its subsequent rise to the lasting rank of island capital. After his triumph in 1787, and because of this accumulation of strategic advantages, Andrianampoinimerina preferred it to his original village, Ambohimanga, in Avaradrano,[5] although this last preserved its sacred character, like the other hills annexed by the king. And perhaps also to mark the rupture of his new kingdom, which was no longer to remain as another *primus inter pares* between the Merina royalties, but to found an innovative political form, which can be described as "national". To legitimize and communicate this innovation to the whole of Imerina (and later, beyond), Andrianampoinimerina used sustained military and coercive force, largely derived from his distributive economic capacity: we must not forget that, in the Malagasy language, one of the forms of the verbal concept "to govern" is *mifehy*, literally, "to bind", "to tie together". But he also had recourse to a parallel ideological construction productive of social cohesion, through alliances, based in turn on interpretations of existing traditions, in particular, on kinship and autochthony. These interpretations went, therefore, from

[5] Avaradrano, literally "North of the water", perhaps referring to the rivers near the capital (Ikopa, Sisaony, Mamba, etc.) and its waters often considered sacred. It designates a current district and formerly an ancient region and then a province (toko) of the unified Merina kingdom, northeast of Antananarivo.

top to bottom, following princely domination, but also from bottom to top: we will come back to this.[6]

11.2.2 The Skyline of Antananarivo: Power on Stage

As for the distribution of groups, holders of collective rights in the new Antananarivo, refounded as the capital in 1787, King Andrianampoinimerina himself was the "designer" of the city by granting the districts where his subjects were to live. By abandoning his ancestral seat as the capital, Andrianampoinimerina acted as a divine king in the Frazerian sense: this choice meant declaring himself as "Great Mediator" (between cosmos and society) throughout the territory of the Kingdom (and all the natural forces involved and connected by its *tompontany*, indigenous lineages) and the people of Imerina. He therefore resized the category of indigenous, *tompontany*, "master or lord of the land", too quickly translated as owner. This category is acquired everywhere in Africa in the logic of the "first occupant", that is to say, the group (living and dead united by an ancestral chain) who made a covenant with the forces, the "souls" of the earth, of places (as they have always existed), a covenant that is continually renewed through ritual. The group is not really the owner but rather a mediator of the "lease", one could say, granted with the land, in particular in relation to foreigners and newcomers (*vahiny)*: land cannot be sold and one the mediating is only kept if the ritual obligations are fulfilled by the autochthonous group. Then the king imposes a second overall mediation for the entire territory, in order to harmonize and optimize the partial mediations of each head of lineage or local communities (*foko*), of each prince. Antananarivo was defined as the place where the confirmation rituals of this multiple alliance (royalty, lineages, land) should be actualized.[7]

This ritualization was visible in all public acts of royalty (such as the *fandroana*, the royal bath, or the enthronement; see Raison-Jourde 1991), but it was especially fixed in the "social landscape" of the city. Andrianampoinimerina decided who he wanted to place in this or that place, just as he was the only one who granted land to those who have satisfied him: then this assignment was and is inevitably linked to the *tanindrazana* (ancestral lands) of the groups settled from elsewhere, that is, the majority, including the elites, *andriana* or *hova*. The sovereign, as absolute master, had decided that humans live on the heights of the hill while the entire marshy plain surrounding the foot of the hill was transformed into rice fields to ensure the security and subsistence of all the subjects. For this very reason, he created the *voromahery* ("eagles"), a group of commoners whom he ordered to supervise this rice cultivation.

[6] For the history of the Merina kingdom, including its unification by Andrianampoinimerina and its transformation into the "Kingdom of Madagascar", see Urfer (2022), where the related scholarly literature is critically reviewed. The book is also valuable for an up-to-date critical view of the island's complete history.

[7] For an introduction to the notion of royalty (and its evolution) among the Merina, as well as to Merina social symbolisms understood in their Malagasy context, see first Raison-Jourde (1991).

Fig. 11.2 Antananarivo, 1890. *Source* Photo by voyager Louis Catat

These *voromahery* with the *ondevo*, "slaves", would then be the first inhabitants of the Lower Town. Until the colonial period, most of the populations of the lower town were *hova*, former free masters, and their slaves, often grouped under the name *mainty* and who became free people over time and, in any case, after the abolition granted by the French conquerors in 1896.

The most impactful message in the purely symbolic register, however, is inscribed on the heights and, above all, on the skyline which has given fame to Tana[8] so far: the hill crowned by the *rova*, the complex that houses the royal palace Manjakamiadana that in turn gives its name to the heart of the *Renivohitra* (see Fig. 11.2). Since the beginning of the nineteenth century, the Rova on the summit (in its own name, as palace par excellence) communicated somewhat lower down (but still high up) with the "square", a real agora located in Andohalo, where the king summoned the people, especially the heads of families, for the *kabary* ("assemblies", "discourses", in this case, "royal assemblies"); nearby a rock, in the style of the Rupes Tarpeia in ancient Rome, was used for executions and saw the first Malagasy martyrs, Amboninampamarinana; at the bottom of the chasm, the royal enthronement field, Mahamasina ("which makes sacred, holy"), linked the Upper Town with the Lower Town (see Fig. 11.1).

[8] Tana, from *tanàna*, "town", is the popular name for Antananarivo (literally, "the city of a thousand").

During the nineteenth century, these symbols would diversify without leaving, therefore, this skyline, read and reread by the *ambaniandro*, the Merina subjects. While the royal houses in the Rova, particularly the Manjakamiadana (finished in 1841, renewed in 1867), even following the changing architectural fashions, drew their legitimacy from the sacredness of the person of the sovereign with all the symbolism that accompanies it, the residence of the Prime Minister, an Andafiavaratra (name designating the group and the residence of the successive Prime Ministers), was to be built on other values and in other conditions (Fig. 11.3). A lineage of Prime Ministers would dominate political life since the time of Ranavalona I (1828–1861), from the order of commoners, the *hova*, a Malagasy bourgeoisie who took power, one could say. His palace, built majestically later (1872) by the will of Rainilaiarivony (last of the Andafiavaratra Prime Ministers, in office between 1864 and 1895, husband of the last three queens), offers itself as a spectacle by exhibiting the market value of all its materials, its dimensions, on a remarkable choice site, slightly below the Rova itself. This presence of Andafiavaratra to the north of the Rova imposes on the eye the existence of the two sources of power, with their own symbols: that of the *andriana*, on the throne, and that of the *hova*, who held the reins of the kingdom. The evolution of a dualism of power throughout the royal period is readable, therefore, on the skyline of the Upper Town, visible from distant lands outside the capital. But the multiplicity of visions of the city cannot be reduced to a double discourse.

The settlement of Europeans during the nineteenth century, such as the self-taught French engineer, and queen lover, Jean Laborde (1805–1878) or the Scottish missionary James Cameron (1800–1875), helped to define a particular architectural style that found its maximum expression in the "double rova" (in wood, 1841, and clad in stone, 1867; see Fig. 11.8). But the newcomers were not going to resign themselves to a role of advisors or technicians: even before the colonial attack, the missionaries, divided and even opposed in various confessions, Catholic, Anglican and Protestant, tried, in vain, to found a real counter-State or a dual State (Raison-Jourde 1991). In any case, we can see how these first Western agents willingly integrated themselves into the royal "city plan", as demonstrated by their competition not only to have the favours of the monarchs, but also to occupy, with their respective seats, the heights and the symbolically significant connection zones of the city: from Faravohitra to Ambatonakanga, from Andohaho to the Rova itself or to Ambohipotsy.[9] Although the Reformed Church (FJKM, local Protestant church resulting from the action of the London Missionary Society) took advantage, the scale of this competition is particularly visible in Andohalo, where we can admire how the new powers of the nineteenth century skirt, it seems as if they are taking hostage, the old area of the royal *kabary*, from the Anglican Cathedral to the Catholic, passing by one of the large churches of the FJKM (Fig. 11.4).[10]

---

[9] Names of well known "hills" in the city core (summits or connecting heights).

[10] There is only one absence between the four churches that will form, after the colonial interlude, the influential Ecumenical Council of Churches of Madagascar (FFKM), dominant until the arrival of the evangelicals in the 90s: the Lutheran Church (FLM), which has scattered from the coasts

Fig. 11.3 Antananarivo, 1862. *Source* Photo by missionary William Ellis

11.2.3 The "Twelve Hills" and the Firazanana*: Crossed Paths in the Capital*

The second axis of the "city plan" drawn up by the Avaradrano faction projects itself outside the city, focusing on the internalization of a notion of capital that one does not hesitate to describe as "national" or even modern, but which, like the skyline, was built on bases that can only be described as traditional.

---

and which, consequently, sits in the middle town (Isoraka, Ambatovinaky) and the Lower Town (67Ha…).

Fig. 11.4 The Rova. *Source* Photo by Helihanta Rajaonarison

Andrianampoinimerina used several ideological instruments to territorialize, to impose, his proposal of capitality for Antananarivo. Perhaps the most "spatial", one could say, is the hierarchization of the ancient *ambonivohitra* through what has been called the ideology of the "twelve hills" or "the twelve sacred hills". But this ideology must be contemplated in a set of alliance measures that refers to what Aidan Southall (1988) had already described with an enlightening terminology, although it may seem an oxymoron: "segmentary state". We will return to this key feature that has survived royalty.

The Twelve Hills instituted by Andrianampoinimerina would constitute the sacred limits of the kingdom, a quality that should intimidate invaders and prevent splits. However, their proximity to Antananarivo suggests that they rather defined the core of the kingdom— a sort of extended capital—and a model of political adhesion. Each of the hills corresponds to a princedom, with an ancestral lineage, *andriana*, as a rule: Nampoina[11] would have married a daughter from each of these lineages—defeated or allied—who remained there. Then all the great lineages of Imerina became linked (fehy) to the monarch. This was done either through a hypergamous marriage which recognized the exceptional superiority of the king (*tsy rafoza ny mpanjaka*, "the king has no in-laws", it is said throughout the island), or through a sacralised allegiance, as with the Mainty enin-dreny (the "Black with Six Mothers"), who could not be

[11] Popular abbreviation of Andrianampoinimerina.

punished with death, *tsy maty manota*. These alliances were accompanied by a strong ritual connection: from the synchronization of the Fandroana (the ritual royal bath at the beginning of the year) to that of circumcisions (the future men of the lineages could only be initiated after the *savatse* of the king's sons had been accomplished), from participation in the ceremonies in Antananarivo to the local guardianship of each of the *sampy*, the royal talismans which did not necessarily coincide with the sacred hills...

It is necessary to note that there is no agreement on a closed list of the "twelve" hills: Ampandrana, Imerimanjaka, Alasora, Antananarivo, Ambohijoky, Ambohidratrimo, Antsahadinta, Ambohimanga, Ikaloy, Ilafy, Ambohimananmbola, and Ambohidrabiby only make up a popular possibility. There are many others who claim this sacredness and host the corresponding cults to this day. This ambiguity clearly shows how the ideology was a mechanism of integration linked to the kingdom unification in Imerina and the number was itself symbolic of plenitude (same number than the *vintana*, fates, destinies, months, of the Malagasy zodiac). But it also shows how this ideology was (and continues to be) used and reappropriated by the different groups to the extent of their opportunities. Despite the absolute discourse of the king and his "ideologues" (see the *Tantara ny andriana*, or Raison-Jourde 1991 in writing on the singularity of the State of Andrianampoinimerina), we are indeed faced with an alliance governed by a reciprocity, asymmetrical certainly, but compulsory, where all parts can take the initiative. Hence, some examples of this autonomy, of this multidirectionality. While circumcision was desynchronized with the fall of Radama II (1863), the official conversion to Christianity (1869) did not erase it, despite its pivotal role in the cult of ancestors; similarly, the cult of royal fetishes[12]survived the royal order to destroy them and they were used by the Menalamba insurgents (1897; see Ellis 1998), and many other reappropriations could be referred on the part of the lineages, the *firazanana*, still very much alive... Although the twelve hills "lost" their political role, often absorbed by the Tananarivian conglomeration and forgotten by the rare urban plans, they have continued to be very popular places of worship, therapy, and gathering, places traced by the lines of several invisible or, rather, invisibilized "city plans", which complement and oppose at the same time that of the king and his avatars (Rakotomalala et al. 2001).

So, we must not forget either that, in the Malagasy language, the verb "to serve" in reference to the actions due to the sovereign, the chief, the superior, is *manompo*, literally "to make, to build, the lord, the master", a linguistic expression that reveals the importance of the bottom-up path in traditional Malagasy powers. The royal city plan was the only one to capture the attention of most of the actors who came after—missionaries, settlers, cooperators—but it is not the only one to continue to act in the new fields of forces in the capital.

---

[12] *Sampy*, large national talismans in *palladium* mode of Imerina, distributed and kept in the name of the king or queen in several villages (i.e. "hills").

11.3 Antananarivo, Capital of Madagascar

The urban and urban symbolism of Antananarivo—and the social (and cultural) projection of its "plan"—are not limited to the city, but first extended to the territory of Imerina unified by Andrianampoinimerina, gradually reaching the entire island, between the nineteenth and twentieth centuries. This insular propagation of the symbolism of Antananarivo led to categorizing the city as "capital", a trait that is now central to approaching any urban imaginary of Tana. However, this "capital" has not been interpreted or instrumentalized in the same way by the different populations of the island, so that it has not given rise to a single symbolism, but rather to an intersection of different visions. These, retroprojected across the city, would weave very different "plans", regardless of what the municipality and state authorities could have established; this cohabitation is not surprising if we consider the extremely powerless of the last ones to impose their criteria, given their enormous budgetary shortcomings. Thus, this pluralism is continually translated into the definition of the city and, therefore, demands to be considered and studied.

11.3.1 Capitality and Pluralism of the Urban Imaginary

The chronology of Antananarivo's rise to the status of capital is part of the modern conception of "capital", which goes beyond the hypertrophies of the city-state (resending to comparison with cases as spectacular as the Roman State and its imperial drift). However, the material survival of its old city, socially connected to the present day, makes visible specific continuities often veiled for the cosmopolitan observer, blinded by the dust of the shanty towns and the omnipresence of landscapes of non-development in the streets of the cities of the "South" (Roca 2014). These continuities imply other chronologies and other cities within the city, beyond infrastructures and architectures. This durability ignores another presumed constant of modern evolution: the deactivation of the sacralising strategies that had characterized the ancient territorial states, premodern ones one might say. Antananarivo thus evokes a diverse genealogy that joins the modernizing European metropolis as well as these traditional "centre of de world" models, those associated to divine royalty, which developed from ancient Egypt to Southeast Asia or the pre-Columbian Andes. Even the enthusiastic commitment of missionaries in the conquest of the moral and urban heights evokes the enchanted urbanism of the ancient city, the old capital, inextricably mixed today with obvious emblems of secularism, of disenchantment of the world.

To understand how these trends coexist as autonomous and non-exclusive dynamics, a diachronic approach is essential, a historical explanation resulting from verification through the available empirical data. In doing so, the first thing that becomes obvious is that Antananarivo is not a "double city", with a traditional, indigenous district or city versus a modern, colonial, foreign district or city. It is not at all a so-called variation of the famous—and simplistic—dichotomy of the "cement city" *versus* the "straw city". There are much more than two logics, two planes, and it is hoped that the exposition of this observation invites readers to question stereotypes about African cities anywhere (v, de Blij 1977; Minyoi 2019; van Noorloos and Kloosterboer 2018; Zoma and Sawadogo 2022).

This multiplicity is largely the opposite of a "plan"; this is particularly true is we conceive the "plan" as the work of a decision-making unit, such as an urban planning office or an ideal identity community, endowed with a supposedly coherent *weltanschauung* and *kultur*. However, even if these logics do not imply or are not in themselves projects as we may understand, they give meaning to urban imaginaries adapted to new conditions; imaginaries actualized in concrete actions (habitats, migrations, network resources, etc.), comprising intentions and, therefore, approaching our notion of project. Imaginaries and actions that undoubtedly intervene in the continuous conformation of Antananarivo, even if it is not easy to distinguish them "aesthetically" in the urban landscape. In fact, Antananarivo's design complex has proven relatively stable, as have some of the fundamental elements of its material structure, with notable continuity since the nineteenth century in the connection of buildings, roads, and quarters or neighbourhoods. Explaining this durability, predicting its immediate future, and challenging the most individualistic visions of a global and urban teleology, may raise questions about the universality of the typification of the modern capital (see, for example, Gottmann 1983; Guzmán et al. 2016).

We propose to organize spatially, territorially, this diversity of nested imaginaries from a sequence of chained geneses of so-called concentric and discontinuous "circular crowns", created by the Merina sociopolitical and economic expansion, from the inner circle where Antananarivo is located towards the periphery of the island. None of these spaces is homogeneous: in fact, only the inner circle and the whole island seen from the outside have given rise to mono-identity discourses, proper and external, debatable and discussed, but persistent. However, we believe that it may be useful to analyse the corresponding groupings that present a certain historical coherence and that emerge in the relations in the capital.

11.3.2 The Inner Circle and the Harmony of Heights Symbology

The inner circle was formed at the end of the eighteenth century with the unification of Imerina under the authority of King Andrianampoinimerina, posthumous name, literally "the noble, the lord, in the heart of Imerina", the land of the Merina people. Although his faction originated from the Avaradrano region (see map), north of Antananarivo, and his domination obscured certain divergent Merina traditions,[13]

[13] See Délivré (1974) in his critique of the Tantara ny Andriana, the "History of Kings" collected by Callet (1908), or again Ellis (1998), when he speaks of the insurrection of the Menalamba.

originating from other sub-regions (the *efa-toko* or *enin-toko*, "four or six parts or districts"), without however completely eliminating them, there is no doubt that urban symbolism has been widely shared. This has contributed to understanding the relief and distribution of the old city as a message of the project around Nampoina. The amplification of Merina spatial logics, socially understandable and observed in Antananarivo, could have been more a question of social adaptation to the crisis of the warlord system than of the presumed simple and unrepeatable genius of an individual, a favourite vision of colonial racist thought to explain African changes.

The distribution in height of the classified groups, of the hierarchical collectives, is consistent with the *ambony/ambany-vohitra* scheme, with its pairs of sociological strata progressively amplified to accommodate the masses arriving from all over the island: (former*) afaka* ("free", autonomous classes) / (former) *ondevo* ("slaves", dependent classes); highlands / coastal; *tompontany* ("master of the land", native) / *vahiny* (foreigner, visitor). Antananarivo then becomes *Renivohitra*, "the Mother of the hills". Even the colonial desecration which meant the transfer of Nampoina ancestors remains, from their tomb in Ambohimanga to Antananarivo, decided by General Gallieni, conqueror and firs general governor, would only reinforce the sacred legitimacy of the capital, as it was before the competition of the Christian churches for the heights. The probable intention was to break the dynastic cult "without breaking the mould", but all this attention only supported the construction of Nampoina, considered proto-nationalist in present-day Malagasy society.[14]

There is no doubt that the people of Imerina eventually accepted the status of capital of Antananarivo, with more or less resistance. This acceptance was associated with the ideological complex of the *ambony/ambany-vohitra* dichotomy and the "twelve sacred hills". However, the operationality of this connection at each point depended on the establishment of multidirectional forces in the pyramidal structure.

In truth, the centripetal force of the discourse of the Avaradrano elite is continually counterbalanced by a myriad of centrifugal forces coming from the ancestral cores of many Merina lineages established in Antananarivo, including those that, transplanted to the Upper Town, have dominated the capital, and the country, for more than two centuries. These ancestral villages, often of a marked rural character, still constitute today ritual and relational poles of the first order, in a society where kinship by filiation groups continues to be a crucial collective cement, even after their legal invisibility during colonization (see Ottino 1998). It is there where the rites of

[14] General Gallieni's disposition combined desecration with humiliation to weaken Malagasy minds. Certainly neither Ambohimanga nor the Rova of Antananarivo were not sacred sites for non-Merina populations, but respect for the sacred character of the royalties even of others was general on the island and the gesture was understood as a sign of who wanted to be the master. Resistance to the attack will be much harder on the coasts, but the observation of widespread attacks, the repetition of the outrage to the ancestors (they even cut off the head of the Sakalava king of the independent Menabe, Toera), the prohibitions of royal cults and ceremonies were able to manage Malagasy solidarity. In the same order of things, the Rova or the Tananarivian skyline, although not triggering allegiances of non-Merina, can produce cohesion in the face of those who come from elsewhere, the non-Malagasy, notably the French. So, they are not discussed as signs of the Malagasy nation (a new concept, just as we do not discuss the capital of Antananarivo), but they have not caused tears at their destruction either, because they are others' heritage.

passage that consecrate belonging to lineages, even clans, are assured, rituals that may connect thousands of members (and more after the generalization of written family registers since the nineteenth century): circumcision, burial, *famadihana* (the famous "second burial"). They house the *fasana*, the collective family tomb, which marks a clan territory (more or less respected by history), the *tanindrazana* ("terroir", "land of the ancestors"), a word which that has been extrapolated to the island territory by the *malgachisants*, the Madagascar pundits, to translate the idea of "homeland"; in fact, under the monarchy, only the sovereigns had their tombs authorized in the royal city. Colonization further reinforced this by prohibiting burials within the municipalities. So, these villages continue to be real focuses of power and social cohesion in permanent, structural interaction with the capital (Blanchy et al. 2006; Rakotomalala et al. 2001). The importance, maintenance, even preservation, of these deep links is what explains the impossible separation of *ambonivohitra* and *ambanivohitra* among the inhabitants of Imerina: the sociosymbolic integration of the capital and its "ethnic territory"—that is to say, where the Merina groups are considered as the *tompontany*, the natives—is a hierarchical process, but a participatory, continuous and not homogenizing one.

On the other hand, as we will see, the non-Merina populations do not seem to have adopted such an ideology, despite their increasing influx towards Antananarivo since the colonial period, with the corresponding intensification of the links between the capital and the Highlands with the rest of the island. For them, the acceptance of the capitality of Antananarivo derives from its incomparable dimension, in the context of a very widespread conscience of a certain island community that does not imply any recognition of its symbolism, its *fomba* (tradition, custom). No needed homogeneity beyond the classification *vis-à-vis* the "non-*gasy*", "foreigners", in the sense of outsiders to the island (coming from *Andafy*), a classification that should be understood according to a segmentary logic, rather than a civic or ethnic one. Result: the convergence of distinct "projects" on the capital. Diverse city projects, much or less parallel, much or less mutually ignored, much or less articulated, in a context of inaction, partiality or impotence of the public administrators and the private investors recognizable in the environment of developing cities (Wachsberger 2009).

11.3.3 Heterogeneity "Disguised" as Anomie of Circular Crowns and Growth of Low Areas

The outer circular, discontinuous crowns (see Fig. 11.10) will also have their urbanist projection (not merely urban) in the capital, but far from the feeling of harmony of the hill model, which can, in a way, falsely evoke the notion of a single or dominant "city plan". That is to say, except for a few elites, notably people of the *fanjakana*, of the

State (see Galibert 2011),[15] non-Merina populations will preferentially settle in low-lying areas or in remote hills. However, and despite certain discourses that assume a common African origin for them (particularly those from the West), stories often tinged with racism (Goedefroit and Roca 2021), these peripheral, "coastal" populations will not be assimilated to the *mainty*—the "blacks", supposedly descended from slaves and dependents, the main occupants of these areas in the nineteenth century—and will not be deployed according to an easily differentiated and hierarchical arrangement, according to the Highlands pattern. They have thus broken with the high-altitude model, even if they seem to adapt to it.

This apparent "disorder" has been linked to their status as *vahiny* ("foreigners", without local *savoir faire*), much more pronounced than among Merina immigrants, which are always possible, even desirable, marital allies, regardless of whether they come from distant Merina regions, such as Vonizongo (to the northwest), Vakinankaratra (to the south), or the edge of the eastern forest. On the part of the Tananarivian elites (Merina *fotsy*, but also their close allies, including the colonizers), there is often added a tacit and unacknowledged presumption of cognitive, organizational and social inferiority somewhat in the pejorative and ethnocentric sense of the French word "peuplade". If the first point has been admitted by all parties, as we will see, the second is simply false. The inclusion of all these Malagasy *vahiny* populations in a rather disconcerting category, that of "*côtiers*", "coastal", was a colonial proposal. It was largely based on a typically diffusionist hypothesis of the settlement and formation of the cultures of the island, applied to a biased and racist reading of the conquest project attributed to Nampoina (*Ny ranomasina no valamparihiko*, "the sea is the limit of my rice field"), and stagnated already at the time of Ranavalona I: quasi-civilized Highlanders (Merina and Betsileo), organized as a dangerous embryo of nationality; decidedly barbarian "*côtiers*", (dis)organized into degenerate and declining kingdoms and other social formations, apparently more manipulable but ignored as collaborators on the path to modernization, excerpt for punctual cases.

As a matter of fact, the so-called "*côtiers*" have not considered themselves to constitute any social unit, nor have they historically acted as such, except when some of them want to oppose what they see as a takeover of the post-colonial state by Merina or Highland factions. The only fairly common category has long been that of *ambaniandro*, "under the day ", an old Merina monarchical terminology applied to subjects, and still used by the inhabitants of the provinces to designate those from Highlands, notably the Merina. "Coastal" diversity contains different historical and adaptive baggage that have strongly contributed to the cultural and symbolic

[15] *Fanjakana*, "government", from the same root (*zaka*) as *mpanjaka*, "king or traditional ruler" is the Malagasy word used to translate the concept of State. The people of the *fanjakana* are, therefore, civil servants, but also people linked to the modernizing action of the colonial and post-colonial states. Most of them are of Merina and Betsileo origin, and this discrimination was even more accentuated in the colonial period. However, a small minority of provincial notables was integrated from the beginning into the indigenous administration and the "evolved" collective, through co-optation strategies known as the "politics of races", which often included training at the Le Myre de Vilers School, created for this purpose in the capital (see Rajaonah, 2002).

heritage of the island, as dominant political formations (such as the Sakalava States in the seventeenth and eighteenth centuries), as kingmakers (advisors with presumed knowledge of the mysteries of power, often coming from the South-East), through their expertise in nature management, often embodied by *ombiasy*, diviners-healers) or as fearsome warriors, jealous of their independence and their uses, occasional mercenaries… In all cases, their societies have had recourse to highly symbolized spatial representations, but their logics, although pooling common elements, haven't come to coincide with each other, nor with the Merina patterns.

As noted, we propose three major "circular crowns" within the island. From the inside out: (1) the territories and their non-Merina populations permanently integrated into the kingdom at the time of Nampoina (notably the Betsileo) by a mixture of coercion and alliance, similar to that used in Imerina, and who came mainly from the Central Highlands; (2) the non-Merina forcibly incorporated following the campaigns of Radama I (with whom the most expansive phase of the kingdom of Madagascar ended), although they may have been allies at the beginning (like Nampoina and Queen Ravahiny, from Boina, in the North-West) and who were never well integrated into the ideological complex of the kingdom (stereotypes of local culture, Christianity, etc.), generating vast no man's lands, escaping any law; (3) the regions and groups which were never subjugated by Antananarivo (south, centre-west) and which will only be integrated into the national project after the colonial victory, having put up the longest and most fierce resistance to the French conquest.

It is important to note that despite the scarcity of studies[16] in this regard, it can be assumed that during the nineteenth century, but also through the colonial period, it was the Merina migrations that fundamentally fuelled the demographic growth of Antananarivo. The population could have increased from a then very considerable figure of 12,000 souls at the death of Andrianampoinimerina (1810), to a range of between 50,000 and 70,000 on the eve of the colonial conquest (1895). Around 1960, when independence was proclaimed, the agglomeration could have exceeded 250,000 inhabitants, including a good part of the 50,000 French who could inhabit the island (more than half of whom would come from Réunion, see Fageol and Garan 2020). Since then, the population has increased enormously to reach around 4 million in 2024 (Wachsberger 2009). Many populations have been "swallowed up" by the Tananarivian conurbation,[17] largely due to an uncontrolled extroversion of the Lower City. Everything indicates that it was at this time when the largest influx of non-Merina into the capital occurred, for economic and professional reasons, redefining the "Lower City".

[16] For various reasons, mainly ideological and power-related, cultural indicators have been seldom taken into account in the already limited sociological surveys that have attempted to describe the society of the capital in recent decades.

[17] You can even see tombs in the urban neighbourhoods, although they are inactive, because a law of 1923, colonial, therefore, but valid until today, prohibits burial in these tombs within the Municipality.

11.3.4 The Recalcitrant Merina Character of the Capital of Madagascar

This predominance of a local demography until the 1960s could explain certain singularities of Antananarivo as a capital. In any case, without denying the importance of this shift in the "pan-insularization" of Tana, what we see is that, during its modernising adaptation, intensified by the French colonial acculturation attempt that was going to remodel it, Antananarivo, seat of a foreign power imposed by force, nevertheless remained a city recognizable under the Merina label. It did not generate either an insular version of the cosmopolitanism typical of imperial capitals, nor the melting pot of a national culture produced by a modern capital that should integrate and represent, or merge, the peoples gathered under the same State. The demographically significant arrival of Malagasy people from "provinces" to the "City of a Thousand" has been therefore taking place in a "foreign" space, a space that the newcomers can barely consider as their own outside the State facilities, paradoxically imposed by colonization, far-off the ordinary new citizens and crowded with Highlands' bureaucrats. Municipal politics continues to be a Merina affair, and, apart from those *côtiers* who work in the *fanjakana* and a few rare entrepreneurs, most of the "provincials" seem to have opted for informal mechanisms for installation and labour integration. Mechanisms in which their own networks, earplug and patronage have played a predominant role. Networks that often have their own "centres" far from the Highlands (Blanchy et al. 2006; Ottino 1998), contributing to weave the "invisible fabric" of the current city, a contribution that may aid to understand the fault lines of certain crises (notably that of 2002, marked by the famous barriers that isolated the province of Antananarivo from the rest of the island, as well as the districts of the capital from each other, see Roca 2023), but also the emergence of new citizen solidarities between groups (see the multiplication of neighbourhoods in the plan of the city, Fig. 11.5).

In fact, the "heights capital" model only seems to have generated replicas to the south of the Central Highlands, in the Betsileo country, the heart of the first outside crown, probably thanks to a similar relief and housing culture, also marked by the search for security, in a context of intensive rice cultivation. But even there, there are distances. Fianarantsoa, the fifth largest city in the country, consciously adopted the pattern (see Raison-Jourde 1991; Mosca 2022). A new city built during the Merina expansion of the nineteenth century, the seat of an administrative division of the "Kingdom of Madagascar" that grouped together (and at the same time subordinated) the old Betsileo principalities and kingdoms, it was later the provincial capital and a connection nucleus with the southeast and southwest coasts. Although Betsileo society had a strategy of adaptation to colonial interference very similar to that of the Merina,[18] identities were not confused (no mixed neighbourhoods or lineage mergers) and neither were the cities (no imitation of the deployment of the statutory

[18] The percentage of people converted to Christianity is even higher than that of Imerina and the city, a stronghold of the Catholic Church, is considered an intellectual focus for the entire island, its name meaning "good education".

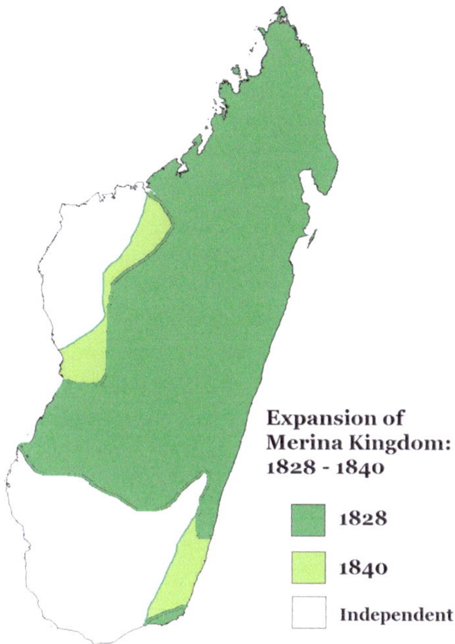

Fig. 11.5 Map of the expansion of the Merina kingdom, around 1840, somewhat idealised, evocative of the "crowns of the capital". *Source* https://es.m.wikipedia.org/wiki/Archivo:Madagascar

groups of Tana). It is therefore not appropriate to consider Antananarivo as the *renivohitra* of Fianarantsoa—and, through it, of the Betsileo country—in the way that an expanded version of the ideology of the "twelve sacred hills" might claim.

It would be even more illusory to advocate for the assimilation of such an ideology outside this region, where the heights do not exist, and have not been, therefore, a symbolic instrument relevant to the spatial hierarchy, neither on the alluvial plains of the West and the South, nor descending the eastern cliff, nor in the forests and bays of the North. One can certainly understand certain well-intentioned attempts to use the "twelve hills" as an instrument of pan-insular ideological cohesion, as it has been tried with the concept of *havana* ("kinship, proximity"). However, we must recognize its distance from local social dynamics, its intellectualism practically confined to certain elites of the capital. Thus, a few years ago, a famous French *malgachisant* went so far as to suggest in a newspaper that Mahabo, the reluctant capital of the conquered Menabe (western Sakalava), was still a "sacred hill" of the new unified Madagascar, carried by the legacy of Nampoina. However, Mahabo, which has a relatively flat relief, today lacks any symbolic representation value in Sakalava society, largely because of its link with Merina expansionism, which makes it more evocative of an abandoned Vichy than of a little Paris of the West Madagascar.

Certainly, other common island cultural traits could be examined, such as the symbolic importance of the relative positions of the elements of the house or the arrangement of the buildings and areas of the city in relation to the cardinal points (and their astrological reading, *vintana*). This is a classic subject of anthropological studies in Madagascar (such as those of Phillipe Beaujard, Jean-Claude Hébert,

Louis Molet or Sophie Goedefroit) and it is undoubtedly something to be considered when approaching the organization of settlements on the island. Nevertheless, here again, what emerges is a picture of great diversity, very much conditioned by the environment and history. For example, in the Highlands, the east and the northeast are generally considered sacred directions, consecrated to the ancestors as mediators with the cosmos, while the north remains the direction of political power. Now, in Tana, it is an impossible mission to try to apply these ideas—and others like them—to the distribution of the hills, the city itself and the populations that would make up the sacred alliance of Nampoina… The conditions of the relief seem preponderant and the Rova, the royal residence rises higher, but to the south of the gathering space of Andohalo… To put another example, in the Sakalava west, although the sacred weight of the east is maintained, the south is the direction of power, in memory of the route followed, according to tradition, by the Maroseraña dynasty, to which is attributed the formation of societies that we would categorise as states. The inversion of ritual behaviours is easy to imagine, an inversion that would reflect the multiplicity of referents of which we were speaking, and which has materialized in a certain way in the forms of incorporation of the coastal people to the Antananarivo metropolitan area.

11.4 Antananarivo, International Capital

This intersection of visions and strategies, of potential or informal projects, would be incomprehensible without including another category of actors, grouped in a fourth space, no longer peripheral but external: these are the foreign populations, who began to flock to Antananarivo during the nineteenth century and among whom the French colonizing contingent has had a singular weight. The colonial conquest had an enormous impact on the evolution and growth of urban agglomerations throughout Africa, including Madagascar and Antananarivo..[19] Starting with the indisputable fact that it was only after the conquest that the capitality of Antananarivo spread in a seemingly irreversible manner to the entire island. However, as we had already anticipated, this imprint is far from having produced a tabula rasa. Not only has it been very limited in terms of forms of sociability, political economy, values and concepts,[20] but in the case of Tana, urban planning has been hampered—or even

[19] On the influence of European models during the colonial period, and their interactions with the Malagasy model in Antananarivo, see Rajaonah (1996), contextualized in Goerg & Huetz de Lemps (2003). As for the validity of the different models in current daily life, as well as for the question of capitality, a fundamental work remains Guérin (2002), through his in-depth approach (Ph.D.) to all dimensions of urban life in Antananarivo and the conceptions (including "city plans") that condition it or derive from it, even the emotional and aesthetic aspects, often very marked in people's relationships with the city.

[20] Despite the decolonial discourse, largely coming from the disconnected diaspora (Afrodescendants "disconnected" by the slave trade), despite the debate between Octave Mannoni and Frantz Fanon on the psychology of the colonized based on the Malagasy case, colonization in Madagascar

eclipsed in some neighbourhoods—by the combination of the lingering centrality of the pre-colonial city and the deficit of post-colonial investment.

11.4.1 On Unexpected Colonial "Alliances": Andafy, the Fourth Crown

Regarding the formation of imaginaries (and projects, necessarily teleological) of and with foreigners in Madagascar, it should be emphasized that there is a strong continuity of this presence in Malagasy history, which has forged a term to designate this fourth crown: *Andafy*, literally "on the other side". The inhabitants of Andafy have had their place in Malagasy identities, either through their origin myths and stories, ancient and modern (Goedefroit and Roca 2021), the conformation of insular socio-cultural forms and ethnicities (Urfer 2022), or throughout often dependent integration into globalization, when Madagascar has participated in the circuits of slave trade, colonization and development. The awareness that everyone is a foreigner on an island if we go back far enough in time (but beware of visions such as that of Ratrimoarivony 2009) does not disavow the decisive—and changing—game between natives and foreigners, *tompontany sy vahiny*, first and last comers, throughout Madagascar.

The colonizers, and then the international development community, chose to settle in the Lower Town (see Fig. 11.4), despite the inclusion of certain urban planning messages in the old plan: notably the immense Andohalo high school (the Gallieni high school of the colonial era), later highlighted—like the entire upper town—by the Hollywoodian sign of Antananarivo on the slope of the hill, and what is sometimes called the "medium town", exemplified above all in Ambohitsorohitra, with the former palace of the governor, then of the president (until Didier Ratsiraka), as well as the main banking headquarters. The choice associated post-colonial elites with the new places of power (Assembly, Senate, businesses headquarters, evangelical churches, etc.). Then the signs of the colonial capital were deployed in the Lower Town, which was redefined following a real declaration of intent, a plan, deployed in the Independence Avenue, former Fallières Avenue, along the Tsaralalana district, where the old railway station, airlines, traditional establishments, the City Hall, the large market are located, ending since the 70s at the Place 13 Mai, a privileged scene of civil society demonstrations in national politics (see Fig. 11.6). This situation, in addition to giving a symbolic ambiguity to the Lower Town—a city of dominant foreigners and "dependent" natives, former slaves, the disinherited, but also coastal

was brief and with a very weak positive impact on most populations (including inculturation). The observation of Malagasy society, as, indeed, of the whole of postcolonial Africa, shows that the popular motto "Decolonize minds" (taken up by the conference of AEGIS—the main network of African study centres in Europe—on education in Africa, "Africa knows", Leiden 2021) should be addressed above all to ex-colonizing societies, to the metropolises, and, to a lesser extent, to their main allies in the former colonies, because they are immersed in a truly plural cultural environment, that has maintained its historical own singularities, that of the masses south of the Sahara.

dwellers and careerists—has largely conditioned the growth and development of the city and, consequently, its modernizing exemplarity.

So, since the consolidation of Antananarivo as the capital, the Lower Town, which houses the demographic dynamism of Tana, is much more than an enclave of poverty and submission. The old and new dependents rub shoulders with the people of the *fanjakana*, with cooperators or investors (even religious, with the strong presence from the 90s of evangelicals, such as the church of Jesosy Mamonjy, or the growing visibility of certain mosques). Far from the obvious hierarchy of the heights, without any perceptible overall planning, the neighbourhoods invaded by the lumpenproletariat are intertwined with new luxury residences or with the rise of large surfaces (a visit to the Ankorondrano quarter, for example, can be revealing in this sense).

Fig. 11.6 Map of Antananarivo. *Source* FTM

The coastal elites willingly fit into this ever-negotiated multiplication, this diversification—from below and from above—of former and simpler "city plan", unitary (kingdom) or double (colony). These peripheral elites choose in the Lower Town the meeting places, profane and sacred, such as the Solimotel for the "notables" of Morondava and the West, or the Lutheran church of 67 Ha—a district "feared" by the bourgeoisie of the capital—for those from the South-East and South-West... It is not because they have no other alternative: it is above all because it is there that their networks are articulated with the economic and political powers, foreigners included, without questioning the predominance of the Highlands in the interface with Andafy, but showing that they don't host a monopoly.

At the same time, part of the power, linked or not to the "Kingdom of Madagascar", remained in the heights: presidential palace, ecclesiastical seats, large banks, JIRAMA,[21] great *Fotsy* families, even a former IRD delegate... It is perhaps the Hôtel Colbert that most embodies this structural ambiguity, this need of the rival founders of the capital to have recourse to each other in order to reassert themselves throughout the island: the "heirs" of the old kingdom and the (neo)colonial power... They are "intimate enemies", as I have called them elsewhere (Roca 2005), paraphrasing and resignifying an expression of the late Paul Ottino ("the intimate foreigner", in reference to the Vazimba). Enclaved in the impossible and contaminated traffic of the average city, next to the old presidential palace, not far from Ambatonakanga, the first parish of the capital (1831), and surrounded by the Karana jewellers,[22] loved and hated by the westernized leaders of the city, the Colbert Hotel maintains a name that seems impossible in these decolonial times. A sign of the deep reasons for the unavoidability until now of the Upper Town which, as has been suggested, poses formidable problems for the planners of today's Antananarivo: the growth of the city in black tide has not been accompanied by a development plan that anticipates current mobility or housing needs to articulate them with the administrative and political functions of the capital.

The Upper Town embodies the African paradox in the face of development: on the one hand, it is an excellent opportunity to make the capital a heritage site and place it on the international map; on the other hand, it is a formidable obstacle to the rationalization of urban planning in the twenty-first century. On the one hand, the Upper Town identifies the capital and the country to the outside world; on the other hand, its well-recognized Merina character makes it fail in all expectations of representing all Malagasy populations, within the island context, as a supposed national symbol. Antananarivo shows sharply the inadequacy of the urban dichotomy "straw city" (African and traditional, presumed not really urban!) *versus* "cement city" (colonial and modernizing, "truly" urban -sic-), to refer to human behaviour, beyond buildings and infrastructures and their alleged functions, if we want to identify the "city plans" at stake in Africa. This is not a minor issue if we consider a horizon that no one dares to deny today: these African cities will be demographically and

[21] JIRAMA (Jiro sy Rano Malagasy), state water and electricity supply company.

[22] Designation of nationals of Indian or Pakistani origin, settled after generations, but most often with an ambiguous status, excluded from Malagasy nationality.

socially dominant in the global community at the beginning of the twenty-second century (Bearak et al. 2021; Statista 2022). In any case, this dilemma for managing African cities is particularly striking, and inspiring, in Antananarivo, because it is not admissible either to demolish the "African city" or to functionally replace it with a new capital, an efficient double (on the model of Brasilia), seeing the conditions of development on the Big Island, but also the social connections materialized in Antananarivo.[23]

11.4.2 A Capital "Homologated" by the International Community?

The only indisputable characteristic of a modern capital is probably that it houses the government of the state in question. However, many well-known capitals have other attributes and functions that various experts have tried to characterize. Let us look at an example and how Antananarivo fits into it. Rossman (2018), following the German historian Andreas Daum, emphasizes that the functions of a capital are those of administration, integration, symbolization of the nation—even performative—as well as that of preserving the common heritage (monuments, history and cultural values considered as their own). While the accomplishment of the first function is not in doubt—apart from the two federalist crises (1991–1992, 2002), in particular the second—few are those who would support the accomplishment of the last three, except with regard to the outside world: either as a place of international representation of the island as a whole, or as a gateway or interface for entry and exit from the island.[24]

To analyse the consequences, and the potentialities (we have already mentioned the conditions and causes), of this lack of adequacy to what is expected of a national

---

[23] In 2018, President Rajoelina launched a proposal to create a new city Tanamasoandro ("city of the sun" or "Sunbeams") in the town of Imerintsiatosika, 25 km west of Antananarivo. Although the project is progressing very slowly, it envisages the installation of the Presidency and the Prime Minister's office, as well as ministries, while the legislative and judicial branches would remain, in principle, in Antananarivo. It seems unlikely that the new city, if it sees the light of day, will be anything other than a satellite of Tana. On the other hand, the deterioration of public services in Antananarivo in recent years contributes to the project being understood as a way for elites to evade their responsibilities and to loosen the immediate pressure of social movements on them. The transfer of the Presidential Palace to Iavoloha by President Ratsiraka may be illustrative of this type of strategy and its limits. The location of the new building (probably inspired by the Rova), 15 km from the city, did not remove power from Tana, nor prevent the Freedom March (1991) followed by the fall of the Second Republic.

[24] The "international representation function" is exemplified by the Mahamasina national stadium, former field of royal enthronement which today hosts the popular national sport teams, either football (the Barea, maverick oxen, returned to the wild)—or rugby (the Makis, perhaps the most well-known lemurs). The "entrance function" is illustrated by the inevitable passage through Ivato, the airport, and through embassies and ministerial offices, as soon as it is a question of going overseas. See Razafimahenina (2022) for an analysis of the geopolitical representativeness of Antananarivo based on its role as a privileged scene of political crises.

capital, we suggest to oppose Rossman's functionalist vision (adapted to the planning of new capitals) to a more empirical or historicist vision (seeking to sort through the diversity existing in the old capitals), as this of Campbell (2000). Around twenty years earlier, this author formulated a capital model through a series of questions provoked by the "reinvention" of Berlin as the capital of a unified Germany at the heart of the European Union. Certainly, Campbell's approach, like Rossman's, presents a strong Eurocentric bias, but it is difficult to deny that, admitted or not, this has been the norm since the 1920s, in the urban planning of the South, of the Third World, of developing countries, of emerging societies, according to the different slang adopted by the international community in each moment. We therefore explicitly confront this bias.

Under this light, Antananarivo, as a capital, appears paradoxically again as a "strange case study". On the one hand, it has the unique character of many capitals in relation to their environment, due to its size and the continuity of its influence, as well as its urban logic, early singular in the island (eighteenth-nineteenth century), or its unexpected aesthetic, modernizing and traditionalist at the same time; it appears unique not only in relation to its national context, but also regarding the whole African subcontinent south of the Sahara. On the other hand, it follows the chronology of modern capitals, but not in the "right place". The advent of the modern capital has been associated with the maturation of the nation-state of European origin, a process which, while starting earlier, would experience a generalized turning point during the nineteenth century, either in Europe itself or in the *criollas* societies[25] produced by modern colonialism. That is to say, the redefinition of Antananarivo as a capital is synchronous with those of Berlin, Rome, Buenos Aires or even Paris, but in Africa, before the advent of colonialism and without resulting from the transmutation of an ancient imperial capital, or even a city-state properly speaking, given its relatively minor importance until its election as capital by Nampoina around 1787.

This initial association with the process of building the Malagasy nation-state would also explain, according to Campbell, that, despite its previous low economic relevance and despite its location far from the coast—a privileged area of exchange promoted by nineteenth century liberalism and colonial needs—Antananarivo joins the ranks of capitals, both political and economic. It is indeed exemplary of the category of capitals where political centrality precedes and stimulates economic centrality. However, Antananarivo does not owe the bulk of its economic power to "state affairs", namely to an presumed excessive differential in the recruitment of formal labour, due to the concentration of the civil service (national and international) and the attraction of foreign investment (as for free zones, the international

[25] "Criollo/a", in Spanish, identifies the descendants of settlers from elsewhere, such as the European "criollos" (descendants of Spaniards, Portuguese, French, English, etc.) or Africans (former slaves) in America. It is not translatable as "creole", a word that has the connotation of mestizo, especially with an African ancestor. In America, as in Australia, with the demographic collapse that followed modern contact with Europe, what was called the "Columbian exchange" (see the works of Kenneth Kipple or Alfred Crosby on this subject), postcolonial societies are predominantly "criolla" societies (very mixed or little mixed), while in Africa or Asia, they are fundamentally indigenous, autochthonous.

services sector or industry). Even if this attraction forces are effective, they are not the only ones: from its conception as capital until today, the city has been the centre of a region that is equally dominant in terms of agricultural production and local commercial dynamism (Dupuy et al. 2020). This was demonstrated during the blockade of Antananarivo province during the 2002 crisis, when food shortages were more acute in cities in other provinces than in the capital.

So, the role of the rural environment in Antananarivo goes well beyond the exoticism of urban rice fields (see Fig. 11.7), transformed into brickyards in the dry sea. Firstly it affects material production and the integration of peasant markets (a crucial constitutive factor in the consolidation of Antananarivo as a capital since the eighteenth century); secondly, it implies culturally defined social ties in relation to multipolar anchors (ancestral villages), defining a network of force lines and visions, of "lived plans" of the city, links which can in no way be explained by just speaking of clientelist dysfunction or ethnicist, or even "tribalist" mystification… Thus, the "proposal" to connect Antananarivo to its rural hinterland goes beyond the contribution of (peri)urban agriculture to food sovereignty. The "rural question" is not only a problem of land use planning and the growth of urban spaces, but an excellent way to begin to draw lessons on future African solutions to global urbanism. The socio-economic uniqueness of Antananarivo and the culture of its political economy constitute a clue to understand that the answer to Campbell's repeated questions on the specificity of capitals cannot be an internationalized presumably consensual model of governance… On the contrary, it suggests a panorama of great diversity by introducing collective subjects of law as such, and not as "sums of individuals". An introduction that can in turn contribute to explaining the deficit of national exemplarity of Antananarivo, a trait required for most typologies of modern capitals.

11.5 Some Thoughts on African Urban Planning to Conclude

Several authors have collected what is a glaring secret: African capitals are both powerful, through the accumulation of resources and decision-makers, and powerless, since they are often the seat of governments and administrations with enormous deficits in the production of services and in the guarantee of the rights and well-being of the populations within their states (Bekker and Therborn 2011). This scenario is crucial to understanding the survival of the solidarities that encourage the "invisible plans" mentioned above. Certainly, some analysts have condemned this urban "informality" in the same terms that they have denounced political informality as corruption or economic informality as irresponsibility towards the common good. Antananarivo proposes a certain number of approaches that could allow us to go beyond these condemnations, which have so far been rather sterile.

First of all, the examination of the evident continuity, sustainability we may even say, of the "Haute Ville plan", prior to the colonial pressure, inevitably leads to the

Fig. 11.7 Partial view of the lower town: colonial quarters, Ampefiloha, Anosy. *Source* Agence Nationale Taratra

admittance that the multiplicity of "invisible" plans also points to "*longue durée*". Thus, this poorly understood "informality" cannot be qualified as a mere undesirable by-product of imperialism, of advanced capitalism, as a dysfunction derived from precariousness… Nor can it be categorized as a "culture of poverty" in the sense of Oscar Lewis. It is above all a long-term adaptive phenomenon, with diverse and sometimes contradictory potentialities, which cannot be dispensed with, given the medium-term incapacity of the State—and of the international community which "shapes" it—to fulfil its functions in a supposedly global democracy. This does not mean that administrations are exempt from their duties, but rather that current challenges call for real participatory approaches. This leads social collectives to act on a case-by-case basis, something that often may involve recognizing indivisible community subjects of rights, which means that they cannot be reduced ("translated") to the international, eurocentric legal order, exclusively funded on individuals. Accepting this situation opens up many possibilities but it also implies enormous challenges because it requires research, creativity and consensus, not only education, mimicry and majority.

It may be useful to recall that, in recent decades, UN agencies such as the WHO, UNESCO or the World Bank have chosen to promote development policies that were once rejected as reactionary, such as the promotion of traditional and complementary medicine or indigenous knowledge, without neglecting, however, granting full and

Fig. 11.8 Lower town: Avenue de l'Indépendance area; see the old market, Zoma, and Faravohitra hill. *Source* Photo by Helihanta Rajaonarison

universal access to scientific medicine or to university training. And they do so even without knowing how to materialize these integrations of knowledge and technologies, aware of the challenge posed by the indivisible "collective property" of many of these potentials, beyond the cognitive difficulties of articulating epistemologies based on different rationalities.

So, why not sharing different urban knowledges and social technologies? Antananarivo—as other African cities—is already an "ongoing experiment". Its history, comparatively contextualized, is rich in lessons, such as those that can be deduced from what is perhaps its central paradox: the maintenance of a very marked ethnic identity, Merina, in a city accepted as the capital by all the populations of the island, the majority of whom are not Merina. Because this acceptance is obvious, there are no alternatives to Antananarivo, and they are not sought, except at the height of certain political crises. But it is also clear that the Malagasy populations continue to perceive themselves and behave (*fomba*) as distinct. And this persistence has resisted more than a century of attempts to co-opt Malagasy culture by an essentialized version of Merina culture: this process, today largely disavowed by science but still popularized, has crystallized in institutions such as the Académie Malgache, created in 1902 in the image of its French counterpart, famous for its universalism.

Thus, elsewhere, one of the authors (Roca 2005) has postulated a sort of "Merina function" of island cohesion, which would be embodied above all in the capital, and which could be considered as a "screen effect" or "membrane". The emergence of a mediating collective, anchored in the island imaginaries that associate and oppose the specific urban planning of the Tananarivian Upper Town with the landscape of the development of the "Lower Towns" across the entire island (Roca 2014), facilitates the continuity of a plural society, even in an unexpected way, and often misunderstood, by settlers, missionaries or cooperators. In the absence of a strong State and sufficient external economic and human investments—which never become "overwhelming", therefore—this screen effect, far from being ethnocidal, opens up "spaces" of autonomy for the continuity of indigenous cultural solutions: these remain "invisible" only to those who persist in looking only through the eyes of a presumed international "community". The mediating collective wants to be demiurgic but remains rather "cultural *degagiste*", in the sense that it manages a passive State that allows the daily predominance of decision-making outside of positive law (constitution, jurisprudence, etc.), but close to the law of the ancestors (*didinrazana, lahatra*). As such, this collective identifies with the foreigner, with global citizenship, and not with a new non-existent national citizen, but it does so without its members having to renounce their respective collective identities: even the Merina contingent—largely in the majority among the people of the *fanjakana* or in the rare middle classes—has not yielded to an acculturative pattern, but has opted for a multicultural existence. No message is stronger to express this cultural pluralism than the well-known urban planning of Tana.

The recent controversial restoration of the Rova, after the never-elucidated fire of November 6, 1995, is revealing. The "liberties" taken by the Malagasy state, without consulting the popular will or the academic expertise, are due both to budgetary difficulties and to the validity of the aforementioned "screen effect" applied to a living heritage, but understood in a very different way by the groups on the island. While UNESCO freezes the world heritage status for lack of historical rigour, the current president, Andry Rajoelina,[26] pushed the alleged continuation of the monarchical project, interpreted as a proto-nationalist "mandate", with the completed construction of an amphitheatre that had never gone beyond the sketch stage in the times of the kingdom, renaming the site as Rova of Madagascar (see Fig. 11.9). Meanwhile, in the irrecoverable inside of the palace, an unexpected "museum of the peoples of Madagascar" has been installed, wrapped in a nationalist glorification of the resistance of the Andrianampoinimerina dynasty in the face of the colonial onslaught. However, it ignores the dark side of the expansion of the Kingdom of Madagascar. The scandal of the historiographical profession or the indifference of the "coastal" population (few from regions outside Imerina had visited the old Rova and few seem likely to visit the new one) have not produced any apparent effects: the government has imposed its criteria, but there is nothing to suggest that this forced establishment of additional layers of "invisibility" can make the invisible planes disappear,

[26] One of the two mayors of Antananarivo (2007–2009), with Marc Ravalomanana (1999–2002), who broke the unwritten rule that prevented the major of the capital from becoming president of the country, by reinforcing a so-called "ethnic" distribution -merina/coastal-, also unwritten, in the summit of state power, modern. The effects of this rupture were a notable instability that cannot be addressed here (Roca 2023).

Fig. 11.9 Rice fields in the city, 2012. *Source* Photo by Helihanta Rajaonarison

although it can delegitimize a little more a State always externally supported (and associated with the exterior by the populations).

This insular functionality of a hierarchical Merina landscape makes the Upper Town inevitable, despite the obvious urban challenges it entails, and which road traffic highlights every day. In this sense, the creation of a hypothetical new capital, if it finally sees the light of day, could hardly be organized as a "substitution", but as a kind of "delegation", a new connection with global society, like Ivato, the airport. It is not that social well-being is secondary (starting with nutrition, hygiene or access to education), but rather that any urban policy aimed at guaranteeing it—and this is a formidable task—will have to rely on this diversity of projects, and of collective subjects involved, if it wants to be inclusive and effective.

It is essential to understand that the hierarchical message of the skyline, of the heights, is identified, but not "accepted", or at least not without interpreting it according to each group involved, not even within Merina society. For most Malagasy populations, the readable hierarchy is that of others, an indigenous visibility that reaffirms the cohesion and the actuality of other indigenous collectives throughout the island, by obliging them to respect the *tompontany* on their own land, opening the way to future reciprocities, which can include mediations with the outside and with the state apparatus (Fig. 11.10).

These singularities do much more than shatter the myth of cities created *ex-nihilo* in Africa (see Martinez in this volume), much more than relativize colonial

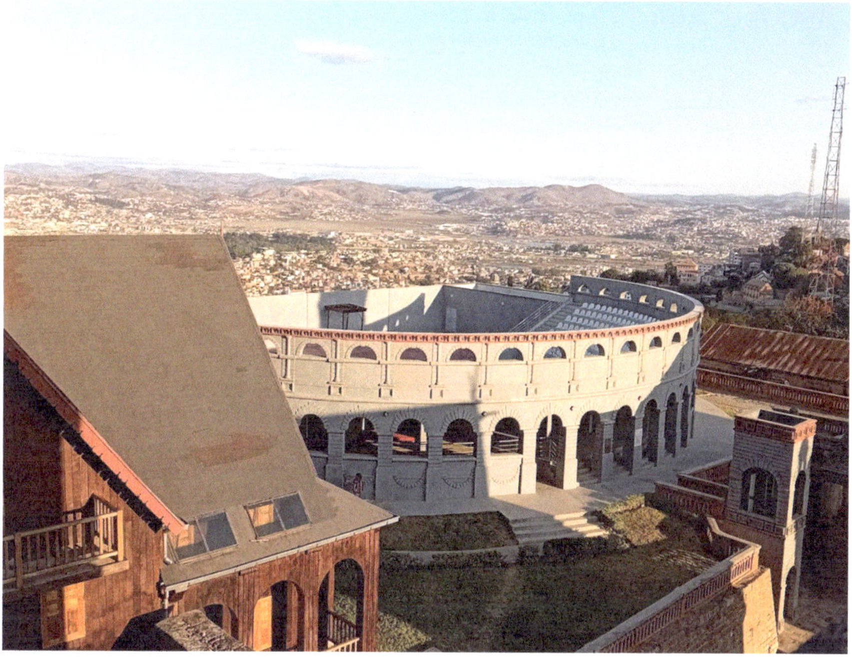

Fig. 11.10 New buildings in the Rova, erected as part of its reconstruction after the fire of 1995, and reclaimed by UNESCO. *Source* Photo by Helihanta Rajaonarison

conditioning in the design of new African cities, supposed to be planned according to the needs of the colonizing populations and nations rather than the colonized (see Minyoi 2019; Zoma and Sawadogo 2022). Favouring hard urbanism (a precolonial "brick city") and "exotically idiosyncratic" architecture, the Tananarivian example focuses on the importance of populations in the city project, these "invisible" plans inviting a positive approach to popular planning, without prior connivance: even if this terminology irresistibly evokes Calvino or Borges for Latin or Mediterranean readers, it is clear that its sources are different, its evolutions are different, and that it calls more for learning than for recognition or homologation.

References

Biao I (2019) Urban planning and the creation of livelihoods in Sub-Saharan Africa in the 21st century. In: Biao I (ed) Learning cities, town planning, and the creation of livelihoods. IGI Publishers, Hershey, pp 55–71. https://www.igi-global.com/chapter/urban-planning-and-the-creation-of-livelihoods-in-sub-saharan-africa-in-the-21st-century/228263

Bearak M, Moriarty D, Ledur J (2021) Africa rising cities. How Africa will become the center of the world's urban future. The Washington Post, upload on 19th November. https://www.washingtonpost.com/world/interactive/2021/africa-cities/

Bekker S, Therborn G (eds) (2011). Capital cities in Africa: power and powerlessness. HSRC Press, Cape Town. https://www.hsrcpress.ac.za/books/capital-cities-in-africa

Blanchy S et al (2006) Les dieux au service du peuple. Karthala, Paris

Bloch M (1983) La séparation du pouvoir et du rang comme processus d'évolution. Une esquisse du développement des royautés dans le centre de Madagascar. In: Raison-Jourde F (ed) Les souverains de Madagascar. Karthala, Paris, pp 265–299

Callet RP (ed) (1878–1881). Tantara ny andriana eto Madagascar. Imprimerie Officielle (traduction française, 1908), Tananarive

Campbell S (2000) Understanding capital cities. In: Cold war metropolis: the fall and rebirth of berlin as a world city. University of Minnesota Press, Minneapolis. Chapter 2. https://websites.umich.edu/~sdcamp/capitals/Ch2.html

de Blij H (1977) Human geography: culture, society, and space. Wiley, New York (11 editions since then, including "the Subsaharan African City Model"). https://www.vaia.com/en-us/explanations/human-geography/urban-geography/african-city-model/

Délivré A (1974) L'Histoire des rois d'Imerina. Interprétation d'une tradition orale. Klincksieck, Paris

Département d'Histoire (s.d.). Bibliographie de Patrimoine Architectural Urbain et Paysager d'Antananarivo (PAUPA). Document Digital, Université Antananarivo

Dupuy S, Defrise L, Lebourgeois V, Gaetano R, Burnod P, Tonneau J (2020) Analyzing urban agriculture's contribution to a southern city's resilience through land cover mapping: the case of Antananarivo, Capital of Madagascar. Remote Sens 12. https://www.researchgate.net/publication/342297987_Analyzing_Urban_Agriculture's_Contribution_to_a_Southern_City's_Resilience_through_Land_Cover_Mapping_The_Case_of_Antananarivo_Capital_of_Madagascar

Ellis S (1998) L'insurrection des Menalamba: Une révolte à Madagascar (1895–1898). Karthala, Paris

Esoavelomandroso FV (1989). Héritage de l'histoire et mode d'urbanisation malgache: Tananarive, dans *Histoire et organisation de l'espace à Madagascar*, dirigé par J. Fremigacci, Cahiers du Centre de Recherches Africaines 7:71–82

Fageol P-E, Garan F (2020). Les Réunionnais à Madagascar durant la période coloniale: espaces de contact et émergence de nouvelles identités (années 1880–1960). In: Arnold M, Duboin C, Misrahi-Barak J (dirs) Borders and ecotones in the Indian ocean: cultural and literary perspectives [en línea]. Presses Universitaires de la Méditerranée, Montpellier. http://books.openedition.org/pulm/6867. ISBN 978-2-36781-422-3. https://doi.org/10.4000/books.pulm.6867.

Galibert D (2011) Les gens du pouvoir à Madagascar : Etat postcolonial, légitimités et territoire. Karthala, Paris

Goedefroit S, Roca A (2021). La "plus belle énigme du monde" ne veut pas être résolue. Le besoin de mémoire multiple à Madagascar. Quaderns-e de l'ICA, desembre [monographique édité par A. Roca & J. Tomàs, Memòria múltiple a l'Àfrica], pp 383–406. https://publicacions.antropologia.cat/quaderns/article/view/371

Goerg O, Huetz de Lemps X (2003). La ville européenne outre-mer [Livre 5 de l'*Histoire de l'Europe urbaine* dirigée par J-L. Pinol, 2 volumes, sur tout le deuxième, De l'Ancien Régime à nos jours. Expansion et limite d'un modèle]. Éditions du Seuil (révisé en 2012; publiée en espagnol en 2011 par Publicaciones de la Universidad de Valencia), Paris

Gottmann J (1983) Capital cities. Ekistics, 50(299):88–93. https://www.jstor.org/stable/43620301

Guérin CF (2002) Vivre à Tananarive. Crises, déstabilisations et recompositions d'une citadinité originale. Géographie. Paris IV Sorbonne, thèse de doctorat. https://hal.science/tel-01705528 Submitted on 9 Feb 2018

Guzmán E, Christie JJ, Bogdanović J (2016) Introduction: the spatial turn and political landscapes of capital cities. In: Guzmán E, Christie JJ, Bogdanović J (eds) Political landscapes of capital cities. University Press of Colorado, Boulder, pp 3–24. https://www.jstor.org/stable/j.ctt1dfnt2b.6

Kamana AA, Radoine H, Nyasulu C (2024) Urban challenges and strategies in African cities—A systematic literature review. City Environ Interact 21. https://www.sciencedirect.com/science/article/pii/S259025202300034X

Mille A (1970) Contribution à l'étude des villages fortifiés de l'Imerina ancien. Musée d'Art et d'Archéologie de l'Université de Madagascar, Antananarivo

Minyoi MK (2019) The historiography of town planning and its shifting paradigms in Sub-Saharan Africa. In: Biao I (ed) Learning cities, town planning, and the creation of livelihoods. IGI Publishers, Hershey, pp 1–32. https://www.igi-global.com/chapter/urban-planning-and-the-creation-of-livelihoods-in-sub-saharan-africa-in-the-21st-century/228263

Mosca L (2022) Fianarantsoa, la capitale du sud de Madagascar: de la ville royale à la ville coloniale. Africa NS/1, 67–90

Ottino P (1998) Les champs de l'ancestralité à Madagascar. Parenté, alliance et patrimoine. Karthala, Paris

Rajaonah FV (1996) Modèles européens pour une ville malgache: Antananarivo XIXe-XXe siècles. dans Coquery-Vidrovitch C. et Goerg O (coord) La ville européenne outre-mer, un modèle conquérant? (XVe-XXe siècles). L'Harmattan, Paris-Montréal, pp 149–162

Rajaonah FV (2002) L'école Le Myre de Vilers (Antananarivo) de 1930 à 1960: politique des races et émergence d'une élite nationale. In: Raison-Jourde F, Randrianja S (dir) La nation malgache au défi de l'ethnicité. Karthala, Paris, pp 147–162

Rajaonarison H (2012) Photographie patrimoniale et contemporaine à Antananarivo. Africultures 88:133–144. https://doi.org/10.3917/afcul.088.0133

Rajaonarison H (2014) Photographie et sociétés à Antananarivo (milieu XIXe-milieu XXe siècle). Thèse d'histoire. Université d'Antananarivo

Rakotomalala M et al (2001) Les ancêtres au quotidien. Usages sociaux du religieux sur les Hautes Terres malgaches. L'Harmattan, Paris

Roca A (2005) "Els merina", en *Avui. Suplement Dominical*: "Cultures del Món". 1–9 (diumenge, 2 de gener)

Roca A (2014) O paisagem como fator de desenvolvimento: refleções comparadas. en: Sampaio J, Duarte Q (eds) Territorios, paisagens e suas dinâmicas. UEMA, São Luis de Maranhão (Brasil), pp 13–49. https://www.editorauema.uema.br/wp-content/uploads/files/2019/04/territorios-paisagens-e-suas-dinamicas-bx-resolucao-1554825728.pdf

Roca A (2015) Metodologies espúries, visions holistes. In: Benet J, Farré, A, Gimeno i J, Tomàs J (eds) Reis negres, cabells blancs, terra vermella. Homenatge al professor d'història d'Àfrica Ferra Iniesta i Vernet. Bellaterra, Barcelona, pp 337–368

Roca A (2023) Ethnicité et non-violence dans la transition démocratique de Madagascar en 2002. Lectures et significations d'une crise. en: Diallo I, Thioub I, Ndiaye AI, Benga N (dirs) Comprendre le Sénégal et l'Afrique d'aujourd'hui. Mélanges offerts à Momar-Coumba Diop. Karthala/CREPOS, Paris/Dakar, pp 229–264

Rafolo A (1989) Habitats fortifiés et organisation de l'espace dans le Vonizongo (centre-Ouest de Madagascar): le cas de Lohavohitra. Institut d'Art et d'Archéologie, thèse pour le doctorat de troisième cycle, Paris

Raison-Jourde F (1991) *Bible et pouvoir à Madagascar au XIXe siècle. Invention d'une identité chrétienne et construction de l'État (1780–1880).* Karthala, Paris

Ratrimoarivony M (2009) *La Lémurie, Terre des esprits: les enjeux spatio-culturels d'une réappropriation de l'identité autochtone à Madagascar*. thèse de doctorat. Université Michel de Montaigne (Bordeaux III), Bordeaux. https://theses.fr/2009BOR30065

Rossman V (2018) Capital cities: varieties and patterns of development and relocation. Routledge, London. https://doi.org/10.4324/9781315735061. https://www.taylorfrancis.com/books/mono/10.4324/9781315735061/capital-cities-varieties-patterns-development-relocation-vadim-rossman

Razafimahenina FAA (2022) Tananarive [1] et les crises sociopolitiques à Madagascar: une réflexion à travers les représentations géopolitiques. dans Regards Géopoliques, v 8, n° 2, pp 2–10 [accroché le 2 Juin]. https://cqegheiulaval.com/2022/06/02/tananarive1-et-les-crises-sociopolitiques-a-madagascar-une-reflexion-a-travers-les-representations-geopolitiques/

Southall A (1988) The segmentary state in Africa and Asia. Comparative Stud Soc History 30(1):52–82. http://www.jstor.org/stable/179022

Statista (2022) Urbanization rate in Africa by country. consulté le 12 Février 2024. https://www.statista.com/statistics/1223543/urbanization-rate-in-africa-by-country/

Urfer S (ed) (2022) Histoire de Madagascar. La construction d'une nation. Hémisphères, Paris

van Noorloos F, Kloosterboer M (2018) Africa's new cities: the contested future of urbanisation. Urban Stud 55(6):1223–1241. https://journals.sagepub.com/doi/full/10.1177/0042098017700574

Wachsberger J (2009) Les quartiers pauvres à Antananarivo: Trappe à pauvreté ou support des individus?. Autrepart 51:117–137. https://www.cairn.info/revue-autrepart-2009-3-page-117.htm?contenu=article

Zoma V, Sawadogo Y (2022) Main characteristics of African cities. J Res Human Soc Sci 10(4):11–17. hal-03653714

Chapter 12
Nador: From Military Barracks to Intermediate City

Salima Ouchen

Abstract This study aims to analyze the historical evolution of Nador, from its humble beginnings to its current development. We will examine the material remains and cultural testimonies to understand the various influences that have shaped the urban landscape of this growing city. Through a methodical approach, we will highlight the major changes that have occurred in Nador's urban planning and architecture. This includes an analysis of the key elements that have contributed to the city's current form, considering various historical and cultural factors. The goal of this study is to provide a clear and detailed view of Nador's evolution, offering an in-depth analysis of its urban and architectural aspects. We hope this exploration will enhance the understanding of Nador's transformation and illuminate the elements that have shaped its contemporary development. This study will reveal how each aspect of the city contributes to its current identity.

Keywords Nador · Military barracks · Urban planning · Spanish colonial architecture · Grid system · Modernization

12.1 Introduction

12.1.1 Overview

Nador is a city located in northeastern Morocco (Fig. 12.1), in the region of the Eastern Rif, bordering the lagoon called "Mar Chica." It is located 150 km west of Al Hoceïma, 80 km east of Berkane, and 110 km south of Taourirt. Its metropolitan area is contiguous to that of Melilla, forming thus an urban continuity. It is situated 200 km from Almería and 60 km from the Algerian border (Fig. 12.2).

Its international airport, Nador-Al Aroui, is the most important in the region, now surpassing Oujda airport in terms of traffic. In 2014, the population of Nador was

S. Ouchen (✉)
Rabat, Morocco
e-mail: atelierarchitecture1999@gmail.com

F. Carrión Mena and S. Rodríguez Alvarez (eds.), *From the City as a Project to the City Project*, The Urban Book Series, https://doi.org/10.1007/978-3-031-94226-6_12

Fig. 12.1 Aerial view of the city of Nador

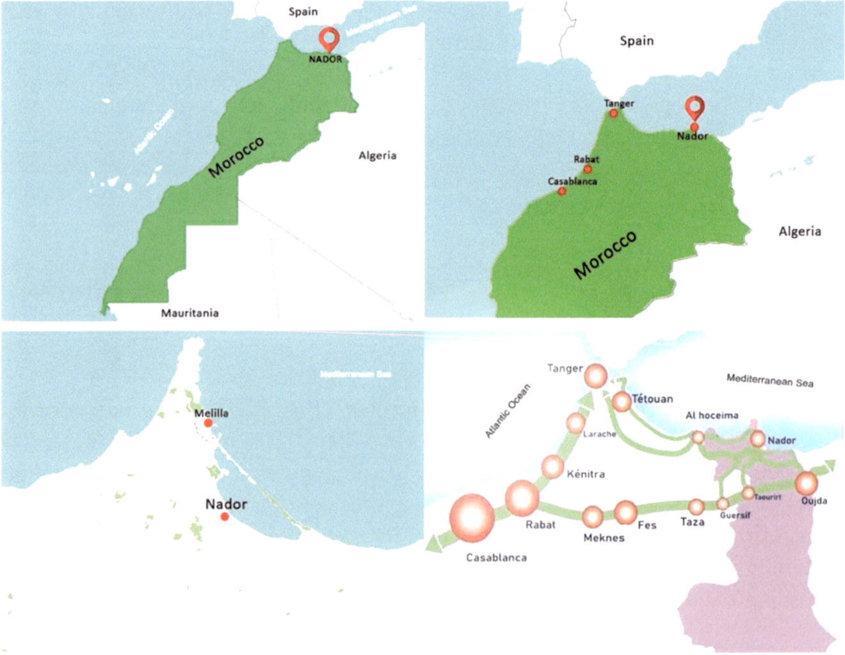

Fig. 12.2 Map of the region of Nador, Morocco

161,726 inhabitants: Now its metropolitan area, one of the main industrial zones of Morocco, has nearly 500,000 inhabitants.

The economy of Nador mainly focuses on steelmaking, with the presence of SONASID, a leader in Africa in this field. As well as modern fishing and international trade, through the port of Beni Ensar. This port is crucial for Moroccans residing abroad, gradually replacing cross-border traffic, locally known as "smuggling," which was done with Melilla. This traffic was interrupted due to the closure of commercial borders in 2014 and 2018.

12.1.2 National Situation of Nador

12.1.3 Strengths

Endowed with a strategic location as a Mediterranean gateway to Europe, backed by natural resources and certain economic potentials, the province of Nador is well positioned to play a leading economic role and constitute a true center of development, both regionally and throughout northern Morocco.

Regarding accessibility, in addition to a relatively well-structured road network, the province of Nador is accessible by sea and air through the existence of the Arroui international airport.

12.2 Nador in the First Half of the 20th Century: The Grid Pattern and the Beginnings of the City

12.2.1 Historical Overview

12.2.1.1 Nador: An Ancient Settlement

The city of Nador has obscure historical origins, with few documents predating 1909. However, there are indications suggesting an ancient human presence in the region, linked to the lagoon known as Mar Chica. A map from 1576 shows constructions in the Nador area, indicating a stable presence. Before 1909, Nador was described as a small scattered village at the foot of the hills, with a mosque called Yamaá El Baida. Travelers from the nineteenth century also mentioned the region, noting the existence of a village called Mezudya. The region included two independent lagoons, Bou Areg and Dzira, used for salt extraction. The dehydration of the lagoon and its connection to the sea were recent events at that time. The Guelayas, inhabitants of the region, organized horse races on the salt flats. These accounts testify to the historical and geographical importance of Nador, rooted in its natural and cultural environment.

Auguste Moulieras, a French author, also describes that the entire region was cultivated, with terraced fields supported by walls. His descriptions of the houses are also noteworthy: isolated villages atop the mountains, a village surrounded by fig trees, scattered huts at ground level, and wells carved into the rock.

On the other hand, this author accurately described in 1895 their market, the Souk Mezzouja or Souk el-Djoumouá (Friday market) at the northwest end of Bou Areg. An important center for commercial transactions where merchants from Melilla gathered. He also clearly mentioned the toponym En-Nadhour (the watcher) as a small locality of 100 households.

12.2.1.2 History and Evolution of the City

General Context:

Nador is the capital of its namesake province, covering an area of 6130 square kilometers. The province is bordered to the east and west by the Moulouya and Neckor rivers, to the south by the Béni Bouyahi mountain range, and to the north by the Mediterranean Sea. Nador occupies a somewhat peripheral position within its province, located on the northeastern edge of the Guelaya region, which includes the Gurugú massif and Cape Trois Fourches. It ranks among Morocco's top ten urban centers. As a Mediterranean city, Nador enjoys an exceptional location along the shores of a vast lagoon, Bou Areg or Mar Chica, at the heart of a rich agricultural region.

This region is currently experiencing rapid spatial growth, visible in the city's urban expansion, with both population and construction surging in recent years. Nador has seen substantial investments in infrastructure, including an international airport, a multifunctional port (for passengers, goods, and industry), a new railway station, as well as expanded roads and industrial zones.

Nador's urban development over the last century has been marked by profound changes, evolving from a small hamlet into a bustling metropolis (Fig. 12.3). In the early twentieth century, the city consisted mainly of a modest urban core, with its first urban plan dating back to 1909. This plan, devised by military engineer Luis Andrade, introduced the "grid" layout—an organized, rectangular scheme centered around perpendicular main streets—reflecting the influence of Spanish urbanism of that time. Over the decades, Nador's population steadily increased, spurring continuous urban expansion.

12.2.1.3 Military Origins of the City and Nador's First Settlement

Before 1909, Nador was a small village with few scattered houses, primarily engaged in agriculture and salt extraction. Following the military campaign of 1909, led by Spanish troops, the military engineer Manuel Pérez Beato oversaw the construction of a large military camp, marking the beginning of the modern city. However, the

Fig. 12.3 History of the city of Nador (Year 500–1900)

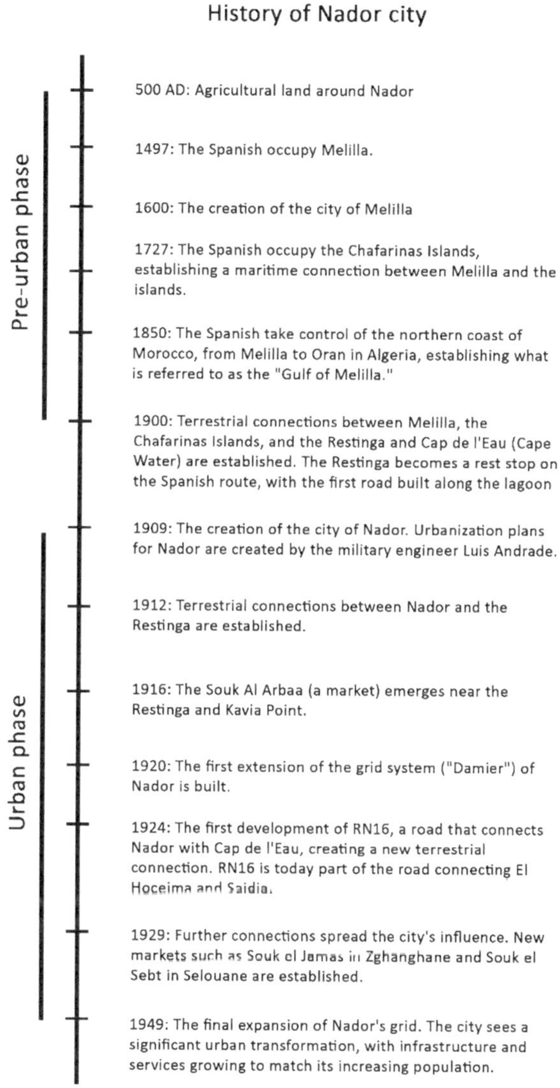

initial development was chaotic, with many newcomers building temporary housing in an informal neighborhood called the "Street of the Stations."

The military occupation deeply affected Nador, destroying part of the village. But by December 1909, reconstruction began with the arrival of the first Spanish settlers. The railway lines, an important element of this time, were already established, connecting Melilla to Morocco with separate train stations for different mining companies.

The urbanization of Nador continued its growth with the increasing influence of European mining companies, which already employed hundreds of local and foreign workers. Despite the initial challenges, Nador became a rapidly expanding urban center, thus marking the beginning of its modern development.

12.2.1.4 Architectural and Urban Diversity: Nador Compared to Traditional Medinas of Morocco

The city of Nador stands out in several architectural aspects compared to traditional Moroccan cities with medinas.

Modern Urban Planning: Unlike the medinas, which often evolved over centuries with narrow and winding streets, Nador was planned according to more modern urban standards, with wider streets and a more orderly design. This is often the result of more recent developments and planned urbanization, especially under the influence of the Spanish colonial administration.

Diverse Architectural Influences: Due to its complex history, Nador showcases architectural diversity reflecting the different cultural and political influences that have shaped the region over time. Examples of Spanish colonial architecture can be found, especially in government and residential buildings constructed during the colonial period. Additionally, elements of modern architecture can be observed in new constructions and recent urban developments.

Recent and Rapid Growth: While medinas are typically historic urban centers that have evolved over centuries, Nador is a relatively young city that has experienced rapid growth during the twentieth century, especially after Morocco's independence in 1956. This rapid growth has led to swift urban expansion, with new residential, commercial, and industrial neighborhoods built to meet the needs of the growing population.

Mix of Styles and Eras: Due to its rapid growth and cultural diversity, Nador features an interesting mix of architectural styles and eras. Historic buildings from the Spanish colonial era, examples of modernist architecture from the 1950s and 1960s, as well as more recent developments characterized by contemporary designs, can all be found.

In summary, the city of Nador is distinguished by its unique blend of architectural styles, modern urban planning, and rapid growth, making it a fascinating place to explore for those interested in architecture and urban history.

12.2.2 Historical Stages

12.2.2.1 The Construction of the New Nador: The Grid (1909–1920)

Under the protection of military positions, notably the Reducto, a vast headquarters was erected on a strategic hill north of the village, supervised by military

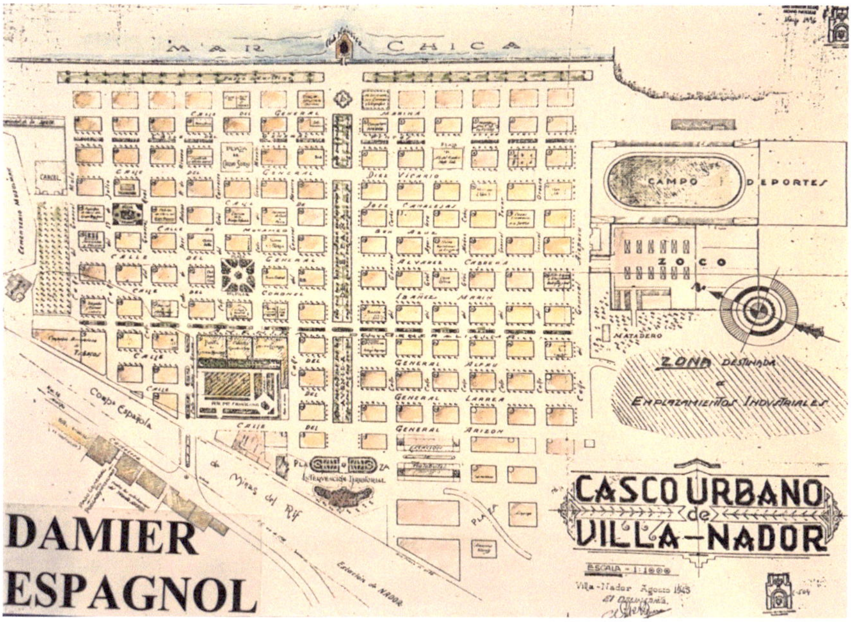

Fig. 12.4 Diagram of the Spanish grid

engineer Manuel Pérez Beato. This project, completed in 1911, spanned an area of 18,000 square meters. Simultaneously, General José Marina reestablished the Mazuza market in January 1910, further enhancing the importance of the new settlement. The arrival of Spanish civilians in December invigorated the village, attracting additional Spanish immigrants. Initially, the settlement was an improvised, unplanned temporary village, but it was soon replaced by the "Grid" upon military orders. Images of this first settlement show an irregular urban layout with poorly built structures (Fig. 12.4).

True civilian settlement occurred in late December 1909, thanks to a project by military engineer Luis Andrade. Now largely forgotten, Andrade was the true architect of an urban plan considered one of the most representative styles of Spain's presence in northern Morocco. His plan was rectangular, consisting of twelve blocks in length and six in width, totaling 68 blocks. The rectangle was intersected by two main perpendicular streets, 40 m long, resembling the cardo and decumanus of a Roman camp. By removing certain blocks, four internal squares were designed, one for each smaller rectangle. The parallel streets along the length measured 20 m, while the cross streets were 10 m wide. However, this initial project quickly transformed due to a growing population, and in 1911, new blocks were added to the west of the rectangle, giving it a trapezoidal shape. The number of blocks then rose to 89. In 1914, the original grid was further expanded, increasing its capacity. By then, there were 103 blocks, 35 more than planned four years earlier. This morphology persisted at least until the 1920s, with further improvements made in the 1930s.

The key advantage of this urban expansion was its relatively large size, allowing for multiple construction phases without reaching saturation. Architectural renovations were also possible: first with wooden buildings, then solid structures, and finally (in the 1930s and 1940s) with the construction of multi-story buildings, marking the end of the city's expansion plan.

The width of the streets also allowed for the planting of trees along the sides, especially on the central avenues, which harmoniously structured the urban ensemble: a large longitudinal avenue (Reina Victoria Avenue, which disappeared when its central space was later created) and a transversal one, España Street, which later became Boulevard Mohamed V. The construction of the town was a long process, with the pace of building heavily dependent on the historical phases tied to Spanish government intervention in Morocco. The grid design dates back to December 1909, when the construction of simple single-story houses and buildings began. At that time, the civil population's size did not demand a larger expansion, and the planned blocks were sufficient to accommodate them. Urban control was initially overseen by military engineers, who were responsible for determining the layout of streets, as well as the designs and facades of buildings.

These new neighborhoods initially housed Spaniards working in various trades and business activities. However, this was not the only residential area in Nador. Before the Spanish intervention, a small hamlet already existed on the hillside to the north of the plain. Later, the installation of the Regulares de Infantería barracks in this area led military engineers to create several housing units for Moroccan soldiers and their families. By 1911, "houses adapted to their customs and traditions" were already under construction, ready to accommodate Moroccan families. That same year, military engineer Manuel Pérez Beato oversaw the establishment of an indigenous school, the first built by the Spanish administration in all of Morocco. This hamlet thus became the origin of Nador's new "Moroccan quarter." Located on the slopes to the northwest of the European settlement, it initially served as the Regulares' quarter. While initially stable, it quickly experienced disordered growth.

The blocks measured 10 m × 5 m, though they could be grouped, and were numbered clockwise: A1, A2, B1, B2, C1, C2, D1, D2, E1, E2, F1, and F2.

12.2.2.2 Urban Evolution of the Grid (1921–1956)

In general, it can be said that the urban development of Nador during the first half of the twentieth century was driven by three main factors. First, it was due to the strengthening and diversification of the civilian population's economic activities, along with the military presence in the barracks. Second, Nador was chosen as the capital of the Khalifian Eastern Region of Morocco, which led to the establishment of all regional administration delegations.

From 1909 to the early 1920s, Nador's social organization included a Junta de Arbitrios, a mixed civilian and military institution responsible for managing urban issues and maintaining infrastructure. On December 22, 1922, a Junta de Servicios Municipales was created, handling urban affairs among other responsibilities. The structure and functions of this Junta were strengthened from 1927 onwards, when the "system of Interventions" was introduced in Spanish administration, making each controller the president of the Juntas. Finally, on March 23, 1934, Nador was officially elevated to city status and renamed Villa Nador. Significant buildings were constructed during this period, including courthouses, a regional prison, a post office, a civil hospital, a dispensary, schools, a market, a slaughterhouse, the Junta de Servicios Municipales building, bajalatos (dignitaries' houses), and many others. It's not surprising that during this time, the population increased from 4699 in 1935 to 8826 in 1940, then to 22,841 in 1953, and 23,443 in 1955.

The Spanish population in the city was always significant and continued to grow between 1909 and 1956. In 1923, there were 1467 Spaniards, increasing to 2810 in 1935 (65% of the total population of 4000), to 5978 in 1940, and to 6050 in 1945 (25% of the total population). By 1953, there were 8895 Spaniards (38%), and by 1955, this number reached 9165 (39%). From these figures, it is clear that the Spanish population played a crucial role in the city until 1956.

At the same time, the Muslim Moroccan population always had a prominent role in the city. It is also important to mention the Moroccan Jewish community, which played a significant role as well. In numbers, the Muslim population was smaller during the early decades but became the majority by the 1940s. The Jewish community reached its peak in the mid-1940s and then gradually declined. These figures reflect the great social and religious diversity in Nador during that period.

12.2.2.3 The Urban Crisis of 1921

Nador's main urban crisis during this period occurred in 1921, coinciding with the disaster of Anoual, when part of the city was destroyed during the war, and some buildings were set on fire. Nador was abandoned, and the entire Spanish population was forced to flee. However, their return was swift, as by 1923, 1467 Spanish residents had already resettled. With the permanent establishment of all Khalifian administrative bodies in the 1920s and some economic activity, Nador began to attract the rural population from its surrounding region. This migratory phenomenon was further intensified by the severe drought that struck northern Morocco in the early 1940s, marking the point when Nador's Moroccan population began to significantly outnumber its Spanish residents.

12.2.2.4 The Main Spaces of Nador in the 1930s and 1940s Were the Main Boulevard and the Seafront Promenade

The major turning point in Nador's history was its elevation to city status in 1934, marking the beginning of its development. Infrastructure projects were launched by the Junta de Servicios Municipales, including the construction of markets, a fish market, a slaughterhouse, and public lighting. A retaining wall was built along the coast of Mar Chica to support the creation of the Paseo Marítimo. The densification of the built space within the Cuadrícula occurred throughout the 1930s and 1940s, with the establishment of various public institutions. The Cuadrícula's expansion to the east led to the construction of new buildings in rationalist and Art Deco architectural styles. Additional works were completed before 1947, such as the construction of municipal buildings, a bus station, a new souk, and the expansion of the Lope de Vega school.

12.2.2.5 In 1949, an Urbanization Plan Was Implemented, Marking the Final Expansion of the Cuadrícula

After all the internal modifications to the grid, it became clear that each population increase would require new expansion areas and the creation of a new urban development plan. Nador functioned efficiently as long as the expansion zone was not saturated. However, its major flaw was the lack of alternative expansion models, which explains why a certain order prevailed around the grid, but not beyond.

In 1949, a study led by a team of architects under Pedro Muguruza Otaño was approved. This team was responsible for drafting general plans for several cities in northern Morocco. After the grid layout in 1909, Villa Nador's urban plan was the first to demonstrate a clear intent to manage the city's growth. The goal was to create a growth model outside the now saturated grid. This plan, similar in spirit to cities like Tetouan, Ksar El-Kebir, Larache, and Al Hoceima, aimed to organize the city through zoning. Different parts of the city would host various modes, activities, and building types in each district or sector (Fig. 12.5).

Facing the existing city, a new southern zone was envisioned: intensive construction and open blocks with various public buildings and open spaces. However, this urban plan was never fully realized, and the city's growth continued in a state of chaos.

12.2.3 Nador in the First Half of the 20th Century: The Grid and the Beginnings of the City.

12.2.3.1 The Grid Today

After Morocco's independence, the colonial administration entrusted architect Eduardo Caballero Monrós with conducting a final urban planning study for Nador, the first city designed by the Spanish administration in 1909.

Spanish architect Fernando Caballero y Vázquez (1914–2007) is known for his contributions to modernist architecture in Morocco during the protectorate period. He played a key role in the design of several iconic buildings, particularly in Casablanca, where his style blended modernist influences with local architectural traditions.

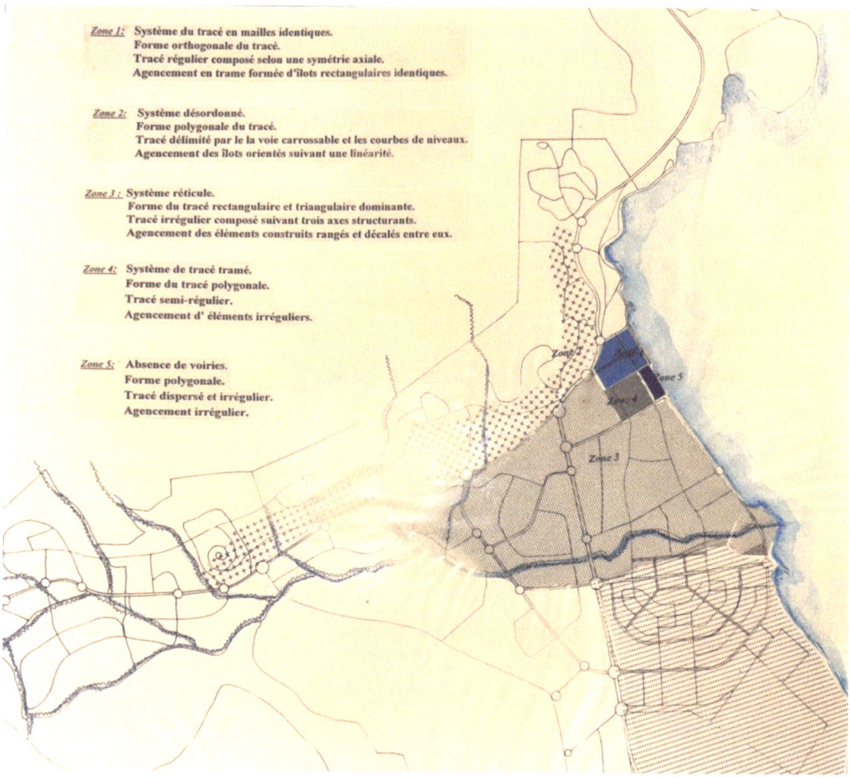

Fig. 12.5 Diagram of the zoning of the city of Nador

His work reflected an innovative approach, aiming to adapt modern architectural principles to Morocco's climatic and cultural contexts.

Caballero participated in the construction of various administrative, residential, and cultural buildings that shaped the Moroccan urban landscape, leaving a visible architectural legacy today. He began working in Nador in 1953 but resigned in 1954 due to work constraints. He returned in 1955, winning a competition for a new project, and was appointed urban planning inspector for the cities of Oujda, Nador, and Al Hoceima until his final departure in 1960. Caballero analyzed Nador's urban problems, highlighting the urban chaos prevailing since the 1950s, with unplanned growth spreading south between the Zeluan road and the coast.

In his 1958 study, Eduardo Caballero provided a detailed portrait of Nador, emphasizing its growing commercial role and importance as an administrative center and transit hub to Melilla. He identified post-independence economic issues, particularly at the Beni Ensar customs, and predicted significant growth for the city, stressing the urgent need for a new port, potentially in Mar Chica or elsewhere along the coast.

Caballero criticized the 1949 urban plan, noting that the city had exceeded its limits since the 1930s. From 1950 onwards, disorganized growth led to the formation of suburbs and slums, which, according to him, would pose a major challenge for the city's future. He also provided detailed demographic data, showing the population distribution across different neighborhoods and villages, highlighting much higher population density in the suburbs compared to the city center:

- In the city center, except for the French station: 1010 dwellings, 73 of which were precarious, housing 1100 families.
- In the suburbs, comprising three neighborhoods: French station and Réguliers district: 55 dwellings, all precarious, housing 79 families. Poblado de Réguliers: 257 dwellings, 43 precarious, and 214 huts, housing 483 families. Poblado Civil: 243 dwellings, 93 precarious, and 82 huts, housing 481 families.

In total, there were 1570 dwellings, 264 of which were precarious and 296 huts, accommodating 2143 families.

Caballero recommended a complete cadastral plan to address land ownership issues in Nador, noting that most properties were unregistered. He also pointed out the lack of municipal or government land, hindering urban interventions.

His urban plan for a city of 35,000 to 40,000 inhabitants envisioned expansion to the south while preserving the souk, the slaughterhouse, and the railway tracks. He advised against residential construction near the coast due to unsanitary conditions, while suggesting the development of the city's hills and the transformation of the Muslim cemetery into a park.

During the second half of the twentieth century, Nador experienced an unprecedented urban explosion. The city's population multiplied, reaching unexpected limits, significantly increasing the built-up area to the point where the old grid became just another neighborhood. What Nador has not lost is its typically Mediterranean character in the urban center. For this city, born in the twentieth century, the grid remains clearly its historic center, where the history of its urban evolution is present in every stone, with archaeological remains from various stages of the city's development still visible. In an era when cities and villages worldwide are highlighting their historical and cultural heritage, these urban elements are a true guarantee of authenticity and rarity for its inhabitants, in contrast to an overly standardized and globalized world.

A part of Nador's history originates in this first neighborhood, where various buildings tied to its unique history can still be seen. Its urban layout and buildings form a shared heritage between Spain and Morocco, built during the period when Spain administered the city, and it is now inseparable from Nador's identity. A city or village cannot live disconnected from its history, as it is what gives it the strength to face the future with confidence.

In 2009, the centenary of Nador's founding was celebrated, though many had forgotten the date. The debate over the preservation of old cities is now more relevant than ever and must shape the objectives of protecting valuable architectural and urban spaces to prevent their destruction and disappearance.

12.3 Architecture in the Grid: From Eclecticism to the Modern Movement

12.3.1 The First Architecture

Nador's humble beginnings were characterized by simple housing, reflecting the precarious conditions of the first settlers in a rapidly developing city. Under strict army regulations, military technicians directed the construction of the first infrastructure. Occupied by Spain in 1909, Nador received a formal administrative framework in 1912 with the consent of the settlers. Spanish military engineers then designed the first urban plan, adopting a simple organization with uniform facades to harmonize the appearance of the buildings.

The first buildings, often single-story, met basic functional needs. Notable historic examples included the church and a few schools, which were the main landmarks of the fledgling city. However, many of these original buildings were demolished over time to make way for more modern structures, erasing much of the visible trace of Nador's first village.

12.3.2 The Eclecticism

Most of Nador's early buildings adopted an eclectic style, influenced by railroad-related projects such as stations, as well as civic and religious architecture (Fig. 12.6). One of the most monumental examples of this eclecticism is the church of Santiago El Mayor, built between 1917 and 1921. This building is a remarkable example of architecture, with an imposing façade marked by twin towers and a pavilion roof, the work of the architect Francisco Serra.

The interior of the church is characterized by a spacious nave covered by a barrel vault and adorned with decorative arcades that give the space a sense of grandeur. Exterior details such as moldings, pediments, and harmoniously integrated windows reinforce the monumentality and overall aesthetic of the building, making the church an architectural symbol of Nador's eclectic period.

12.3.3 Modernist Trends in Nador

In Nador, the influence of the modernist movement was relatively limited, largely overshadowed by the architectural development in Melilla. The modernist works in Nador were generally modest, characterized by decorative details such as floral moldings on openings and balconies, likely imported from or inspired by Melilla. These balconies, typical of the style, often featured curved wrought ironwork or floral motifs.

Fig. 12.6 Iconic building from the eclectic style era

The Club of Hunting and Fishing, despite being partially mutilated and at risk of disappearing, is considered one of the most significant examples of modernist architecture in the city. Even in its dilapidated state, it continues to testify to the impact of this architectural movement in Nador, marking an important chapter in the city's urban history.

12.4 Architecture in the 1920s and 1930s

12.4.1 Neo-Arabic and Art Deco

Art Deco in Nador manifests in three distinct trends. The first merges Art Deco elements with Andalusian influences, represented by architects such as José Larrucea and Manuel Latorre. The second takes a more rationalist approach, akin to modern architecture, illustrated by Francisco Hernanz and Manuel Latorre. Finally, a later aesthetic inspired by aerodynamic lines and machine aesthetics is introduced by Manuel Latorre.

The first trend combines Neo-Arabic characteristics with Art Deco elements, particularly in the civil constructions of the colonists. The Villa Nador hospital, designed by José Larrucea Garma, embodies this fusion with its rational lines and subtle Neo-Arabic decorative touches.

12.4.2 Art Deco and Rationalism

Nador adopts a distinctive architectural identity with Art Deco and Rationalism, thanks to the arrival of architects from the Civil Construction Service. Francisco Hernanz and Manuel Latorre are the key architects behind many buildings in Nador, with Latorre being the more prolific of the two.

Hernanz, known for his rationalist and Deco work, designed several schools and the Nador prison, characterized by a cubic design and vertical openings reminiscent of Art Deco. Rationalist architecture is distinguished by changes in design and innovations in construction typologies, evident in the use of new architectural languages and distinctive elements. Latorre's work in this vein includes the local intervention in Guelaya, inaugurated in 1940, which showcases a rationalist composition with cubic volumes and horizontal lines, marked by two vertical pillars that emphasize the main façade, which later became the headquarters of the Moroccan Royal Gendarmerie.

12.4.3 The Aesthetic of Machinery and Aerodynamic Designs

Between 1939 and 1942, Manuel Latorre designed the Flechas Navales refuge (now the Nautical Club), an extraordinary building inspired by aerodynamic principles, elevated on stilts above the Mar Chica. It resembles a boat moored at the port, reminiscent of river steamers with its prominent smokestack. This work, one of the key examples of this style in Nador, still dominates the waterfront promenade and has been restored.

This building reflects the urban changes of the 1940s in Nador, highlighting the axis of the waterfront promenade along the Boulevard. This area was characterized by architecture linked to Rationalism and Art Deco, featuring notable chalets, such as the one belonging to the regional intendant, which has unfortunately been demolished.

Another building with aerodynamic shapes, located at the end of the boulevard, echoes the aesthetics of machinery, albeit heavily transformed from its original appearance (Fig. 12.7).

12.4.4 Art Deco Design: Craftsmanship and Architecture

Nador showcases high-quality doors that reflect the Art Deco style, spanning a broad period from the 1930s to the 1940s. These doors, found in both Art Deco buildings and others, illustrate the layering of styles in various contexts.

The double-leaf doors are intricately carved, often featuring decorative wooden and wrought iron elements that accompany the upper oculi. The wood displays tubular and undulating forms, sometimes adorned with decorative scales, while the

Fig. 12.7 The Restinga Fort

oculi adopt various polygonal shapes, contributing to a break from straight lines and disrupting conventional logic.

12.4.5 *Rifian Architecture*

In the Rif region, military architect Emilio Blanco Izaga introduced Rif architecture, particularly in Villa Alhuceimas, aiming to provide the Rif community with a distinctive architectural identity. Manuel Latorre, in collaboration with Blanco Izaga, designed two works in this style in Nador, despite the city not being located in the Rif: a fondouk and a mosque.

The fondouk, surrounded by modern buildings, retains Rif characteristics such as sloped walls, vibrant colors, and geometric patterns, while the mosque adopts an Arabesque aesthetic with a characteristic dome and minaret, albeit altered over time.

Latorre successfully integrated Rif aesthetics while concealing some of his primary contributions. Unfortunately, these buildings have been demolished over time, erasing a significant part of Nador's architectural heritage.

12.4.6 The Sgraffito Architecture

The epilogue of Art Deco architecture is reflected in what we refer to as sgraffito architecture. This trend is characterized by the use of rationalist or aerodynamic design in its structure (square belvederes with right angles, curved or straight balconies, etc.), but with an envelope based on stucco surfaces adorned with decoration. Stucco serves as a façade finish that simulates marbled tones and shades, showcasing high quality and elegance with reddish, blue, or green colors.

The examples known in Nador exhibit reddish tones, positioning these works in the early 1940s if we follow the evolutionary pattern seen in Melilla and the city of Al Hoceima. Additionally, these stucco surfaces are decorated with highly schematic floral sgraffito motifs in a Deco style, forming bands across balconies, windows, walls, and the entire surface. The decoration is achieved through sgraffito techniques on the stucco surface, shaping the decorative ornamentation.

12.5 During the 1940s, in Nador, There is a Trend Toward More Classical and Baroque Architectural Programs

In the eastern regions of the Rif and Kert during the 1940s–1950s, architecture experienced a decline in the avant-gardism previously embodied by Francisco Hernanz. Only Manuel Latorre maintained a metropolitan level, yet his work reflected more of a continuity from the 1930s rather than serving as a precursor to autarkic architecture.

In Nador, Latorre reinforced the urban fabric with key buildings along major roadways, such as the Regional Intervention building and the Nautical Club. However, regional architectural programs remained modest compared to those in western Morocco, such as in Tétouan.

Throughout the 1940s, Art Deco lost popularity in favor of new currents. The departure of Francisco Hernanz from Nador, coinciding with the onset of the Spanish Civil War, did not necessarily lead to a political shift, but architectural forms tended toward greater sobriety and classical monumentalism.

12.6 New and Old Models in Architecture During the 1940s and 1950s

12.6.1 Social Housing

In Nador, some monumental buildings combine Art Deco with traditional elements like baroque consoles, while others incorporate architectural innovations such as the patented "seagull wing" concrete system. However, the demolition of certain structures has led to the loss of their original essence.

Later, the city witnessed the emergence of buildings adopting a picturesque language with regional elements, such as cornices and tiled roofs, while maintaining white facades. The transition from the 1940s to the 1950s was marked by the construction of new residential buildings, some retaining traditional features while others anticipated modern architecture.

12.7 The Triumph of the Modern Movement: Claudio Verdugo

In the second half of the 1950s, the modern movement took root in the eastern region of Morocco, marking the end of Spanish architectural influence. The widespread use of reinforced concrete, exemplified by the roof of the Rif cinema in Nador, signifies this evolution. Eduardo Caballero Monrós, active in the region until 1960, embodies this transition toward rationalism and the modern movement.

In Nador, iconic buildings such as the airport control tower and the Rif hotel reflect this modern aesthetic, characterized by white facades, cubic shapes, and the boldness of reinforced concrete. Claudio Verdugo, a Spanish architect residing in Morocco, brings a brutalist vision marked by the functionality and visual strength of concrete.

Verdugo completed several projects in Nador, including a vocational training center in 1960, the first of its kind in independent Morocco. His other educational projects, like the School of Masters and Hassan II School, addressed the growing needs of the Ministry of National Education.

In the 1970s, notable constructions included the new post office building, which has undergone significant transformations, as well as the city's hospital, which was completed in 1984, and the official headquarters of Amalato along the waterfront. The latter, a remarkable example of rationalism, contrasts with traditional architecture but has experienced renovations that altered its original appearance.

12.8 Nador Beyond the Grid: The Military Space

Before 1940, Nador's urban structure was characterized by its grid-like or checkerboard expansion, along with military barracks lining the city to the north and southwest. In the 1940s, the Regular Cavalry Quarter extended along the Zeluán and Monte Arruit roads, featuring a small defensive garrison near the aerodrome. Although these military constructions have since been dismantled, they hold significance for Nador's urban history, with important barracks also built in nearby areas such as Segangan, Tauima, and the seaplane base at Atalayón. Together, these military structures form a significant unit and constitute an essential part of the region's contemporary urban heritage.

12.8.1 Origins of the Regulares Infantry Quarter no. 2 in Nador Before 1940

Following the establishment of the new city of Nador and its grid plan for civil urban expansion, the Army decided to construct a significant complex of military buildings on the hills north of the city, along the road leading to Melilla. The initial projects date back to 1911, with a plan developed by engineer Manuel Pérez Beato for the Nador Barracks, which included various military units. These military installations were influenced by subsequent urban growth, with some structures disappearing and others being transformed into headquarters for the Moroccan Royal Armed Forces.

The first barracks, built according to the original 1911 design, were simple wooden structures. Over time, they were replaced by masonry buildings, with some original elements still remaining despite partial losses incurred during the Battle of Annual. The cavalry squadron quarter, one of the most significant in northern Morocco, gradually took shape, with modifications and renovations carried out by military engineer Luis Castroverde in 1923.

Today, these barracks, although in a state of abandonment and occupied by marginalized housing, bear witness to early 20th-century military architecture. They feature eclectic decorative details centered around the framing of openings and cornices, reflecting the architectural style of the period and the historical significance of military presence in Nador (Fig. 12.8).

12.8.1.1 The Barracks on the Right Side of the Road

The construction of the barracks along the Melilla Road commenced concurrently with the 1911 project, with wooden barracks quickly erected. By 1913, an extension of the infirmary was added to accommodate the growing military population. Following the destruction in 1921, a reconstruction period ensued.

Fig. 12.8 Zegangan barracks

Starting in 1923, with the installation of the Regulars No. 2 forces in the adjacent barracks, a functional division between the two areas was established, with the main road serving as a definitive separation. Dependencies for the Guardia Civil were also set up in this zone, including a stable projected in 1923. The military barracks were arranged from northeast to southwest, while later-built housing blocks were oriented north–south, resting on older military structures.

The façade of the barracks adopts a post-1923 historicist aesthetic, featuring two polygonal towers, a semi-circular arch entrance, and a perimeter walkway with battlements. This monumental entrance distinguishes it within the current urban fabric, justifying its preservation in future interventions. Other surviving elements include the buildings, the barracks, the swimming area, and various artisanal features such as hydraulic tiles, which further reflect the architectural heritage of the site.

12.8.2 Changes in the Military Structure During the 1940s

12.8.2.1 The Walls of the Infantry Regiment no. 2 Neighborhood

During the 1940s, the military facilities in Villa Nador were completed, encompassing the barracks for the Regular Infantry and Cavalry units. Several military engineers, including Luís Sicre Marassi and José Ramírez Ramírez, along with

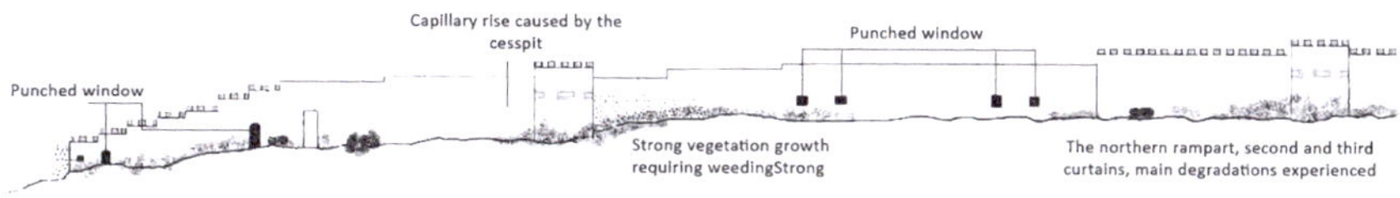

Fig. 12.9 View of the ramparts of the military quarter

architects like Manuel Latorre Pastor, oversaw the architectural renovation of these military towns over a span of more than fifteen years.

During this period, the barracks for the Regular Infantry No. 2 took shape, featuring walls adorned with battlements and cubic towers. Unfortunately, the current condition of these walls is highly degraded, with significant fragments lost over time (Fig. 12.9).

12.8.2.2 The Refurbishment of the Infantry Regiment no. 2 District: Neo-Arabic Architectural Models and Proposals

In the Regular Infantry No. 2 quarter, the construction of new buildings reflects a stylistic diversity. Some structures adopt arabesque forms, while others adhere to the prevailing rationalist style of the era.

The Muslim school, designed by military engineer Luís Sicre Marassi in 1940, exemplifies this trend with simple yet repetitive arabesque shapes, highlighting the increasing significance of the Regulars' quarter within Nador.

Other projects incorporate Arab-Andalusian elements into their decor, such as the Non-Commissioned Officers' Residence, designed by engineer José Ramírez Ramírez in 1941, and the Officers' Residence. These buildings blend decorative Arab details with a rationalist aesthetic, creating an eclectic ensemble.

The infirmary and the health depot, also in a neo-Arab style and designed by José Ramírez Ramírez in 1947, feature a façade that embodies two styles: one with Arab elements for the officers and another more rationalist section for the troops, reflecting an eclecticism dictated by the building's function.

These structures are located in a prominent area of the quarter, emphasizing the creation of a monumental entrance adorned with elegant ornamental details such as tiles and stucco. The stairs and balustrades add a sense of elegance to this military space, often regarded as one of the most thoughtfully designed quarters in the eastern colonized zone.

12.8.2.3 Reform of the Regular Infantry no. 2 Quarter: Rationalist Models and Proposals

During the first half of the 1940s, many pavilions of the military installation were reconstructed, including those for troops, services, and other utilities within the quarter. Several engineers, including Modesto Sánchez Llorens, Juan Nunell Ortega, and Maury Carvajal, oversaw these projects. Between 1942 and 1944, this ambitious construction program was completed, revitalizing the military structure.

The pavilions from this period feature façades characterized by elements indicative of a new architectural style, with moldings, horizontal details, and horizontally opening windows. The works primarily followed the 1942 design, built upon the earlier structures created in 1911. In total, eight new pavilions were constructed, alternating between two-story and single-story designs, employing concrete and new construction techniques to provide improved conditions for the soldiers.

12.8.3 The Cavalry Regulares Quarter: The Epilogue of Arabesque Forms

In military construction programs, renovations are not limited to the Infantry Regulares quarter but also include a new quarter for the Cavalry Regulares.

Located outside the Cuadrícula, along the road from Nador to Tauima, Zeluán, and Monte Arruit, this quarter has benefited from several construction projects led by engineers such as Maury Carvajal in 1944 and José Ramírez Ramírez in 1945 and 1948, followed by work from Noreña de la Cámara in 1951 and Eduardo Colomer Molinos between 1954 and 1955.

The aesthetics of these facilities blend neo-Arab elements, evident in façade details such as raised arches, with more modern trends like horizontal windows and simple façades. Inside the quarter, a water tower designed by Santiago Noreña de la Cámara in 1953, resembling a minaret, becomes an architectural landmark in the city.

12.8.4 The Aviation Barracks

The development of the Nador aerodrome began on September 8, 1920, to replace the Zeluan aerodrome, which had become inoperative after the events of 1921. Its location, closer to Nador than Zeluan, was designated as Tauima. Established in the 1930s, the aerodrome's structure has retained its original rectangular shape despite subsequent renovations in the twenty-first century. Its main façade features two cylindrical towers adorned with historicist elements such as crenellations. Inside, the barracks reflect the military architectural styles used in Spain during the twentieth century in Morocco. While some pavilions have preserved their original eclectic architecture, others have been modernized over time. Several of these pavilions are still in use by the Royal Moroccan Armed Forces.

12.9 Nador Today

Architecture:

The architecture of Nador is shaped by a variety of factors, including its history, culture, and contemporary needs. One can expect a blend of traditional and modern architectural styles. Historical buildings may reflect Arab-Andalusian influences, while new constructions could incorporate contemporary elements.

Urban Planning:

Nador is a city located in northeastern Morocco, along the Mediterranean coast. It has experienced significant urban development over the past few decades. Local authorities have likely implemented urban planning initiatives to manage this growth, focusing on land use, infrastructure management, and enhancing the quality of life for residents.

12.9.1 Urban Challenges in Nador: Complexities and Prospects

The urban challenges facing Nador are numerous and varied, reflecting the complexities of a rapidly developing city. Here are some of the most pressing challenges:

- **Unplanned Urban Expansion**: Rapid population growth and accelerated urbanization can lead to unplanned urban expansion, with the emergence of informal settlements and inefficient land use.
- **Inadequate Infrastructure**: The rapid population increase can strain existing urban infrastructure, including roads, water and sanitation networks, public transport systems, schools, and hospitals.
- **Transport Management**: Traffic congestion and mobility issues can pose significant challenges, necessitating investments in efficient and sustainable transport systems as well as alternative mobility solutions.
- **Affordable Housing**: The rising demand for affordable housing can make it difficult for many residents, particularly low-income populations, to access decent living conditions.
- **Social Inequalities**: Socio-economic disparities between different neighborhoods can lead to social inequalities, resulting in unequal access to basic services such as education, healthcare, housing, and infrastructure.
- **Waste Management**: Urban growth can lead to increased waste production, requiring effective waste management practices to prevent pollution and protect the environment.
- **Resilience to Natural Risks**: Due to its elevation and proximity to the Mediterranean coast, Nador is vulnerable to natural risks such as floods, storms, and earthquakes, necessitating resilient urban planning to mitigate risks and protect residents and infrastructure.
- **Environmental Sustainability**: Promoting environmental sustainability within urban development is crucial for preserving natural resources, reducing carbon emissions, and mitigating the effects of climate change.

By addressing these challenges in an integrated manner and adopting a participatory and inclusive approach, Nador can progress toward sustainable and resilient urban development that enhances the quality of life for its residents while preserving its natural environment and cultural heritage.

*12.9.2 **A Royal Initiative***

According to the 2014 census, the province of Nador has a population of 565,426, representing 24.4% of the total population of the Oriental region. Its urban population stands at 392,623, which corresponds to an urbanization rate of approximately 70%. With a population density of 175.5 inhabitants per square kilometer, Nador benefits from a strategic location facing Spain and close to Europe, further enhanced by its natural resources and economic potential. This geographical situation allows it to play a major economic role and become a development hub in northern Morocco. Its significance was underscored in King Mohammed VI's speech in 2003 in Oujda, which outlined the directions for the development of the Oriental region. Since then, the province has witnessed the realization of numerous projects that have transformed the region both economically and socially, paving the way for promising new horizons.

"Concerned about concretely demonstrating our high solicitude for this region, which possesses significant potential and industrious, highly motivated human resources, we have decided to launch a Royal Initiative for the development of the Oriental Region. This initiative, centered around four pillars, aims to stimulate investment and promote the creation of small and medium enterprises by the youth. It also seeks to equip the region with the necessary basic infrastructure and prioritize the encouragement of large economic projects. Furthermore, the initiative aims to promote education and training while fully embracing the principles of solidarity. To ensure effective implementation, we have outlined specific mechanisms for financing, monitoring, and evaluation.

In order to alleviate the unemployment affecting youth categories in the Region and to create conducive conditions for work and self-employment, we have decided to initiate this Initiative with an initial allocation of 30 billion dirhams, to which many institutions will contribute to finance business projects and guarantee the loans granted to them. This operation will be overseen by the Regional Investment Center."

Excerpt from the Royal Speech.

Pronounced in Oujda, March 18, 2003.

12.9.3 Urban Strategies and Policies

Launched in 2008 by His Majesty King Mohammed VI, the Marchica Lagoon development project aims to enhance the natural riches of the Nador region through the creation of seven thematic cities around the lagoon. This investment of over 2 billion euros is part of a sustainable development strategy, aligning with other major regional projects such as the Nador Med West port complex and the Ain Beni Mathar thermosolar power plant. The project's first phase involved the depollution of the lagoon, generating local jobs and promoting the regeneration of the marine ecosystem.

12.9.4 The Marchica Lagoon Project

The first phase of the project involved the depollution of the Marchica Lagoon, requiring an investment of approximately 1.5 billion dirhams (150 million euros). The depollution efforts had a significant social impact by mobilizing local labor, which has since been trained in environmental professions. The creation of a one-kilometer-long and 300-m-wide channel facilitated the rejuvenation and purification of the marine environment, promoting the restoration of the marine ecosystem and the return of migratory birds (Fig. 12.10).

Fig. 12.10 View of the Nador Lagoon from Mount Gourougou

12.9.4.1 Project Overview

- **Environmental Rehabilitation:**

  Lagoon decontamination
  Planting and landscaping
  Ornithological Park

- **Economic, social, and cultural development:**
  - Promotion of the artisanal fishing and aquaculture sector
- **Territorial development:**
  - New city of Beni Ensar
  - Restructuring and urban requalification of under-equipped areas
- **Opening new territories for urbanization and creating new activity hubs:**
  - Atalayoun City
  - Two Seas City
  - New City of Nador
  - Fishermen's Village
  - Flamingos Bay
  - Marchica Sport
  - Marchica Orchards

This ambitious project generated 15,000 jobs during the construction phase, including 4700 direct jobs and 10,300 indirect jobs. Once the sites are fully operational, a total of 65,000 jobs will be created, comprising 30,000 direct and 35,000 indirect positions. To successfully execute this large-scale initiative, it's important to note that a public limited company, Marchica Med, was established by decree on March 5, 2008, with a capital of 500 million Dirhams, equally shared between the state and the Hassan II Fund. For each of the seven projects within the program, a specialized subsidiary will be created, with private shareholding participation.

The first subsidiary responsible for developing Atalayoun City has been established, according to Said Zarrou, President of the Board of Marchica Med. He emphasizes that this structure allows the state to supervise and ensure the proper execution of such a large program in compliance with established standards.

This initial city has been designed over an area of 45.2 hectares according to ecological standards, prioritizing renewable energy sources such as wind and solar power, along with the recycling of wastewater. It includes a residential area with a 370-room hotel, 650 villas and riads, and 2230 apartments, totaling 14,000 beds. The project also features two marinas, a golf academy, and retail shops. Construction began in 2009 and was completed in 2014.

The next project, Two Seas City, spans 15 hectares along the lagoon. It consists of 320 villas with a total of 1920 beds, 193 apartments providing 772 beds, and 280 hotel rooms with 560 beds. Plans also include the construction of two marinas, a market, and a research center.

Atalayoun City

The Atalayoun Peninsula, named after the Amazigh word "Talaouine," meaning "water sources," offers breathtaking views of the Marchica Lagoon. The marina allows boaters to stay at the 4-star Atlas Marchica Hotel or the 5-star Port Hotel, benefiting from the services of the Golf Academy.

Atalayoun City features 650 villas and 2230 apartments, with the first phase including 149 apartments near the Golf Academy and three hotels of varying standards, providing 370 rooms for a total of 740 beds.

The Atalayoun Golf Resort, boasting an 18-hole golf course and a Golf Academy, is perfect for both amateur and professional players. Designed by Ron Garl in collaboration with SKS Studio, this international-standard golf course has attracted the Jason Floyd Academy for training programs. Accommodations around the Atalayoun Golf Resort include the 5-star Port Hotel, the 4-star Atlas Marchica Hotel, and a boutique hotel, all of which have been constructed.

Two Seas City

Two Seas City embodies the innovative challenge of the Marchica program, merging economic development with environmentally friendly tourism. This eco-resort spans 120 hectares and balances international comfort standards with sustainable construction between the lagoon and the Mediterranean Sea. The complex features four resorts inspired by their immediate surroundings.

New City of Nador

Nador City will be restructured in line with the overarching theme of the Marchica project, "Between City and Nature," while rationalizing the current urban situation:

- Harmonization of the urban fabric.
- Restructuring of the three urban centers: Nador, Beni Ensar, and Arekmane.
- Eradication of substandard housing.
- Establishment of a multimodal transport network.
- The new city of Nador will include hotel infrastructure offering 32,000 beds.
- Marchica Med, the agency responsible for developing the Marchica project, has outlined a multimodal mobility policy for the region in collaboration with the ONCF (National Railway Office):
- Renovation of the road network.
- Tram-Train connection from Laaroui Airport to Beni Ensar Port.
- Alternative transport modes such as seaplanes, cable cars, funiculars, and lagoon transport.

Fishermen's Village

Near the residential neighborhoods located to the east of Nador, the Fishermen's Village has been constructed on the current village of Kariat Arekmane–Larbaa Orekman in Amazigh. The new port serves both as a fishing port and a marina for pleasure boats. The accommodation infrastructure includes 6000 beds.

Flamingos Bay

Flamingos Bay combines a wide range of accommodations, facilities, and services aimed at lake tourism, catering to sports enthusiasts and golfers. The accommodation infrastructure offers 29,000 beds.

Marchica Sport

Marchica Sport spans 14 hectares, featuring various facilities designed for amateur athletes, high-level sportsmen, and professionals from different disciplines. It provides a range of training and high-performance sports facilities, with accommodation for 5000 beds.

Marchica Orchards

Situated in the natural park of the lagoon, Marchica Orchards aims to attract tourism focused on mobile homes and the rental of cottage-style villas across 14 hectares. The accommodation infrastructure will be substantial, with a planned capacity of 12,000 bed equivalents. This marks the seventh and final phase of the grand Marchica project.

Nador WestMed Industrial Zone

Objective: To create a modern and competitive industrial zone that attracts national and foreign investments, promotes economic diversification, and creates jobs.

New Nador West Med Port

This project aligns with the National Port Strategy set for 2030, which aims to enhance Morocco's market share in international maritime trade and cruising, integrate the port system into the regional transport network, contribute to the regional balance of the Kingdom, promote social and human development, and support the competitiveness of economic activities.

The new Nador West Med Port is part of the Oriental Pole, one of six poles defined in the national port strategy. The Nador West Med port complex will cater to the following activities:

- Container transshipment traffic.
- Oil product traffic for national needs and transshipment.
- Roll-on/roll-off, bulk, and specialized bulk traffic for national and regional needs.
- Project Site the Nador West Med port complex is strategically located at the Bétoya Bay, on the western side of Cape Three Forks, less than 250 miles from the Strait of Gibraltar, directly facing the major East–West maritime routes for container and oil product traffic.

12.10 Conclusion

1.1 Summary of Key Findings

The analysis of Nador's urban evolution in the twentieth century reveals a fascinating journey marked by profound and ongoing transformations. Initially structured according to a grid layout imposed by Spanish colonial influence, the city serves as a valuable testament to that period. Although this organizational model remains present, it has struggled to keep pace with rapid demographic growth. Consequently, Nador has experienced accelerated urban expansion, necessitating the creation of new residential, commercial, and industrial neighborhoods to meet the needs of its burgeoning population.

Nador's evolution reflects a complex interaction between Spanish and Moroccan influences, both of which have shaped the city's identity. This cultural duality underscores the importance of preserving architectural heritage while addressing the challenges posed by modernization. The celebration of Nador's centenary in 2009 highlighted not only the city's achievements but also the contemporary challenges related to its rapid development. Thus, it becomes imperative to find a balance between urban growth and cultural preservation to ensure a sustainable and harmonious future.

12.10.1 *Perspectives for the Future of Urban Planning in Nador*

The future of urban planning in Nador is filled with opportunities for innovative and sustainable transformation. In response to the challenges posed by rapid urbanization, it is crucial to adopt an integrated approach that encompasses strategic planning, investment in infrastructure, and a strong commitment to environmental sustainability. Implementing advanced technological solutions and environmentally friendly practices will be key to supporting this growth in a sustainable manner.

Simultaneously, active participation from local communities is essential for creating a living environment that meets citizens' needs while preserving the city's cultural identity. By promoting social inclusion and community engagement, Nador can develop dynamic and vibrant urban spaces. By leveraging its historical strengths and addressing its challenges with a proactive vision, Nador can strive to become a model of sustainable urbanism, thereby offering its residents a high quality of life while respecting and enhancing its rich historical heritage.

12.10.2 *Actions and Recommendations*

12.10.2.1 Heritage Rehabilitation and Conservation

Protection of Heritage Elements: Establish non-construction servitude zones that are either landscaped or planted to protect valuable heritage sites.

Encouragement of Restoration: Foster the restoration of historical buildings, adapting them for new uses such as cultural spaces, colonial barracks, churches, or institutions that respect their historical character.

12.10.2.2 Revitalization of Historic Neighborhoods

Urban Development Projects: Promote urban development projects that incorporate local architectural styles in historic neighborhoods, including former military camps and elevated fortifications, by using local materials and sustainable construction techniques.

Ecologically and Culturally Inspired Materials: Use construction materials that are ecologically balanced and culturally inspired by the rich heritage of the region (Mediterranean, maritime, Andalusian battalion style, etc.), while ensuring aesthetic alignment with the natural landscape (e.g., Cale, Gourougou).

Local Resource Utilization: Leverage local natural and heritage resources to create a distinctive architecture that preserves regional identity, avoiding standardization and monotony in Nador's urban landscape. Examples:

- Shell Utilization: Use shells from the Rif beaches (e.g., Tcharana, Cala Blanca, Tibouda, Bocana Beach) for façade ornamentation and decoration.
- Pebble Use: Employ pebbles from the beaches for flooring and wall base cladding.
- Natural Stone Integration: Incorporate local natural stone (such as Gourougou stone) in façade walls and interior panels.
- Regional Inspiration: Draw inspiration from the unique characteristics of the region for future architectural projects.

12.10.2.3 Consideration of Site Opportunities

Architectural Arrangement: Organize the architectural design of the urban façade along the water front (Lagoon Marchica, Cale).

Preservation of Natural Elements: Preserve natural elements that maintain interesting landscape qualities by implementing sightlines (overlooking the lagoon, islands, mountains, etc.).

12.10.2.4 Integration of Local Identity in Modern Urbanism

Urban Design: Integrate architectural elements and motifs inspired by local craftsmanship into future urban development.

12.10.2.5 Training and Awareness

Architectural Training: Raise awareness among architects working in the region about the values of local heritage through training and architecture competitions adhering to local guidelines.

12.10.2.6 Sustainable Development and Space Management

Creation of Public Spaces: Develop public spaces that are accessible to all, utilizing ecological management methods (e.g., local flora such as Gourougou pine, cacti, and Mediterranean perennials).

12.10.2.7 Tourism Promotion

Digital Promotion: Utilize digital tools to promote Nador's cultural and historical heritage, attracting tourists and investments.

Event Leverage: Capitalize on regional events to showcase local heritage (e.g., project images of architecturally significant military sites and historical monuments on event signage and banners, or organize festivities featuring monumental gateways or beach ramps).

12.10.2.8 Conclusion

Nador must strive for a future where heritage preservation and sustainable urban development coexist harmoniously. The city should focus on maintaining its cultural identity while adopting ecological practices. By valuing its unique resources and encouraging ambitious projects, Nador can become a model of balanced and attractive development, enhancing the quality of life for its residents and attracting visitors and investors alike.

Bibliography

Bravo Nieto A (2006) El Protectorado Español en Marruecos: La arquitectura y urbanismo en la zona del Rif

Capel H (2006) *Colonialismo, arquitectura y ciudad: El urbanismo del Protectorado*

López Carvajal C (2015) El impacto del colonialismo español en el norte de Marruecos: Nador y su desarrollo urbano. Revista de Historia Urbana y Territorial

López Viguera MD (2013) La ciudad colonial en el Rif: Nador y su evolución urbana. Español en Marruecos

Ortega Pérez N (2012) El urbanismo del Protectorado Español en Marruecos: Modelos y realidades. Boletín de la Real Academia de Bellas Artes de San Fernando

Mateos JS (2011) La arquitectura colonial en el Rif: Nador y la construcción de la identidad urbana. Revista de Estudios Marroquíes

Reyes Gómez JC (2010) La expansión urbana de Nador durante el protectorado español. Hespérides. Anuario de Estudios Almerienses

Souto AL (2015) El protectorado español en Marruecos: Desarrollo urbano y políticas coloniales en Nador. Cuadernos Geográficos

Part V
Latin American Cities

Chapter 13
Colonial Cities: Vulnerability and Contingency

Rodrigo Castro Orellana

Abstract This work examines the type of rationality that would be at stake in the process of founding colonial cities in the Americas. It seeks to establish that the conquest of the city during the sixteenth century would not be the realisation of a supposedly classical-Cartesian reason, nor the condition of possibility of the modern dream of a total order of reality. On the contrary, the dynamics of urbanisation obeyed to practical criteria and situational needs that tried to respond to the vulnerability faced by the conquerors. This hypothesis is demonstrated by the description of the phenomenon of the nomadism of Hispanic urban complexes studied by Alain Musset. The research shows that the colonial city was the result of a series of supervening facts and that the colonising power was always conditioned by contingent realities and specific practices of Spaniards and Indians.

Keywords Colonial city · Colonisation · Utopia · Contingency · Nature

13.1 Introduction

The process of urbanisation that has taken place in the Americas since 1492 is unique in human history in terms of its complexity, speed, and scale. In <100 years, between the end of the fifteenth century and the first half of the sixteenth century, more than 360 cities were founded in Mexico and Peru alone (Calvo 1995; Terán 1989). This urban network spread from one end of the continent to the other, from North America to southern Chile, crossing different geographical locations and confronting different indigenous cultures. In this sense, Hispanic colonial power, and the desire to build cities seem to be closely linked. In fact, the first gesture of possession of the new territories in the Indies always consisted of the founding of a city, a rite marked by an enormous symbolism that followed that ancient Greco-Latin tradition that made the urban fact the constitutive aspect of the civilising process. In this rite of taking possession, a space that was supposedly "without owner" (*res nullius*) was

R. C. Orellana (✉)
Complutense University of Madrid, Madrid, Spain
e-mail: rodrigocastro@filos.ucm.es

F. Carrión Mena and S. Rodríguez Alvarez (eds.), *From the City as a Project to the City Project*, The Urban Book Series, https://doi.org/10.1007/978-3-031-94226-6_13

transformed, because of a legal process, into a property with a just title (Lucena Giraldo 2006, p. 35). To be legitimate, this act required the discoverer to take a handful of land, drink water or even shout in the new region. The notary would then record these gestures and the town crier would publicly testify to what had happened. Finally, the appropriate masses were celebrated, the streets and plots of land were laid out, and the first cabildo (town council) was formed.

However, the city did not function exclusively as a local tool for the appropriation of space, but also defined a global strategy of colonisation throughout the so-called New World. Initially, "it was a place of provisioning, rest and decision-making", but later "it became the nucleus of stabilisation and irradiation of Spanish colonisation" (Lucena Giraldo 2006, p. 30). The urban network functioned as a platform to produce new conquest initiatives and the consolidation of those already underway. It could be said that the colonial city developed as a fundamental instrument of conquest, a fact of which the continent's indigenous populations were fully aware. This is illustrated by a story attributed to the Araucanian chieftain Lautaro, who is said to have said to his companions in 1556 as they set out to destroy Santiago: "Brothers, know that what we are going to do is to nip in the bud where these Christians are born, so that they will never be born again" (Ramón 2000, p. 33). The city was thus recognised as a key part of colonisation and was linked to an easily identifiable materiality, which consisted in trying to assimilate the unknown through the urban management of the territory (management of unknown places, control of new environmental conditions, government of an enigmatic and hostile indigenous population, etc.). In this dynamic, we cannot exclude the role of the symbolic dimension, especially if we consider the precariousness of most of the urban centres founded during the colonial period and the consequent need to strengthen and secure the bonds of an incipient community that perceived infinite threats around it. Ángel Rama is therefore right when he states that "American cities have been consigned to a double life since their origins" (Rama 2009, p. 44). The physical order of material construction, with its specific contingency, its ups and downs, its circumstances, and uncertainties; and, on the other hand, the order of signs, which is more tied to certainty and stability, and which is decisive at the time of each realisation (2009, p. 45). Nevertheless, it seems to me that the priority that Rama gives to the symbolic level in the founding of Spanish–American cities, as well as the content that he attributes to the ideal structures that would have intervened and conditioned sixteenth-century urban planning, should be re-examined.

13.2 The Ordered City and the Cartesian Cogito

In his 1984 work *La ciudad letrada*, Rama devotes a first chapter, entitled "La ciudad ordenada" [The ordered city], to the study of the Latin American city as the manifestation of a specific dream of the conquering intelligence. His main argument is that the European impulse to transform the city according to a set of abstract ideals would have found in the New World a "unique opportunity" to materialise in a place thought to be devoid of its own values, that is, as a kind of "tabula rasa" where it would have been possible to initiate a new history of the city and of man (Rama 2009, pp. 32–33). From an urban point of view, America would have represented the real possibility of carrying out a kind of singular *creatio ex nihilo*; it would have been understood as the starting point for the beginning of a new epoch.

The Uruguayan author's hypothesis on this point is crucial: the *conquistador-urban* act would not reproduce or replicate the existing models of the European metropolis but would move the ideal or the horizon of the future, which lay as a slumbering power in the Old World, to its effective realisation in a new cosmos that was beginning to write its history. The utopian structure of the founding of the colonial city would thus be revealed. The New World would contain a force that would ignite the imagination and dreams of the Castilians, leading them to build urban centres that would be completely subject to the authority of an ideal city. According to Rama, the content of this ideal would obey the general principle of what he calls "ordering reason" (2009, p. 36). The word "order" would thus have been central to the cultural aspirations that migrated to the new continent. Desires for symmetry, fears of hierarchy and dreams of classification also crossed the Atlantic.

As an example, Rama recalls the instructions given by the King to Pedrarias Dávila in 1513 regarding founding cities:

> When you have seen the things that are necessary for the settlement of the places, and have chosen the most profitable site and the one in which most of the things that are necessary for the pueblo are to be found, you are to distribute the lots of the place to build the houses, and these are to be distributed according to the qualities of the people and to be given in order at the beginning; so that when the plots are done, the pueblo will seem ordered, both in the place that is left for the square, as well as the place where the church will be, and in the order that the streets will have; because in the places that are newly made by giving the order at the beginning without any work or cost, they are ordered and the others are never ordered (2009, p. 37).

The future of "order" to which this text refers would correspond, according to the approach of *La ciudad letrada*, to the dream of a geometric reason that would later find its philosophical development in the work of Descartes. Here, Rama draws a unique parallel between the ideal conception of reason, which orders and measures according to the principle of a universal *mathesis*, and the characteristic design "a chord and a ruler" of the urban plan in most Latin American cities (Rama 2009, p. 39). The order expressed in the grid of the colonial city would be the manifestation of a reverie of thought imposed before any practical realisation (2009, p. 40). In the same way that the city is the product of an ordered and mathematical reason in its appropriation of an unknown land, of a wild and untamed nature, colonisation as a whole could be understood as an enterprise oriented towards the negation of any otherness. From the point of view of intellectual history, however, this parallel that Rama draws between the design of colonial cities and what Foucault called the *classical episteme*, that is, a discursive regime characterised by taxonomic classification and geometric continuity, is hardly plausible. One wonders how the Castilian conquistadors could have thought in Cartesian terms a century before the French thinker published his *Discourse on Method* (in 1637).

Enrique Dussel has developed an argument that seeks to clarify this supposed relationship between the conquest of America and Cartesian thought. His approach corrects this historical nonsense of *La ciudad letrada*, while preserving the essence of Rama's argument. To this end, he introduces a distinction between a first modernity, identified with the Christian, humanist and Renaissance *ethos*, whose hegemonic power would have been Spain; and a second modernity, which would begin in the seventeenth century with the decline of Hispanic power and the rise of new powers such as Holland, England, and France (Castro-Gómez 2005, p. 49). The first modernity would produce a type of subjectivity that can be characterised as centred on the "conquistador-ego", which relates to otherness (the Indian) from a logic of exclusion and domination. This *ego conquiro* would be "the proto history of the constitution of the ego cogito" (Dussel 1994, p. 47), that is, the meaning of the Cartesian cogito would have to be understood in relation to an unquestioned ideal of subjectivity that precedes it: the colonising/conquering self or the imperial self (Grosfoguel 2007, p. 64). For Dussel, between the conquering moment and the Cartesian moment there would be a perfect continuity corresponding to the development of a fundamental device of modernity: the "colonisation of the life-world (*lebenswelt*)" (Dussel 1994, p. 48).

From this perspective, the urban conquest of the sixteenth century would not be the materialisation of a supposedly classical-Cartesian reason, but the condition of possibility of the modern dream of the total order of the real. The act of conquest would produce the negating incorporation of the other into the dominant totality (Dussel 1994, p. 41), which prefigures the claims of a *ratio* that seeks to subjugate nature by eliminating all discontinuity and contingency. The construction of the colonial city could therefore be seen as the practical confirmation of the ideal structure of the *ego conquiro*, and thus as the first milestone in a process of unilateral rationalisation aimed at consolidating an imperial power without counterweights. It is important to stress the full importance and scope of Dussel's approach. His thought, and in particular his interpretation of the history of the conquest and colonisation of the Americas, has had an enormous influence on the project of Latin American liberation philosophy and also on the so-called decolonial theories. The basic thesis is to establish an identity between Western European modernity, as a total, uniform and homogeneous system, and the *colonisation of life*. This would translate into an imperial power that effectively designs spaces and subjectivities, subsuming any otherness or difference, a power that would have played a decisive role in the urban practice of the colonial period. My purpose here is to explore some recent contributions of historical studies to the Spanish colonial city, with the understanding that such an exercise shows the need to reconsider Rama's specific analysis of this configuration of the city, as well as Dussel's general description of the Hispanic imperial power as a hegemonic system.

13.3 Compensation, Mimesis, and Contingency

The colonial city is not the result of the "free flight of an imagination" (Rama 2009, p. 32) that dreams of a taxonomically ordered space. Rather, it is the product of a series of urban planning decisions based on various principles that were considered scientific at the time. Considerations such as the location and spatial orientation of the city, the effects of the climate or the health conditions of the environment were largely taken from Vitruvius' *De architecture*, a work that had been widely disseminated in Castile thanks to the commentaries of Diego Sagredo in his *Medidas del romano*, published in Toledo in 1526. The text *De Regimine Principium* by Thomas Aquinas also had a significant influence, stating that a good prince should ensure his fame by building cities that would endure over time, something that required attention to a series of location criteria that focused mainly on the problem of environmental health. In this, Aquinas, like Vitruvius, followed to the letter the Hippocratic canon of the *Treatise on Airs, Waters, and Places*, which established an essential relationship between an unhealthy climate and moral degradation or the relaxation of people's customs (Musset 2011, p. 107). According to Hipocrates, "when one arrives in an unknown city, one should be concerned with its position: how it is situated in relation to the winds and the rising of the sun" (1997, p. 40). In the same way, when confronted for the first time with a building space, the conquerors considered it necessary to give priority to concerns about health and environmental conditions rather than to any possible geometric administration of the new territories.

However, it is clear that the Hispanic American city is generally characterised by a plan ordered according to the grid model. The question, then, is whether we should attribute this design to a supposed pre-figuration of Cartesian rationality, as Rama and Dussel suggest, or whether we should rather explain this geometric layout of the colonial city based on a series of old theories and as the result of a series of supervening events that forced urban planning practice. According to Alain Musset, the grid plan refers to mediaeval traditions such as those found in the walled cities of south-west France, or "bastides". The design of the "bastide" ("bastida", "nova bastida", or "nova population") made the square the element that divided the land into plots, all of which were defined by the orthogonal grid of streets (or "carreyras"). This morphology of rectangular streets with rectangular blocks is so old that it can be traced back to the urban plans of Hippodamus of Miletus in the fifth century BC. In Spain, the orthogonal layout appears in cities under Roman occupation, such as León, Zaragoza, or Zamora, and is put into practice from the eleventh century onwards with the creation of new urban centres linked to the Reconquest (Musset 2011, pp. 44–45). For example, one of the paradigms of this design was the camp of Santa Fe, founded in 1491 by the Catholic Monarchs in the context of the siege of Granada, inspired by the ideas of Vitruvius and the Roman castros or *castrum* (Lucena Giraldo 2006, p. 68). Certainly, the great theorists of the Italian Renaissance, such as Alberti, Filarete, or Martini, also decisively inspired Hispanic architects and urban planners, but their influence in the American case can be considered later and not related to the basic urban plan of the colonial city. It should not be forgotten that Alberti, in his *De re aedificatoria* (*On the Art of Building*), proposed a radial urban plan, that Sforzinda—the ideal city dreamed of by Filarete—was a city in the shape of an eight-pointed star and completely walled, or that Martini's utopian city reproduced the spatial organisation of the human body. All of them were clearly far from the grid. Leon Battista Alberti's *De re aedificatoria*, published in 1485, is the most important Renaissance treatise on urbanism. In this context, it is interesting to note how Alberti rejects the idea of building an ideal city and defends a model of urban planning based solely on historical concreteness and practical considerations (Seta 2002, p. 40 ff).

In any case, the role played by all these theoretical traditions in the design of the colonial city may have been more limited than one might initially suppose. The grid, in this sense, was imposed primarily by the realities of the situation, which demanded an orderly and hierarchical distribution of plots and buildings. This spatial organisation of the city provided orientation, reference points, stability and security for a fragile and incipient community that was just beginning to develop. In this context, the key question would be the cultural representations available to the first Castilian colonisers, which would lead us to understand urban design as the expression of a compensatory mechanism aimed at transplanting Spain's own original order into a new and unknown reality. This could explain, among other things, the conquistadors' habit of giving the cities of the New World the same names as those of Spain. In the Americas, we find cities with the names Trujillo, Valladolid, Salamanca, Córdoba, Guadalajara, Cuenca, Mérida, León, and so on. In fact, in this search for similarity, Seville was the urban reference par excellence. Every city in the New World sought to conform as closely as possible to the political and legal structure of the Andalusian city (Musset 2011, p. 317). In the case of Lima, the capital of the viceroyalty of Peru, this even led to urban projects that attempted to copy Sevillian gardens or houses to the millimetre (2011, p. 318). The psychological necessity demonstrated by mimesis makes it possible to establish a link between the act of founding the Spanish-American city and the concrete conditions of vulnerability faced by the first conquistadors. This link would be much more significant and relevant than the role played by the theoretical traditions of Greco-Latin and mediaeval urbanism. Both elements: the conditions of vulnerability and the urban concepts of the time, offer us the appropriate historical context to explain and understand the phenomenon of the colonial city in the sixteenth century.

Musset argues that the circumstances faced by the first colonisers played a fundamental role in determining the spatial organisation of cities. This was the dynamic that characterised the struggle against the indigenous population in the first half of the sixteenth century. In fact, the way in which the natives were fought could explain why the Hispanic American cities were not founded as walled enclosures. Faced with adversaries armed with bows, maces or spears, an open urban layout with straight streets facilitated cavalry charges and the use of long-range firearms (Musset 2011, p. 48). Later, when the threat of war came more from Europe, it became necessary to build walls and towers to make the fight against cannons and rifles more effective. The walled colonial city would therefore appear at a later stage in the colonisation process. According to the architectural historian Cesare de Seta, the first fortress and walled enclosure in Cuba was built in 1540 as a result of pirate raids and invasions by the English navy. Other examples of this are Cartagena de Indias, which is shown on plans as a walled city from 1593, or San Juan de Puerto Rico, which was not fortified until the end of the sixteenth century (Seta 2002, p. 119).

In short, the founders of colonial cities had to make urban planning decisions based on their specific ways of representing the world, much more attached to practical criteria and circumstantial needs than to strictly theoretical criteria, and even less to regulations derived from a previously imposed rational model. A final proof of this is the fact that the first regulation attempting to regulate the urban situation in America dates to 1573. These were the so-called "Ordinances of Discovery, New Settlements and Pacification", issued during the reign of Philip II, at the height of the first phase of colonial expansion. These ordinances set down on paper a series of practices that had already been applied in reality (Seta 2002, p. 55) and that were well known to the Crown through a long tradition of letters, chronicles, relations, and so on. There was no solid, consistent, and a *priori conquest urban mechanism*, but rather a modality of construction and management of the city directly linked to contingency.

13.4 Cities on the Move

The link between the founding moment of colonial cities and the problem of contingency can be confirmed if we pause to analyse the phenomenon of nomadism in Hispanic American urban complexes, extensively studied by Musset in his *Ciudades nómadas del Nuevo Mundo* (*Nomadic Cities of the New World*). Certainly, the conquistadors founded cities with the aim of dominating and populating a new territory, in accordance with the aspiration described by Thomas Aquinas that a city should last and endure over time. However, the numerous relocations of cities during the colonial period show the difficulties encountered in achieving these objectives and allow us to observe the fragility that characterised the colonisation process. This phenomenon of urban relocation was not an exceptional or isolated event, but a reality of great magnitude that crossed the American continent (almost 200 cities were relocated, some of them two or three times). These urban displacements should not be confused with other events such as the abandonment of urban centres or the relocation of indigenous populations by religious orders in the sixteenth century. According to Musset, a relocation involved "both the migration of the inhabitants and the reconstruction of all the structures, buildings and services of the original city" (Musset 2011, p. 28). In most cases, the relocations took place within a radius of 10–20 km (although there are exceptions, such as San Lorenzo El Real in New Mexico, which was moved 50 km from its original site, or Esteco in Argentina, which was moved 80 km), and affected some regions to a greater extent, such as Central America and northern South America (Colombia and Venezuela). In the early stages of the city's life, the relocation could be decided and resolved without major conflicts, but 10 or 50 years after the first location, the circumstance of displacement exacerbated all kinds of political tensions within the community.

There are several reasons for this shift in urban centres. First of all, there is the complex relationship that the colonial city established with nature, which manifested itself in the great difficulties that the Castilians had in understanding the peculiarities of the territory and coping with its vagaries (Musset 2011, p. 212). Nature in the New World was presented to the conquistadors as a set of deceptive landscapes, of which there was no adequate or precise knowledge. For example, landscapes covered with luxuriant vegetation were, almost counter-intuitively, totally inhospitable to agriculture or the cultivation of essential European products such as wheat. On the other hand, the regular natural disasters (floods, volcanic eruptions, earthquakes) that occurred in America increased the sense of insecurity that plagued the colonial cities. The problem of telluric activity, for example, illustrates the limits of the conquest of reason in this urban context. The theories of the time valued sandy soils as less prone to the effects of earthquakes (Musset 2011, p. 77), on the assumption that an earthquake would be the consequence of pressurised air circulating beneath the earth's crust and trying to find spaces to escape to the surface. This would explain why many cities (Guatemala is an emblematic case) have been repeatedly destroyed, moved, and rebuilt as a result of earthquakes.

However, this experience of an incomprehensible and frightening nature was part of a perception of the healthy and the unhealthy which, as we have already seen, was a priority in the colonisers' representation of the ideal places to find their cities and settle their populations. Thus, one of the main arguments for the relocation of a city was the search for a more favourable climate, that is, one that was more similar to that of Spain (Musset 2011, p. 109); and the attempt to locate in places that were not subject to the threat of earthquakes, an unattainable objective since, as we know, such phenomena are due to faults that affect very large geographical areas (Musset 2011, p. 86). Such relocations were therefore often completely useless in terms of the problems they were supposed to solve and, logically, this failure increased scepticism towards supposed expertise, intensified recourse to astrology or superstition and reinforced the diagnosis of catastrophes in terms of divine punishment resulting from the dubious morals of the city's inhabitants. At other times, the expulsion of a city was the result of incessant conflicts with the indigenous inhabitants of a particular area (Musset 2011, p. 179). In this context, the term "colonial period" sometimes risks implying that the processes of conquest were consolidated and completed during the sixteenth century, which is not necessarily the case. Until the end of the eighteenth century, the colonising towns on the northern frontier of New Spain were repeatedly attacked by hostile indigenous populations (Musset 2011, p. 188). This was also the case in the so-called "Indian Flanders", the war against the Araucanians in southern Chile, which continued well into the nineteenth century. In addition, conflicts with the Indians broke out regularly in various regions of the Americas, sometimes precisely because of the confiscation of land and the relocation of a town (Musset 2011, p. 389). During the sixteenth and seventeenth centuries, cities such as San Luis in New Spain, Comayagua in Honduras, La Concepción del Bermejo in Argentina, Santa Cruz in Bolivia, or La Imperial in Chile were abandoned as a result of the indigenous threat (Musset 2011, pp. 179–191). It should be added that, from the second half of the sixteenth century, this danger of possible attack by the natives was compounded by the problem of pirate plundering, a circumstance that explains a significant number of urban relocations from the coast to the interior, mainly in southern Mexico and Central America.

However, despite the existence of this set of objective facts that could justify a change of urban location, colonial Hispanic American societies faced the decision to relocate in a particularly traumatic way, especially when many years had passed since the founding of the city. The debates about relocation involved enormous intra-community disputes, which also indicate the precariousness of the legal and political order that governed the city, and the very real possibility that the discussions could degenerate into open and explicit conflict (Musset 2011, p. 255). According to Musset, two logics usually clashed: that of the living and that of the dead. The logic of the living was that of the proponents of relocation, who wanted to ensure the continuity and permanence of the city by developing it on new foundations. The logic of the dead, on the other hand, was that of the opponents of relocation, for whom the work and life of their ancestors, the memory of their deeds and the physical or spiritual presence of their tombs made the soil of the city sacred (2011, p. 266). In this context, the absence of a single position on the part of the Church increased the tensions at stake, as each sector claimed to be the most appropriate expression of Catholic orthodoxy (2011, p. 269). Certainly, the debates and arguments for or against the transfer could conceal other interests and speculations of an economic nature that were at stake in the permanence of the city or its eventual displacement. Above all, however, these debates showed that for its inhabitants the city was more than just an accumulation of structures and material goods. Indeed, the fragility of the buildings revealed the fictitious nature of the unity of the community itself. Colonial societies were as fragile as their urban buildings were precarious and volatile.

For all these reasons, the relocation projects were extreme situations that exposed the reality of the cities, revealing societies divided and segmented by interests that were difficult to harmonise: Indians and Spaniards, Creoles and peninsulars, rich and poor, urban dwellers and peasants, clerics, and laymen (Musset 2011, p. 246). Above all these divisions, however, was the decisive opposition between the Crown or its representatives and the Creole population. Sometimes the king himself had to be convinced of the necessity of the transfer, sometimes of its inconvenience. In this context, arguments were used that could be considered as important precursors of a Creole consciousness that was beginning to claim a local identity (Musset 2011, p. 283). A good example of the latter is the stubborn resistance of some Mexican Creoles to the Crown's attempt to relocate Mexico City following the great flood of 1629 (Musset 2011, pp. 284–297). What for the Crown, seen from a distance, was the solution to a technical and financial problem, was for a significant number of Mexico's inhabitants rather the expression of a deeply symbolic collective failure of their projects and ways of life.

13.5 Fragility and Utopia

The idea of the "tabula rasa" as a way of characterising the founding instance of the Hispanic American city does not seem to us to be a satisfactory way of describing the complexity of the process of designing and constructing the colonial urban network in the New World. Upon their arrival in the Americas, the conquistadors were confronted with a set of exceptional and radically new circumstances for which they had neither conceptual frameworks nor technical tools capable of offering effective solutions. It was never possible to intervene in space as if it did not impose its own determinations, and it was never possible to do so from a supposedly ideal rationality, devoid of ancestral beliefs and customs, fears, and emotional needs. In this sense, it can also be said that the colonial city does not obey the materialisation of a utopian urbanism. First of all, from the point of view of the original design of the main American urban centres. As we have already noted, Renaissance utopianism proposed an ideal city enclosed within a walled perimeter, an enclosure that resembled a fortress. This isolation from the outside world can be seen in the urban model of Filarete's Sforzinda or, in literary terms, in the later works of Moro or Campanella. Campanella's *City of the Sun*, for example, is divided into seven circles, each with four gates facing the cardinal points. If someone wanted to enter the city, he would have to be able to cross each of the circles with their corresponding ditches, in a kind of system of increasing difficulty (Campanella 1980, pp. 143–144). Obviously, the impossibility of entering the *City of the Sun* expresses the abysmal distance that would exist between the ideal city and the real world, but at the same time it indicates that the utopian space must fold in on itself, since it embodies the consummation of all human aspirations and projects. It is not possible to enter the *City of the Sun* easily, just as it makes no sense to leave it.

In contrast, the colonial cities of the first half of the sixteenth century were not built as fortified centres but were completely open to the outside world. We have already reflected on the importance of the nature of the confrontation with the indigenous population in the choice of this model, but perhaps it is also necessary to consider that this openness reveals a different relationship with nature. In utopian cities, precisely because they are the result of a "free flight of the imagination", any expression of nature is purified. This does not mean that the structure of urban utopia is limited to the mere expression of fantasy. It is one thing for utopia to be written with the power of the imagination, and thus to have the capacity to bend, disrupt or reinvent the "natural given"; it is quite another to consider the power of the utopian narrative to penetrate reality and even to guide some kind of practice. From the first point of view, individual desires and interests are subsumed under the imperative of the common good, becoming and contingency disappear as the result of a perfect administrative order, and uncertainty gives way to the infallible rule of knowledge. All this is alien to the colonial city, which seems more like an artefact constantly exposed to an unknown and ungovernable nature, an apparatus that has been unable to transform nature into an external power or an element subject to its domination. There are undoubtedly several aspects that the colonial city shares with the narrative of ideal cities, such as the layout of long, rectilinear streets or a certain universalisation of the urban plan. It should not be forgotten that on the island of Utopia it was enough to know one of its cities (Amaurota) in order to know them all, because they were so similar. Something similar happens in the colonial urban network of the New World, where the grid layout is insistently reproduced. In this case, however, it is the reality of Hispanic American cities that influences the utopian rather than the other way round. Remember that Raphael Hitlodeo, the central figure in Moro's utopian account, was supposedly an explorer who got separated from Vespucci during the latter's voyage to the Indies and ended up on the island of Utopia. There is therefore a link between the voyage to the New World and the Utopian text. Erwin Walter Palm offers an interesting example of this overdetermination of utopian cities by American cities. This is the scheme of the ideal city formulated by Albrecht Dürer in 1527 in his treatise on fortification. According to Palm (1951), this ideal city project was decisively influenced by the publication of Cortés' Second and Third Letters in Nuremberg in 1524, together with the plan of the city of Tenochtitlan. Our thesis, then, can be summarised in the idea that the utopian city is a consequence of the American conquest and urbanisation, and not the other way round.

In any case, one of the main arguments against the existence of an essential relationship between colonial cities and utopianism is offered by Musset's approach to nomadic cities. Logically, the destruction of a city and its eventual relocation represented an extraordinary opportunity for architects and urban planners to modify their plans and models, to free themselves from the constraints imposed by inadequate ideas, to formulate new alternatives, and finally to have the chance to "draw *ex nihilo* the outlines of an ideal city" (Musset 2011, p. 407). However, this utopianism never took place in the context of the displacement of Latin American cities. Urban planners and architects were forced to respect and reproduce the urban and social realities already established in the American territory. In this sense, the colonial process of urban planning was a repetition rather than a radical innovation. The vulnerability and precariousness of the colonising experience reinforced the reproduction mechanisms of discourses and practices already known to the Spaniards (ideas about the construction of cities, legal concepts, religious principles, among others), as well as the desire to belong to a common identity provided from afar by the Crown. In fact, there is an enormous immobility of urban plans and a conservatism of designs, which is not due to the determination of the Crown or its supposed iron control over the Ibero-American space, but to the resistance of the inhabitants of the cities, who faced the misfortune of relocation (430). The displacement of the city represented a threat to the stability of the community, which saw its perception of insecurity and precariousness increase. Therefore, any change added to this situation, such as the substantial transformation of the urban plan, could only be welcomed as the seed of new concerns. This would explain why local elites rejected radical changes to the original city in order to preserve and guarantee the fragile internal equilibrium of colonial society (Musset 2011, p. 434).

Musset presents an interesting case that illustrates how, until the end of the eighteenth century, the basic principles established by the colonial city and made explicit in the ordinances of 1573 continued to significantly determine the urban landscape.

This is the case of the relocation of the city of San Pedro de Riobamba in Ecuador, following the earthquake that destroyed it in 1797. Faced with the possibility of designing a new city from scratch, the Audiencia of Quito chose the Spaniard Bernardo Darquea to draw up the plan for the future city. In Spain, Darquea had collaborated with Pablo de Olavide in the construction of the new towns of Sierra Morena around 1770, a truly biopolitical project that aimed to build a model society that followed the reformist principles of the Enlightenment: machine production, efficient performance, and the usefulness of the individual (Vázquez García 2009, pp. 43–53). The project for the new city of Riobamba that Darquea submitted to the authorities for consideration was a plan concentric in radius on a square perimeter that almost completely dispensed with right angles (Musset 2011, p. 434), a model that radically departed from the grid layout of the original city. The proposal did not allow for any kind of equivalence between the old layout, with its social segmentation and hierarchies expressed in urban space, and the future city organised as an ideal metropolis. Darquea's project was, as expected, rejected on the grounds that it threatened the social, political, and moral stability of the city (Musset 2011, p. 436). In 1799, work began on the construction of the new San Pedro de Riobamba, following the classical plan of the colonial era.

13.6 Conclusions

The multiplicity of transfers of Hispanic American cities highlights the fragility of the urban planning process in the New World, but also the persistence of the colonisers in continuing to cling to the model of the city as the only means of coping with contingency. However, this phenomenon corresponds to a much broader and more complex context. The existence of a vulnerable society that sought to consolidate itself in a territory saturated with previous conditioning factors and disconcerting features that radically questioned the forms of knowledge and representations available to the colonisers. All this, as will be understood, problematises the value and meaning of certain narratives about the colonial city that have been deployed in Latin American thought, such as the image of a rationally and geometrically planned city or the figure of an ideal city that would be built on the basis of the violence involved in defining the colonised territory as an immaculate space. All these symbolic expressions have served to underpin an interpretation of history that denounces the colonial power of European modernity, a reading of our past that runs through and underpins some decisive theoretical approaches such as the programme of the philosophy of liberation or the decolonial turn.

In fact, these projects have been formulated in a way that makes them completely dependent on an idea of history that establishes the existence of a hegemonic, totalising and excluding power and, moreover, describes its monolithic continuity from the Hispanic colonial past to the present. In this sense, the observation of the profound contingency that defines one of the key aspects of Hispanic colonisation, namely urbanism, suggests the need to explore other ways of approaching the study of cultural and historical relations between Europe and America. As we have already pointed out, two models of historiographical work are juxtaposed here. The philosophy of liberation and the decolonial approach depend on a totalising concept of history that establishes the existence of a single colonial logic (the symbolic violence of the coloniser, the Eurocentric *ratio*, the *coloniality of power*, etc.) that emerges from the sixteenth century to the present and that cuts across different layers of social reality: subjectivity, the construction of knowledge, social relations, institutional practices, and geopolitical dynamics.

From my point of view, using the tools of Foucauldian archaeo-genealogy, it would be more appropriate to opt for a study of practices in their specific contexts, which would allow us to unravel strictly delimited logics that can belong to a micro-, meso-, or macro-political level, without implying the existence of a global and univocal rationality that links them. From the perspective of the assumptions of historiographical contextualisation and with regard to the specific case of the colonial city, the first model minimises all the aspects that make up a context (rationalities, beliefs, experiences) and tries to emphasise the radically new character of the event. In this way, an original meaning is introduced into the genesis of colonial urbanism that cannot be explained by anything previous or by any context. It is about the production of the *ego conquiro* and the establishment of the totalitarian power of modern reason, something that was already present in the utopian and geometric design of the American city in the sixteenth century.

On the contrary, from the point of view of the approach I am proposing, the colonial city would be a particular mechanism that obeys a series of conjunctural rationalities that come together in it and that try to respond to the challenges posed by a complex reality. What is of interest, therefore, is to know what made the emergence of the colonial city and its particular precariousness possible, with the understanding that this is the only way to clarify the necessarily limited horizon of what this event implied for the future (for example: the opening of utopianism). As Jay (2011) argues, occurrences that can be qualified as events are rather rare and difficult to identify. Certainly, a contextual approach would have to be alert to the intrusion of what would be pure possibility into history, but it would also have to operate with caution against the *invention of the event*, which conceals a *transcendentalising* of the past and a useful ideological justification for the present.

The colonial city was the result of a series of events that demonstrated time and again that the colonising power was completely conditioned by the concrete realities imposed by the place and by the specific practices of both Castilians and Indians. In a sense, the colonial city was always being built and unbuilt, as if it were fulfilling the cursed cycle of Sisyphus. Perhaps it is in this precariousness, which denies the foundational nature of an omnipotent, unilateral, and imperishable power, that lies the path that must be followed in order to achieve greater intelligibility of today's Latin American urban landscapes, which are subject to new forms of instability and fragmentation.

References

Calvo T (1995) L'Amérique ibérique de 1570 à 1910. Nathan
Campanella T (1980) La ciudad del sol. In: Imaz E (ed) Utopías del renacimiento. Fondo de Cultura Económica
Castro-Gómez S (2005) La poscolonialidad explicada a los niños. Editorial Universidad del Cauca
Dussel E (1994) 1492: El encubrimiento del Otro. Hacia el origen del «mito de la Modernidad». Plural
Grosfoguel R (2007) Descolonizando los universalismos occidentales: el pluri-versalismo transmoderno decolonial desde Aimé Cesaire hasta los Zapatistas. In: Castro-Gómez S, Grosfoguel R (eds) El giro decolonial. Reflexiones para una diversidad epistémica más allá del capitalismo global. Siglo del hombre editors, pp 63–78
Hipócrates (1997) Tratado de los aires, las aguas y los lugares. Gredos
Jay M (2011) Historical explanation and the event: Reflections on the limits of contextualization. New Liter Hist 42(4):557–571. https://doi.org/10.1353/nlh.2011.0043
Lucena Giraldo M (2006) A los cuatro vientos. Las ciudades de la América Hispánica. Marcial Pons
Musset A (2011) Ciudades nómadas del Nuevo Mundo. Fondo de Cultura Económica
Palm EW (1951) Tenochtitlan y la ciudad ideal de Dürer. Journal De La Société des Américanistes 40:59–66. https://doi.org/10.3406/jsa.1951.2542
Rama Á (2009) La ciudad letrada Fineo
Ramón A (2000) Santiago de Chile (1541–1991). Historia de una sociedad urbana. Editorial Sudamericana
Seta C (2002) La ciudad europea del siglo XV al XX. Istmo
Terán F (ed) (1989) La ciudad hispanoamericana, el sueño de un orden. Centro de Estudios Históricos de Obras Públicas y Urbanismo
Vázquez García F (2009) La invención del racismo. Nacimiento de la biopolítica en España, 1600–1940. Akal

Chapter 14
Historical Context of the Q'ente and Santa Rita De Q'ente Properties (Historic Sanctuary of Machupicchu) and Its Implications for the Machupicchu Property

Alex Ivan Usca Baca

Abstract The first antecedents regarding the ownership of the Q'ente and Santa Rita de Q'ente properties come from the Sillque hacienda. During the colonial period (seventeenth–eighteenth century), the Sillque hacienda, owned by the Betlemitas, covered extensive territories located on the left bank of the Ollantaytambo and Machupicchu districts. In the mid-nineteenth and early twentieth centuries, the usufruct of the lands of the Sillque hacienda was accessed through various mechanisms such as inheritance, donation, purchase and sale, leasing, among others, which caused the fragmentation of the Sillque hacienda. In 1944, the Sillque hacienda (also called Cutija or Primavera), owned by Tomasa Ferro de Abrill, was divided among her heirs. As a result of this division, lot No. 4 or Q'ente property was awarded to José E. Abrill Vizcarra, who in September 1944 transferred it by purchase and sale to the Zavaleta Flores family. This transfer did not include the main archeological monuments located on the left bank of the current Historic Sanctuary of Machupicchu. In 1969, the Q'ente and Santa Rita de Q'ente properties were expropriated by the Agrarian Reform Law and registered in the name of the Peruvian government. Since 1977, the government has been assuming the legal defense and has demonstrated through historical and legal analysis that the transfers made during colonial and republican times were illegitimate and affected the intangibility and inalienability of the lands of the current Historic Sanctuary of Machupicchu.

Keywords Estate · Agrarian reform · Allocation · Adjudication

A. I. U. Baca (✉)
Machu Picchu National Archeological Park, Cusco, Peru
e-mail: alexusca25@gmail.com

F. Carrión Mena and S. Rodríguez Alvarez (eds.), *From the City as a Project to the City Project*, The Urban Book Series, https://doi.org/10.1007/978-3-031-94226-6_14

14.1 Background of the Ownership and Usufruct of the Sillque Hacienda[1] (1698–1935)

Advances in historical research, based on the available documentary information regarding the usufruct of the lands of the Sillque hacienda between the seventeenth and twentieth centuries, suggest the existence of different types of ownership, in addition to its permanent commercialization and litigation for its possession between civilians, Spaniards and religious.

During the colonial period, there were five haciendas on the right bank of the Vilcanota basin (i.e., Mascabamba, Huatabamba, Phiri, Tanccac and Chillca) and five on the left bank (i.e., Pachar, Cachicata, Sillque, Cutija-Utquibamba and Mescay) (Glave and Remy 1983, p. 183). Of these haciendas, Sillque and Cutija would include lands of the present Historic Sanctuary of Machupicchu. Subsequently, during the eighteenth to twentieth centuries, these haciendas went through a process of fragmentation and transfer of dominion through inheritance, sale-purchase and lease, which led to the division of the parent haciendas and consequently the birth of others.

In this context, by 1698, the Sillque hacienda was donated by the priest Juan Centeno to the Betlemitas. The lands of this hacienda were located mainly on the valley and were suitable for corn production. In addition, the hacienda had other areas with natural pastures and cultivated land in the higher areas (Glave and Remy 1983, p. 315).

In 1714, Priest Joseph de la Soledad, prefect of the Almudena Convent of Cusco (owner of the Sillque hacienda) and Isidro Juárez de Bera, on behalf of the other heirs of Francisco and Nicolás Juárez (owners of the lands of Wayllabamba, Qesqa, Churo, Pampaqhawana, Matara, Guairon and Waqoto, among others) presented a petition for the measurement and composition of their lands. At the time of the measurement, these lands had the following boundaries: *to the southeast, the Ankasqocha lagoon, the Aqoq'asa and Tanqaq Piray apacheta; to the southwest, the Salkantay and Palqay mountains; to the northwest, Intiwatana, Wayruroq'asa, Waynapicchu, the Vilcanota River, the Kusichaka ravine up to Qesqa and the Ankasqocha apacheta.*[2]

In 1725, the lands (i.e., Q'ente, Waqoto, Churo and Wayllabamba, among others), which had been given in trust to Captain Mateo Velásquez Cobarrubias in 1668, were sold by the descendants of Gabriel Mariño to the Betlemitas for 2430 pesos. The latter integrated them into the Sillque hacienda.[3] However, Glave and Remy (1983, p. 336) indicated that the acquisition of the Q'ente lands was for 2000 pesos and took place in 1728.

In 1784, there are several claims about the Q'ente property and its parts, besides being decisive for the total legalization of the properties of the Betlemitas. Friar

---

[1] The Sillque hacienda, between the seventeenth and mid-twentieth centuries, was one of the main haciendas, located on the left bank of the current districts of Ollantaytambo and Machupicchu.

[2] ARC. College of Sciences of Cusco (Colegio Ciencias), Silque documents, book 12, part 11, ff. 7v-12, 1635–1722.

[3] ARC. College of Sciences of Cusco (Colegio Ciencias), leg.14ª, cuaderno 9, ff. 14v-28, 1629–1745.

Josef de San Miguel, prior of the convent of La Almudena, presented a petition in which he indicated: *"my convent is the owner of lands named Quenti, Guato, Churo, Guayllabamba, Guairurocasa, Puerto Salcantay, Champicasa, Yntihuatana, Pichoc, Guaynapichoc and other names, which it owns by virtue of the titles that in due form it presented which are composed of the composition that it made of them with His Majesty Don Diego Sanabria Catcorrayo [...] and of the confirmations of the diligences practiced."* In addition, he requested that *"it be ordered that I be given this testimony authorized in public form [...] as this document has been presented and the titles expressed therein have been demonstrated, the present notary Miguel de Acuña will give this party the testimony it requests of said titles".*[4] It is evident that the Betlemitas legally justified their possession of the Q'ente property (which included, within its limits, the *llaqta* of Machupicchu). In December 1784, Friar Manuel de la Encarnación, in representation of the Betlemitas, took formal possession of Q'ente property and the Pitupukio property, *made up of the lands of Pitupukio, Wayruroq'asa, Intiwatana and others with boundaries at the confluence of the Vilcanota and Ahobamba rivers, with the Picchus knife and continuing through Yunkapata (Intipata) to the Salkantay snow-capped mountain.*[5]

In this context of the post-colonial period, precisely the republican period after the Independence of Peru, Simon Bolivar decreed the expropriation of ecclesiastical property in favor of hospices and schools. Thus, the Sillque hacienda, owned by the Betlemitas, was given to the College of Sciences of Cusco. However, it was later awarded to Marshal Agustín Gamarra as a reward for his support to the independence cause: a Supreme Resolution of May 28, 1832 ruled the granting of the possession of this hacienda in favor of Gamarra when he was president of Peru.[6] In the document, the appraisal of each of the properties that make up the Sillque hacienda was indicated, such as: *"Pampa de Sillque, San Isidro, Patapata, Cutija, Camicancha, Cachiccata, Miscae, Chamana, Guaccoto, Quente, Carpamayu, Guayllabamba, Chillipahua, Rocococha, Chuya, Muyucancha, Ocoruro, Chancachuco, Hanccascocha, Pocpo, Coyabamba, Pampacahuana, Chaquil and Vilque".*[7] In the respective appraisal, no information was provided regarding the real extension of the hacienda. Also, no information was clearly indicated about boundaries. Upon the death of Marshal Gamarra, the Sillque hacienda was inherited by his son, Andrés Gamarra. In 1847, having contracted excessive debts with Colonel Ramón Nadal, Andrés Gamarra was forced to sell it by public auction in favor of the latter for 42,000 pesos.[8] The transfer of the Sillque hacienda included within its boundaries important production areas located in different ecological levels, as Llanos refers: "*...it included: Cachicata, Chillipahua, the pastoral lands at the height of Sallkantay, today property of the Sondor hacienda, Limatambo, the current Sillque hacienda, Cutija and the lands*

[4] ARC. Cabildo del Cusco, Causas Civiles, leg. 55, cuaderno 9, ff. 1-1v, 1783–1784.

[5] ARC. College of Sciences of Cusco (Colegio Ciencias), Silque documents, book 12, part 14, f. 22, 1635–1722.

[6] ARC. Notarial Protocol. Mar y Tapia, Pablo del, protocol 166, ff. 352–366, 1828–1835.

[7] ARC. Notarial Protocol. Mar y Tapia, Pablo del, protocol 166, ff. 355–359v, 1828–1835.

[8] ARC. Notarial Protocol. Araníbar, Pedro José, protocol 13, exp. 6, 1847–1848.

Fig. 14.1 Extension of the Sillque hacienda at the end of the eighteenth century. *Source* Google 2023; own elaboration

of the sugarcane fields that extended to the Santa Teresa bridge, now property of Huadquiña..." (Llanos 1926, p. 233).

This reference shows that during the republican period, the Sillque hacienda covered a large area that included a large part of the land that is currently part of the Historic Sanctuary of Machupicchu and the Buffer Zone (Fig. 14.1). Among its boundaries, it not only included valley floor lands, but also dry and irrigated lands, places and settlements, which in times before the republic belonged to indigenous societies who used the land through the principles of reciprocity and redistribution.

In the middle of the nineteenth century (1860), Jacinto Alegría, usufructuary of a large part of the lands on the left bank of the current Historic Sanctuary of Machupicchu, sold the Intiwatana hacienda to Santiago Angulo and his wife Rufina Guillén, for 450 pesos. The boundaries of the Intiwatana hacienda were: *to the south-east, the Sillque and Kusichaka hacienda; by the heights, Waqoto, Mediocucho, Wayllabamba, Llullucha, Huch'uywayruro, Toqorumiyoq, Cedro Cortado and Salkantay; by the Vilcanota River, Intiwatana, Picchu, Hatunpicchu, Huch'uypicchu, Carmenga, Pepino, Chachabamba, Durasnuyoq, Champiq'asa and Q'ente.*[9] In February 1862, Santiago Angulo and his wife Rufina Guillén sold the Intiwatana hacienda (for the

[9] ARC. Notarial Protocol. Gárate, Carlos, protocol 102, exp. 29, ff. 245v–247v, 1860–1862.

same value at which it had been acquired two years earlier) to Colonel Ramón Nadal, who integrated it into the Sillque hacienda.[10]

The lands acquired by Ramón Nadal helped to consolidate his position as the main usufructuary of the Sillque hacienda; however, in 1864, in his will, he declared his children Julián, Alejandro, Antonia and Adeodato Nadal as his universal heirs. The first two died at an early age and the last two divided their assets as follows: Antonia received the Lucre hacienda and obraje and Adeodato obtained the Sillque hacienda.

In 1896, the lands of the Sillque hacienda were still in the possession of the Nadal family, who after the death of Adeodato Nadal executed a deed of extrajudicial division of the assets of the Sillque hacienda, which defined that the lands would be distributed as follows: Adeodato's widow, ***Genara Suarez****, received Camicancha, Ankasqocha, Chancachuco, Chaquil, Wayrancalle, Cutija, Pampaqhawana, Palqay and Churumayu.* ***Ramón Nadal*** *obtained the lands of Manchaybamba, Cachicata, Patapata Chico and Churu.* ***Alejandro Nadal*** *got Mansanayoq, Vilque, Coyabamba and Intiwatana.* ***Eduardo Nadal*** *got Chillipahua, Roccoccocha, Turpay, Pacopata, Wacakancha, Chiara and Pocepoc, Wamankancha, Capillapata, Anapahua and Jacas.* ***Enriqueta Nadal*** *was left with the main hacienda of Silque, Patapata Grande, Qesqa, Rocobamba, Qarpamayu, Meskay, Willkaraqay, Waqoto, Wayllabamba and Chamana.*[11] During this period and until the beginning of the twentieth century, land transfers were made between the heirs of Adeodato Nadal and also with third parties,[12] leading to the division of the Sillque hacienda (Fig. 14.2).

In 1904, Mariano Ignacio Ferro and his wife, María Laureana Vizcarra de Ferro, began to acquire several properties on the left bank of the Vilcanota River from the Nadal family. These properties made up the Cutija hacienda and included the sectors of Q'ente, Santa Rita, Intiwatana, Wayllabamba, Pampaqhawana, Primavera, Lourdes and Patapata, among others. In 1921, María Laureana Vizcarra stated that she was the owner "*... of half of the Cutija hacienda [...], composed of the fractions: Cutija, Manchaybamba, Kcamicancha, Santa Rita and Intimatana [Intiwatana], all with their adjacent i estancias...*".[13] The property acquired by the Ferro-Vizcarra conjugal partnership was transferred to their heirs Manuel Esteban, Felipa, María Salomé, Saturnina and Tomasa. In the same year, 1921, Mariano Ignacio Ferro, in his capacity as executor of his wife, and as an advance payment of the legitimate, gave the Primavera hacienda[14] to his daughter Tomasa Ferro de Abrill for the sum of 50,000 soles.[15]

---

[10] ARC. Notarial Protocol. Gárate, Carlos, protocol 102, exp. 29, ff. 921v–922, 1860–1862.

[11] ASUNARP. Zona Registral X. Sede Cusco, partida 02016781, t. 1, f. 60, as. 2, 1898.

[12] ASUNARP. Zona Registral X. Sede Cusco, partida 02016781, vol. 28, f. 219, as. 34, 1908; t. 40, f, 92, as. 48–48v, 1910.

[13] ARC. Notarial Protocol. Alosilla, José, book 390, ff. 421–423, 1921.

[14] Colonial documents from the seventeenth century about the Sillque hacienda refer to the place name Cutija, which will survive until 1921, when it changed its name to Primavera.

[15] ASUNARP. Zona Registral X. Sede Cusco, partida 02016781, t. 64, f. 227, as. 63, 1922.

Fig. 14.2 Division of the Sillque hacienda among the heirs of Adeodato Nadal, 1896. *Source* Google Earth 2023; own elaboration

In 1935, by means of a resolution of July 31, 1935, the brothers José Luis, Carlos Alejandro and Julia Lourdes Abrill Ferro were declared heirs of Tomasa Ferro de Abrill and the first two became, together with their father, owners of the Primavera hacienda[16] (Fig. 14.3).

14.2 Fractionation of Hacienda Primavera (Cutija)

On August 18, 1944, in the property division trial followed by the heirs of Tomasa Ferro de Abrill, the Primavera (Cutija) hacienda was divided as follows: *José Emilio Abrill Vizcarra was awarded lot 4 or Q'ente; José Luis Abrill Ferro was awarded lot 2 Primavera and the Pampaqhawana punas; Carlos Alejandro Abrill Ferro was awarded lot 3 or Camicancha; and Julia Lourdes Abrill Ferro received lot 1 (Chuyamayu),*[17] which was not located in the understanding of the Primavera hacienda, but in Huayopata (La Convención) (Fig. 14.4).

With respect to lot No. 4 or Q'ente, by court order in 1944, the following sections were identified: *Q'ente, Santa Rita, Intiwatana, Cedrobamba, Punas Wayruro,*

[16] ADRAC. Agrarian Reform, leg. 311, exp. 2008, 1974.

[17] ADRAC. Affectation. Leg 311. 2008 File. Year. 1974–1990. Fs. 14–27.

Fig. 14.3 Cutija-Primavera hacienda before the division between the Abrill-Ferro family. *Source* Google Earth 2023; own elaboration

Fig. 14.4 Division of the Primavera (Cutija) hacienda among the heirs of Tomasa Ferro de Abrill, 1944. *Source* Google Earth 2023; own elaboration

Fig. 14.5 Q'ente property or Lot No. 4. *Source* Google Earth 2023; own elaboration

Matara, Mesada and Palqay. In addition, this property included all the forests, headwaters of mountains that are located around the ruins of Machupicchu.[18] The boundaries of the Q'ente hacienda extended to the north with the Vilcanota River, to the east with the Camicancha and Pampaqhawana haciendas, to the south with the La Estrella hacienda, having the Salkantay knot and crest as a dividing line, and to the west with the Huadquiña hacienda, having the Aobamba gorge as a natural boundary; the surface extension of the hacienda covered a total area of 22,000 ha (Fig. 14.5).

With respect to the Q'ente fraction, where the *llaqta* of Machupicchu is located, file 452, book 3, located and analyzed by Rogger Ravines (1975 [2012]), is of great importance. According to the information contained in it, in 1933, José Emilio Abrill Vizcarra asked the Peruvian State for the expropriation of his lands and two years later for compensation, which, through Ministerial Resolution No. 2511 of the Ministry of

[18] ADRAC. Affectation. Leg 311. 2008 File. Year. 1974–1990. Fol. 20.

Instruction, recognized the importance of the expropriation of the entire Primavera hacienda and referred the document to the Ministry of Development for the corresponding procedure. After various complex procedures and opinions, Supreme Resolution No. 3975 of December 6, 1944 declared the creation of the Ollantaytambo National Park to be of public utility and authorized the Peruvian National Archeological Board to follow the expropriation file of the Primavera hacienda owned by Abrill Vizcarra (Ravines 2012 [1975], pp. 86–89).

In September 1944, the Q'ente fraction (composed of the sections Q'ente, Santa Rita, Intiwatana, Cedrobamba, Matipata, with all its comprehensions and the punas of Wayruro, Matara, Mesada and Palqay) was sold by José Emilio Abrill Vizcarra to Julio Zavaleta Flores and Rosa Zavaleta Álvarez for 63,000 soles. However, it was noted that the sale did not include the five most important archeological sites: *Machupicchu, Waynapicchu, Phuyupatamarka, Sayaqmarka and Wiñaywayna.*[19] After the transfer of the Q'ente fraction in favor of the Zavaleta Flores family, they proceeded indirectly to usufruct the lands suitable for agricultural activity, which were generally located on the valley floor of the Kusichaka and Vilcanota rivers.

14.3 Agrarian Reform in Predio Q'ente and Santa Rita De Q'ente

The situation of the land ownership regime at the beginning and until the middle of the twentieth century in Peru, and particularly in the Cusco region, was characterized by the dominance of landowners over agrarian production. This situation served as the basis for the Peruvian State to generate a political project aimed at changing the traditional agrarian structure. Thus, by Supreme Decree No. 44 of August 5, 1966, the entire department of Cusco was declared an Agrarian Reform Zone. Three years later, on June 24, 1969, with the enactment of Agrarian Reform Law No. 17716, the process of land affectation and adjudication began under the guidelines of the law.

The process of Agrarian Reform consisted of two phases: **the affectation or appropriation**, which consisted of the expropriation of the haciendas and its delivery to the State. And the **adjudication**, which consisted of the transfer of the hacienda lands in favor of the feudatories previously qualified as beneficiaries. The reform process helped to put an end to inequality in land ownership and to the abuse of the hacienda owners, thus achieving "*an integral process and an instrument of transformation of the agrarian structure of the country*" (Rueda 1970, p. 1).

The landowners had to comply with the rules of the decree, which required them to submit a sworn statement, including the property titles, location plans and number of feudatories, among other data. Once this requirement was fulfilled, the General Directorate of Agrarian Reform and Rural Settlement (DGRA-AR) proceeded to

[19] AGN. "Testimonio de escritura pública de compra-venta del predio Q'ente granted by J. Emilio Abrill Vizcarra in favor of Julio Zavaleta Flores and Rosa Zavaleta Álvarez. Before the notary public Rosendo A. Fernández," ff. 15–21, 1944.

prepare the total or partial land affectation plan, stating the extension and reasons for the affectation of the areas under consideration. If there were no grounds to the contrary, the DGRA-AR, under the protection of article 149° of the TUC of DL No. 17716, was in charge of "*the actions of transformation of the structure of rural land tenure, with legal personality, only for the effects of the affectation, acquisition, expropriation and adjudication of rustic lands*" (Eguren 2009, p. 77).

In Machupicchu district, the agrarian reform process became effective in the former haciendas of Q'ente, Santa Rita de Q'ente and San Antonio de Torontoy. The expropriation process in these haciendas was based on the provisions of articles 17, 19 and 21 of DL No. 17716, which provided for the total or partial affectation of lands that were exploited by feudatories, small tenants, subtenants and other non-owner farmers, on the condition that the allocated landholdings did not exceed three times the family agricultural unit determined for each zone.

In this context, the Agrarian Reform procedure of the ***Q'ente*[20] *and Santa Rita de Q'ente properties or haciendas,*** located on the left margin of the current Historic Sanctuary of Machupicchu, was carried out separately by properties, as a result of the extrajudicial division carried out by their owners, which is why the procedure and technical reports of the Agrarian Reform Law were developed under the protection of the legislation of Legislative Decree 17,716.

On October 21, 1969, Rosa Zavaleta Álvarez, as possessor of the ***Q'ente hacienda*,**[21] in compliance with Article 50, paragraph a) of Law 17,716 of the Agrarian Reform Law, presented her sworn statement in which she stated: A simple copy of the notarization of the holographic will granted by Julio Zavaleta Flores; a notarial deed of division of assets granted on September 23, 1965; a notarial copy of the deed of sale dated September 12, 1944; a literal certificate of ownership of the Sillque property; and the information regarding the number of feudatories, surface area in hectares, agricultural production and plans. This information was analyzed by the DGRA-AR, and the pertinent technical studies were carried out, qualifying and indicating the areas subject to affectation.

On September 24, 1973, the legal advisory office of the DGRA-AR issued Report No. 1636-73-AJ-XI, detailing the legal study for the titling of the Q'ente hacienda. Two aspects stand out from this document. First, the Q'ente hacienda was originally owned by José Emilio Abrill Vizcarra, having acquired it in 1944. On September 12 of the same year, Abrill Vizcarra decided to sell it to Julio Zavaleta, stating that "*... this sale does not include the payment of the indemnities that are still pending before the government for the expropriation of the Inca cities Machupicchu, Huayna Picchu,*

[20] In 1944, Q'ente hacienda included the Q'ente, Santa Rita, Intiwatana, Cedrobamba, Wayruro, Matara, Mesada y Palqay, Machupicchu, Phuyupatamarka, Sayaqmarka, Waynapicchu, Wiñay-wayna and Matipata sectors, with a total area of 22000 ha.

[21] The process of affectation and adjudication of the Q'ente property was quite complex for three reasons: (a) the existence of important evidence of immovable cultural heritage; (b) the presence of feudatories; and (c) the persistence of the heirs of the Zavaleta family, who were not satisfied with the expropriation of their land.

Huiñayhuayna, Sayac Marca, Phuyupatamarca, currently owned by the State...".[22] Secondly, in the will of Julio Zavaleta Flores dated November 15, 1966, it is indicated that his assets, including the Q'ente hacienda, had passed through an extrajudicial distribution carried out with his spouse Rosa Zavaleta Alvarez.[23] As a result of this distribution, the estate was divided as follows: "*...to Rosa María Zavaleta Álvarez de Zavaleta, in the property "Oquente," which has the following boundaries: On the north with the Vilcanota River, throughout its course; on the south, with the high peaks of the Matara puna, which separates it from the Palccay puna; on the east, with the Camicancha and Pampa ccahuana hacienda, punas of the Primavera hacienda, being the natural boundary that rises from the snowy Salccantay...within this boundary is assigned to Rosa M. Zavaleta Álvarez the following zones: Kusichaka, Tarayoq, Huayllabamba Huayruro and Matara; and for the husband and wife Julio Zavaleta, he is assigned the fraction of the same Q'ente hacienda, which are within the following boundaries: On the north the Vilcanota river, from kilometer nine hundred and twenty-three of the Santa Ana railroad, which is in front of the bad pass of Escalerayoq, until kilometer one hundred and twenty-two; on the south, with the Estrella hacienda, having as dividing line the knot and crest of the snowy Salccantay; on the west with the Huadquiña hacienda, having as natural limit the ravine and the Aobamba river up to the mouth of the Vilcanota river; on the east the high peaks of Huayruro, Matara and Palccay, within the boundary, there are the following places: Paccaymayo, Santa Rita, Huiñaywayna, Phuyupatamarka, Machupicchu, Huaynapicchu, Inkaraqay, Intiwatana, Rompe, Matipata, Sayaqmarka, Yanaqocha, Mesada and Palqay*"[24] (Fig. 14.6). We infer that the extrajudicial division carried out by the spouses was a legal strategy to avoid the total expropriation of the property; however, the agrarian reform procedure of the Q'ente property was carried out separately and by property, as detailed below.

On December 10, 1973, the DGRA-AR presented a technical report on the ***Q'ente property***, detailing that the total area of the property was 6936 ha, of which 5369 ha were uncultivated land that would revert to the Peruvian State's domain, and the land reclamation quota was determined at 1567 ha, which included cultivated land, natural pastures, forest area, installations and constructions. In addition, this property included 827 ha of snow-capped mountains and archeological zones (Machu Q'ente, Wayna Q'ente, Tiendachayoq and Patallaqta)[25] with an area of 7.60 ha. Likewise, at the time of the different technical inspections, the lands of the Q'ente property were being used indirectly by Julio Carlos Zavaleta Zavaleta (6.80 ha) and by feudatories (27.60 ha).

[22] AGN. "Testimonio de escritura pública de compra-venta del predio Q'ente granted by J. Emilio Abrill Vizcarra in favor of Julio Zavaleta Flores and Rosa Zavaleta Álvarez. Before the notary public Rosendo A. Fernández," ff. 15–21, 1944.

[23] ADRAC. Affectation and adjudication, leg. 15, 1970–1975, exp. 204, ff. 1–3.

[24] ARC. Emilio Muñiz Notarial Protocol. 1965. Protocol N° 60, folio N° 1837v to 1841. "Testimonio de la escritura pública de División y Partición de Bienes".

[25] The archeological zones identified in the Q'ente property were based on the reports of the Cusco's Directorate Archeological Board (Patronato Departamental de Arqueología del Cusco).

Fig. 14.6 Division of the Q'ente property or lot No. 4. *Source* Google Earth 2023; own elaboration

On January 22, 1974, after the technical report, the owner Rosa M. Zavaleta Alvarez made observations on the plan of the land affectation quota (1567 ha), stating that within the reversion area there are other lands that could complement the family agricultural unit (UAF) for the feudatories, and that the data of previous qualifications of feudatories were made in a unilateral manner. Due to these observations, it requested the DGRA-AR to qualify it as a beneficiary of the Agrarian Reform Law. However, through report No. 179-74-DA-PARA-ZA-XI of the Sub-Department of Agrarian Reform, the technical reports of affectation were ratified, and the land was expropriated.

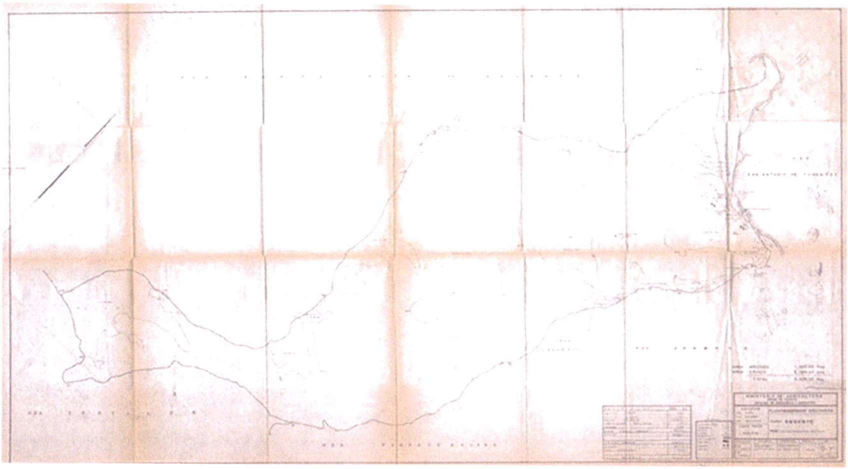

Fig. 14.7 Q'ente property, with a surface area of 6936 ha, including affected area (1567 ha) and uncultivated area (5369 ha)

On May 15, 1974, pursuant to Directorial Resolution No. 0365-74-DZA-XI, the DGRA-AR and D.S No. 0444-75-AG (April 29, 1975) approved the definitive plan for the affectation of the Q'ente property with an extension of 1567 ha, distributed among the following sectors: Matara, Wayllabamba, Q'entemarka and Tarayoq. Articles 17, 19, 21 and 65 of the Agrarian Reform Law state that the land **will be affected in its entirety**. This statement was supported in items 14 and 16 of the technical report presented by the DGRA-AR on December 10, 1973. Once the lands of the Q'ente hacienda were expropriated, the State, represented by the DGRA-AR, took possession and proceeded to register them in entries 145 and 148 of entry No. 02016781 of the National Superintendence of Public Registries (Sunarp).[26] With respect to registry entry No. 145, the DGRA-AR, through Supreme Resolution No. 0322-75-AG-DZ-XI of July 22, 1975, became the owner of the uncultivated land comprising an area of 5369 ha. In the same way, it is clear from the entry No. 148, that the DGRA-AR became the owner of the 1567 ha, due to the sentence of June 28, 1976 dictated by the land judge of Quillabamba (Fig. 14.7).

In December 1975, the DGRA-AR, in compliance with articles 67° and 77° of the TUC of DL No. 17716, stipulated that the adjudications would be made by the General Directorate of Agrarian Reform and Rural Settlement in favor of peasants without land or who possessed insufficient land. In addition, the norm stipulated that the adjudications would be made only to cooperatives, peasant communities, agricultural societies of social interest and natural persons, previously qualified (Caballero and Alvarez 1980, p. 39). However, this did not apply in the area of the current Historic Sanctuary of Machupicchu, since the land and buildings of the Q'ente hacienda were

[26] ASUNARP. Zona Registral N° X-Sede Cusco, item 02016781, t. 179, f. 301, as. 145–148.

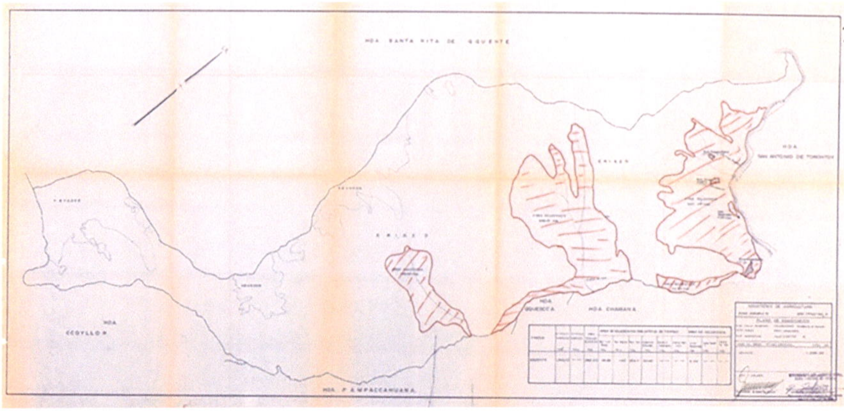

Fig. 14.8 Q'ente property, with details of the polygons corresponding to the affected areas (sectors: Matara, Wayllabamba, Tarayoq and Q'entemarka)

awarded to the Qquente Peasant Group.[27] After studying the feudatory qualification documents in accordance with Article 84 of the Agrarian Reform Law,[28] the legal advisory office of the DGRA-AR issued RD No. 1136-75-DZA-XI on December 22, 1975, in which 33 feudatories of the Q'ente hacienda were qualified as Agrarian Reform beneficiaries.

On January 28, 1976, through RD No. 280-76, the DGRA-AR proceeded to adjudicate the lands of the Q'ente hacienda in favor of the Qquente Peasant Group[29] (1563 ha), distributed among the Wayllabamba (680.51 ha), Tarayoq (44.80 ha), Q'entemarka (641.29 ha) and Matara (196.40 ha) sectors. The adjudication was carried out through purchase-sale contract No. 0040-76 of January 31, 1976, where the adjudicators agreed to pay the amount of 2,091,390 soles for land, plantations and machinery[30] (Fig. 14.8).

At the same time, the Agrarian Reform process of the **Santa Rita de Q'ente** hacienda took place between 1969 and 1976. On October 1, 1973, after carrying out a pertinent study of the file presented by the owners, the legal advisory office

[27] During the Agrarian Reform implementation process, peasant groups emerged as an organizational model. Originally they had been conceived as transitory associative forms, which grouped smallholders who had been awarded land, but the goal was to integrate them into cooperatives. In the case of Historic Sanctuary of Machupicchu, the attempts of the allottees to constitute themselves as communities or peasant groups remained in process, due to the recognition of the Historic Sanctuary of Machupicchu as cultural-natural patrimony of the Nation.

[28] Art. 84. To be admitted as an applicant for the adjudication of Family Agricultural Units, it was required: to be older than 18 years old, to be head of family, to be a peasant, to reside in the land of adjudication and not to be owner of other lands.

[29] ADRAC. Affectation. Province of Urubamba. District of Machupicchu, leg. 320, 1976, exp. 4, f. 187.

[30] ADRAC. Affectation and adjudication of Agrarian Reform, leg. 14-B, 1973–1993, exp. 196, ff. 52–55.

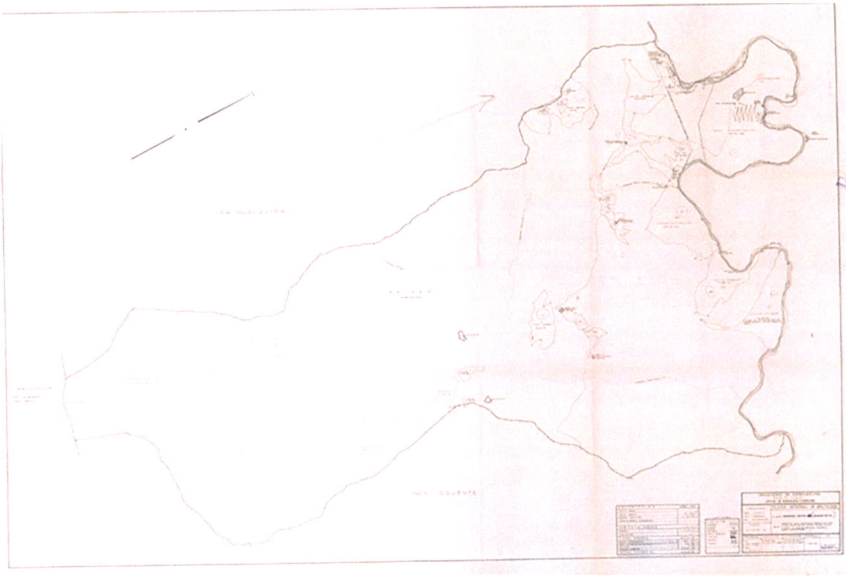

Fig. 14.9 Santa Rita de Q'ente property, with a surface area of 12,770 ha, including affected area (1134 ha) and uncultivated area (11,636 ha)

of the DGRA-AR issued report No. 1652-73-AJ-ZA-XI, detailing the scope of the legal study for the titling of the hacienda. The information in the file of the owners' affidavit was the same that had been presented for the Q'ente hacienda, with the only difference being that the owner-heirs clarified that *"the State is only the owner of the ruins, but the area occupied by these ruins corresponds to the testator [...], owners in condominium, who according to the affidavit give it the name of Santa Rita de Qqente..."*.[31]

On December 7, 1973, the DGRA proceeded to carry out a technical report of affectation, where it was specified that the total surface of the hacienda according to the plan was 12,770 ha Fig. 14.9 of which 11,636 ha were considered as uncultivated land, which passed to the domain of the Peruvian State; the surface of snow-capped mountains and lagoons was estimated at 2022.40 ha; the archeological zones was 129.66 ha; the Machupicchu hydroelectric area was 60.80 ha and the total area to be affected was defined as 1134 ha. Likewise, inspections carried out by technical personnel from the Ministry of Agriculture established that the Santa Rita de Q'ente property was occupied and worked by feudatories in an area of 21.30 agricultural hectares and by the anti-crests Saturnino Delgado Moscoso and Cornelia Delgado Tintaya, who occupied 201.90 ha of natural pasture.

On May 15, 1974, by means of RD No. 0364-74-DZA-XI, the DGRA-AR declared the 1134 ha of the Santa Rita de Q'ente hacienda to be affected. Subsequently, on

[31] ADRAC. Affectation and adjudication of Agrarian Reform, leg. 15, 1970–1975, exp. 203, ff. 1–4.

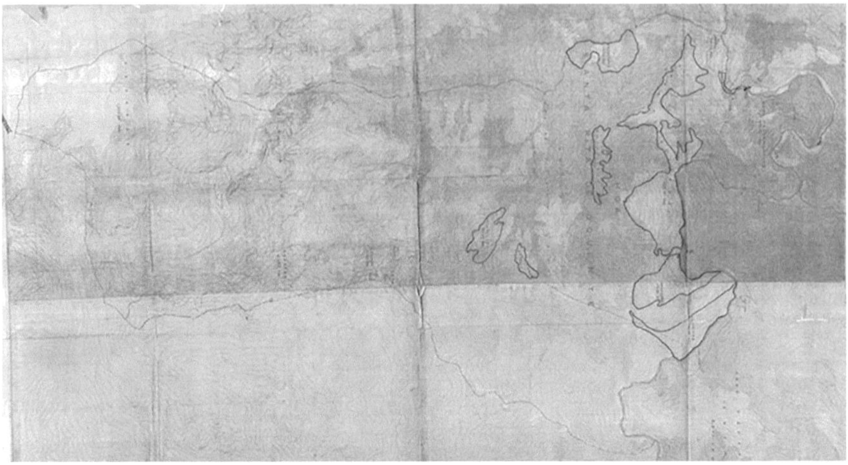

Fig. 14.10 Santa Rita de Q'ente property, with areas marked as affected and adjudicated

November 28, 1974, the Ministry of Agriculture promulgated SD No. 1207-74, which approved the definitive plan of affectation. Once the lands of the Santa Rita de Q'ente hacienda were expropriated, the Peruvian State took possession and proceeded to register them in the name of the General Directorate of Agriculture and Rural Settlement.[32] In this regard, registry entry No. 142 states that, by Supreme Resolution No. 0321-AG-DZ-XI dated July 22, 1975, the DGRA-AR became the owner of 11,636 ha. Along the same lines, registry entry No. 143 states that, by judgment of the land court of Quillabamba on September 27, 1976, an area of 1134 ha was expropriated in favor of the DGRA-AR, which consisted of eleven sectors (Fig. 14.10).

On April 8, 1976, after reviewing the affidavit files presented by the feudatarios and having verified in the field the possession and veracity of the data, the DGRA-AR decided (through RD No. 0286-76-DZA-XI) to benefit fourteen feudatarios. Based on the stipulations of Article 83°, the DGRA-AR and the Huayllabamba-Huayna Q'ente Peasant Group, represented by Paulino Herrera Sayco and Valerio Sayco Ttica, signed on June 24, 1976 the purchase and sale contract No. 1046-76, where the purchase and sale adjudication of 1182 ha was arranged, as well as the plantations of the described property in favor of the Huayllabamba-Huayna Q'ente Peasant Group"[33].

In 1989, the National Institute of Culture (INC) asked the DGRA-AR to rescind DS No. 1207-74-AG and DS No. 0444-75-AG. This was due to the fact that the Q'ente and Santa Rita de Q'ente haciendas were located inside the Historic Sanctuary of Machupicchu, which did not correspond to current legislation regarding the protection of cultural heritage. For this reason, on September 10, 1991, the DGRA-AR issued decree DS No. 036-91-AG, which left the legal effect of the protection

[32] ASUNARP Zona Registral N° X-Sede Cusco, item 02016781, t. 140, f. 451, as. 142–143.

[33] ADRAC. Affectation and adjudication, leg. 14-B, 1973–1983, exp. 196, f. 57.

of the Q'ente and Santa Rita de Q'ente haciendas without legal effect and entrusted their protection to the INC.[34] As a result of this decree, Julio Carlos Zavaleta and Rosa Zavaleta Álvarez, by means of a document dated October 10, 1991, requested the Quillabamba Land Court to annul the entire expropriation process of the Q'ente and Santa Rita de Q'ente properties. That is, the annulment of the contract between the DGRA-AR and the Q'ente and Santa Rita de Q'ente Peasant Groups and the cancellation of the registry entries in the name of the DGRA-AR. In December of the same year, the land judge of Quillabamba granted the petition of the Zavaleta family and declared null and void everything that had been done in the expropriation process, ordering the cancellation of the property registration in favor of the DGRA-AR and that, once the cancellation had taken place, real and effective possession be given to the owners of the expropriated land. Based on this provision, the Zavaleta family proceeded to register the resolution in entries 156 and 157.[35] In 1993, the legal Directorate of the National Institute of Culture appealed the ruling of the Quillabamba Land Court, a request that was granted by Resolution No. 36 of April 30 of that year, which declared the nullity and invalidity of the proceedings. In 2001, the Mixed Chamber of the Superior Court of Cusco revoked the resolution issued by the Mixed Judge of Urubamba and ordered the cancellation of the existing registry entries in favor of the Zavaleta family. Finally, the Peruvian State regained ownership of the Q'ente and Santa Rita de Q'ente property.[36] Consequently, the Peruvian State, through the Decentralized Ministry of Culture of Cusco, continues to administer the cultural heritage of the Machupicchu Historic Sanctuary in accordance with the regulations and the expropriations carried out by the DGRA-AR.

14.4 Implications Regarding Machupicchu Property

After the application of the Agrarian Reform Law in the area of the current Historic Sanctuary of Machupicchu, claims were filed by the heirs of the Zavaleta family, who sought to reclaim the property affected by the Agrarian Reform on certain legal and factual grounds. Faced with this situation, the Peruvian State, from 1977 to date, has been assuming the legal defense and demonstrating that the plaintiffs are not the owners of the Q'ente and Santa Rita de Q'ente properties because, before the enactment of the Agrarian Reform Law, there was already a vast legislation aimed at protecting and safeguarding the cultural heritage.

Since the dawn of Independence, the DS N° 89 of 1822 already considered that the archeological monuments belonged to the Peruvian nation. On June 13, 1929,

---

[34] El Peruano (1991, p. 99863).

[35] Archive of the Superior Court of Justice-Cusco (ACSJC) (3° Civil Court. Judicial process 002228-2005-0-1001-JR-CI-03, t, II, ff. 50–55; claim Zavaleta family, owners of Q'ente and Santa Rita de Q'ente).

[36] The cancellation was registered in entries 181 and 182 of the record N° 9603 of the Real Estate Registry of the Public Registries of Cusco.

the Peruvian State enacted Law No. 6634, which in its Article 1° indicates: *"The monuments existing in the National Territory prior to the time of the Viceroyalty are property of the State.* The *right of the Nation over said monuments is inalienable and imprescriptible"* (Ministry of Education 1929, p. 4). Likewise, in Article 22, it states: *"The Government will provide what is necessary for the conservation and repair of the ruins and archeological sites that are declared national monuments by special laws and in any case of the sites of Saqsayhuamán, Ollantaytambo, Machupicchu..."* (Ministry of Education 1929, p. 8). On the other hand, regarding expropriation processes, article 5° states: *"If the archeological properties referred to in the preceding articles were located on private property, the State could expropriate said properties in accordance with the law, in the superficial extension sufficient for their conservation and the scientific explorations to which they lend themselves"* (Ministry of Education 1929, p. 5).

On December 6, 1944, in order to conserve and preserve the archeological heritage, the Peruvian State ordered the creation (through Supreme Resolution No. 3975) of Ollantaytambo National Park, which included the district of Machupicchu (established on October 1, 1941 through DL Nº 9396). Thus, it was declared of public utility and the Patronato Departamental de Cusco was authorized to proceed with the expropriation procedures (Ravines 2012, p. 88).

On January 31, 1969, the Cusco's Directorate Archeological Board, based on Law No. 6634, provided for the delimitation of archeological areas, creating the Machupicchu National Archeological Park with an area of 10,724 ha, whose boundaries were: on the east, the Kusichaka, Chamana and Willkaraqay ravines; on the southwest, Wayruro and Yanaqocha; on the west, the Huadquiña hacienda; and on the north, the Vilcanota River[37] (Fig. 14.11). Since the delimitation of the PANM, the Peruvian State, through the competent institutions, began to carry out actions aimed at the conservation, restoration and research of such an important area.

Twelve years later, DS No. 001-81-AA of January 8, 1981, created the Historic Sanctuary of Machupicchu with an intangible area of 32,592 ha. As of that date, the Ministry of Agriculture (with jurisdiction over natural protected areas) and the Ministry of Education (with jurisdiction over cultural and archeological issues through the then National Cultural Institute [INC]) assumed control of the sanctuary. Then, on December 9, 1983, the Historic Sanctuary of Machupicchu was inscribed by UNESCO as a mixed heritage site on the World Cultural and Natural Heritage List, with registration number 54.

The legal norms regarding cultural heritage and added to this the analysis and interpretation of historical documentation helped to demonstrate that in colonial times and part of the republic the lands of the indigenous people were inalienable because this was prescribed by the compilation of the Indian laws regarding the distribution of indigenous lands. These could not be sold to Spaniards nor to other people but they belonged to the common indigenous, which passed from parents to children, from generation to generation and in the absence of them for their community or *ayllu*. In

[37] ADRAC. Affectation and adjudication of the Agrarian Reform. Province of Urubamba, leg. 320, 1976, exp. 3, ff. 12–20; adjudication.

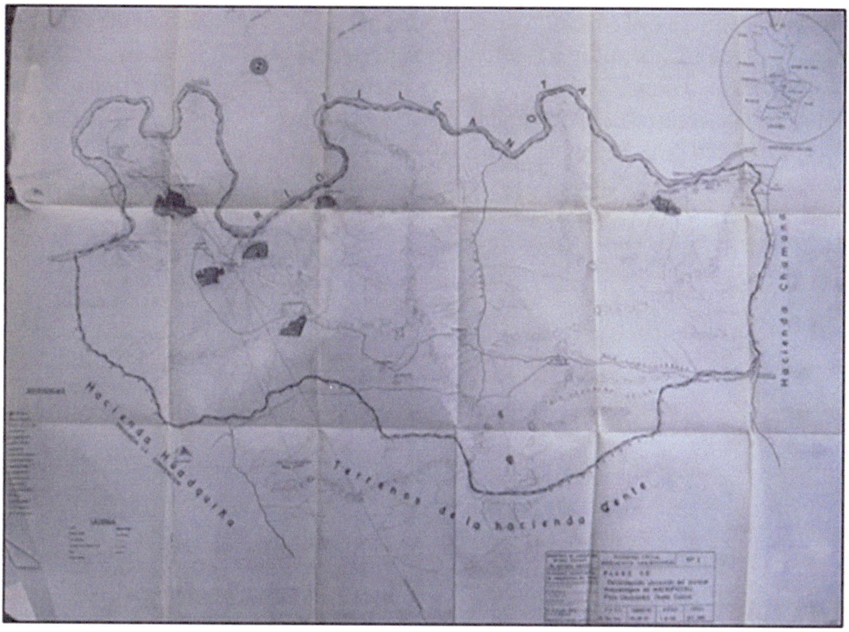

Fig. 14.11 Delimitation of the PANM, 1969 (ACDC 1968)

this perspective, the lands of the Historic Sanctuary of Machupicchu, especially those that correspond to the ex-land of Q'ente and Santa Rita de Q'ente, from their origin at the beginning of the conquest, were part of the lands assigned to the indigenous people that had been transferred from the area of Cusco on October 4, 1572, to the area to found the town of San Francisco de Victoria de Vilcabamba. Being so, in merit to the legislation of Indians law No. 16, title XII, book IV, these lands assigned to indigenous people were given to them with inalienable character and by no case they could be sold or alienated to third parties. This condition was maintained throughout the colony and even until the beginning of the republic, specifically until 1849, so the donations and subsequent sales of these lands by Mariano Santos, Jacinto Alegría, Santiago Angulo, Ramon Nadal, Vizcarra Ferro, Abrill and Zavaleta were illegitimate and illegal.

The Peruvian State's defense has demonstrated and sentenced that the Zavaleta family is not the owner of the Q'ente and Santa Rita de Q'ente properties and of the archeological monuments located in the aforementioned properties. For this reason, the Superior Court of Justice through the 3rd civil court sentenced the following resolutions: (a) ***Resolution No. 237 of August 2, 2013***, referred: The Peruvian State is the owner of these properties having acquired them by expropriation through the General Directorate of Agrarian Reform and Rural Settlement (DGRA-AR). This is evidenced by the judicial proceedings followed with File No. 10-76, before the land judge of La Convención and File No. 76-75, before the land judge of Quillabamba, both of which ended with the declaration of ownership in favor of the

Peruvian State. (b) ***Resolution No. 473 of January 23, 2019***, declared the cassation appeal unfounded, therefore the sentence of Resolution No. 273 was confirmed. (c) ***Resolution No. 477 of January 13, 2020***, resolved to the effect that the first instance sentence, second instance sentence and cassation sentence be registered in the registry entry No. 02016781. On January 20, 2020, the Superior Court of Justice-Third Civil Court sent resolutions 237, 473 and 477 to the head of the Registry Zone No. X, Cusco Headquarters, so that he could arrange the registration of the judgments.

14.5 Conclusions

The transfer processes of the lands located on the left margin of the current Historic Sanctuary of Machupicchu were illegitimate because the lands of the indigenous people were inalienable. In addition, we must highlight the legal corpus that was promulgated since the beginning of the republic as the Supreme Decree No. 89 of April 22, 1822, stated that: *"the monuments that remain from the antiquity of Peru are a property of the nation, because they belong to the glory that derives from them.*" In this sense, the archeological monuments of Machupicchu, Waynapicchu, Wiñaywayna, Sayaqmarka and Phuyupatamarka, which are the subject of the aforementioned claims, were included in this Decree, so that before September 12, 1944 (date of the sale made to Julio Zavaleta Flores) the Peruvian State was already in possession of the aforementioned archeological monuments.

Between 1896 and 1965, the usufruct of the lands of the Sillque hacienda went through a process of transfer between the Nadal, Ferro, Abrill Vizcarra and Zavaleta families. In 1944, by court order, it was determined that the Cutija (Primavera) hacienda, owned by the Abrill Vizcarra family, would be divided among the descendants of Tomasa Ferro de Abrill (José Emilio, José Luis, Carlos Alejandro and Lourdes Abrill Vizcarra). The heirs were divided in four lots: Chuyamayo or lot No. 1 (Huayopta-la Convención), primavera and punas Pampaqhawana or lot No. 2, Camicancha or lot No. 3 and Q'ente or lot No. 4, this last lot was awarded to José Emilio Abrill Vizcarra, who decided to transfer it to the spouses Julio Zavaleta Flores and Rosa M. Zavaleta Álvarez, whose estate included the sections of Quente, Santa Rita, Intihuatana, Cedrobamba, Matipata, Matara, Palcay, Huayruro, among other places, leaving on record that the sale did not include the payment of indemnification that is still pending before the government for the expropriation of the Inca cities: **Machupicchu, Waynapicchu, Wiñaywayna, Sayaqmarka, Phuyupatamarka, currently owned by the State**.

In 1965, the Q'ente parent property was divided into two properties, Q'ente and Santa Rita de Q'ente. We infer that the extrajudicial division of the Q'ente hacienda was due to the State's agrarian reform policies, which were oriented to the structural change of land tenure (landowners, feudatories, etc.). This policy led the former landowners to maintain the right of usufruct of the land in accordance with the law. However, in an effort to avoid expropriation, the landowners resorted to legal strategies that led to the division of the Q'ente property. In spite of this, the Peruvian

State, through the General Directorate of Agrarian Reform and Rural Settlement (DGRA-AR), proceeded with the affectation of the Q'ente and Santa Rita de Q'ente properties, a procedure that was carried out separately.

Considering that the procedure was developed by properties (Q'ente and Santa Rita de Q'ente), the DGRA-AR carried out the cadastral survey of each of the properties. A reading of each of the plans ratifies the supreme decrees of affectation and adjudication. In the case of the Q'ente property, the total area was 69,636 ha, of which the affectation quota was set at 1567 ha (DS No. 0444-75-AG) and 5369 ha of uncultivated land went to the Peruvian State, represented by the DGRA-AR. The Santa Rita de Q'ente property covered a total area of 12,770 ha, of which 11,636 ha of uncultivated land became the property of the State, and the quota was set at 1134 ha (DS No. 1207-74).

References

Primary Sources

Regional Archive of Cusco (ARC)

College of Sciences of Cusco, Silque documents, book 12, part 11, pp 1635–1722
College of Sciences of Cusco, leg. 14a, notebook 9, 1629–1745
Cabildo del Cusco, Causas Civiles, leg. 55, cuaderno 9, 1783–1784
Notarial Protocol. Mar y Tapia, Pablo del, protocol 166, 1828–1835
Notarial Protocol. Araníbar, Pedro José, protocol 13, exp. 6, 1847–1848
Notarial protocol. Gárate, Carlos, protocol 102, exp. 29, 1860–1862
Notarial Protocol. Alosilla, José, book 390, 1921
Notarial Protocol. Emilio Muñiz. Protocol No. 60. 1965

Archive of the National Superintendence of Public Registries-Cusco (ASUNARP).

Zona Registral X. Sede Cusco, entry 02016781, t. 1, as. 2, 1898
Zona Registral X. Sede Cusco, entry 02016781, t. 28, as. 34, 1908; t. 40, as. 48–48v, 1910
Zona Registral X. Sede Cusco, entry 02016781, t. 64, as. 63, 1922

Archive of the Regional Directorate of Agriculture of Cusco (ADRAC).

Affectation. Leg 311. Expediente 2008. Year. 1974–1990
Affectation and Adjudication. Report No. 1652–73-AJ-ZA-XI. Estudio Legal de Titulación del predio rustico denominado Santa Rita de Qquente. Leg. 15. File No. 203. Year. 1970–1975
Affectation. Leg 311. 2008 File. Year. 1974–1990

Secondary sources

Caballero J, Álvarez E (1980) Aspectos cuantitativos de la Reforma Agraria, 1969–1979. IEP, Lima
Eguren F (2009) La Reforma Agraria en el Perú. In: Debate Agrario, p 44
Glave L, Remy M (1983) Estructura Agraria y Vida Rural en una Región Andina Ollantaytambo entre los Siglos XVI y XIX. Centro de Estudios Bartolomé de las Casas, Cusco
Llanos L (1926 [1923]) Las Ruinas de Salapunco. In: Revista Universitaria, p 51
Ravines R (2012) Machu Picchu. Monumento Arqueológico. in: Boletín De Lima 34:167
Rueda G (1970) Ley de Reforma Agraria. Art. 1° del D. L. 17716. In: Nueva Legislación sobre Reforma Agraria. Editorial El Perú y Sus Leyes, Lima

Chapter 15
Tenochtitlan: Agriculture, Urbanism and Aztec Houses in the Sixteenth Century

Jorge González Aragón Castellanos

Abstract The settlement of Tenochtitlan Island was initiated by two groups coming from the ethnic group of Chichimecas who made pilgrimages from northeastern present-day Mexico. The Aztecs began their settlement at Tenochtitlan in 1325 according to Aubin Codex, 1576, National Library of France, and The history of the Mexicans as told by their paintings, 1530, Library of the University of Texas at Austin; it is an agricultural community, generate a city and a dominant empire in much of Mesoamerica (Paul Kirchhoff coined the concept of Mesoamerica in 1943, to designate the region shared by various ethnic groups with a similar cultural development, from the Panuco River in the middle north of Mexico to the Motagua River between Nicaragua and Costa Rica, at the time of contact with Europeans in the sixteenth century.); in 1521, they are defeated and conquered by the alliance between European warriors and those from indigenous peoples of the region, especially the Tlaxcalans. Even before this violent encounter and still during the sixteenth century, there is a pattern of settlement of Aztec tradition that is identified in indigenous cartography called Aztec codices: 1. *Plano Codex in Maguey Paper*; 2. *Codices-plans of sixteenth and seventeenth century houses in Mexico City Tenochtitlan.* There are still few references that are currently held to the city and the houses of Aztec tradition in the ancient city of Mexico-Tenochtitlan because the archeological findings are almost unknown and prevent a reconstruction of the urban and architectural spaces of tradition, on the other hand, European manuscript documents from New Spain marginally mention the indigenous building.

Keywords Tenochtitlan · Aztecs · Agriculture · Urbanism · Housing

J. G. A. Castellanos (✉)
Autonomous Metropolitan University - Xochimilco, Mexico City, Mexico
e-mail: jgonzaleza@correo.xoc.uam.mx

F. Carrión Mena and S. Rodríguez Alvarez (eds.), *From the City as a Project to the City Project*, The Urban Book Series, https://doi.org/10.1007/978-3-031-94226-6_15

Fig. 15.1 *The Pilgrimage Strip*[1] Tenochtitlan glyph and Tlatelolco glyph. *Source* First page of the *Codex Pull of the Pilgrimage*

15.1 Origin of the Aztecs and the City of Tenochtitlan

Below are some brief notes on the origin of the settlement of Tenochtitlan Island with its two sectors: Tenochtitlan to the south and Tlatelolco to the north (Figs. 15.1 and 15.2).

A. It records the beginning of the pilgrimage located in the north-east of present-day Mexico, in the mythical place of Aztlan, a small island within a lake, from there comes a priest rowing a canoe bound for Cerro Torcido or Hueycolhuatepec, which has a cave where the god Huitzilopochtli appears speaking in abundance to guide the priest and his Aztec tribe in the year 1 tecpatl (year 1116 of the European calendar[2]), to the new land to be populated.

[1] *The Pilgrimage Strip* is a Mesoamerican codex in the form of a folding screen containing 22 plates.

[2] Galarza and Libura (2012), p. 19.

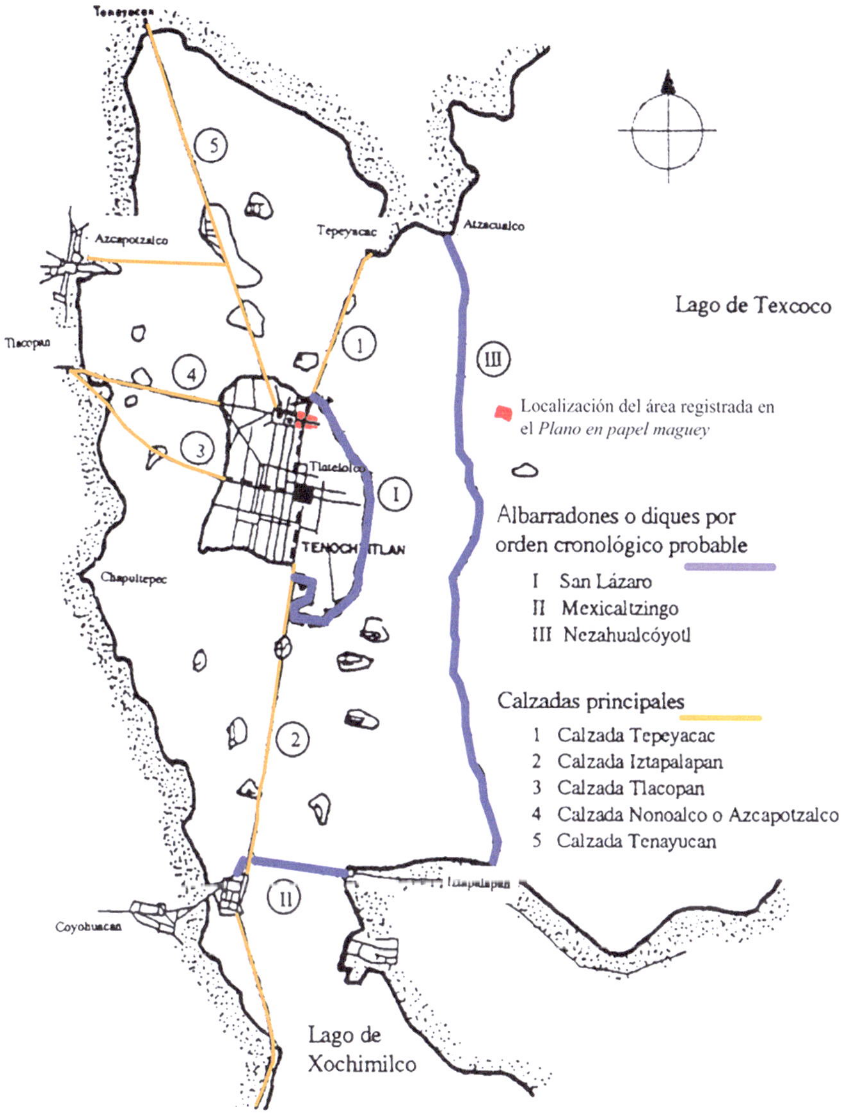

Fig. 15.2 Tenochtitlan Island in the sixteenth century

B. Detail of page 19 of *The Pilgrimage Strip* in which we observe the Cerro del Chapulín or Chapultepec west of the Island to be designated as Tenochtitlan. In Chapultepec, the Aztecs remained for 20 years, where they celebrated in 2 Caña the rite of new fire, performed every 52 years; in 1325, they set out in boats on the island of Tenochtitlan, just 11 km away, with the lake in between to found the city.

C. Detail of the first page of the *Mendocino Codex*. Glyph of Tenochtitlan (tetl = stone, nochtli = nopal) the place where nopales abound among the stones, so-called by the Aztecs, the eagle that perches on the flowering nopal, which emerges among the stones, is not considered in the reading of the glyph, perhaps, by grammatical economy. However, to date, it is the national emblem, although distorted because the eagle has a snake, pointed out by some authors as the symbol of sin, that is, an imposture from the vision of the conquerors. Tenochtitlan had as his first ruler or Huey Tlatoani Tenochzin, the great lord or chief of the Aztecs, also called Mexica Tenochcas upon his arrival on the Island.
D. *Glyph of Tlatelolco appearing in the Codex Tlatelolco.* It means the terrace or plain on a large sandy promontory. Tlatelolco is located in the northern part of Tenochtitlan Island and comes from the same Aztec or Mexica ethnic group, they are also known as Mexica tlatelolcas and occupied that place around 1337 (Valle 1994, p. 21). México-Tlatelolco was twin city with Mexico-Tenochtitlan from 1337 until 1473, the year it was subjugated by the aztecs of Tenochtitlan and later fell under spanish rule following the conquest in 1521 (Monjarás-Ruiz 1987, pp. 21–22).

15.2 Tenochtitlan's View from a Corpus of Aztec Tradition Codices

The characteristics of a part of the city and the houses of Tenochtitlan Island were determined with the study of the *Plano Codex on maguey paper* corresponding to an urban sector located east of Tlatelolco and north of the ceremonial center of Tenochtitlan, located in the current plinth of Mexico City and adjacent areas; we also present the architectural reflection on a set of 34 codices-plans corresponding to 33 homes[3] located mostly in the neighborhoods around the Spanish urban trace established by the conquistadors around 1522.

With the study of the corpus indicated, we observe the urban and architectural culture Tenochca, linked to the lake condition of the region, with first-class solutions, in particular, for its outstanding hydraulic works,[4] in addition, the building experiences of the Aztecs have functional, spatial and constructive patterns and codes very different from European ones; their quality can only be appreciated in the cultural, temporal and relationship of the works with the ecological environment and meeting the needs of the population.

The aforementioned codices record the way of making city and house-rooms of Aztec tradition, we study them with the theory and method developed by Joaquín

[3] There are 2 plans that were made for only one house, that is why there are 34 plans for 33 houses.

[4] Several authors maintain the thesis that among the agricultural cultures of the time, those that had the control and development of hydraulic works stood out, which allowed the generation of surpluses and the creation of despotic states with the rise of the militia, education and more elaborate cultural expressions. Angel Palerm also supports this thesis and carried out relevant research on the hydraulic works in the Mexico Basin in the sixteenth century.

Galarza in a wide work published mainly in France and Mexico. Galarza has shown that each of the pictographs with their shapes, colors and arrangement in the plastic picture has a glyphic content that was translated into the language of origin: Nahuatl. The author cited mentions:

> … the "text-images" in these manuscripts were actually small stories in Nahuatl still made by Tlacuilos, indigenous painters-writers who had preserved their ancestral system of writing by painting[5]

We are interested in presenting the sixteenth-century city of Tenochtitlan and the single-family and multi-family houses-rooms. It is a reading exclusively from the perspective of the architect-urbanist[6] and not all the elements present in each of the indicated codices are taken.

15.3 The Plane Codex on Maguey Paper

Surely, the codex that most accurately represents the traditional indigenous city is the so-called Plano on maguey paper or Partial Plano of Mexico City (Fig. 15.3); it is a large panel of 238 × 168 cm. painted on amate paper (made from the bark of a Mexican tree of the same name and not on maguey paper as its name suggests), it is deposited in the Chamber of Codices of the National Museum of Anthropology in Mexico; the hypothesis proposed by Joaquín Galarza Regarding its date of realization, he considers it uncertain because it has had several additions until the middle of the sixteenth century and it is possible that it has begun before the Spanish conquest.[7] The plan is a document that maintains the traditional pictorial characteristics of the Aztecs, that is, of stories of various themes through expression in Aztec-Nahuatl glyphs and records a part of the city of Mexico-Tenochtitlan, located east of Tlatelolco Square.[8] On this plane, it can be seen that in the sixteenth century, the urban trace of this area was reticular and homogeneous, barely altered by the hydraulic works that allowed the city to be made in a lake area. This area of the city registers around 420 housing units, mainly intended for a group of indigenous warriors and is also a cadastral plan that identifies the location and owners of each of the plots. As far as the urban issue is concerned, the only thing that remains to be added is that the following components of the urban structure could be identified in the aforementioned plan:

---

[5] Joaquín Galarza and Jorge González Aragón, Exhibition "La vivienda azteca en la ciudad de México en los códices-planos de los siglos xvi–xvii" (Spanish for The Aztec dwelling in Mexico City in the codices-plans of the sixteenth and seventeenth centuries), Museo Nacional de Arquitectura, Mexico, September 1995.

[6] This topic was developed under the guidance of and in continuous conversation with Carlos González Lobo for almost a decade.

[7] For an analytical and interpretative description from the urbanistic point of view, see González Aragón (1993).

[8] According to researchers Manuel Toussaint, Federico Gómez Orozco and Justino Fernández 1938 (1990); on the other hand, architect Luis González Aparicio made a hypothesis that locates this part of the city to the west of Tlatelolco Square (González-González and Cué 2006).

Fig. 15.3 Schematic drawing of the Plano Codex on maguey paper. *Source* by the author, drawn by Claudia García Jiménez

Original classification: *Plane Codex on maguey paper*. Original size: 168 × 238 cm. Guarded in the Chamber of Codices, National Library of Anthropology and History, CDMX.

Material support: Amate paper (not on maguey paper as its name says). The Aztec-Nahuatl Codex recounts a multiplicity of themes including a list of Aztec rulers and their periods of government, dialogs between leaders, wars and the urban cadastre that precisely defines the territory, plot and urban structure, housing units and their owners among other elements.

15.3.1 Hydraulic Works

i. Similar to the grandeur of pre-Hispanic hydraulic works in the Mexico Basin, Plano records hydraulic works of various order that have been classified by their magnitude and pictographic representation as follows: slurry or dikes to contain the waters of Lake Texcoco, wide channels for water conduction, channels for the irrigation of agricultural chinampas and channels for the circulation of canoes (Fig. 15.4).
ii. Road system. The urban sector alluded to had a system of roads with different hierarchies and arrangement in the whole, the roads are identified by their brown color and black contours throughout their journey, some in the central part of the route have footprints. According to the magnitude of the road its location and the pictographic representation we classify them as follows: intercity roads (communicate this urban area on the islet of Tenochtitlan with the mainland), main roads (which structure and communicate the urban complex and link with the interurbans), secondary roads (lower trajectory) and neighborhood roads that communicate small sets of plots (Fig. 15.3)
iii. Chinampas. In the Plano, it is observed that the urban complex is made entirely of chinampas (artificial portions of soil anchored with local tree trunks called hollows to the lakebed), it looks like a floating city because of the amount of water found there. Chinampas were used for the construction of houses, buildings intended for worship, roads and, to a large extent, for intensive agricultural irrigation cultivation. The chinampas for agricultural use were cohabiting with the housing areas, they were part of them; their location was throughout the urban complex and the proportion they occupied was close to 50% (Fig. 15.4)

A. *Hydraulic works* identified according to their hierarchy and function in the urban complex according to their representation in glyphs: dams, canals and spring.
B. *Road system* of intercity, primary, secondary and neighborhood roads identified according to their representation in glyphs.
C. *System of chinampas* or portions of land anchored to the bottom of the lake by tree trunks "expects." The chinampas are surrounded by irrigation canals for intensive farming.

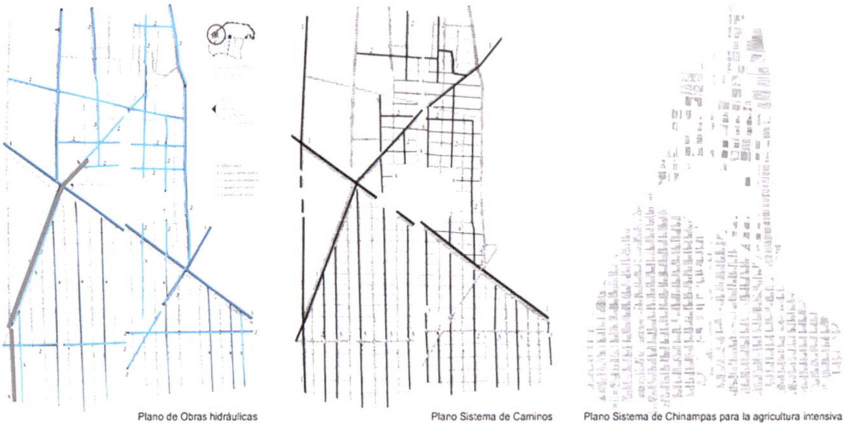

Fig. 15.4 Hydraulic works, road system and Chinampas for agricultural cultivation

i. Parceling. It has a homogeneous orthogonal layout only altered by the arrangement of relevant hydraulic works such as dams and main channels made for the control and use of the waters. The plots are regularly 90%: they are based on a scheme where each plot corresponds to a housing unit that is integrated directly into the pedestrian roads and the canals for canoes (Fig. 15.5). The housing unit was composed of the plot on which the house was erected, the agricultural chinampas (portions of land anchored to the lakebed), usually a group of seven with irrigation canals on the sides for intensive agricultural production, the canals and roads of various kinds that bounded the housing unit on all four sides.

ii. Pre-Hispanic housing and religious buildings.[9] Here you can see a housing building with the code EC, it is from an indigenous nobleman and is characterized by having on the frieze a black ribbon with a white circle, apparently appears headlong although what indicates is that its cover "looks" to the east. The pre-Hispanic religious buildings are located as indicated: three in the eastern zone, two "looking" to the west and the other to the east, in the western zone there is a building and "looks" to the west. (Fig. 15.5)

iii. Some urban and architectural elements with their associated glyphs. Below are some of the components registered in the Plane that present various glyphic representations and that served as the basis for the classification of the components of the urban structure, in the case of roads, canals, housing units and buildings of pre-Hispanic worship (Fig. 15.5).

[9] The housing unit in pre-Cortesian buildings is the concept defined by Linda Manzanilla, archaeologist, researcher at UNAM and member of the Colegio Nacional.

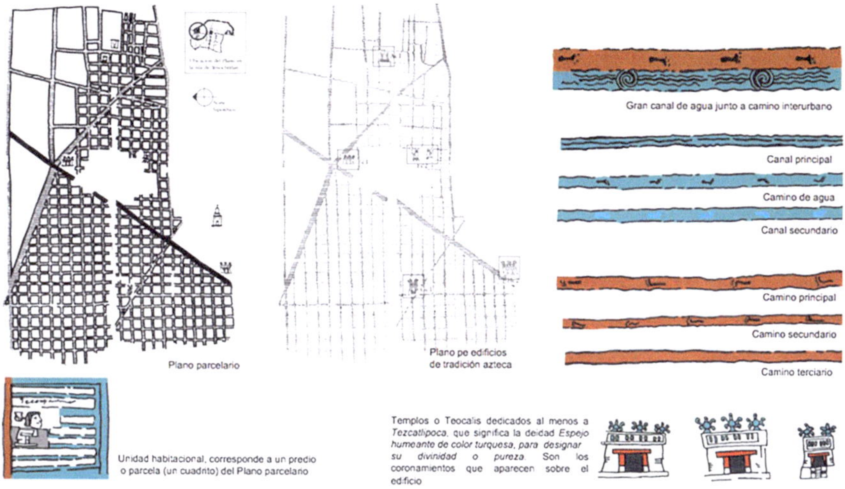

Fig. 15.5 **a** Plane of parcel, **b** plane of buildings of Aztec tradition, **c** hierarchy of roads and canals, **d** housing unit on the plane on maguey paper and **e** teocalis dedicated to the deity Tezcatlipoca, which safeguards young warriors, coinciding with this sector of the city occupied by warriors, who are identified by the knotted hair of the occupants of each of the plots

15.4 Aztec Homes[10]

With the analysis of 34 codices-plans of indigenous houses of the sixteenth and seventeenth centuries elaborated by local painters, in order to establish a typology of spaces and constructive elements that shows the physical organization of housing in the ancient city of Mexico-Tenochtitlan and an interpretation of the sociocultural aspects linked to it (Fig. 15.6).

The original plans of the houses are in the General Archive of the Nation in Mexico. The corpus analyzed is a valuable architectural testimony for the ease of having such a broad group of documents that can be compared and have reliable information. In the case of the plans of the houses are documents created with the cadastral rigor that was required to present it before a New Hispanic court during the sixteenth and seventeenth centuries and initiate a process that allowed to confirm the possession of the owner or the transfer of the house from one person to another, usually from an indigenous owner to another indigenous, and in some cases to a Spanish.

It should be noted that the documents analyzed are plastic and that they were prepared according to the traditional pictographic conventions of the Aztecs, so each of the glyphic elements is part of the architectural reading we make.

[10] For an extensive consultation of the 34 codex-house plans, see: Gonzalez Aragon (1996).

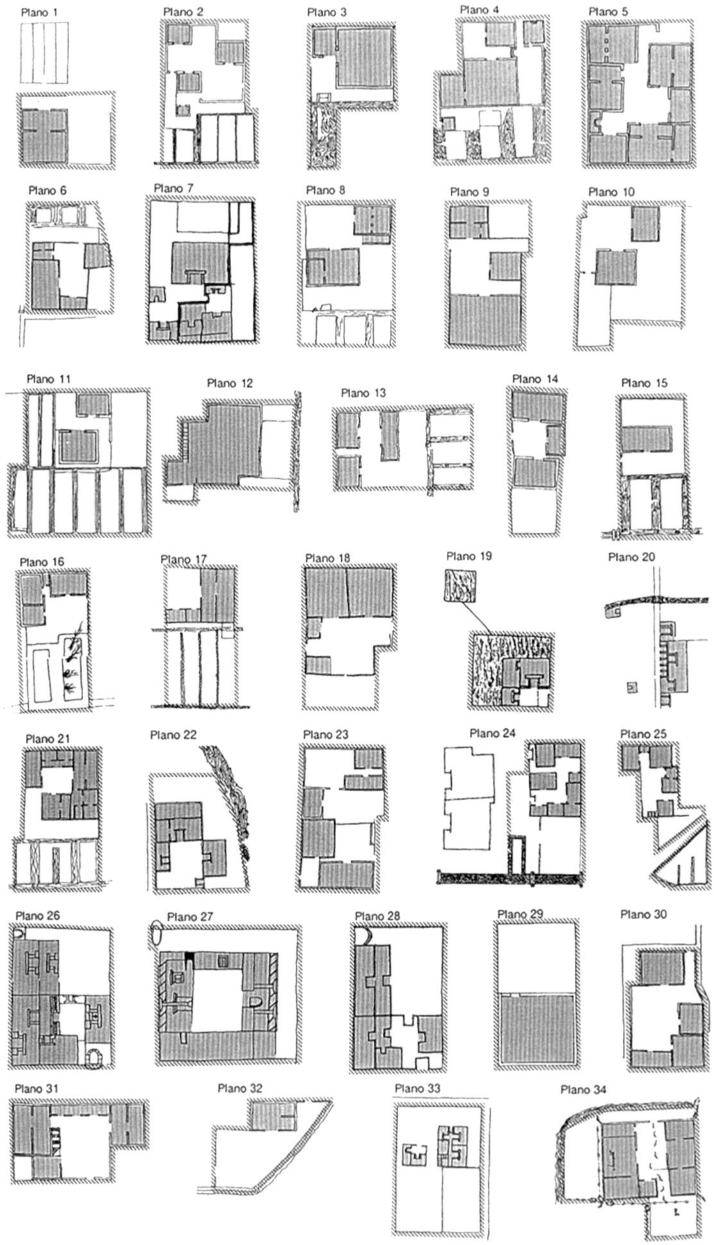

Fig. 15.6 34 Codices-plans schematized by the author, in these plans the built areas stand out[11]

[11] Idem.

15.4.1 The Houses

The indigenous house has characteristics that allow us to perceive a domestic atmosphere very different from what we are used to today:

i. The constant association of agricultural plots and chinampas to the house with the canals for irrigation and circulation of canoes. The presence of these spaces in houses located in the central area of Tenochtitlan indicates the close relationship of domestic activities with agricultural ones on eminently urban grounds.
ii. The association of built areas, the courtyard, agricultural chinampas and the plot make up a consistent unit that can be called a housing unit.
iii. The presence of the courtyard is the central element around which the built spaces are organized.
iv. A balanced and sometimes loose relationship of open and cultivated spaces in relation to those built.

15.4.1.1 Placement of the Plot in the Housing Complex

The most relevant feature of the plot (area occupied by a housing unit) in its integration with the immediate urban context is independence: the plot is integrated into the urban context through roads or canals and is generally independent, that is, there is a separation marked by natural elements, roads and walls between both plots, a fact that allows the housing unit to maintain total independence from its neighbors and integrate into the spaces. public: roads and canoe circulation channels.

15.4.1.2 The Spatial Components of the Housing Unit

The quarters As for the uses of the rooms, they were identified: the cihuacalli or fourth of the women that was intended for the preparation of food; the rooms for total rest that were used mainly as bedrooms, the scarce interior lighting privileged that manual activities were carried out outside the buildings; there are a large number of rooms with lobbies with two and up to three internal divisions. The files accompanying the plans mention that some of the houses have rooms intended for sacred images or the dead. Finally, it is worth mentioning that due to the size of some rooms and their location inside the house, it is likely that they were small spaces intended for the storage of work or household objects.

As for the organization of the space that the rooms have, there are them with a single space or single rooms; with two double spaces or quarters and with three triple spaces or quarters. The rooms generally have access to the center of one of the walls, according to the previous classification, several quarter modalities are presented (Fig. 15.7).

The yard. It is the open central space that is made up of the arrangement of the buildings. The presence of the courtyard is found in all the cases analyzed; however,

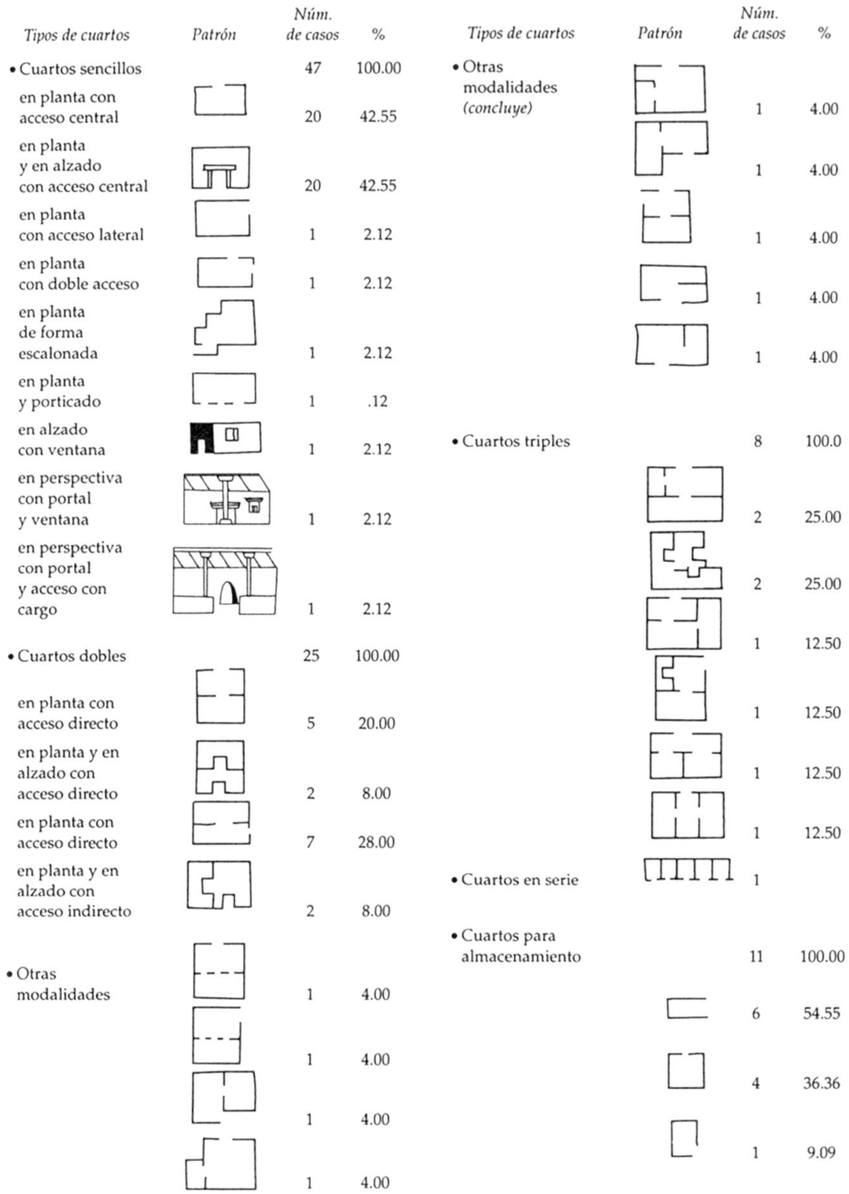

Tipos de cuartos	Patrón	Núm. de casos	%
• Cuartos sencillos		47	100.00
en planta con acceso central		20	42.55
en planta y en alzado con acceso central		20	42.55
en planta con acceso lateral		1	2.12
en planta con doble acceso		1	2.12
en planta de forma escalonada		1	2.12
en planta y porticado		1	.12
en alzado con ventana		1	2.12
en perspectiva con portal y ventana		1	2.12
en perspectiva con portal y acceso con cargo		1	2.12
• Cuartos dobles		25	100.00
en planta con acceso directo		5	20.00
en planta y en alzado con acceso directo		2	8.00
en planta con acceso directo		7	28.00
en planta y en alzado con acceso indirecto		2	8.00
• Otras modalidades		1	4.00
		1	4.00
		1	4.00
		1	4.00

Tipos de cuartos	Patrón	Núm. de casos	%
• Otras modalidades *(concluye)*		1	4.00
		1	4.00
		1	4.00
		1	4.00
		1	4.00
• Cuartos triples		8	100.0
		2	25.00
		2	25.00
		1	12.50
		1	12.50
		1	12.50
		1	12.50
• Cuartos en serie		1	
• Cuartos para almacenamiento		11	100.00
		6	54.55
		4	36.36
		1	9.09

Fig. 15.7 Identification of the different spaces of indigenous housing obtained from a Corpus of 34 codices-plans

although the courtyard has a universal character in the houses, that of the Aztec house has its own peculiarities; it is generally delimited by buildings on two and even three sides, closing the other sides with "blind" walls of solid materials or with vegetable fences.

Another feature is that the rooms around the courtyard have their accesses outdated with each other, that is, in relation to virtual axes of composition, the generality is the asymmetry of the accesses of the rooms seen in plan, which allows greater visual intimacy for the occupants of them and accentuate the public use of the courtyard.

Chinampas for intensive farming. It should be mentioned that chinampas are portions of land that rose artificially above the level of the lake's waters and were used both for domestic construction and for civil works or for agriculture, in Mexico we generally use the term chinampa to refer to portions of land surrounded by ditches and dedicated to agricultural cultivation. About 40% of the properties analyzed had chinampas for agriculture, the agricultural condition of the housing units, as said, is present in almost all of the cases analyzed, either by the presence of chinampas or by the crop plots.

The rectangular shape of the chinampas is homogeneous in the territory of the Mexico Basin although their dimensions vary depending on the site.

Main orientation of homes. In most homes, we find that accesses to the east or west predominated, a facet that comes from a pattern of habitability in which the rising or setting of the sun is the determining element in the orientation in terms of lighting and comfortable temperature of the rooms of the house (Fig. 15.8).

15.4.1.3 Indigenous Measures

The architecture has been linked to anthropometric measurements, the system applied in the domestic architecture of Aztec tradition is based on the measurements of the Aztec body of the sixteenth century that had an average height of ~160 cm (Fig. 15.9) and that due to that condition has proportional characteristics delimited from some parts of the human body and according to the average indigenous size, this system was used both for the design and for the cadastral record of the housing units.[12]

15.5 Conclusions

Around urbanism.

a. The evidence that the indigenous people were highlighted in the elaboration of plans and maps recording stories and information properly territorial: environmental, agricultural, urbanistic, architectural, and also incorporating scenes or historical accounts of political, social and economic life.

[12] The Aztec measurements of the sixteenth century were consulted in the text by Matías Alonso (1984).

Fig. 15.8 Four codices-plans of Aztec houses in the General Archive of the Nation, Mexico. Perspectives interpreted by the author and illustrated by Luis G. Aragón Medina are added

b. The urban pattern made in the Tenochtitlan area recorded in the Plano on maguey paper responds to certain geometric relationships. However, its characteristics are more closely linked to the adaptation of the settlement to the site; to the various techniques of carrying out civil and domestic work that allowed them to integrate into the lake without hurting it and make the most of the resources it offered through the consumption of the fauna and flora of the place; the technique of the chinampas for intensive agricultural cultivation and the construction of levees, canals and roads for the control of the level of the lake's waters in the urbanized area.
c. Through the observation of the city's networks of roads and canals, the hierarchy of these urban elements and the continuity that exists to integrate canals and simple roads to more complex ones are clearly delimited.
d. There was careful care in the registration of the property and in the realization of civil works, which is manifested in the later data that were incorporated into the Plano on maguey paper (palimpsestos). This forces us to think about a city

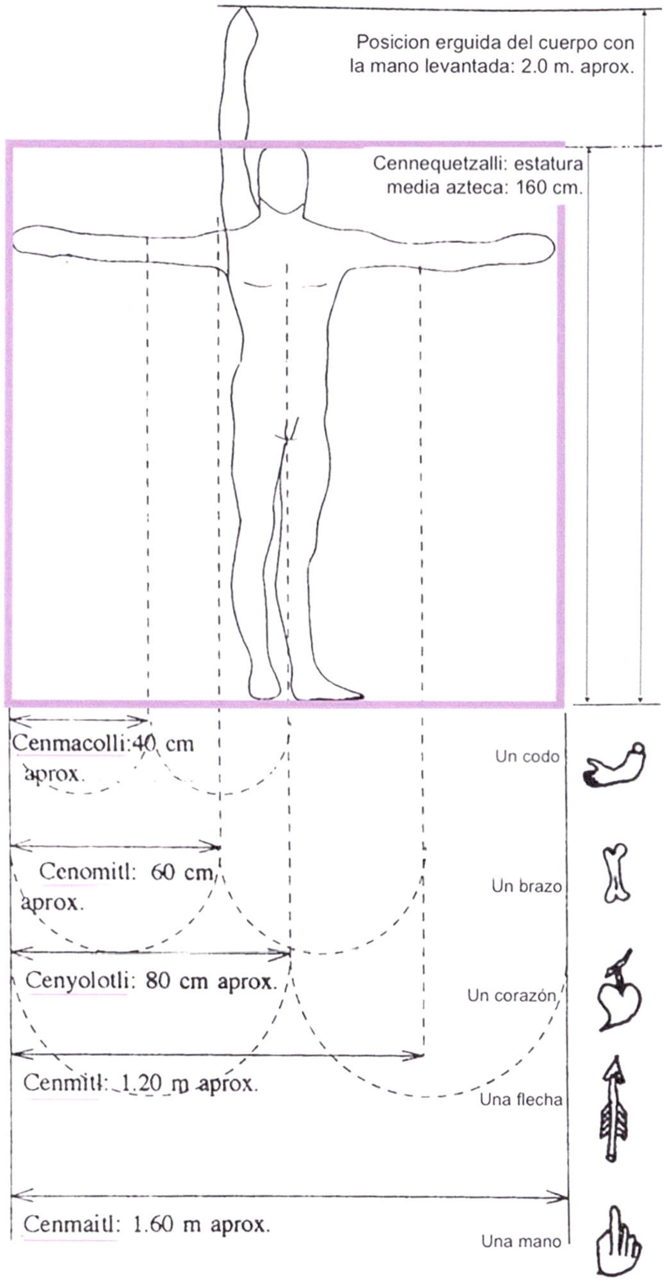

Fig. 15.9 Model of the human figure with the Aztec representation of the measurements based on the geometric description of Marco Vitrubio around 15 BC and drawn by Leonardo Da Vinci around 1490. For our particular case, the author of this text decided to elaborate this emblematic model to clearly represent and understand the indigenous measures of Aztec-Nahuatl tradition used in the measurement of houses and land

management system that directed the growth of the city and the realization of civil works.

Around the homes.

Previously, the typological characteristics of the spaces of the Aztec tradition house were presented with which we can confirm that the designs presented common patterns: simple and complex with a wide diversity of spaces for rest, coexistence, food preparation, storage and artisanal and agricultural production that indicate comfortable living conditions in large and ventilated spaces; also a mastery of the anthropometric scale used in housing designs and as a method of measurement for the cadastral records of the housing unit. In fact, through the analysis of the codices-plans we can perceive an architectural and urban culture of advanced Aztec tradition for the time that cannot be valued from the dominant patterns of other cultures of the same period, although from the satisfaction of the economic, social, functional and habitability needs of houses and the city in their relationship with the ecological environment.

References

Galarza J, Libura KM (2012) Para leer la Tira de la Peregrinación, México, Ediciones Tecolote

Galarza J et al (1981) 20 planos urbanos del siglo XVI en la ciudad de México. Centro de Investigaciones y Estudios Superiores en Antropología Social [inédito]

González Aragón J (1993) La urbanización indígena de la ciudad de México. El caso del Plano en papel maguey. Universidad Autónoma Metropolitana, Unidad Xochimilco, México

González Aragón J (1996) La casa de tradición azteca en la Ciudad de México. Siglos XVI y XVII". (Tesis de Maestría). Universidad Nacional Autónoma de México, México. https://repositorio.unam.mx/contenidos/65639

González-González CJ, Cué L (2006) Pasado y presente de la región de Tenochtitlán. La obra de Luis González Aparicio. Con notas y comentarios de C. Javier González y González. México. Grupo DANHOS

Matías Alonso M (1984) Medidas indígenas de longitud. México. Centro de Investigaciones y Estudios Superiores en Antropología Social (Cuadernos de la Casa Chata no. 94)

Monjarás-Ruiz J et al (1987) Tlatelolco rival de Tenochtitlan. INAH- UDLA Puebla, Mexico

Valle P (1994) Códice Tlatelolco, México, INAH—BUAP

Chapter 16
The Plan of Tenochtitlan and the Symbolic Dominance of the Holy Roman Empire During the Diet of Nuremberg (1524)

Daniel Astorga-Poblete

Abstract The first European-published depiction of Tenochtitlan, produced in Nuremberg in the year 1524, symbolized the political and religious supremacy of the Holy Roman Empire over the Aztec capital. This map, incorporated within a Latin translation of Hernán Cortés' accounts, depicted Tenochtitlan along with its ceremonial center, surrounding lake, and notable edifices. This publication functioned as both a geographic and symbolic instrument to assert Habsburg authority during the Diet of Nuremberg amidst political struggles and deliberations concerning religious reforms following the Edict of Worms. The map articulated hierarchical relationships, accentuating the Habsburgs' divine right and dominion over the Mexica. Furthermore, the design of the map paralleled Renaissance urban layouts and indirectly addressed contemporary discourses regarding idolatry amid the Reformation. Ultimately, it constituted a potent medium linking territorial conquest to imperial ideology.

Keywords Map of Tenochtitlan · Habsburg dominance · Nuremberg imperial diet · Conquest of Mexico

16.1 Introduction

The first image of Tenochtitlan published in Europe was printed by Friedrich Peypus in Nuremberg in 1524 (Fig. 16.1). Included as a woodcut of the book *Praeclara Ferdinandi Cortesii de Nova Maris*, a Latin translation of the *Second Letter* and *Third Letter* of Hernán Cortés, the plan presented the recently discovered Tenochtitlan and a map of the Gulf of Mexico. The city was shown in a bird's-eye view before falling to the Spanish-Tlaxcaltecan alliance, with the ceremonial buildings at the center of the image and surrounded by the Laguna of Mexico and Lake Texcoco. Adding the

D. Astorga-Poblete (✉)
University of La Serena, La Serena, Chile
e-mail: daniel.astorga@userena.cl

F. Carrión Mena and S. Rodríguez Alvarez (eds.), *From the City as a Project to the City Project*, The Urban Book Series, https://doi.org/10.1007/978-3-031-94226-6_16

Plan of Tenochtitlan to the *Praeclara* opens the question regarding the context of production, political motives, and strategic utilization of the image in the book's context (Chartier 1997). Particularly, the history of the image's acquisition by the Nuremberg printer indicates the importance of the context in the production of the Plan's final version. As critical cartography proposes (Pickles 2004; Piper 2002), maps are not just mere reproductions of a neutral space but rather embodiments of power. Considering maps as documents establishing control over territory, land, and subjectivities, we aim to understand how the Plan displays the Imperial dominion over this new territory. Power is involved not only in the production of the image but also in its materiality. In this scenario, the Plan of Tenochtitlan presents a reality of the Aztec capital in the context of a complicated relationship between Charles V and the Holy Roman Empire. Therefore, we propose two arguments. First, the Plan printed in Nuremberg was based on an indigenous prototype acquired in the Americas that traveled from the Central Valley of Mexico to the German city where the imperial Diet was set to take place. Thus, we will establish the passage of a native image of Tenochtitlan to Seville, Valladolid, and then to the hands of the court of Archduke Ferdinand in Nuremberg, where Johannes von Revellis took the task of arranging its publication along Cortés's accounts. The transference of native artifacts from the fallen Tenochtitlan to the Diet sets up the locus of enunciation of the final version of the image. Second, the imperial Diet of 1524 activates a series of dichotomies and hierarchies in the Plan, projecting power and dominance over the Mexica while commenting on the political struggles between the Emperor and the German electors regarding the Edict of Worms. The Plan articulates two elements to establish the territorial and religious control of Charles V over the Mexica and the Holy Roman Empire: the banner of the Emperor and the stone idol at the ceremonial center. On the one hand, the banner displays the dominance of the Habsburgs in Tenochtitlan, emphasizing Catholic symbolisms and the heavenly alliance with Charles, discarding any variation of the Christian faith beyond the Emperor's support. On the other hand, the stone idol mobilizes gender systems of sovereignty between the "conquered" Mexica and the Habsburgs. By setting up the stone idol at the center of Tenochtitlan, the Plan emphasizes idolatry in a context where the public adoration of imperial relics was challenged by the Nuremberg council.

16.2 The Authorship of the Image

Although the Plan of Tenochtitlan has been heavily studied over the past century, just a few scholars have argued that the image was based on a native prototype. Mundy (1998) proposed that the Plan was made by indigenous hands, an argument that is also supported by Hill Boone (2011) and Matos Moctezuma (2011). Mundy and Hill Boone indicate that the first printing of the Aztec city in Nuremberg in 1524 was based on a native image of Tenochtitlan brought to the Imperial city. There, it was reworked by an anonymous artist who made the final design that ended up in the woodblock. Mundy (1998) proposed that the city plan presented some details that were difficult

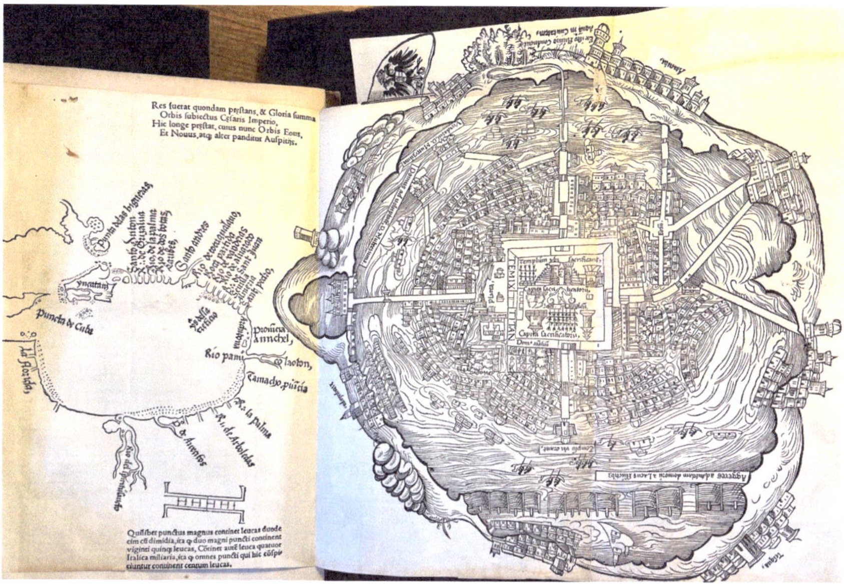

Fig. 16.1 Friedrich Peypus, Map of Tenochtitlan and the Gulf of Mexico, 1524, woodcut, 34 × 48 cm. *Source* Vienna, Österreichische Nationalbibliothek

to depict by German artists acquainted with descriptions of Tenochtitlan found in Cortés's *Second Letter* and *Third Letter*. According to Mundy:

> The cleft between the temples of the two gods aligned with the rising sun of the equinox; on these days, the temples appeared to channel the sun itself on its ascent. Only recently excavated, these temples have long been the site of speculation. Cortés neglected to mention the dual nature of the central pyramids, and other European accounts and pictures present it fancifully. (p. 16)

Hill Boone (2011) agrees with Mundy regarding the astronomical motivation of the sun at the equinox between the two temples. Hill Boone uses the research of Aveni and Gibbs (1976) to explain the importance of the sun between the temples and its relation with the story of Tenochtitlan's foundation. This information was not present in either Cortés's *Praeclara* or the accounts published in Europe related to Tenochtitlan.

Discarding the idea of a native hand inspiring the Plan, some scholars have assigned the authorship to a certain Martin Plinius. The first time this name appears in the discussion is in Ola Apenes's book *Mapas Antiguos del Valle de México* (1947). Thanks to a conversation with the historian Federico Gómez de Orozco, who was doing research in Germany at that time, Apenes argued that a German engraver named Martin Plinius was the man behind the woodcut of the coast of the Gulf of Mexico and the plan of the city. He also mentioned that Plinius was active from 1510 to 1536.[1] Different scholars mentioned Martin Plinius's name by quoting

[1] Apenes (1947), p. 20.

Apenes' book.[2] However, neither Apenes in his book nor Gómez de Orozco, Toussaint, and Fernández's book *Planos de la Ciudad de México* provided a reference to this artist.[3] Moreover, there is no indication of a Martin Plinius in the collections of German artists, engravers, and woodcutters prepared by Bartsch (Strauss 1978–2025), Geisberg (Strauss 1974), Muther (1884), Dodgson (1903), and the series *The New Hollstein's German engravings, etchings and woodcuts, 1400–1700*(Beaujean et al 1996–2025). The name of this artist does not appear either in the Nuremberg State archives, the Nuremberg City Council, or in the Imperial archives in Vienna. Considering these facts, there is no indication that Martin Plinius existed and/or was the designer of the woodcut or the cutter of the block.[4]

Other scholars have attributed Dürer as the artist behind the plan of Tenochtitlan. Palm (1951) argued that the German master was the designer of the woodcut. We know that the German master saw different Mexican artifacts when he stayed in Brussels as part of the court of Margaret and that by 1524, he was in Nuremberg, where he was visited by Frederick, the elector of Saxony who was part of the Nuremberg Diet. However, Dürer never mentioned seeing a map, book, or indigenous painting, and Charles V did not take any item sent to him by Cortés after 1522 to Brussels. Moreover, the German master would not have allowed a design made by himself to be published without his monogram.[5]

Peypus may have used a previously printed image of Tenochtitlan, an idea proposed by Gresle-Pouligny (1999). Nonetheless, there are no indications that Peypus could have used a previously published image. Cortés's accounts were published before the Nuremberg edition, with none of these prints incorporating images of the Aztec capital. The Conquistador wrote his *Second Letter* on the October 30, 1520, and arrived in Spain in the Autumn of 1521 (Thomas 2020, p. 593). The letter was first published by the German-born printer Jacobo Cromberger in Seville on November 8, 1522.[6] Cromberger's edition reached different parts of Europe and excited readers eager for information about the new lands found in the Indies. Thus, George Coçi in Zaragoza did not hesitate to copy Cromberger's edition and published a new version almost a year later. The news of Tenochtitlan expanded quickly. By the end of 1522, Michiel de Hoochstraten published in Antwerp two abbreviated

---

[2] The name of Martin Plinius as the artist behind the woodcut is mentioned by Vargas (2000, p. 25); Mundy (1998, p. 29); Matos Moctezuma (2009, p. 186); Hill Boone (2011, p. 43) who also acknowledges the missing information of the engraver; Jiménez (2014, p. 247); Martínez (2015, p. 319); Ferman (2022, p. 16).

[3] Toussaint et al. (1938), p. 98.

[4] Hill Boone is the only scholar that raised a question regarding Plinius. In a footnote on her article on the Plan of Tenochtitlan, she mentions: "I have not been able to find other information about this engraver" (2011, p. 43).

[5] Pon (2004), pp. 141–154.

[6] Since the royal authorization that designated Cortés as general captain and governor of New Spain was only effective on the October 11, 1522, it is reasonable that Cortés' father, Martín Cortés, waited until the imperial council was favorable to his son on the matters related to Diego Velázquez to publish the letter. The *real cédula* that named Cortés as governor and general Captain of New Spain was issued in Valladolid on the October 15, 1522. The *cédula* stipulated that Cortés would be the governor, leaving behind the conflict between him and Velásquez (Martínez 1990, pp. 343–346).

versions in Flemish and French, and Andrea Calvo prepared an Italian version that was published in Milan around the same time. Whereas the *Second Letter* took almost two years to get published, Cortés's *Third Letter* was published less than a year later by Cromberger. Although the exciting news about the mighty Aztec city reached Europe quickly and astonished the readers, none of these publications added a woodcut of the city of Tenochtitlan.

16.3 The Itinerary of the Image from Valladolid to Nuremberg

Arguably, the native image of Tenochtitlan traveled along Cortés's *Third Letter* on May 22, 1522, from the Central Valley of Mexico to Seville. At that moment, three ships were sent, but only one arrived safely. While Juan de Ribera and Antonio de Benavides' La Rábida was able to evade the attacks of Jean du Fleury, Alonso de Ávila and Antonio de Quiñones' ships were not that lucky and lost a part of the treasure to the French pirate. On October 1522, De Ribera and Benavides' ship arrived in Seville with part of the treasure, the *Third Letter,* and other documents to the Council. The treasure brought in La Rábida caught the attention of the imperial court, which was able to see them in Valladolid on November 1522, a month after De Ribera and Benavides's arrival. De Ribera was interviewed by Peter Martire d'Anghiera, Marino Caracciolo, Gasparo Contarini, and Tomaso Maino. According to Martire, they were able to see different artifacts brought from the Indies, as well as a Tlaxcalan who danced with his ritual clothing. Martire was not only interested in these treasures but also in some maps De Ribera brought with him. The Italian described two: one painted in cotton cloth, thirty feet long and narrow (Martire d' Anghiera, f. XXXIIIv 1530).[7] After commenting on the regions that this map presented, Martire described a second map:

> Post grādiorem mappā paulo minorē vidimus aliā: sed nō minus allicietem animos nostros, cum suis templis ac pōtibus civitatem ipsam Tenustitanā. Ea incolarum manu cum lacunis depict cōplectebatur. (Martire d'Anghiera, f. XXXVr 1530).[8]

The map of the fallen city described by Martire echoes the woodcut of Tenochtitlan printed in Nuremberg in 1524. Both images projected Tenochtitlan with its temples,

[7] This map showed: "the plains and provinces whether vassal or hostile to Moctezuma… as well as the lofty mountains which completely surrounded the plain" (Martire d'Anghiera 1912, pp. 198–199).

[8] "After seeing the large map I have mentioned, we were shown a somewhat smaller one but which interested us just as much. It is a native painting representing the town of Temistitan, with its temples, bridges, and lakes. (Martire d'Anghiera 1912, p. 201). In his account to the Venetian senate, Contarini also gives a small reference to the city with this image in mind: "This city is marvelous in size, site, and artifice, placed in the middle of a lake of salt water, which surrounds about two hundred miles, and at one end it joins with another lake of fresh water; however, it is not very deep, and the water rises and falls twice every day, as it does here in Venice. Some roads that go to the city have their foundations in the lake." (Contarini 1840, p. 53).

causeways, and aqueducts, though Martire is keen to mention that the one he saw was "native."

Part of the treasure brought in La Rábida and copies of the *Second* and *Third Letters* of Cortés traveled from Valladolid to Nuremberg on the August 20, 1523, as a gift to Ferdinand, who would preside over the Diet by the end of that year. As Martín de Salinas commented on his letters to the Archduke and Gabriel de Salamanca, Ferdinand was keen to receive news from the new lands found in the Indies.[9] Martín de Salinas informed that Henry du Hemricourt departed Burgos for the imperial city with the treasure, though the inclement weather did not allow the agent to arrive earlier. He encountered the Archduke in Nuremberg no later than the January 12, 1524, as the inventory of chaplain Martín de Paredes described.[10] In that inventory, Paredes wrote that he received some indigenous feather-work ("plumarios") and shields from New Spain while staying at Nuremberg. Those artifacts appeared among the treasures shipped in La Rábida and described by Cortés in his inventory. Martire d'Anghiera also mentioned them along with the maps De Ribera presented.[11] Paredes' registry of these artifacts did not explicitly mention the reception of maps, as these inventories were designed exclusively for jewelry, clothing, and weaponry acquired by Ferdinand. Nonetheless, the inventory mentioned that Ferdinand received at least "three books from New Spain,"[12] which were transported by Antonio Calvo on the August 24, 1524, in Vienna, a couple of months after Hemricourt's arrival in Nuremberg and Ferdinand's departure from the imperial city to Austria. It is unclear whether Hemricourt was carrying Mesoamerican codices like an amoxtli or books printed in Europe related to the events of New Spain. Usually, among the Spaniards that encountered these amoxtli, the Spanish word used to refer to the folded indigenous pictographic manuscripts of the Central Valley of Mexico was "libro" (book). Thus, it seems that Hemricourt not only carried with him the feather-work, shields, and clothing but also other things such as Mexican codices (which sometimes included "maps" like folio 2r of the *Codex Mendoza*) and copies of the *Second* and *Third Letters* of Cortés.[13]

---

[9] Salinas (1903), p. 312.

[10] *Rationes* (1521–1524), p. 197.

[11] Martire d'Anghiera (1912), pp. 195–198. This description coincides with the inventory of Cortés written in 1522. Part of those items also appeared in the treasure sent to Ferdinand (*Rationes* 1521–1524, pp. 197–199).

[12] *Rationes* (1521–1524), pp. 49–50.

[13] In the prologue of the *Praeclara*, Pietro Savorgnano mentions that he has received the same letters written in Spanish that Cortés sent to Charles informing him about the lands found (Cortés 1524a, b, pp. IIv–IIr). Since Ferdinand was interested in the events of Cortés in the Central Valley of Mexico, Charles sent copies of the letters of Cortés; however, it is unclear whether he sent manuscript copies or the previously published editions from Cromberger. Nonetheless, we are inclined to consider that it was the former. Indeed, the *Codex Vindobonensis S.N. 1600*, held in Vienna, compiled manuscripts copies of all the letters written by Cortés. According to Woodrow Borah (1962), the codex was prepared in Vienna for Ferdinand probably around 1528 using manuscript copies written before the summer of 1528 (p. 91). It is reasonable to think that copies of the *Second* and *Third Letter* sent to Ferdinand were used in the Nuremberg edition and the *Codex Vindobonensis S.N. 1600*.

Someone within the court of Archduke Ferdinand brought the letters and the plan to Friedrich Peypus. According to Peter Martire, in an account written in 1524 to Pope Clement VII, the recently elected bishop of Vienna, Juan de Granada, was involved in the publication.[14] After he arrived in Vienna, Juan de Granada decided to change his name to "Johannes von Revellis" a German variation that sought to favor his Spanish figure in a hostile environment to Iberians. This is the name appearing on the front page of the Latin translation of Cortés's accounts printed in Nuremberg. Revellis was an important figure in the court of Archduke Ferdinand by the end of 1523. While he was the head of the Church in the Austrian city, he assumed the role of almoner, confessor, and court preacher of Ferdinand until 1530 (Koretz 1970, p. 45; Pfohl 2020, p. 137). Thus, his closeness with Ferdinand, his name in the frontispiece of the *Praeclara*, and Peter Martire's information to Pope Clement VII point out that Revellis had access to Cortés's accounts and treasure sent to the Archduke while he was in Nuremberg presiding over the imperial Diet. Since Martire d'Anghiera's description of the "native map" coincides with the plan printed in Nuremberg, this indicates Hemricourt was carrying some fascinating maps of the Yucatan peninsula and a plan of the Aztec fallen city, those that De Ribera had shown to the Italian chronicler in 1522. The presence of the plan of Tenochtitlan and the map of the Gulf of Mexico in this German edition, unseen in the previous ones, suggests that Revellis used the same cartographic documents brought by De Ribera to enhance his edition of Cortés accounts.

16.4 The Plan

The final version of the Plan presents the city of Tenochtitlan to the reader as a city island surrounded by the Laguna of Mexico. At the center of the image, the artist placed an idol holding two serpents, a sculpture that, for some, represents the goddess Coyolxauhqui (Hill Boone 2011, p. 35) or Coatlicue (Mundy 1998; Gresle-Pouligny 1999). Above the idol, the artist placed the Templo Mayor with its two pyramids in honor of the gods Huitzilopochtli and Tlaloc. To the left of the Templo, a tzompantli holding human skulls was depicted, and to its right, what seems to be the Temple of Tezcatlipoca (Matos Moctezuma 2009, p. 94). Below the idol, a bigger tzompantli was drawn surrounded by two buildings that could be the Calmecac, the school for the Aztec nobility, and the Templo del Sol to the left and right of the viewer, respectively (Matos Moctezuma 2009, p. 94). Outside the religious center, the Plan indicates buildings related to the political and social life of Tenochtitlan: the palace of Moteuczoma Xoyocotzin and the zoo of the city to the lower left of the ceremonial center; the "pleasure house" and the garden of Moteuczoma in the upper-left quadrant; the city of Tlatelolco indicated by the inscription "forum" in the upper right quadrant, which refers to Tenochtitlan's twin-city as a commercial center; and the pier inscribed as a temple in the lower quadrant.

[14] Martire d'Anghiera (1912), p. 212.

As Battcock explains, the Tenochca cosmos materialized in different buildings within the city, linking the sacred with a specific spatial location, which transpires in the Plan (2017, pp. 70–71). Outside the amphibious city of Tenochtitlan, the artist presented Nezahualcoyotl's dike, which controlled the mixing of salty water from Lake Texcoco with the sweeter water of the Laguna of Mexico. Tenochtitlan built another barrier, Ahuitzol's dike, which also ran from north to south as Nezahualcoyotl, though it was not depicted in the Plan. However, causeways and aqueducts like the Causeway to Tlacopan and Tenayuca and the Aqueduct from Chapultepec were presented to illustrate the connection between the city and the shore of Lake Texcoco. Only three other cities are mentioned besides Tenochtitlan: two of them belonged to the Triple Alliance: "Tesqua" (Texcoco) and "Atacuba" (Tlacopan), while "Iztapalapa" (Ixtapalapa) might have been included since it was the first city visited by Cortés in his entrance to the Valley of Mexico (Ferman 2022, p. 105).[15]

16.4.1 The Eagle and the Idol

Although the viewer faces the Plan with the West side at the top, it seems that the intention of the designer was for the viewer to spin the image 180º as the toponyms of the cities around Lake Texcoco, the indication of some key buildings and sailors in the lake are only perceived correctly if we change the orientation of the page. Likewise, Matos Moctezuma has addressed the problem of the imprecision of the ceremonial center, which is shown with the Templo Mayor at the West instead of East, the real location in Tenochtitlan (2011, p. 186). In addition, the tzompantli are inverted in their disposition. The bigger tzompantli should be to the West instead of East, and the smallest to the North. Spinning the ceremonial center at 180° coincides with its orientation with the rest of the geographical sites. Moreover, if we look closely at the ceremonial center, we can notice that the base for this complex is inverted as the perspective is drawn as if we were looking at it from the back and not in front of the viewer.

The orientation of the Plan highlights the dichotomic relation between the Habsburgs and the Mexica, expressed by the banner of Charles V and the stone idol in the ceremonial center. While the banner might indicate where Cortés placed his forces in Tacubaya, where he prepared the final attack on Tenochtitlan, it also emphasizes a hierarchical relation with the stone idol. Thus, the editor positioned the final view of the Aztec capital with the banner at the top, allowing the viewer to appreciate the coat of arms correctly while inverting the ceremonial center from its original disposition to highlight the opposition of two forces in Tenochtitlan represented by the double-headed eagle and the stone idol, underscoring the dominance of the European force.

[15] The conformation of the Triple Alliance derived from the military aid between Mexico-Tenochtitlan and Texcoco, with minor participation of Tlacopan, to confront Azcapotzalco (Battcock 2011, p. 27).

The double-headed eagle represents the Holy Roman Empire and the coat of arms of the Castilian and Burgundian territories of Charles (Damen and Meer 2022). According to Filipek (2023), the eagle has been used by kings and emperors as a symbol of power, strength, and endurance. Usually, it is associated with heaven since the eagle is the bird that carries the emperor's body to the celestial site (17). The eagle's extended wings in pulpits in medieval times indicate the majesty and power of the word of God. Its bicephalic characteristic denotes the dual sovereignty of the Emperor both in secular and in religious matters (Filipek 2023, p. 18; Hodaj 2019, pp. 160–161).

Interestingly, during the Middle Ages, the battle between the light and Christian spirit and the forces of evil and darkness is allegorized in the confrontation between the eagle and the serpent. While the former expresses the heavenly power, the snake carries the attributes of earthiness and the underworld. Thus, the victory of the eagle over the snake refers to the dominance of heavenly power over evil. Although other contemporary plans of city sieges depicted battles and warfare, as in Niklas Meldemann's Siege of Viena, the almost solitary plan of Tenochtitlan does not seem to be interested in showing how Hernán Cortés confronted the Tenochcas. While the *Praeclara* explained how the conquistadors entered and destroyed the Aztec city, the plan displayed the dispute on a symbolic level. In this dichotomic and hierarchical configuration, the double-headed eagle represented the Holy Roman Empire and Charles, while the decapitated stone idol holding two serpents portrayed the Tenochcas. The banner of the Habsburgs placed at the top of the plan aligns with the eagle's power. It is a Catholic and heavenly strength, symbolizing the dominance of spiritual and earthly dimensions with its double-headed configuration ready to attack the evil and sensual forces of the decapitated idol, which lacks rational control. The double-headed eagle of the Habsburgs allegorizes the light that would take the city of Tenochtitlan, while the central figure holding the snakes shows the power of the darkness and the death spree consolidated in the skulls surrounding it.

The opposition of the double-headed eagle and snake, Christians, and idolaters also establishes the control of life and death in the Aztec city. The banner is placed to the west and outside the Laguna of Mexico, near the forest of Chapultepec. This is the place that brought sweet water to Tenochtitlan. The provision and control of water supplies was the expression of power and political force (Wilson 2012), which in the plan of the Aztec city is dominated by Cortés's forces. The aqueduct of Chapultepec and the causeway to Tlacopan began at the west gate of the Coatepantli (the snake wall at the Templo Mayor) and ended up at the hill of Chapultepec (Carballal and Flores 1989, p. 74). As the plan points out, Tenochtitlan brought water from Chapultepec to the city through this aqueduct. The western side of the Laguna of Mexico is seated at a slightly higher elevation than the eastern. During pre-Hispanic times, the island of Mexico was fed by numerous rivers and springs from the western part of the Valley, bringing sweet water to the Laguna. Control of the salty water from Lake Texcoco and the freshwater from the west of the Laguna of Mexico was critical for Tenochtitlan (Mundy 2015). It allowed the balance of water irrigation to the chinampas and prevented the flooding of this agricultural system and the city through the dikes. Nonetheless, the plan of Tenochtitlan highlighted the aqueduct as the source of drinkable water necessary for the population's life, leaving aside its other functions. Indeed, the image focuses on the distinction between sweet and salty water, life and death. While the control of sweet water is aligned with the double-headed eagle, the idol extends its dominion over salty water circulating through Tenochtitlan.

The relationship between the banner of the Habsburg and the stone idol also mobilized gendered systems of sovereignty (Bendall 2023, p. 43; Gagné 2017, p. 130). In the case of the plan, the artist articulated a power and hierarchic relation between the double-headed eagle as a representation of the male invader in opposition to the female native invaded and conquered. Although the decapitated idol does not project explicitly a gender to the modern viewer, contemporary iterations of the plan, as seen in Bernardino de Viano's edition printed in Venice (1524) and Benedetto Bordone's *Isolario* (1528), depicted a female idol at the ceremonial center. Thus, the stone figure in the contrapposto position indicates lasciviousness and carnality while holding two serpents as extensions of its body, as Cleopatra was represented occasionally during the Renaissance. The female figure sits below the masculine double-headed eagle that opens its wings to attack. In this case, the banner not only indicates the military campaign but also announces the disposition of the male conquistador to hunt and conquer the female city. As Elizabeth Reid has commented concerning the Italian Wars, the Italian city of Genoa was personified as a dependent mistress and ruined woman by artists in the Genoese entries (2021, p. 83). Similarly, Tenochtitlan is condensed into the sensual stone idol at the mercy of the invader.

16.4.2 The Idolatrous Center: The Diet and the Relics

The artist behind the plan did not leave aside a commentary on the religious situation in Tenochtitlan. Indeed, the inscriptions highlighting the temple where Mexica sacrificed their subjects, the tzompantli, and the stone idol clearly stated the difference between Christians and Tenochcas. Nonetheless, the stone idol and the tzompantli hinted at a complicated discussion during the Diet of Nuremberg in 1524 regarding the adoration of images and the exhibition of relics. The Diet was the general assembly of the emperor of the Holy Roman Empire with the representatives of the different semiautonomous territories that were part of it. During the 1520 s, the Diet served as a forum for negotiations to advance a general church reform that confronted the emperor and papal delegates with the prince-electors. The Diet of Worms of 1521, the first presided over by Charles as emperor, discussed Luther's ex-communication. The Edict of Worms, signed by the emperor on the May 25, 1521, declared Luther an outlaw, though it was not enforced as many princes felt uncomfortable with the imperial intervention in their internal affairs. They were concerned the imperial cities could activate an open revolt if Luther was imprisoned (Close 2018, pp. 128–129). After the Diet of Nuremberg in March 1523, the participants agreed to reunite again by the end of the same year to address the problem of Luther and the enforcement of the Edict of Worms.

Delayed, the Diet of Nuremberg began on the January 12, 1524, and Luther's point of view on the veneration of images and relics as idolatry was felt hard by the papal delegate Lorenzo Campeggi. The Nuremberg city council decided not to celebrate the traditional rituals of Easter week, with no ritualized public display of the imperial relics challenging a papal bull established in 1452. This act defied the figure of Ferdinand and Campeggi, with no support from the German princes. As Carina Johnson explains: "Explicitly, the ban acted on the council's religious convictions that rituals involving the veneration of material objects or the use of sumptuous display in religion were problematic" (2011, p. 93). The veneration of relics, as a practice that extended to the veneration of images and other practices, such as the use of indulgences, reaffirmed the church's superstition, corruption, and avidity (Johnson 2011, p. 93). Neither Archduke Ferdinand nor Campeggi were able to oblige the Council to display the relics nor enforce the Edict. At the end of the Diet, the Council declared the meeting of a national religious council later that year in Speyer despite the counterarguments of Ferdinand and Campeggi. The Council's decision was received badly by Charles and Pope Clement VII, with the former declaring the decision null and void in the following months.

Thus, the idol lapideum and the tzompantli comment on the adoration of images and relics. As city portraits during the Renaissance projected the invisible character of a city and the interior life of an urb (Maier 2012, p. 714), the plan of Tenochtitlan reflects on the idolatrous configuration of the Aztec capital. Indeed, the projection of Tenochtitlan depicted by the Nuremberg artist is hierarchical, where the civic and religious center dominates the rural and profane space (Urroz 2022, p. 142). By circularly depicting the city, the German Briefmaler establishes Tenochtitlan as a medieval European city. As Baschet (2009) argues, the city center presents the church and the cemetery. Outside this nucleus, houses and buildings were erected, and beyond this, cultivated land, forests, and other conquered territories were established (Baschet 2009, p. 371). Following this scheme, at the nucleus of the inner life of Tenochtitlan, the artist portrays the ceremonial center and the tzompantli along the stone sculpture, signaling a parallel to the church and the cemetery while highlighting the difference with Christianity by inscribing the word "idol." Outside the ceremonial center, houses and agricultural lands are depicted, and beyond the island, the subject cities. Nonetheless, as the plan of Tenochtitlan comments on the exhibition of the skulls and the stone sculpture as an idolatrous practice, it connects it with the denial of the Nuremberg council to display the relics during the Diet of 1524. Therefore, if the inner life of the Aztec city was based on their idolatrous behavior, the imperial city of Nuremberg was defined as aligned with Reformation ideals of pious life. As Johnson explains regarding the Nuremberg edition of Cortés's Letters, for the reform-minded reader, the description of temples and the portrayal of idols and images that Mexica worshiped accentuated the idolatrous aspects of Aztec religious practices (2011, p. 94).

16.5 Conclusion

The publication of the Plan of Tenochtitlan in Nuremberg in 1524 not only served as an aid for the reading of Cortés's *Praeclara* and a geographical representation of the Aztec capital but also functioned as a powerful and symbolic tool for the Habsburg during the Diet of Nuremberg. Nonetheless, the image successfully projected Nuremberg's perception of Habsburg's dominance over the Mexica, illustrating the hierarchical positioning of the double-headed eagle and the stone idol. By presenting these elements, the Plan ensured their readers the dominance and superiority of the Holy Roman Empire over the Aztecs. While previous studies have focused on the geographical and indigenous elements of this city view, a deeper contextual analysis reveals the broader implications of its creation and publication. The context of the Imperial Diet of Nuremberg in 1524, with its concurrent discussions on the Reformation and the political climate of the Holy Roman Empire, significantly influenced the map's design and the messages it conveyed. However, the image mobilized gender relations and religious differences that portrayed European control over Tenochtitlan, the artist behind the final version of the image successfully aligned the religious and political agendas of Nuremberg.

Acknowledgement This research has received financial support from Project ANID Anillo ATE220054 "Heritage, Space and Gender" (Chile), the Newberry Library (USA), and the John Carter Brown Library (USA).

References

Anders F (2001) Die Schätze des Montezuma. Museum für Völkerunde

Anonymous (1520) Ein Auszug ettlicher Sendbrieff dem aller durcheleüchtigisten grossmechtigiste[n] Fürsten und Herren Herren Carl… Printed by Fridericum Peypus

Anonymous (1521–1524) Rationes cottidianarum expensarum. 1521–1524 imperatoris Caroli V. a camerario hispanico conscriptae. Österreichische Nationalbibliothek

Anonymous (1522) Newe Zeitung von Dem Lande das die Spanier Funden Haben ym 1521 Iare Genant Jucatan… Printed by Matthes Maler

Anonymous (1523) Ein Schöne Newe Zeytung so Kayserlich Mayestet aus India… Printed by Melchior Ramminger

Anonymous (1960) Codex Vindobonensis S.S. 1600. Cartas de relación de la conquista de la Nueva España escritas por Hernán Cortés al Emperador Carlos V y otros documentos relativos a la Conquista, años de 1519–1527. Akademische Druck U. Verlagsanstalt

Apenes O (1947) Mapas Antiguos del Valle de México. UNAM

Aveni A, Gibbs S (1976) On the orientation of pre-Columbian building in central Mexico. Am Antiq 41(4):510–517

Aveni A et al (1988) Myth, environment, and the orientation of the templo mayor of Tenochtitlan. Am Antiq 53(2):287–309

Baschet J (2009) La civilización feudal: Europa del año mil a la colonización de América. Fondo de Cultura Económica

Battcock C (2011) La conformación de la última "Triple Alianza" en la Cuenca de México: problemas, interrogantes y propuestas. Dimensión Antropológica 18(2):7–30

Battcock C (2017) La tenochtitlan de alvarado tezozomoc. Telar 18:43–60
Beaujean D (2010) Hollstein's German engravings, etchings and woodcuts 1400–1700: Hans springinklee. Sound and Vision Publishers
Beaujean D, Biller JH, Bleyerveld Y, Hofbauer C-M, Röver-Kann A, Mielke U, Seelig G, Tipton S, Hollstein FWH (1996) The new Hollstein: German engravings, etchings and woodcuts, 1400-1700. In: Jacoby JW, Luijten G, Zijlma R, Knauer M, Kaulbach H-M, Diefenbacher J, Leuschner E, Messling G, Turner S, Leeflang H, Bevers H, Wiebel C, Bartrum G, Leesberg M, Schoch R, Tanner P (eds) Sound & Vision Interactive, Sound & Vision Publishers
Bendall S (2023) Female personifications and masculine forms; gender, armour and allegory in the Habsburg-Valois conflicts of sixteenth-century Europe. Gend Hist 35(1):42–67
Borah W (1962) The Cortés Codex of Vienna and Emperor Ferdinand I. The Americas 19(1):79–92
Carballal M, Flores M (1989) Las calzadas prehispánicas de la isla de México. Arqueología 1:71–80
Chartier R (1997) On the edge of the cliff: history, language, and practices. The John Hopkins University Press
Close C (2018) The imperial diet in the 1520s. In: Whitford D (ed) Martin Luther in context. Cambridge University Press, Cambridge, pp 125–134
Contarini G (1840) Relazione di Gasparo Contarini ritornato ambasciatore da Carlo V, letta in senato a di 16 Novembre 1525. In: Alberi E (ed) Relazioni degli ambasciatori Veneti al senato, vol 2. Tipografia e Calcografia all'insegna di Clio, pp 9–73
Cortés H (1522a) Carta de relación enviada a su S. majestad del emperador nuestro señor por el capitán general de la Nueva España llamado Fernando Cortés. En la cual haze relación de las tierras y provincias sin cuento que han descubierto nuevamente en el Yucatán… Printed by Jacobo Cromberger
Cortés H (1522b) Des marches, îles et pays trouvés et conquits par les capitaines du tres illustre… Charles Ve. Printed by Michiel de Hoocstraten
Cortés H (1523a) Carta de Relación. Printed by George Coçi
Cortés H (1523b) Nove de le isole and terraferma nuevamente trovate in India. Printed by Andrea Calvo
Cortés H (1524a) Praeclara Ferdina[n]di. Cortessi de noua maris oceani Hyspania narratio sacratissimo. ac. Inuictissimo Carolo Romanoru[m] Imperatorio Semper Augusto, Hispaniaru[m], [etc] Regi Anno Domini. M.D.XX. transmissa: in qua continentur plurima scitu, [et] admiratione digna circa egregias earu[m] p[ro]uintiaru[m] vrbes… potossimu[m] q[ue] de celebri ciuitate Temixtitan variisq[ue] illi[us] mirabilib[us], qu[a]e lege[n]te[m] mirifice delectabu[n]t. Trad Pietro Savorgnano, Printed by Fridericum Peypus
Cortés H (1524b) La praeclara narratione di Ferdinando Cortese della Nuoua Hispagna del Mare Oceano, al sacratissimo… Printed by Bernardino de Viano
Cortés H (2013) Cartas de Relación. Editorial Porrúa
Damen M, Meer M (2022) Heraldry and territory: coats of arms and the representation and construction of authority in space. In: Damen M, Overlaet K (eds) Constructing and representing territory in late medieval and early modern Europe. Amsterdam University Press, pp 243–276
Dodgson C (1903) Catalogue of early German and Flemish woodcuts preserved in the department of prints and drawings in the British Museum. British Museum
Feest C (1995) The collecting of American Indian artifacts in Europe, 1493–1750. In: Kupperman K (ed) America in European consciousness. University of North Carolina Press, pp 324–360
Ferman Cruz F (2022) El Mapa de Núremberg: Un Acercamiento a la 'Visión Española' de México-Tenochtitlan. Secretaría de Cultura-INAH
Filipek S (2023) The King of birds and the bird of kings: about the symbolism of the Eagle in culture, belief and art. Asian J Soc Sci Stud 8(2):17–24
Gagné J (2017) Collecting women: three French kings and manuscripts of Empire in the Italian wars I. Tatti Stud Ital Renaiss 20(1):127–184
Gregg R (2009) City views and the Habsburg and Medici courts. Depictions of rhetoric and rule in the sixteenth Century. Brill

Gresle-Pouligny D (1999) Un plan pour Mexico-Tenochtitlan: les representations de la Cite et l'Imaginaire Européen (XVIe-XVIIIe Siècles). L'Harmattan
Hill Boone E (2011) This new world now revealed: Hernán Cortés and the presentation of Mexico to Europe. Word Image J Verb vis Enq 27(1):31–46
Hodaj E (2019) The cult of the eagle metamorphosed- a semiotic approach. In: International conference. Vlora, pp 159–164
Jiménez A (2014) El Mapa de Hernán Cortés. Memorias de la Real Academia Sevillana de Ciencias. Real Academia Sevillana de Ciencias, pp 235–257
Johnson C (2011) Aztec regalia and the reformation of display. In: Bleichmar D, Mancall P (eds) Collecting across cultures. Material exchanges in the early modern Atlantic world. University of Pennsylvania Press
Keunecke H (1985) Friedrich Peypus (1485–1535). Zu leben und werk des Nürnberg buchdruckers und buchhändlers. Mit einem kurztitelverzeichnis seiner drucke. Mitteilungen des Vereins Für Geschichte der Stadt Nürnberg 72:1–65
Kopallik J (1894) Regesten zur geschichte der Erzdiöcese Wien. Druck von K. Goraschek
Koretz S (1970) Das niederländische element am hofe Ferdiands I. Universität Wien
Maier J (2012) A 'true likeness': The Renaissance city portrait. Renaiss Quart 65:711–752
Martínez J (1990) Documentos cortesianos, vol 1. Fondo de Cultura Económica
Martínez J (2015) Hernán Cortés. (Electronic ed.) Fondo de Cultura Económica
Martire d'Anghiera P (1530) De orbe novo Petri Martyris ab Angleria Mediolanensis protonotarij Caesaris senatoris decades. Printed by Miguel de Eguía
Martire d'Anghiera P (1912) De Orbe Novo: the eight decades of Peter Martyr D'Anghera. Trad. Francis Augustus MacNutt, vol 2, The Knickerbocker Press
Matos Moctezuma E (2011) Reflexiones Acerca del Plano de Tenochtitlan Publicado en Nuremberg en 1524. Caravelle 76:183–195
Matos Moctezuma E (2009) Tenochtitlan. Fondo de Cultura Económica
Mundy B (1998) Mapping the Aztec capital: the 1524 nuremberg map of Tenochtitlan, its sources and meanings. Imago Mundi 50:11–33
Mundy B (2015) The death of Aztec Tenochtitlan, the life of Mexico City. University of Texas Press
Muther R (1884) Die Deutsche Bücherillustration der Gothik und Frührenaissance (1460–1530). München & Leipzig
Palm E (1951) Tenochtitlan y la Ciudad Ideal de Dürer. Journal De La Société des Américanistes 40:59–66
Pfohl J (2020) The court chapels of the Austrian line (I): from emperor Ferdinand I to Emperor Matthias. In: Weaver AH (ed) A companion to music at the Habsburg Courts in the sixteenth and seventeenth centuries, vol 4. Brill, pp 131–175
Pickles J (2004) A history of spaces. Routledge
Pietschmann H et al (2005) Alemania y México: Percepciones mutuas en impresos, siglos XVI–XVIII. Universidad Iberoamericana
Piper K (2002) Cartographic fictions. Rutgers University Press
Pon L (2004) Raphael, Dürer, and Marcantonio Raimondi: copying and the Italian Renaissance print. Yale University Press
Reid E (2021) Gendering political relationships in Genoese ceremonial entries. Sixteenth Cent J 52(1):79–110
Salinas M (1903) El emperador Carlos V y su corte según las cartas de don Martín de Salinas embajador del infante Don Fernando. Real Academia de la Historia
Strauss W (1974) Max Geisberg the German single-leaf woodcut: 1500-1550. Hacker Art Books
Strauss W (1978–2025) The illustrated Bartsch: German masters of the sixteenth century. Abaris Books
Strauss W (1984) The illustrated Bartsch: Erhard Schön and Niklas Stöer. Abaris Books, New York
Thomas H (2020) La conquista de México: Moctezuma, Cortés y la caída de un imperio. Planeta
Toussaint M et al (1938) Planos de la Ciudad de México. UNAM
Urroz R (2022) México-Tenochtitlan: modelo de ciudad renacentista. Antropol Am 7(14):125–146

Vargas G (2000) La Nueva España en la Cartografía Europea, Siglos XV–XVI. In: Mendoza H (ed) México a través de los Mapas. Plaza y Valdés, pp 15–32

Wagner H (1929) Three accounts of the expedition of Fernando Cortés. Printed in Germany between 1520 and 1522. Hisp Am Hist Rev 9(2):176–212

Wilson A (2012) Water, power and culture in the Roman and Byzantine worlds: an introduction. Water Hist 4:1–9

Part VI
The Classic City

Chapter 17
The Political Community in Classical Urban Thought (Utopia)

Rubén C. Lois-González

Abstract This chapter examines the traditional notion of the classical city, associated with the model of the ancient Greek city. This urban example was considered the example of rationality, respect for public space and space designed to serve citizens. However, classical Greek civilization was slave-owning and excluded women from enjoying the most valued meeting places. Greek urban planning was idealised as the birthplace of democracy, of the community sense of the city and a major market area. The Greek city decisively influenced the one developed in the Roman Empire, in a different context of political unity and subordination to Rome. In any case, both the idealised classical Greek city and the Roman one are a product of Eurocentric historiography, which must be qualified. There was an admirable classical Greek city, despite the inequalities and injustices associated with this civilization, but the Chinese, Indian and American cities of antiquity must also be considered classical. In all of them, there are urban elements of enormous interest, which allude to the classical myth of the primaeval city, which until a few decades ago was only considered for Europe and the Mediterranean.

Keywords Classical City · Agora · Ancient Greece · History of urban planning · Citizen · Slavery

17.1 On the Notion of the Classical City

The application of critical theory to the study of cities and urbanisation has completely transformed the interpretation of urban history since the late twentieth century. The traditional evolutionary and cumulative explanation regarded the Greek city as the origin and foundation of everything that came after. This was always a Western-centric reading of the past, placing its millennial origins and the foundations of plural communal life in the Mediterranean. In Greece and throughout this region, the appreciation of both residential and public spaces, along with the beauty in

R. C. Lois-González (✉)
University of Santiago de Compostela, Santiago de Compostela, Spain
e-mail: rubencamilo.lois@usc.es

F. Carrión Mena and S. Rodríguez Alvarez (eds.), *From the City as a Project to the City Project*, The Urban Book Series, https://doi.org/10.1007/978-3-031-94226-6_17

their forms and economic functionality, was established (Morris 1994; Benevolo 2006). However, this historical interpretation has come under scrutiny due to the rise and rationale of decolonial theory. Additionally, numerous studies on the history of cities have proliferated in Asian, African, and American countries, showcasing the significant importance of non-Western contributions to the formal design of cities and the principles of urban planning (Vance 1977).

While acknowledging this renewal in the theory and history of the city, it cannot be ignored that the example of the classical city, associated with the ancient civilisations that flourished in the Middle East and the Mediterranean, holds great relevance both as a model and in its internal organisation. The first examples of urban habitats were tested in the Anatolian Peninsula, Mesopotamia, and ancient Egypt. In Catalhuyuk, Turkey (dating from 6600 to 5700 BC), urbanisation appears to have preceded the formation of a truly organised civilisation based on territorial power (Soja 2005). In fact, this settlement, which may have reached 10,000 inhabitants, along with others, suggests, according to archaeologists, that the birth of the city preceded the development of mature control structures. It was urbanisation, driven by the need to facilitate trade and certain artisanal advancements, that spurred historical progress, contrary to the long-held belief that it was the other way around (Jacobs 1969; Soja 2005). Regarding Egyptian or Mesopotamian cities, we only have fragmentary documentary references, but it is undeniable that Thebes and Memphis functioned as regional capitals of Ancient Egypt, serving as centres of power and hosting several thousand inhabitants. Similarly, in Mesopotamia, the examples of Ur and Babylon achieved a level of popularity that continues to resonate in the present. In both cities, political and religious power spaces are identified through temples, palaces, fortresses, and ziggurats, alongside extensive areas reserved for intense residential functions, combined with various economic activities.

Traditionally, these urban examples from the valleys of the Tigris, Euphrates, and Nile were taken as the origin of the city on a global level. However, research conducted in other parts of the world and archaeological excavations have shown that, alongside the emergence of urban phenomena in the Middle East and the river valleys of the Indian subcontinent, cities also developed elsewhere. All these cities were expressions of the power of the period and the progressive development of economic activities that were not strictly agrarian (storage of goods, crafts, trade, accounting, etc.) (Caves 2005; Orum 2019). Similarly, in America, the flourishing of capitals and major centres indicates the existence of other autonomous hubs of urbanisation (Lara 2022). Therefore, the linear view of the city originating from the riverine civilisations along the Mediterranean, and later continued in Greek and Roman civilisations, does not hold. Similarly, the idea that urban forms, such as regular, orthogonal layouts or the establishment of symbolic centres and residential neighbourhoods are exclusively creations of these Western predecessors, is also unfounded. Despite all this, it cannot be denied that the classical Mediterranean city marks a before and after in the history of urbanism. Furthermore, the innovations of Greek urbanisation were transferred to Rome, passed in one form or another to the Middle Ages and the Renaissance, and were imposed, with varying degrees of pressure, in America and several colonial cities on the African and Asian coasts.

Regarding the classical city, it has often been emphasised that the Greek *polis* constitutes its most perfect model. In our view, this is a conventionalism that we will uphold. On the one hand, Greek urban centres inherited the legacy of the cities of Egypt and Mesopotamia, although they modified the central spaces reserved for power. Temples are less important, as are fortresses and palaces, while public spaces for citizens and *acropolis*es emerged with strength. In this way, urban planners better understood the morphological and structural elements that facilitate the lives of their inhabitants (dwellings, street layouts, neighbourhoods, the public market, etc.). Ultimately, it is a finished example of a city. On the other hand, the *polis* spread throughout the Mediterranean. They were established on numerous islands, in Sicily, southern Italy, and on the current Turkish coast, among their preferred founding territories (Finley 1963; Graham 1999). Finally, the Greek city transmitted its heritage to the Hellenistic cities, which were created in Asia beyond Persian territory and in the Nile Valley. These cities, known as *Alexandrias*, exhibit certain elements of evolution compared to their predecessors, yet they represented the dissemination of Hellenic urban concepts to new regions (Wiemer 2013).

It has often been argued that the Roman city embodies elements of continuity with the classical Greek model. To some extent, the Roman Empire made significant advancements in creating a truly integrated system of cities, introducing important public works within the urban fabric, and effectively responding to the urbanisation drive by constructing new popular residential neighbourhoods (Wilson 2011). However, the symbolic and structural elements of classical Greek urbanism are much more original and, for this reason, have been taken as the first major references of the city in classical antiquity. The central example of the Greek city will be studied, although we will also provide some relevant data about Roman cities.

The chapter is structured into five parts and some conclusions. In the first part, the focus will be on the characterisation of classical antiquity as a slave-owning society. The contrasts between the power groups and a large portion of the population, who were subjected and had no personal rights, will be reflected in the production of urban spaces and the resulting city morphology. The second section centres on presenting the different models of the city and their evolution over time. Undoubtedly, the example provided by classical Athens or the Greek colonies in Asia Minor and Sicily will constitute the purest typologies of the urban planning proposals of the period. In the third part, we will delve into the Greek city to complete a series of analyses initiated in the previous section. From Greece, we will move to the Hellenistic period, which goes beyond the few years that the Macedonian kings led an expansive Greek world. Concerning Rome, the main changes brought about by the Empire will be listed, both in terms of the generalised growth of city sizes and the importance gained by major public works within them. The fourth section addresses the influence of the classical city on the history of urbanism. This includes morphological impacts, with the example of orthogonal plans; structural impacts, with the individualisation of the centre; and functional impacts, with the separation of commerce and market areas within the built fabric. Finally, brief conclusions will be drawn.

17.2 Classical Greek Antiquity: Slave-Owning Society and the Origin of the West

The symbolic importance of the classical Greek city in the history of urbanism has been associated with two fundamental traits. These traits provide, on one hand, a justification for its existence and functioning and, on the other hand, significant value in defining the concept of the Western urban phenomenon. Starting with the social and economic causes that justified the birth and consolidation of the Greek *polis*, it is generally agreed that it emerged within the context of a slave-owning society. This perspective, heavily influenced by the historical materialism of Marxist thought, was accepted by the most renowned specialists in the history of classical Greece, such as Finley (1963), Vernant and Vidal-Naquet (1988), and Vidal-Naquet (2000). For these authors, the existence of slave relationships in agrarian work, domestic service and certain commercial activities enabled the formation of a privileged class, that of the citizens. This class developed the idea of the *polis*, created its institutions, fostered public debate forums and cultivated the arts and philosophy. In essence, they constructed a classical notion of the city, with the *agora* as the symbolic centre of the urban area, a space for the freedom of debate, always close to the market and the port, where the economic heartbeat of the city thrived (Lois-González 2019).

The Greek city, the *polis*, depended on the existence of a rural environment, which would generate the necessary production surpluses for a significant portion of landowners to act as an idle and dominant class in the nearby urban area. The citizens constituted the foundation of the *polis*, an exceptional urban model in antiquity, as they are credited with the invention of politics. This resulted in the relative distribution of the powers and attributions of the state among a collective of individuals who possessed civil rights (Finley 1963). These individuals were a clear minority in the city, as they were exclusively men (women and children were excluded from power), and their number was significantly lower than that of slaves and foreigners (the *metics*). In any case, this traditional perspective of the Greek city, although accurate in its core elements, requires nuance in several areas. On one hand, it is an interpretation focused on the model provided by classical Athens in the fifth century BC, which is by no means representative of the entire diversity of *polis* that existed in the Hellenic peninsula, Asia Minor, Sicily, and southern Italy over centuries. On the other hand, recent research has introduced new elements regarding the Greek social structure, which is more complex than a simple description of a slave-owning society would suggest.

A set of documentary sources and archaeological excavations show that the formation of the *polis*, in the usual sense of the term, occurred in the eighth century BC, although some evidence places them as early as the end of the ninth century BC. Their existence would continue until the Hellenistic period, as the territorial expansion led by Alexander the Great in the second half of the fourth century BC disrupted the entire pre-existing urban system. The *polis* first emerged in the Greek, strictly Hellenic space, but the dynamics of civilisation led to the founding of new *polis* on the Aegean islands, the current Turkish coast, and the southern half of Italy. Some of these urban examples of newly planned cities, such as Miletus or Priene, became prominent references of classical urbanism due to the regularity of their layout and the internal order of their structure (Morris 1994; Pateta 1975; Caves 2005). Therefore, a common socio-economic base for the *polis* expanded over a relatively wide area of the northern half of the Mediterranean, justifying their influence on later Western urbanism.

Regarding the social structure that sustained the permanence of the Greek city, some clarifications are needed. While the slave-based foundation of the socio-economic structure justifying the existence of the *polis* is not disputed, numerous recent studies have explored the diversity of living and working conditions of the slaves. For instance, the conditions and origins of slaves living in rural areas and engaged in agricultural tasks do not seem to closely resemble those of slaves living in the cities, where they typically performed domestic service tasks or worked in commerce and urban businesses. In rural areas, a significant number of slaves came from free families who lost their status due to debt. In contrast, among the slaves in the *polis*, there were many foreigners who had been captured in some military conflict (Wood 1988; Fernández Arancibia 2013). Other foreigners not subjected to slavery, the *metics*, usually held occupations related to commerce and manufacturing. Their economic conditions could be good, but they lacked political rights. In recent years, debates have intensified regarding the actual importance of small landowners or owners of modest businesses in Greek cities (Carledge 1993). It is generally believed that their significance must have been greater than what classical Marxist-inspired theories defended throughout the twentieth century. Thus, the specificity of the slave-based *polis* would be mitigated by a significant presence of free men who worked to survive and barely participated in the government and public life of the city, unlike the citizens. A final note on social life in the *polis* refers to the lack of spaces designed for the female population, who, like slaves and children, lacked rights.

As can be deduced, the *polis* is linked to a differential use of space, depending on the social group in question. In any case, this organisation of classical Greece allowed for the consolidation of a type of city and urban lifestyle that would be taken as a reference for the West. This is due both to the territorial expansion of Greek urbanisation under Alexander the Great and Hellenism, which extended a type of layout born in the Hellenic Peninsula beyond Persia and into North Africa, and to the interpretation that considers the Roman city as an heir to its predecessors in Athens, Euboea, or Miletus. Additionally, in the *polis*, citizens developed philosophy, theatre, mathematics, and geography. Thus, the notion of the city as an innovative centre, so ingrained in European culture, begins with this recognition of the fundamental role played by classical Greece.

17.3 City Models: Evolution Over Time

The accepted idea that the classical *polis* is the origin of Western urbanism has always been because Greece fully created an urban lifestyle, restricted to a minority of citizens, but projected throughout history. The city is the space of community, politics, governance, debate, human interaction, and innovation. Additionally, the original model of the *polis* in Greece expanded through colonisation, and all examples of European urbanism have succeeded in distant places thanks to their conquest and domination (Lois-González et al. 2012). Generally, Greek cities were modest in size: their most notable examples housed several tens of thousands of people, while most did not even reach this demographic threshold. In fact, the history of traditional urbanism has assigned an exemplary role to Greek cities, which today would be merely provincial capitals or regional hubs. This is because Western urban theory has praised these cities for their ideals of individual freedom, democracy, and participatory governance, viewing them as places open to innovation and with populations inclined to host cultural events and public debates. In any case, citizens with full rights in the *polis* numbered only in the hundreds or a few thousand, so the traditional arguments of urban history tend to suffer from a pronounced tendency towards idealisation, reinforced by the great prestige that the culture of classical Greece has always possessed.

In relation to the importance of the Greek city as the first great model of Western urbanism, it is worth highlighting the existence of a recognisable centre of the *polis*, distinctly separated from the spaces reserved for the port or residential sectors. This central space is consistently depicted in all descriptions and archaeological evidence of cities from this civilisation, spanning from the Archaic period (eighth century BC) to the Hellenistic period (third century BC). In the urban centre, two emblematic places have always been distinguished: the *agora* and the *acropolis*. These spaces appear to have been integrated into the conventional history of urbanism, with one area expressing the civil power of the city and its economic strength through the market, and another reserved for religious uses, whether temples, cathedrals, mosques, or synagogues. It matters little whether these areas are close together or relatively separate; what is emphasised is how the *polis* demonstrates the concept of the city as a privileged space for the social or political community, whether it is dedicated to business, worship, or debating urban governance.

From a political standpoint, throughout Greek history, there is a tendency to contrast the democratic *polis*, like fifth-century BC Athens as the prime example, with aristocratic ones, with Sparta as the reference. In this classification, once again, there is a certain idealisation of the existing political systems. In fact, many cities were governed by aristocratic elites, while in others (far fewer), the role of citizens was greater. In the former, temples would have more significant symbolism, whereas in the latter, the *agora* would reinforce its prominence. In any case, archaeological evidence has allowed the establishment of a set of common patterns across the urban world, varying only according to the size of each settlement: thus, Athens was a fully developed city, with all the elements of an important centre, while in Sparta, the traces of urbanity were quite diluted. Therefore, the Athenian case is one where the spaces reserved for the market and citizen gatherings are perfectly recognisable, with the Panathenaic Way connecting to the *acropolis*. This model is highly regarded in the history of Western urbanism, but it had less widespread prominence than is often attributed to it.

Another undoubtedly fundamental aspect in recognising the Greek *polis* as the first great historical reference of Western urbanism is the fact that entire cities were designed from scratch in some colonies of Asia Minor. In this way, the classical city becomes the first example of comprehensive urban planning, with the examples of Miletus and Priene being indisputable milestones (Morris 1994; Caves 2005). In these colonies, the *agora* and the *acropolis* occupy a central space, defining the urban centre, always close to the market place and port infrastructures. The plan is then completed with relatively extensive residential sectors, where the different sizes of the blocks and their formal regularity could corroborate the existence of certain processes of residential segregation. In short, the centre and residential areas, differentiated in the layout according to sectors, constitute two individualised parts that are constantly repeated in European and Mediterranean-based urbanism.

As a final feature of the legacy that the classical Greek city left in the Western world throughout its centuries of existence, it is worth highlighting that the urban planning materialised in some colonies was not the result of pre-established academic regulations, but rather a practical response to the concrete needs of founding new cities. The absence of large power structures and the prominence acquired by the collective decisions of the citizens explain why there were no specific planning authorities within the government but rather general principles assumed about what the *polis* should be. Therefore, the creation of colonies is more a consequence of citizen-driven urbanism, now highly regarded, than of specific dictates from any strong, centralised power, which was absent in classical Greece but eventually asserted itself in the Hellenistic period.

17.3.1 Greece, as an Example of the Classical City

Encyclopaedias, manuals, and dictionaries related to the history of urbanism often cite Greece as a recurrent model of the ancient city, a precursor to everything that developed afterwards. From a morphological and structural perspective, the example provided by the *polis* differentiates a series of elements that are always repeated: the *acropolis*, the port, the wall, the residential quarters, and the industrial areas (Morris 1994; Caves 2005). Moreover, the Greek city is located in a Mediterranean environment, implying the predominance of communal life over domestic life. In fact, the major landmarks of the urban core—the *agora*, *acropolis*, and port—provide a favourable framework for sociability and exchange among citizens and workers. The family home is of lesser importance; only the open interior courtyards, which seek to take advantage of sunlight, hold some interest as a precursor to the much more appreciated Roman houses and mansions in the history of urbanism. Experts agree that the predominance of public spaces over private ones is evident in the urban layout, where the placement of the *agora*, temples, gymnasiums, and other leisure spaces stands out, while residential areas filled the rest of the city (Lois-González 2019).

The symbolic and religious centre of the polis was its *acropolis*, typically situated in a prominent topographical position where the city's temples were concentrated. As often noted, the architecture of the temples emphasised the importance of their exterior space, as they were rarely entered. For this reason, from an artistic perspective, the value of the buildings was concentrated on their façade and sides, with the decoration of the pediment and the forms and motifs of the columns. The temples were typically regular and balanced, somewhat mirroring the plans that characterised many of the newly planned Greek urban colonies. Given their location, most Greek cities had port facilities that would be considered rudimentary today, given the modest size of the ships that engaged in trade across the Mediterranean. Nevertheless, the association of port and city appears here for the first time in history and would serve as a reference for the conceptualisation of Western urbanism for more than two millennia. Alongside the port, commercial activities flourished, intertwining the city and mercantile activity once again (Benevolo 2006; Orum 2019).

There is a certain consensus that the *agora* represents the most important contribution of classical Greek urbanism to the concept of the city. The *agora* is the civic centre where the main public buildings are located, such as the *bouleuterion* (the seat of the assembly responsible for the daily affairs of the *polis*) and the *prytaneion* (reserved for offices and administrative functions). Additionally, it is situated near the market square and usually the port, thus linking the symbolic centre to the economic foundations of the city (Lois-González 2019). Above all, the *agora* transcends its significance as an urban element and becomes the term used to describe the place where public debate, discussion, and discourse on matters of general interest occur. The *agora* is, therefore, the essence of Greek democracy, specifically Athenian democracy, though, as explained above, only a minority of citizens, supported by a slave-owning society and excluding women, could participate and play a clear role in the debates. This idealised, somewhat falsified, concept of the city has persisted for a long time in European and Western urbanism, where parliaments, government buildings, and even revolutions occur in civic centres, always favourable to expressions of freedom, although these images are often somewhat contrived.

From a formal standpoint, the *agora* offers other valuable attributes for urbanistic thought. Essentially, the *agora* is a public square, and the notion of a central city square emerges in this historical period. Additionally, the square allows for better perception of the buildings situated around it. Therefore, if these buildings are intended for governance, the exercise of power is highlighted by the perspective offered by a large open space. A characteristic often highlighted about Greek civilisation is the absence, or lesser relevance, of palaces in their cities. This was due to the very structure of society, which was led by a restricted nucleus of citizens or aristocrats, but never linked to a single-person power, at least until the Hellenistic period. Along with the *acropolis*, a notable feature of the *agora* is its accessibility at the city centre. In fact, this indispensable element of classical urbanism is the origin of what would later become the Roman forum and, by extension, the central squares of medieval cities in Western Europe (Morris 1994). Finally, within the *agora*, a typically present space were the *stoas*, colonnaded walkways where citizens would gather and exchange opinions. Associated with the *agora* and located nearby were the main public baths of the cities, often known as *balaneia*.

To conclude this enumeration of the constitutive parts of the Greek *polis*, understood as the central model of the classical city, we must briefly mention the residential quarters and the city wall. Regarding the former, archaeological excavations reveal compact and fairly dense spaces, where houses with interior courtyards rarely exhibited significant dimensions. The home was a secondary place for the Greek citizen, and there are scarcely any notable examples of extensive mansions. For the other social groups, modesty and monotony were the norms in neighbourhoods with high levels of occupation. Finally, the city walls cannot be compared to those of the Roman Empire in terms of size and tendency towards monumentality. They served to delimit the urban area and played a certain defensive role, especially when conflicts between the *polis* became widespread in the fourth century BC.

17.3.2 The Legacy of the Hellenistic and Roman World

The Western history of urbanism, frequently evoked, establishes a continuity between the model provided by the classical Greek *polis* and the Roman city. However, it is often noted that the three centuries of the Hellenistic period serve as a temporal and evolutionary bridge between the Greek and Roman worlds. On one hand, the notion of the city-state, characteristic of Hellenic civilisation, disappears after Alexander the Great. An empire is established, encompassing a vast, interconnected territory where cities become part of a hierarchical and networked system, distinctly different from the previously discussed independent *polis*. Secondly, there is a diffusion of the Greek-origin urban model towards the East. Hellenism is characteristic of the Hellenic Peninsula, Asia Minor, and southern Italy, but it also extends to Egypt and, notably, further east, covering many territories in Asia and reaching nearly to the borders of the Indian civilisation. Consequently, the major cities incorporate many elements of the preceding national civilisations and function to some extent as capitals for a monarch or governor of a part of the Empire (Hauser 1950).

A significant novelty of Hellenistic society is the relaxation of the citizen's role and status. In a unified territory, their political and social significance loses much of its original meaning, partially replaced by local oligarchies formed from previous indigenous aristocracies that have also lost some of their power, along with provincial rulers and high officials. In an increasingly interconnected Empire, merchant groups reinforce their prominence in a context where the power of money is greater. As a result of these changes, there is a growing appreciation for architectural monumentality in cities that, despite everything, continue to display their Greek origins (Poette 1920).

Among the elements that became widespread in the Hellenistic city, the library and the museum stand out culturally. Indeed, the legacy of this period is often associated with its most emblematic monument, the Library of Alexandria in Egypt. At the same time, urbanism reflects the power of the monarch, the governor, and the dominant oligarchies through buildings such as temples, altars, public fountains, theatres, baths, and gardens, which sometimes served as burial places for prominent families. In this context of expressing power through grand constructions, it is notable that a characteristic element of the polis, the colonnaded street, reaches its greatest prominence. The tendency towards monumentalism and the exaltation of power dominates everything, and in this setting, large public works gain increasing importance (Mumford 1961). Undoubtedly, this is a clear precursor to what would later characterise the Roman city. In Hellenism, the social relevance of citizens becomes clearly diminished, as manifested by the disappearance of the agora, while the economy and commercial exchanges drive urban progress. Numerous examples of cities from this period emerged practically from nothing, making the planning of perpendicular, rectilinear streets very common, as seen in Alexandria located in the Nile delta, the most famous city founded by Alexander the Great. Finally, the importance of panoramic views and perspective is reinforced, consistent with the period's taste for monumentalism (Pateta 1975).

Roman civilisation, particularly during its imperial period, inherited the legacy of the classical Greek city, but it was nuanced by the Hellenistic experience. In this sense, Western historiographic narratives have traditionally drawn parallels between the transmission of cultural, religious, and artistic principles, as well as social organisation, from the Greek world to Rome, and the developments in the specific domain of urbanism and urbanisation (Morris 1994). However, the Roman Empire exhibits a territorial dimension unknown until that era; it maintained centralised control from the city of Rome over vast areas, ranging from the Iberian Peninsula to the edge of Syria, and from mid-England to the Sahara Desert. The extent of this civilisation justifies the emergence of a genuine system of cities, heavily influenced by the Empire's political and administrative division. This places Roman cities in stark contrast to the *polis*, despite attempts to establish continuity in urbanism. The most important Roman cities were provincial or diocesan capitals, connected by a network of roads essential for maintaining the Empire's cohesion. These cities served crucial functions as centres of power and major economic hubs, receiving agricultural or mineral products from their respective territories and sending them on to Rome, from where other goods and merchandise were received (Morley 2011; Wilson 2011).

The Roman city continued to maintain the regular grid pattern of its Greek predecessors, particularly the examples provided by the colonies. Two main roads, the *cardus* and the *decumanus*, organised the city's layout and intersected perpendicularly. One ran from north to south, and the other from east to west. At the point of intersection, the forum was established, resembling the Greek and especially Hellenistic tradition by concentrating the main public and religious buildings of the city. Its importance as a symbolic centre was intentionally emphasised, utilising the monumentality of the constructions and the significance of the public spaces as meeting places for citizens. This design followed a Mediterranean logic in urban planning. However, in Rome, the major architectural landmarks during this period were monumental public works or places designated for collective entertainment and leisure (Giedeon 1961).

In most studies on the history of Western urbanism, the cities of the Roman Empire are always distinguished by three major architectural landmarks: the theatre, the amphitheatre, and the circus. The theatre, inherited from the Greeks, reflected the citizens' interest in the performance of comedies and dramatic plays. The circus, on the other hand, was large enough to host horse races and was typically located on the outskirts of the city. The amphitheatre, usually more monumental, was designated for gladiatorial combats and violent spectacles. The tall façade of the amphitheatre symbolised the isolation of a place within the city where, exceptionally, bloody spectacles could take place. Regarding major public works, the importance of the roads, aqueducts for transporting water over long distances, fountains, and bridges has already been mentioned. The cohesion of the Empire, essential for its functioning, along with the distinctly pragmatic nature of Roman urbanism, granted immense importance to engineering projects, which advanced notably throughout the period (Welch 2007).

As for the other elements present in Roman cities, the walls gained significant monumentality. They were designed to isolate the city from its surroundings, where indigenous populations could pose problems to the established power, which was always concentrated in the main cities. Some of the Empire's walls have been preserved almost intact, such as those in Lugo (Spain), or have retained their monumental gates, like those in Trier (Germany). The walls distinguished the city from the countryside, maintaining a classical tradition that continued throughout the medieval period. In addition to the walls, many cities had ports that were more developed than those of the Greek and Hellenistic predecessors, as trade increased significantly across the Mediterranean and other more challenging seas like the Atlantic. Some cities, such as Rome, promoted independent port cities, with Ostia being a prominent example, indicating the importance of mercantile activities throughout the period. Lastly, Roman houses were built around a central courtyard, with a pond in the middle, which served to organise the rest of the rooms. Archaeological excavations have provided a thorough understanding of the residences of Roman citizens, which are always presented as an advanced model of Mediterranean housing. In this regard, natural disasters like the destruction of Pompeii by the eruption of Vesuvius have made it possible to study numerous details of urban residential spaces in Imperial Rome (Lugli 1940; Crema 1959).

From a historiographical perspective, Roman civilisation was pivotal in introducing urban life and cities to large parts of Western Europe and North Africa. This development was considered crucial, as it was the Europeans who would later spread their urbanisation typology to most continents. As previously mentioned, cities became essential to the functioning of the Empire: they received products generated by agriculture or mining in their dependent territories and sent them on to Rome or other central regions of this cohesive territory. Cities were founded throughout the Iberian Peninsula (where there were already some earlier examples of Phoenician, Carthaginian, or Greek origin), in present-day France, much of England, western

Germany, the Netherlands and Belgium, Libya, Algeria, Tunisia, and Morocco. With Rome, it is possible to describe a civilisation with a global reach, as it encompassed a significant portion of the known Western and Mediterranean world at that time. Within the Empire, a genuine hierarchical urban system was articulated, based on the role each city played in the existing political-administrative division. At the same time, the urban layout model was similar, as most major centres were newly founded. Rectilinear layouts, the centrality of the forum, and the prominence of large constructions and public works were consistently found in most provincial capitals and other territories throughout the areas occupied by this civilisation (Morley 2011; Wilson 2011).

During this long period, a primary association between the city and progress was established, which would persist throughout the history of urbanism. Innovations reached the main centres, such as a unified language, the bureaucracy responsible for territorial governance, the typology of temples and principal constructions, and the cultural expressions of the time. This association of urbanity with innovation, stemming from the existence of an interconnected system of centres, has remained a constant in global cities up to the present day. Finally, Roman cities experienced a degree of continuity in the medieval period in Europe. This was not only because some of the main buildings were repurposed several centuries later but also because the roles played by the city walls in defining urban areas, the importance of centrally located temples and basilicas, and the maintenance of certain public works were inherited from Rome. These elements would reappear in Muslim cities from the sixth to the tenth centuries and in the Christian cities that emerged in the following years.

17.4 Repercussions on the History of Urbanism

Without a doubt, Europe created the conventional way of making history, which became popular during the Enlightenment and the bourgeois revolutions. Through the narration of the past, the aim was to analyse a constant progress of humanity century by century. As time progressed from one century to the next, production would increase, technical advances would be recorded, the urban and commercial world would advance, and the general welfare of people would improve, except during a limited number of crisis periods, which were also considered somewhat exceptional. For urbanism, both this perspective and its Western origin have been reflected in a history of cities that starts tentatively from the civilisations of the Near East but extols the Greek *polis* as the origin of everything. The symbolic value of the centre, the city of citizens, and public spaces for the community continued in Rome, building the medieval urban experience and later the more modern periods. Based on this, with its successes and significant falsehoods, the dynamics of global urbanisation are still understood today. The linear evolutionary history was questioned by Marxist-inspired materialism, which focused its explanations on characterising modes of production. Here, again, the slave-owning society emerges as the classical model of ancient history (as opposed to the vagueness that characterises the Asiatic mode of production), and the urban past continues to be exemplified by the classical city model embodied by the Greek *polis* and the Hellenistic and Roman cities.

Undoubtedly, although the Renaissance exaltation of the classical has been fading towards the present, the weight of historical materialism on the European interpretation of urbanism has continued to be in force. This fact has somewhat negative consequences. A critical theory of the nineteenth century, which emphasises social inequalities, reaffirms a fundamental inequality: between the Western city since Greece, always considered central, and the remaining regions of the world, interpreted as peripheries to a greater or lesser extent. Only the recent decolonial theory, postcolonial in the world of English expression, has begun to undermine these partial readings of the history of urbanism.

In relation to the entrenched idea that Greek and Roman cities constitute the model of the classical city, several questions can be posed. Firstly, as mentioned earlier, the central space of these cities was reserved for citizens and the general public, an attribute that has been upheld throughout history. People would gather in these places either to meet and debate or to praise the established power. The collective use of the symbolic centre of capitals is a constant throughout history. The variations associated with this space are numerous: promenades, whether colonnaded or not, and open and visible squares in more participative societies; monuments, palaces, and temples in eras where power seeks to project and reaffirm itself through mass gatherings around its most emblematic buildings (Panofski 1956; Rossi 1966). Another urbanistic approach arises from constructive elements or city layouts that are disconnected from their central purpose: large squares that remain empty for most of the year, while the population clusters in densely populated market areas, ports, or networks of narrow, commercially vibrant streets. In any case, the dialectic between the forms, symbols, and uses of the city was established in antiquity and is emphasised in treatises on the subject.

A fundamental notion is that the origins of European urbanism can only be understood in the Mediterranean context. This explains the immense importance that has always been placed on outdoor spaces, which take advantage of high levels of sunlight and can be used by large numbers of people. This tradition differentiates Western urbanism from others, such as that of northern China or Japan, where the network of relatively enclosed houses, narrow streets, and alleys holds greater significance. It is true that another model of city with Mediterranean origins, the Islamic city, features a collection of inward-facing houses. While family privacy is preserved, within the homes, courtyards play a prominent role, and the market areas (the souk) or the surroundings of the main mosques serve as public spaces primarily enjoyed outdoors.

Indeed, the use and control of these public spaces in the city indicate which social groups are important or hegemonic. In this regard, the Greek *polis* is extremely clear: the agora is presented as the quintessential place reserved for citizens. There, public matters were debated, assemblies held, or philosophical discussions conducted. This space always reflected the power of this group and their control over the structures of governance. Of course, public squares and promenades were also occupied by those driving the economic vitality of the city, such as traders and merchants, as well as by the slaves who bore the burden of the less pleasant yet essential activities. This somewhat idealised dominance of public spaces was progressively replaced in the Hellenistic and Roman worlds by the centralised power of monarchs and governors, who expressed their authority through monumental buildings and significant public works. The square, the agora, was no longer the central place where the assembly opined and made decisions. Instead, they became spaces from which to observe the magnificence of the rulers, expressed in grand palaces, temples, theatres, and other constructions that always stood out for their size.

A final element of the classical city with significant influence on subsequent urbanism is the definition of a city structure, where the centre is already distinguished from the residential neighbourhoods on the map. Major populations were internally organised through principles of internal space classification. Those areas more accessible and visible from a distance defined a clearly established centre. Meanwhile, residents occupied their homes in predetermined neighbourhoods or ones that had developed spontaneously over time. This specialisation of urban structure is perfectly observed in the newly founded Greek colonies and the Roman cities in the provinces. Prior to this, the urban precedents of ancient Egypt or Mesopotamia did not clarify these fundamental guiding principles for the history of urbanism. This is also why the classical Greco-Roman models are considered so important.

17.5 Conclusions

The general analysis of the classical Mediterranean city, established in Greece nearly three thousand years ago, presents an ambivalent significance. On one hand, Greek *polis* and Roman cities have been fundamental reference points for subsequent urbanism. On the other hand, the Eurocentric and Western explanatory model of the past reveals numerous aspects open to criticism, which have been increasingly highlighted in recent times. Therefore, in this chapter, we have aimed to be cautious: while acknowledging the legacy's significance in many respects, we also highlight that the principles of Greco-Roman urbanism have been praised with interpretations that deliberately elevate their historical importance. It is evident that urban planning, the creation of central religious and secular spaces, the integration of ports into the urban layout, and the importance of city walls are well documented in these civilizations. Consequently, their study as classical examples holds undeniable relevance. However, many of these elements are not exclusive to Greco-Roman cities; they also emerged in China, certain river valleys in India, Mexico, Central America, and the Andes. The model of the ancient city examined here is but one of the various possibilities that a diverse world offers. Therefore, in the future, studies that interpret ancient Asian or American cities as creators of urban development models should be strengthened. In this way, their historical discrimination with respect to the hitherto undisputed Greek/Mediterranean/European example, which should not be based on its traditional formulations, would be corrected.

Another aspect that has been emphasised is the relationship between urban centrality and the organisation of power. The variations within the Greco-Roman world are significant and have therefore been discussed. Contrary to the idealisation of fifth century BC Athens as a model of civic urbanism, current historiographical consensus considerably nuances this interpretation. The supposed Greek democracy was, in reality, the exercise of power by a male minority; furthermore, aristocratic regimes were more numerous throughout the classical period. From Athens and other civic *polis*, the agora remained a model of a symbolic centre and public square, used for trade and a wide range of discussions. In contrast, the Hellenistic and Roman worlds already reflected the importance of the image of power through monumental architecture and grand public works. This is a clear precedent for medieval European and Mediterranean cities, though not so original compared to the urbanism found in the Americas or East and South Asia.

References

Benevolo L (2006) Storia della Città (4 volumi). Laterza, Bari
Carledge P (1993) Classical Greek agriculture I: recent work and alternative views. J Peasant Stud 21:127–136
Caves RW (ed) (2005) Encyclopedia of the City. Routledge, London
Crema L (1959) L'Architettura Romana. Sei, Torino
Fernández Arancibia L (2013) El límite entre libertad y esclavitud: conceptos e ideologías de los amos de la Antigua Grecia. Revista Electrónica Historias Del Orbis Terrarum 10:112
Finley M (1963) The ancient Greeks: an introduction to their life and thought. Viking Press, New York
Giedeon S (1961) Architecture and the phenomenona of transition. The three space conceptions in architecture. Harvard University Press, Boston
Graham AJ (1999) Colony and mother city in ancient Greece. Manchester University Press, Manchester
Hauser A (1950) De socialgeschichte der Kunst and literature. Verlagsbuchhandlung, Munich
Jacobs J (1969) Economy of cities. Penguin Books, London
Lara F (2022) Structures and typologies of latin American and Caribbean cities. In: González-Pérez JM, Irazábal C, Lois-González RC (eds) The Routledge handbook of urban studies in Latin Ameruca and the Caribbean. Routledge, London, pp 92–116
Lois-González RC, González JM, Escudero LA (2012) Los espacios urbanos. El estudio geográfico de la ciudad y la urbanización. Cátedra, Madrid
Lois-González RC (2019) Agora. In: Orum AM (ed) The Willey Blackwell encyclopedia of urban and regional studies, vol I. Willey-Blackwell, Hoboken, pp 21–24
Lugli G (1940) I Monumenti antichi de Roma e Suburbio. Bardi, Roma
Morley N (2011) Cities and economic development in the Roman empire. In: Bowman A, Wilson A (eds) Settlement, urbanization and population. Oxford University Press, Oxford, pp 143–161
Morris AEJ (1994) History of the urban form. Before the industrial revolutions. Routledge, London
Mumford L (1961) The city in history. Penguin Books, New York
Orum AM (ed) (2019) The Willey Blackwell encyclopedia of urban and regional studies. Willey-Blackwell, Hoboken
Panofski E (1956) Studies in iconology. Routledge, London
Pateta L (1975) Storia dell´Arquitetura. Antologia critica.: Fabbri, Milán
Poette M (1920) Introduction à l´urbanisme. Boivin, Paris

Rossi A (1966) L'architettura della città. Marsilio Editore, Padova
Soja E (2005) Mesogeographies: on the generative effects of urban aglomerations. In: Text of conference of TCP annual distinguised lecture: territory, culture and politics. Department of Geography, University of Newcastle, Newcastle Upon Tyne
Vance JE (1977) This scene of man: the role and structure of the city in the geography of western civilization. Palatine (Illinois) Harper's College Press.
Vernant S, Vidal-Naquet P (1988) Mythe et Tragédie. Tome 2. La Découverte, Paris
Vidal-Naquet P (2000) Les Grecs, les historiens et la démocratie. La Découverte, Paris
Welch KE (2007) The Roman amphitheatre: from its origins to the Colosseum. Cambridge University Press, Cambridge
Wiemer H-U (2013) Hellenistic cities: the end of Greek democracy? In: Beck H, A companion to ancient Greek government. Willey and Sons, Hoboken, pp 54–69
Wilson A (2011) City sizes and urbanization in the Roman empire. In: Bowman A, Wilson A (eds) Settlement, urbanization and population. Oxford University Press, Oxford, pp 161–196
Wood EM (1988) Peasant-citizen and slave. The foundations of the Athenian democracy. Verso, London

Part VII
The Industrial City

Chapter 18
The Dialectics of Industrial-Immigratory City

Marcelo Corti

Abstract This chapter expresses various propositions to understand the industrial city as a technical and political project: (1) The characterization of "immigratory" due to its coincidence with the beginning of the great migration from the countryside to the cities. This process is developing both in countries where the process of industrialization can be clearly identified and in those that accompanied this process from a dependent or subordinate position. (2) The boulevard, the large urban park, the neighbourhood as an urban unit and mass housing as a social and political issue constitute the characteristic contributions of this period. (3) Urban planning as a differentiated discipline founded on scientific bases is largely a product of these processes, as well as an instrument for their organisation and control. (4) Guided means of transport and other infrastructure resulting from the same industrialisation process (lighting, telegraph, sanitation, etc.) structure the space and time of this city. (5) The subsequent emergence of the automobile and its infrastructure, particularly the motorways, generates tensions and transformations that prefigure and enable the transition to the contemporary globalised metropolis. (6) Just as the industrial-immigratory city extended (and did not destroy) the historic city in its different expressions, the contemporary globalized metropolis develops on the physical and social base of the city of the nineteenth and twentieth centuries.

Keywords City · Industry · Immigration · Nineteenth century · Urbanism

18.1 Introduction

François Ascher (2005) categorized cities into three periods that, with some modifications, can be summarized and characterized as follows:

M. Corti (✉)
National University of Córdoba, Córdoba, Argentina
e-mail: marcelo.corti@unc.edu.ar

F. Carrión Mena and S. Rodríguez Alvarez (eds.), *From the City as a Project to the City Project*, The Urban Book Series, https://doi.org/10.1007/978-3-031-94226-6_18

- One, the first, of enormous extension, from the first cities in the Neolithic to the beginning of the nineteenth century. This historic city encompasses everything from Athens and Rome to Kyoto and Jerusalem, the medieval European cities, the Arab city, the cities of colonization in America and, in general, all those that emerged and developed before the Industrial Revolution. They are cities with different functions (military, administrative, religious, service centres for rural areas, ports, etc.), generally of limited size, whose internal mobility is pedestrian or animal traction and whose fundamental public spaces are the square and the street. A feature that distinguishes them is their legibility, facilitated by the monumentality of their most significant buildings, the succession of their representative spaces and the clarity of their edges.
- With the Industrial Revolution, a second period begins, that of the industrial city, which not only develops in those cities where industry flourishes but in all those affected by this early process of globalization: commercial enclaves, raw material supply centres, etc. In this sense, what defines these cities is not so much the industry and its factories (except, of course, in the cities that lead this process, with London and Manchester at the forefront) but the strong immigration from the countryside (or in our American cities, from abroad) to the city. Buenos Aires, for example, has developed all the characteristics of this type of city since the end of the nineteenth century, without significant industrial production appearing until the great European wars forced—or allowed—a process of import substitution. Something similar could be said of the "hundred cities for the pampas" (Pesoa Marcilla 2012), created at the same time in the rural interior. That is why in this article we have decided to call this stage of urban history the **industrial-immigratory city**. As we will see, these industrial-immigratory cities are characterized by their accelerated expansion over their rural peripheries, the appearance of guided means of transport as a form of internal and regional mobility and by the commercial avenue and the park as significant public spaces.
- A third period, which has been developing since the second half of the last century, corresponds to the contemporary globalized metropolis. Its features are the extreme territorial dispersion, the predominance of the private automobile as a means of internal and external mobility, the expansion of the service sector in traditional centres and in peripheral expansions, the resulting polycentrality and the appearance of substitutes for public spaces such as the shopping mall, social networks or digital platforms.

These three cities do not replace each other nor do they develop exclusively: it is likely that any city in the world, whatever its scale, has features of each of them. Even recently created cities have the features of historic cities, even cities where there were never industries have the features of industrial-immigratory cities, and even underdeveloped cities have some aspect (even if it were just a precarious internet network) that connects them to economic and social globalization. Therefore, we will analyse the industrial cities that are the subject of this article from two dimensions:

- their urban characteristics,
- their persistence, transformation and opportunities today.

18.2 Characteristics of the Industrial City

As we have seen, the essential characteristic of the industrial city is **its expansion and the growth of its urban area**, in some cases through planned expansions (the best-known examples of which are those of Barcelona with the Cerdà Plan of 1859 and the New York grid of Manhattan outlined by the Commissioners' Plan of 1811); in others through subdivisions and urbanizations developed by private actors; in others, encompassing surrounding villages or towns, in an early process of metropolization. This growth generated the need for new forms of mobility that would allow for internal connectivity within cities. The answer was **guided means of transport**: local and suburban railways, trams and, somewhat later, the underground. This generated a double movement: the transport axes connected the existing centres with the peripheral expansions and with nearby cities, and opened up the possibility of new urbanisations in the interstices and rural spaces in between. According to Herce Vallejo (2013):

> The railway stations in the municipalities on the outskirts of large cities also generated urbanisation in their suburbs and the tram lines to the train stations were often the axes of the consolidation of these suburbs.
>
> Whether in the public or private sector, the capacity of the stations to regenerate city conditions, that is, to generate added value in the land around them, would soon begin to be exploited. Thus, the early purchases, from 1873 onwards, of large plots of land in the Swiss valley of Flon by a company called Société des Boulevards, linked to the Lausanne-Ouchy railway company, are documented.
>
> Many others have not been able to be documented due to the difficulty posed by the secrecy typical of urban development companies. But among the well-known ones, we can mention the Rieselfeld neighbourhood in Fribourg, the Lake Forest and Riverside in Chicago or the Forest Hill in New York, and the Central Railway in the Argentine Pampas, or those linked to the construction of the Union Pacific in the USA, have already been mentioned. In London, at the beginning of the century, working-class neighbourhoods began to be built along the axes of the metropolitan railway. Between 1900 and 1914, the city government built 17,000 dwellings for this purpose in depressed neighbourhoods and 11,000 on the outskirts, along the railway lines, at the borders of the county

A by-product of this new rail mobility is the **city of neighbourhoods**. Although neighbourhoods and districts have always existed in large historic cities, they were generally simple administrative or factual divisions, in some cases linked to landmarks (generally religious: the "parishes" of a certain church or temple) but always dependent for their activities on the attributes and services of the historic centre. The neighbourhood of the industrial-immigratory city, instead, is a socio-spatial unit developed around a commercial and service centre, generally formed around a train or subway station or a tram stop. Schools, squares, perhaps a hospital, neighbourhood institutions such as libraries or union headquarters, cinemas and theatres, sports clubs and other facilities provide attributes of urbanity and socialisation and in many cases form a proud neighbourhood identity. London at the end of the nineteenth century and Buenos Aires at the beginning of the twentieth century, for example, are cases in which football clubs are formed in neighbourhoods, which are often incorporated into institutional names: Chelsea, Tottenham, West Ham, Fulham; Chacarita Juniors, San Lorenzo de Almagro, Tigre, Colegiales, Lanús, Banfield, Quilmes and Defensores de Belgrano.

The expansion of the urban area and population growth cause a great distance from nature. Overcrowding and the lack of sanitation infrastructures simultaneously generate very serious health problems, such as the cholera or yellow fever epidemics that decimated entire cities in the nineteenth century. A specific response to both problems is the emergence of the **park** as a characteristic public space of the industrial-immigratory city; it is no longer a question of the square as an agora or forum but of an area of extraordinary dimensions (sometimes equalling the surface area of the city's built fabric) that reintroduces and reinterprets nature within large cities. Its origin may derive from generous donations from monarchs or great lords or from concessions in new urban openings and expansions. Boulogne and Vincennes at the western and eastern ends of Paris, Central Park in the middle of Manhattan, Santa Lucia Hill in Santiago or Chapultepec in Mexico City are examples in this sense. Sometimes, large parks are the accompaniment or culmination of large developments carried out by public or private actors: Regent Park as the culmination of Regent St. in London, Park Güell in Barcelona, Parque Sarmiento donated by Miguel Crisol in his development of Nueva Córdoba in Argentina. These parks have, in addition to their hygienic and landscape-oriented objective, a social function of representation and homogenization of customs and "decorum" among the working and popular classes; Adrián Gorelik (1998) has studied these trends in *La grilla y el parque*, about Buenos Aires.

The promenade is also an unavoidable function of the other great creation of public space in the industrial-immigratory city: the **avenue or boulevard**. And it is also a real estate product of the bourgeoisie that generated or took advantage of the Industrial Revolution. The most emblematic examples are the Haussmannian boulevards of Paris and the avenues of the Eixample in Barcelona. The Parisian boulevards are the product of the transformations of the prefect Haussmann, based on the demolition of the existing medieval fabric and the opening of representative connections between the large squares and monumental buildings of the city. Although the story about this urban renewal usually focuses on the repressive component of the operation (opening roads that allow the rapid movement of people), the Hausmannian city project transcends this mere aspect and introduces a modern criterion of the city, appropriate to the new social and productive conditions. Immortalised in literature by Baudelaire and Benjamin, the boulevard is a real estate product that today we would consider environmentally sustainable due to its urban compactness; however, its origins were determined by the deficiencies of the time… Without lifts, the logic of high-rise buildings indicated the first level as a noble plane, the residence of the upper bourgeoisie or the aristocracy, often the developers of the buildings. The successive levels of height were intended for different layers of the middle classes and the poorly air-conditioned mansards for the lowest strata of domestic service (precariousness of the means of preserving and preparing food made necessary them to be located immediately next to their employers). The ground floors are used for general access and for the location of shops, bars and restaurants, taking advantage of the large circulation of people through these densely populated neighbourhoods and thus completing the desirable rental scheme.

A similar case is that of the avenues and promenades of the Ensanche de Cerdà, whose construction differs from those of Paris because it is carried out on an expansion over free land, on the rural periphery. The modernist houses of Passeig de Gracia, for example, reproduce that scheme of spatial and social distribution. The avenues, especially the largest ones, introduce the concept of segregated routes for different means of transport and constitute true devices of comfort and urban sanitation, from their sanitary conduits below ground level (in the case of Paris, even with "pneumatic" systems for distributing correspondence that anticipate the immediacy of the telephone and the Internet) to their trees of landscape interest, climate regulators and oxygen renewers. Cerdà also establishes a principle that is becoming more and more valid: at least half of the street surface must be used for sidewalks and pedestrian areas in general.

These achievements of the industrial-immigratory city coexist with difficulties or obvious problems inherent to its condition. Industry requires large surfaces and cannot withstand the competition for real estate rent with other uses such as commercial or residential. Therefore, it tends to be located in places with low land costs, such as the outskirts of cities or less well-off neighbourhoods. Large factories generate **urban barriers** that affect connectivity and degrade their surroundings. Something similar is produced by the railroad, which in its interurban connection arrives with stations located on the very edge of consolidated centres; the railway tracks and grids divide the urban areas that grow later, or establish barriers with respect to coastal areas (as do other facilities, such as wastewater treatment plants or thermoelectric plants). Le Corbusier himself (1962) considered the contemplation of rivers in cities to be "an idle pastime" and imagined their banks as the ideal place to locate the services that make the flourishing of industry possible.

The overcrowding of working-class neighbourhoods, the emission of gases, smoke and polluting effluents by "paleotechnical" industry—according to Mumford's (1971) classification—and other factors (such as the manure produced by animal-powered transport) generate environmental pollution that is at the origin of numerous health crises.

And the conditions of exploitation of the labour force from rural immigration generate a miserable situation for working-class and popular housing, pointed out by political and ideological forces as diverse as Engels' study (1845) on the situation of the working class in England or the encyclical Rerum Novarum by Pope Leo XIII (1891). This situation is the basis of another urban product of the industrial-immigratory city: **housing as a political, social and architectural issue**.

To address this problem, proposals and actions arose from various fields. Christian churches in various countries promoted initiatives linked to charity and philanthropy, generally with very affordable or directly symbolic rents. Some private entrepreneurs built company towns for their employees and families. But the most forceful and effective response was the public production of low-cost or subsidized housing by national, subnational and municipal governments. These state initiatives range from the neighbourhoods built by socialist or social-democratic administrations in the South District of Amsterdam, the *hof* of red Vienna or the *siedlungen* of Frankfurt to the massive programs of the Soviet Union or the Federal Housing Administration (FHA) in USA. In practice, the provision of mass public housing was one of the pillars of the Welfare State and Keynesian capitalism. As we will see, the discussion on types of housing, their form and grouping were one of the key points of the theoretical formulation of Modern Architecture and its urban thought. It should even be noted as a political contribution of modernity the placing of the issue of housing for the working and middle classes within the sphere of attention of architectural culture.

All these environmental and social problems generated by the growth of industrial-immigratory cities explain the contemporaneity of this stage with the emergence of urbanism as a specific discipline. There has always been urban thought and theory (from Aristotle in his Politics to Renaissance writers such as Alberti or the Laws of the Indies with which the cities of the Spanish colonization in America were founded), but it was in the mid-nineteenth century, coinciding with the empirical evidence of the problems that were emerging in the cities of the Industrial Revolution, that treaties and theoretical bodies appeared that aspire to build knowledge and proposals for the solution of urban conflicts from a particularized knowledge. Again, we must mention Cerdá, whose General Theory of Urbanization of 1867 systematizes the concepts applied in the Barcelona Eixample and is probably the first text where the word *Urbanismo* appears in the Spanish language to refer to the discipline that addresses the issues of the city with a method of scientific pretensions. This urban planning for industrial-immigratory cities continued, among other instances, with the appearance of the first zoning regulations in Germany and USA, in a still very primitive instance linked to the provision of adequate turning radii for fire trucks or the regulation of the heights of new skyscrapers based on criteria of sunlight and ventilation, or even to ominous racist measures, such as the prohibition of laundries in the elegant neighbourhoods of San Francisco to prevent the settlement of Chinese families who ran that business in the city.

The City Beautiful movement initiated in Chicago by Daniel Burnham with the World Exposition of 1893, and foreshadowed in the plans of Benoit for La Plata in 1882 and Aarão Reis for Belo Horizonte in 1893, constitutes a formalistic attempt to organize the industrial-immigratory city; it reached its highest realization (and possibly its greatest disappointment) in Marion and Walter Griffin's plan for the capital of the new Australian state, the boring Canberra, in 1910.

More profound in their conception, the Garden City movements (proposed by Ebenezer Howard 1898) and the American regional economy of the 1920s and 1930s, centred on "neighbourhood units," proposed complete alternatives to the evils of unplanned urban expansion. The Garden City proposed new cities that would unite the qualities of the country and the city and include industry in suitable environmental conditions. Although it had few concrete realizations, such as Letchworth in England, it influenced and still influences such contrasting realizations as the new European cities of the second post-war period or the suburban "privatopias" that emerged in USA and today spread throughout much of the world (although these communities of specific interests or closed neighbourhoods are far from constituting integral and integrative experiences like those proposed by Howard). The neighbourhood unit, of which Radburn (New Jersey) is one of the most successful examples, proposed instead an orderly and reasonably self-sufficient expansion of the suburbs, with continuity of the urban area and good links to their pre-existing urban surroundings. The book Cities of Tomorrow, by Hall (1996), presents a detailed and well-founded analysis of these and other stages of twentieth century urban planning.

Other interesting variants were, for example, the Linear City of Soria and Mata in Madrid, founded on an avenue where a tram organizes the elementary mobility; the residential nuclei are joined perpendicularly to this directional axis. Today, it is a neighbourhood integrated into the city. This same conception, perfectly adapted to mobility by guided means, was developed in projects such as the Pampus Plan by Van der Broek and Bakema in the Netherlands and, in practice, in the city actually built in metropolises with good railway infrastructures; The latter is, for example, the case of the northern, western, southwestern and southern corridors of the metropolitan area of Buenos Aires.

In the field of urban planning (as we saw, a discipline formalized in the period we analysed), the most notable example is Sir Patrick Abbercrombie's Greater London Plan, drawn up in the final years after the Second World War. It had a great influence on subsequent urban planning, establishing a green ring to contain urban growth, which leads outwards from this ring to new cities connected to the centre of the metropolis by motorways and railway lines.

In Anglo-Saxon countries, the tradition of urban planning as a discipline of physical formation of the territory was incorporated into the so-called "urban design", while the social and economic aspects were absorbed by "planning". Modern architecture, for its part, proposed solutions such as Tony Garnier's apt Industrial City in 1901, the developments on the territory of the Soviet "de-urbanists" of the 1920s or the "three human establishments" of Le Corbusier in 1959. The Athens Charter of the International Congress of Modern Architecture (CIAM) of 1933 summarises the predominant proposals of the avant-garde movement: strict separation of residential, recreational, productive and circulatory functions, urban space as a mere plane of emergence of sculptural buildings well sunned and ventilated and, an inevitable consequence of these choices, the predominance of the automobile as a means of mobility. In fact, the separation of functions is in practice the urban application of the Fordist production model and its assembly line, ironically described in Chaplin's Modern Times or René Clair's For Us, Freedom. But the application of these principles has its greatest realization (the strictest separation of functions between the monumental civic axis and the residential wings) in Lucio Costa's 1959 plan for Brasilia, which is not an industrial city but a new national capital. As in the cases cited above (La Plata, Belo Horizonte, Canberra), the ideas of ordering the industrial immigratory city are concretized in cities whose function is clearly political-administrative.

A first disciplinary reaction to the problems of orthodox modern urbanism is proposed by Team X from the 1950s; without denying the contributions of modernity, its members claim the mixture of functions of the traditional city and social life in urban spaces described as the street and the square. From other perspectives, Jacobs' (2013) criticism of Robert Moses' urban renewal in North American cities and Lefebvre's (1969) criticism of French bureaucratic planning is harsher. But we are already in the 1960s, with its critical movements both against capitalist consumer society and against the authoritarianism of the socialism that really existed east of the Iron Curtain; the seams of industrial society are already creaking from left and right and other social, economic and political paradigms are already appearing that will constitute new inputs for the production of the city.

18.3 Post-Industrial Intervention in the Industrial City

However, a good part of the production of the city after the 1960s, after the oil crisis of 1973 and, in general, after the "30 glorious years" of the Welfare State (1945–1975) and the heyday of Fordist industry, is aimed at correcting the problems of the industrial-immigratory city or taking advantage of the opportunities generated by its decline or transformation. In general terms, we will call this production "post-industrial", aware that such a name is only useful to encompass a series of concurrent phenomena and processes, such as the transformation of industrial production modes from Fordism to Toyotism (Coriat 1993), the crisis of "real socialism", productive and financial globalization, the tertiarization of the economy, the precariousness of employment, the impact of new information and communication technologies, etc.

In the field of urban planning, these processes are manifested in the adoption of so-called strategic planning (a model originating in the military sphere and later adopted in the business sphere) to replace the normative and regulatory planning that had predominated in the second half of the twentieth century. One of the key points of strategic planning is the abandonment of totalizing pretensions regarding the development of the city and the adoption of a vision of the city in parts, sometimes assimilated to the conception of traditional Chinese medicine with the metaphor of **urban acupuncture**: an intervention on a delimited point of the city whose impacts transcend the area of operation and reach nearby or even distant areas. The Great Urban Project (GUP), also called "large operations or interventions" or simply "urban projects", is a type of intervention characteristic of this model. In another text (Corti 2015), we have commented on this:

> In general, they are linked to the obsolescence of some forms of production and their consequent impact on the urban configurations associated with them; [...] Contemporary Large Urban Projects are the physical expression in cities of the processes of globalization, financialization and economic relocation and of the new forms of business organization. They generally have the following characteristics:

- They are located especially in central areas or in their immediate periphery.
- They are developed on large, vacant plots or with obsolete or abandoned uses: ports (one of the most common cases), large industries, barracks, airports, etc. [...]

A typical case of GUP is the transformation of large factories and industrial establishments whose production or production system is already obsolete, or whose company decides to change its line of business or location due to better labour and raw material cost ratios elsewhere in the world, or even whose original location in peripheral areas of the city has been overtaken by consolidated urban development and has become urban rental areas more suitable for residential and service locations than for industrial use.

The renovation of the former Pirelli headquarters in Milan, for example, is a particular case of this type of large urban project. The decision to abandon traditional production activity on the La Bicocca site, 5 km from the historic centre of Milan, was based on the new conditions of production and business organisation that made its facilities obsolete, but also on the exceptional relative conditions of its location, which by the 1980s was already completely surrounded by consolidated urban areas. The company then decided to undertake an urban-real estate operation, transforming the headquarters into an area of advanced university research and teaching, with the aim of attracting young professionals to whom a variety of residential products are offered on the same site. Among other collaborations provided to the project, the Milanese Municipality built the Teatro degli Arcimboldi on the site, linked to the prestigious Scala. For its part, the Università degli Studi di Milano chose the Bicocca (at the suggestion of the Municipality) to locate its new headquarters dedicated to innovative research. The physical project was entrusted to the most important architectural studio-firm in Italy, Vittorio Gregotti e Associati, winner of the competition held in 1987 among twenty major international studios. Although the conjunction of a single client and a single architect results in a formal image and an urban fabric affected by a certain monotony, an interesting network of public spaces and a respectful recovery of the industrial building heritage can be seen. A similar case occurred in the other great Italian industrial city, Turin, although it was resolved with a different operation: the transformation of the former Fiat car production plant, Lingotto, into a commercial and cultural centre. As part of the intervention, a large pedestrian suspension bridge was built, the Footbridge or Olympic Arch, which crosses the railway access grid to Porta Nova station, connecting Lingotto with the area of the Olympic Village of the 2006 Winter Games; as we will see, the intervention on large railway infrastructures is also a type of operation characteristic of this new conception of urban planning and management. In Paris, the former Citroën car factory, situated in a prime location in the 15th arrondissement near the left bank of the Seine, was transformed into a large urban park, the Parc André-Citroën, which opened in 1992. A similar fate befell the former La Fundidora metallurgical factory in Monterrey, Mexico, which was transformed into a large park for the 2007 UNESCO World Forum of Cultures, which it hosted. The factory's chimneys and some large machinery were left in the new park as commemorative sculptures in homage to industrial heritage. A similar approach, but on a much larger scale, was used in Duisburg-Meiderich, Germany, the heart of the steelmaking axis of the Ruhr Basin during the 1950s–1980s. Once abandoned, it was subject to environmental clean-up and the public park of Landschafts park was created, which opened in 1991. The park pays homage to the site's previous use, the steel and coal production plant, thus recovering the memory of its industrial past.

In neighbourhoods characterised by the density of factories, warehouses and logistics sheds in large industrial cities such as Toronto or New York, the obsolescence or abandonment of these uses left a park built with evident constructive solidity, good environmental performance and a certain spatial quality. A product that began to be developed in the 1970s to take advantage of these conditions was the loft or, in general, the recycling and enhancement of these old factories for upper-middle class or professional residential uses, commercial, recreational-cultural, etc. The area of New York neighbourhood of Soho, later extended to nearby neighbourhoods as Tribeca, became an effective or symbolic model of this trend.

Another type of urban operation on obsolete factories is the one carried out in the @22 district in Barcelona. The operation combines two types of urban policies focused on a plot-by-plot renovation: the reconversion of a declining industrial area, the Poblenou neighbourhood, the industrial heart of Barcelona at the end of the nineteenth century and a good part of the twentieth century (it was called "the Catalan Manchester") and the creation of a cluster of economic activities, in this case aimed at attracting information and communication technology companies with the resources of the city's international image, its urban quality and the immediate proximity to the beaches on the Mediterranean. The district was planned in the 1990s, together with the nearby Forum of Cultures, in an attempt to recover the momentum of the 1992 Olympics. It is a successful process of urban transformation, although not so much in the form imagined as an area of technological innovation or an incubator for ICT companies. Instead, there are plenty of hotels, universities and real estate projects, but it is easier to find healthy survivors of the old industrial times (mechanical workshops, warehouses) than innovative companies. There are also some iconic chimneys, remnants of that "Catalan Manchester", but this fascination with chimneys as a vestige of the old industrial past seems to often conceal the destruction of a valuable heritage, such as that of the Can Ricart factory. Better luck was that of Can Framis, an old textile factory recycled as a museum of contemporary art by the Vila Casas Foundation, which features a courtyard surrounded by the two surviving industrial warehouses. On the other side of the city, in the Bordeta district, another large abandoned factory, Can Batlló, was the subject of a collective self-management project that includes a social housing project managed by Sogeur and financed by Triodos, the "ethical" bank. The community operation has thus become a benchmark for bottom-up citizen management in Barcelona.

A laudable attempt at sustainable urban transformation of an "industrial cemetery" is the Eixo de Tamanduatehy project, in the most dynamic industrial metropolis in Latin America, the gigantic agglomeration of São Paulo. It was not promoted by the central city of the agglomeration, but by a peripheral municipality, Santo André, an industrial and working-class hub that has suffered from an intense process of industrial decentralization since the 1970s. This creates enormous urban voids and causes a decline in its capacity to attract investment, as well as an increase in unemployment and urban violence. In this way, the area was transformed into what Borja and Muxi (2003) describe as a traditional industrial region, "suffering from both the traditional deficits of the peripheries of large centres and the more recent ones due to the relocation of activities". In an attempt to improve its quality of life and reactivate its economic activity, the municipality set out at the end of 1997 to define a strategic project that would unite the city and the ABC (the group of municipalities Santo André, São Bernardo and São Caetano), on the axis of the Tamanduatehy River. The project is proposed in synergy with other economic and urban revitalization projects. It covers an area of 12.8 km^2 and brings together and plans efforts for urban, economic, social and environmental requalification of the area with criteria of inclusive urbanism and economic and social development. A participatory planning process had already identified the Tamanduatehy River axis as having great development potential, due to, among other factors, its privileged metropolitan location within the third largest Brazilian consumer market (the Greater ABC Paulista), its accessibility and the coexistence of large empty areas available for new development, consolidated urban infrastructure, qualified labour and a set of urban, tax and economic laws favourable to development. The type of investment programs, uses and even the companies that would intervene in the different urban operations were the same as in any other large Latin American city: hypermarkets, business centres, private universities, etc. However, the territorial implementation of these operations and the conditions of negotiation between the public and private sectors were foreseen to be substantially different from those of the usual "landing" of decontextualized megaprojects.

Among the urban operations and partnerships carried out, there are, for example:

- Pirelli City: a kind of "Bicocca" in São Paulo; this project (carried out on land and establishments owned by the Italian company) allowed the opening and boulevardization of streets, the construction of a public park and a municipal school.
- Industrial Operations 1 and 2: allowed the redevelopment of Av. Industrial, recreating a new local and regional development hub with the implementation of service and commercial ventures: the ABC Plaza Shopping Mall, a business centre, a housing complex and the Ibis e Mercure hotel centre. These operations are identified as drivers of local transformation by inducing new ventures and the reconversion of old abandoned industrial warehouses and under-used areas into areas reintegrated into the city and the population, generating employment, income and quality of life.
- Regional Logistics Centre: consists of the implementation of one of the largest dry ports in Latin America, jointly between the private companies COSIPA/MRS Logistics and the Prefecture of Santo André.
- UniABC Partnership: result of the Urban Operation and partnership (cooperation) between the University of Gran ABC and the Prefecture of Santo André, allowed the urban-landscape restructuring of another section of Av. Industrial. The installation of the university transformed a degraded area with abandoned warehouses into a new academic centre of regional importance.
- Partnerships with the hypermarkets Pão de Açucar and Carrefour, in exchange for the execution of green areas and public facilities.

An analysis of the status of the Eixo de Tamanduatehy project (or, to be more exact, group of projects) shows the negative consequences caused by the unexpected loss of a very strong and committed public leadership with the project, such as Mayor Celso André. The project is currently at a standstill. In other cases, abandoned industrial establishments are being repurposed for traditional industrial use, albeit adapted to new production procedures and processes, such as in the La Cantábrica industrial park in Morón, Buenos Aires.

Without a doubt, the greatest degradation suffered by an industrial city is clearly that which caused the decline of automobile production in Detroit, the "Motor City" of the twentieth century and the cradle of "Fordism." Detroit is today a city that is trying to overcome its fiscal bankruptcy, depopulated to less than half of the almost two million inhabitants it housed in the 1950s (its golden age), and buildings in ruins in most of its pericentral neighbourhoods (the inner city). The urban solution proposed today is expressed in the plan of the Detroit Blight Removal Task Force (a name that is already significant in itself: a planning office whose task is to remove debris, rust and dust), which aims to demolish 40,000 buildings in the space of 5 years and ruralize the land they occupy today. The environmental component that in other cities is assigned to urban compactness, sustainable mobility and energy conversion, in Detroit's blight removal involves the correct disposal of asbestos and the reuse of recoverable materials. It is difficult for us to imagine, from a place in the world where the problem is cities that grow without a plan, without control and without resources, the logic of a city where people no longer want to or can live. But the case of Detroit can be an interesting example of a Fordist city in which post-industrial solutions do not serve to solve the problems generated by the industrial crisis. The Great Urban Project (caricatured in the Delta City designed to protect Detroit in the classic 1980s film Robocop) has failed and in its place the task of removing the ruins has been entrusted to urban planning.

Other important urban operations carried out in the last decades of the twentieth century and the beginning of the current one have been directed towards the recovery of coastal strips in central urban areas. The most significant and widespread examples are those in which the urban recovery and renewal of old ports is carried out. In this case, the obsolescence of port use coincides with that suffered by Fordist factories, although the impossibility of adapting the old facilities to the large surfaces required for the location of containers (according to the new logistics and maritime transport scheme) seems instead to be a product of technological evolution.

The first great examples of urban-port recovery are located in American cities: the old port of Baltimore, the Quincy Market in Boston, the Pier 17 in New York and Pier 29 in San Francisco. These are operations that seek to help revitalize urban centres degraded by the combination of the "white flight" of the second post-war period and the destruction caused by the Robert Moses-style motorways. They rely on the festive potential of nautical imagery and its facilities; they promote recreational, gastronomic and cultural uses, very much based on the recovery of the built heritage but also, of course, on the urban-real estate valorisation of the surrounding area.

Much more ambitious was the urban recovery of the old port area of the Thames (the Docklands), in London, which was also a test of Margaret Thatcher's neo-liberal economic policies applied to urban development. In order to attract investment, a total deregulation of urban planning regulations was carried out, requiring only that the developer companies mitigate some environmental or transport impacts produced by their undertakings. The consequence of these actions was the formation of a chaotic urban plot and fabric and an elitist and segregator socio-economic profile. For the 2012 Olympic Games, a sector located to the south-east of the Docklands was renovated, highlighting the clean-up of a large environmental liability generated by old, highly polluting industries.

Another significant work in this regard was the urban recovery and integration of the southern end of the port of Rotterdam (Kop van Zuid), which brought together some of the best Dutch architects in a series of works of great individual interest, such as the Erasmus Bridge by Ben Van Berkel. In Australia, the operation of the Melbourne Docklands, at the mouth of the Yarra River, stands out; in the other great Australian city, Sydney, the operations to recover port areas on the bay are characterised by the extreme concentration of high-rise buildings. This is the case of Barangaroo, the most recent of these projects, very aggressive towards the successful urbanity of the nearby neighbourhood of Millers Point.

Perhaps one of the best examples of port recovery is that of the old Port of Genoa, which in 1992 underwent a major operation to mark the fifth centenary of Christopher Columbus' voyage. The most important intervention in this project was the simple demolition of the walls that separated the port from the city and the connection and expansion of the central area. The demolition of a motorway that runs through the area was also considered, but given the difficulties in finding an alternative motorway, it was decided to keep it; however, due to its elevated layout and the relative transparency of its support structure, the infrastructure does not constitute a barrier to the connection between the city and the port.

The recovery of the Nervión Estuary and the new cultural, administrative and commercial centre in the Abandoibarra area of Bilbao has too often been summarised as the construction of the Basque headquarters of the Guggenheim Museum, the extraordinary building by Frank Gehry. Bilbao's particular location in relation to large Spanish and European cities largely explains the focus on tourism as an alternative economic base to that of the large nautical industry that was in decline towards the last quarter of the twentieth century. The museum is located at one of the extreme points of the nineteenth-century expansion completion operation zone with a block-shaped area in which the Iberdrola Tower stands out, the skyscraper designed by César Pelli (a classic of contemporary GPUs); at the other end is the Euskalduna Jauregia Conference and Music Palace, designed by Federico Soriano and Dolores Palacios. Bilbao is possibly the clearest example in the world of the fusion between show business culture and post-industrial urban renewal.

In Latin America, the recovery of the Old Puerto Madero as an extension of the historical centrality of Buenos Aires is particularly noteworthy. In Rio de Janeiro, Porto Maravilha adds to the reconversion of the traditional port area the operation of clearing a coastal highway, transformed into a boulevard with a tram connected to the city centre. In Valparaíso, the future of its waterfront is still under discussion, which depends largely on the location chosen for the expansion of its port terminal.

Another type of production characteristic of the industrial-immigratory city that is currently undergoing transformation is the large complex of public housing. In the case of London, where these operations characterized the urban policies after the Second World War, the transformation has questionable social and political consequences. In her book Big Capital, Minton (2017) reveals the political-economic structure that explains the visible city and its social consequence: the displacement of thousands of Londoners from their city and the enormous difficulty for working, middle-class and professional sectors to access housing in the British capital and in other cities in the United Kingdom.

The book combines testimonies of the lives of people displaced from the neighbourhoods where they spent their entire lives with an impeccable description of the apparatus by which the expulsion was implemented and the tremendous increase in land values that underpins this process. How, for example, does one manage to demolish an entire neighbourhood of social housing such as Heygate in Southwark Council to make way for a luxury complex such as Elephant Park, whose apartments none of the former residents can even dream of accessing? For Minton, this is a new politics of space in which global capital completely reconfigures the territory based on rates of return on property value that far exceed those that productive activity can support.

In the beginning, it was the anti-modernist rhetoric expressed by Charles Jenks in his famous argument regarding the Pruitt-Igoe complex in St Louis that fuelled the bad reputation of the post-war council housing complexes, which at the beginning of the 1980s were home to a third of London's population. This interpretation of architectural and urban criticism was combined with Thatcher's discourse of individual effort and entrepreneurship to get out of the Welfare State. The Right to Buy programme, designed to "transform proletarians into property owners", allowed the dismantling of the housing structure by giving tenants access to ownership. This operation had an implicit trap: the housing units were sold but not the land on which they were located, which remained municipal. Forced purchase of housing was applied in this century to demolish it and make way for extremely luxurious interventions, dedicated to a global elite (the "alpha elite"), presumably linked to money laundering operations. The "owners" received a payment that did not consider the value of the land and whose amount, even combined with life savings, was not enough to buy an equivalent property, not even in the new luxury development but in the same district. Families are thus forced to move to peripheral neighbourhoods, losing social capital, networks of relationships, access to work and urban services that they had achieved in their original neighbourhood. In the districts that receive them, the same mechanism is reproduced and the economic filter displaces families and households outside London to other cities where the process begins again, in a devastating domino effect. The municipalities that thus recover the land from the former social housing do not even capitalize enough to take advantage of the process for urban regeneration that allows them to maintain their residents; the sale to the development corporations barely covers the cost of the sum of forced purchases.

The Housing Act of 1949 promoted integrated neighbourhoods with a social mix. The device of expulsion and social cleansing that currently replaces it generates personal and family situations almost similar to those portrayed by Charles Dickens in the nineteenth century. This process cannot be separated from the disillusionment that led to the Brexit victory in the 2016 referendum. Terry and Brenda, former residents of Heygate confined to Sidcup, an hour and a half from central London by public transport, voted Leave even though they were not sure whether or not it was the best option. "But then I thought that the Labour municipal government of Southwark kicked us out of our house and did nothing to protect us. They left us with no options…", says Terry, and voting to leave the European Union thus became an option.

In contrast to this London process, other European examples instead postulate the regeneration of the complexes and the permanence of their inhabitants. This is the case, for example, of deFlat in Amsterdam, Sluseholmen in Copenhagen and, above all, the extensive public housing stock in Vienna or Berlin, cities that according to Minton base their urban policy "on not resembling London". In the same vein, we can cite the project by Lacaton and Vassal, Frédéric Druot and Christophe Hutin architecture for the transformation of 530 housing units into three modernist social housing buildings in Bordeaux. Built in the early 1960s, the 'Cité du Grand Parc' sector is a district with more than four thousand low-income housing units. The buildings, which are the subject of this renovation project, are structures of ten to fifteen floors; following legal disputes due to the poor state of maintenance of these buildings, the municipality finally decided to undertake an extensive project to renovate and reclassify them. Due to their privileged location, these structures have enormous potential to rescue the self-esteem of the neighbourhood and its residents.

Another type of operation to transform the industrial immigratory city is the intervention on railway and road infrastructure. A very particular case is the Zone d´Aménagement Concerté (ZAC) of the Austerlitz station in Paris, where a railway area on the left bank of the Seine is being recovered for urban use. The decisive operation in this case is the creation of an elevated platform over the railway grid, defining a new urban "ground level". The operation connects with the new National Library and, through a pedestrian esplanade over the Seine, with the ZAC of the Bercy Park (former wine cellar area).

In Turin, the Spina Centrale is a thirteen-kilometre long boulevard resulting from the burying of the railway section between Zappata (very close to the aforementioned Lingotto) and Rebaudengo, in four successive sections. This work allowed the connection of the urban fabric, the creation of new avenues, public spaces and prestigious cultural facilities and the consequent real estate appreciation of the plots facing it. It is worth noting that this type of operation has such early antecedents as the opening of New York's Fourth Avenue, carried out on the underground railway line that ends at Grand Central Station.

Using another criterion, we can add to this category the intervention in New York's High Line on the elevated railway network that connected the meatpacking establishments in the Meatpacking District, converted into a refined linear park.

Finally, it is also worth mentioning the dismantling or burying of many urban highways "à la Robert Moses", typical of the fifties and sixties, a recognition of the negative impacts of these infrastructures on historically formed urban fabrics. An emblematic case is the "Big Dig" in Boston, so called because of the magnitude of the excavation works that involved the burying of the city's central highway and its conversion into an urban avenue with parks and real estate development on its edges. We have already mentioned the case of the Rio de Janeiro Periférico in Porto Maravilha; also well known are the dismantling of the elevated highway over the Cheonggyecheon River in Seoul, recovering it as an urban promenade, and the burying of the highway on the edge of the Manzanares River in Madrid, a fundamental component for the generation of an extensive linear park and numerous urban and real estate operations.

18.4 A Legacy Under Discussion

In this article we have expressed various propositions to understand the industrial city as a technical and political project:

- We have added the characterization of "immigration" due to its coincidence with the beginning of the great migration from the countryside to the cities. This process is developing both in countries where the process of industrialization can be clearly identified, in particular the one initiated after the so-called second Industrial Revolution, and in those that accompanied this process from a dependent or subordinate position, as suppliers of raw materials and/or as recipients of products manufactured in the central economies.
- The boulevard, the large urban park, the neighbourhood as an urban unit and mass housing as a social and political issue constitute the characteristic contributions of this period to the constitution of cities.
- Urban planning as a differentiated discipline founded on scientific bases is largely a product of these processes, as well as an instrument for their organisation and control.
- Guided means of transport and other infrastructure resulting from the same industrialisation process (lighting, telegraph, sanitation, etc.) structure the space and time of this city, supported by the historical city formed, of which in the great majority of cases it is its extension and continuity.
- The subsequent emergence of the automobile and its infrastructure, particularly the motorways, generates tensions and transformations that prefigure and enable the transition to the contemporary globalised metropolis that has become the predominant model since the last third of the twentieth century.
- Just as the industrial-immigratory city extended (and did not destroy) the historic city in its different expressions, producing qualitative transformations on the pre-existing urbanization, the contemporary globalized metropolis develops on the physical and social base of the city of the nineteenth and twentieth centuries. Even within the framework of productive transformation processes that often imply the obsolescence of equipment and installations from the industrial era, this fixed capital, this built asset, is the support that conditions and makes possible a good part of the urban interventions of the twenty-first century.

References

Ascher F (2005) Los nuevos principios del urbanismo. Alianza, Barcelona Original, Paris
Borja J, Muxi Z (2003) El espacio público. Electra, Barcelona
Cerdà I (1867) Teoría general de la urbanización y aplicación de sus principios y doctrinas a la reforma y ensanche de Barcelona. Imprenta Española, Madrid
Coriat B (1993) Pensar al revés. Trabajo y organización en la empresa japonesa. Siglo XXI. Madrid
Corti M (2015) La ciudad posible. Guía para la actuación urbana. Café de las ciudades. Buenos Aires
Engels F (1845) Die Lage der arbeitenden Klasse in England. Otto Wigand, Leipzig
Gorelik A (1998) La grilla y el parque. Espacio público y cultura urbana en Buenos Aires, 1887–1936. Universidad Nacional de Quilmes, Buenos Aires
Hall P (1996) Ciudades del mañana. Historia del urbanismo en el siglo XX. Ediciones del Serbal, Colección La Estrella Polar. Barcelona. 1996 (Original: Londres, 1988)
Herce Vallejo M (2013) El negocio del territorio. Alianza, Madrid
Jacobs J (2013) Muerte y vida en las grandes ciudades. Capitán Swing. Madrid. Original, Nueva York
Le Corbusier (1962) La ciudad del futuro. Infinito. Buenos Aires (Original: 1924)
Lefebvre H (1969) El derecho a la ciudad. Península. Barcelona. (Original: Le droit a la ville. Editions Anthropos, París. 1968)
León XIII (1891) Rerum Novarum. Encíclica papal. Vaticano
Minton A (2017) Big capital: who is London for? Penguin Books
Mumford L (1971) Técnica y civilización. Alianza, Barcelona
Pesoa Marcilla M (2012) Cien ciudades para la Pampa: idea, técnica y construcción de la forma urbana en las nuevas ciudades del siglo XIX de la provincia de Buenos Aires. Revista Iberoamericana De Urbanismo 7:3–17

Chapter 19
Haussmann and the Disruption of the Apparatus of Capture in Mid-Nineteenth Century Paris

Antoine Paccoud

Abstract Haussmann's public works in Paris in the mid-nineteenth century have been at the center of a historiographical debate on the rationales for what is considered to be one of the most ambitious, single minded (and heavy-handed) examples of 'city improvement' in the modern world, a process that has been called 'haussmannization'. In this chapter, I draw on previous work on Haussmann's daily planning interventions to show that this concept covers over the fact that, between 1853 and 1859, Haussmann drew on Saint-Simonian ideas to force through the entrenched influence of property owners over planning. What haussmannization thus obscures is that the transformation of Paris under Haussmann saw the playing out of an essential conflict at the heart of urban development—that which centers on the private capture of housing and land—at a time in which the degree of housing capture was extreme. To provide the theoretical tools to understand how Haussmann's planning practice sought to break the hold of property owners over the development of the city, I use the Deleuzoguattarian notion of the 'apparatus of capture' as a means to isolate three facets of property owner domination over housing: control, extraction and production.

Keywords Paris · Haussmannization · Urban planning · Housing · Apparatus of capture

19.1 Introduction

This chapter offers a reinterpretation of my published work on Haussmann and the transformation of Paris (Paccoud 2016; 2019; 2023).

The work of Baron Haussmann and Emperor Napoleon III during the Second Empire (1853–1870) defined much of the Paris we know today: the wide, tree-lined and monumental boulevards with uniform facades and the perspectives they create.

A. Paccoud (✉)
LISER, Esch-Sur-Alzette, Luxembourg
e-mail: antoine.paccoud@liser.lu

F. Carrión Mena and S. Rodríguez Alvarez (eds.), *From the City as a Project to the City Project*, The Urban Book Series, https://doi.org/10.1007/978-3-031-94226-6_19

Haussmann began his career in public administration in 1831 and had served as Prefect, the highest central government representative in a Department, in Gironde and Yonne before his appointment as Prefect of the Seine Department by Napoleon III in 1853. Under his tenure as prefect, the city's transformation was profound, ranging from the most mundane of planning activities (such as street leveling and widening, sewer construction, enforcing building standards) to the most monumental (the boulevards, the architecture, and the perspectives). The almost two million inhabitants in the urban region saw their city change before their eyes in less than two decades. What few realized however, is that in this rapidly changing environment, the planning apparatus and its regulatory power were the locus of a fierce opposition between Haussmann and city's powerful property owners.

Haussmann's public works have been at the center of a historiographical debate on the rationales which pushed forward what is considered to be one of the most ambitious, single minded (and heavy-handed) examples of 'city improvement' in the modern world. There have thus been strong critiques attached to the process of 'haussmannization': the forceful imposition of order over a city (Pinheiro 2002; Scott 1982), proto-gentrification (Engels 1942; Lefebvre 2000; Merrifield 2014; Smith 1996), or the consolidation of urban capitalism (Benjamin 2002; Harvey 2006). Work on French urban history has however spearheaded a reconsideration of certain aspects of Haussmann's work in Paris, most notably François Loyer's study of the Haussmannian architectural system (Loyer 1987), and Florence Bourillon's work on Parisian industrial quarters which nuances claims of worker displacement (Bourrillon 1996).

In this chapter, I will use my previous work on Haussmann's planning practice, or in other words the daily decisions taken by Haussmann in his carrying out of planning interventions, to show that the concept of haussmannization tends to cover over some interesting aspects of the public works. Indeed, while Haussmann oversaw the transformation of Paris between 1853 and 1870, his tenure can be split into two distinct periods: between 1853 and 1859, Haussmann drew on Saint-Simonian ideas to force through the entrenched influence of property owners over planning; after 1859, losses in the courts handed back planning power to property owners (Paccoud 2016, 2019). Little is known of Haussmann's planning practice in the first period: contemporary critiques emerged after the end of his tenure (Marx' *Civil War in France* and Zola's *La Curée [The Kill]* in 1871, Engels' *The Housing Question* in 1872). Based on an unexplored archival resource, I argued that Haussmann had a strategic approach to the use of planning law designed to align the public works with the interests of Saint-Simon's industrial class over those of idle property owners in a context of extreme wealth inequalities (Paccoud 2019). Indeed, in 1847 Paris, the richest 10% of decedents held 98.3% of total wealth, with 55.8% of total wealth being held by the richest one percent; in comparison, these were, respectively, at 66.9 and 23.7% in 1994 (Piketty et al. 2006).

What the concept of haussmannization thus obscures is that the first decade of the transformation of Paris under Haussmann saw the playing out of an essential conflict at the heart of urban development—that which centers on the private capture of housing and land—at a time in which the degree of housing capture was extreme.

To provide the theoretical tools to understand how Haussmann's planning practice sought to break the hold of property owners over the development of the city, the next section presents the Deleuzoguattarian notion of 'apparatus of capture' as a means to isolate the three facets of property owner domination over housing: control, extraction and production. This is followed by a summary of the state of planning law before Haussmann's arrival and of how he used it to fashion his planning practice. Drawing on epistolary exchanges between Haussmann, property owners and the Minister of the Interior, the next section shows how this planning practice disrupted property owner domination over housing in an attempt to foreground the primacy of city-wide interests over those of particular property owners. Before concluding, the chapter then uses these findings to discuss how the opposition between Haussmann and property owners is covered over by the concept of haussmannization, and how a different comparative gesture could do it justice.

19.2 Housing as an Apparatus of Capture

Gilles Deleuze and Félix Guattari introduced the notion of 'apparatus of capture' in their joint 1980 book *Capitalisme et Schizophrénie II: Milles Plateaux* [1987 for the English translation]. For Deleuze and Guattari:

> what forms the apparatus of capture are two operations always found in the convergent modes: direct comparison and monopolistic appropriation. And the comparison always presupposes the appropriation: labor presupposes surplus labor; differential rent presupposes absolute rent; commercial money presupposes taxation. The apparatus of capture constitutes a general space of comparison and a mobile center of appropriation (Deleuze and Guattari 1987: 444).

The production of housing represents an ideal configuration of capture precisely because it is at the intersection of the three 'heads' of the apparatus of capture: land, work and currency. The production and the consumption of housing thus involve the three forms of direct comparison (of land, of work and of objects exchanged via the market) and the three forms of monopolistic appropriation (absolute rent, surplus labor and the issuance of currency).

19.2.1 Land

The direct comparison of land plots produces a differential rent, while its monopolistic appropriation produces an absolute rent. Differential rent derives from the fact that some pieces of land are more productive than others, or are in a better location, which allows their owners to draw from them a higher rent. Absolute rent takes place at a higher level: since land is by its very nature limited and non-reproducible, its monopolistic appropriation provides landowners with the ability to set the general level of land prices, thereby generating rent linked only to their position as owners.

This absolute rent is what grounds the general market value of land, over which play out differences linked to location and productivity.

The monopolistic appropriation of land is predicated on its owner being able to exercise full control over it, which can be summarized in the statement: 'no-one can take my property away from me'. Land hoarding (Paccoud et al. 2022) is an instantiation of this principle, but it more generally manifests in the protection of property rights, i.e., in the idea that private ownership supersedes public interest considerations in the use of a particular land plot or building. Absolute rent thus emerges because land, or housing, is in the hands of a minority aware of its value; but it also requires full control over property, e.g., the certainty that the owner will have the final say over what is done with it. Without this certainty, absolute rent is threatened.

19.2.2 Work

In the domain of 'work', there is a direct comparison of activities that enables the measurement of productivity. This comparison serves the entrepreneur by allowing the optimal management of the labor force. Surplus labor, like absolute rent, occurs at a higher level: whatever the productivity of the workers, the entrepreneur always extorts a part of what has been produced. There is always a difference between what is produced with a certain expenditure of labor and what the remuneration for that labor can buy of what is produced. This is the difference between use value (what has been produced as such) and exchange value (the price of what has been produced once it arrives on the market). The monopolistic appropriation of labor allows the entrepreneur to establish the general level of profit, i.e., the extent to which exchange value exceeds use value.

The operation of surplus value in the field of housing is of course visible in the profit margins of developers that originates in the construction process. The focus here is on another type of process: the extraction of rent by the owners of property, which stands in a homological relationship to surplus value extraction. In both cases, there is a reliance on a particular type of violence, what Deleuze and Guattari call 'lawful violence' (*violence de droit*). Citing Marx, they explain that the capitalist

> Does not limit himself to taking away or stealing, but extorts the production of a surplus value, in other words, he first contributes to the creation of that from which he takes away.... A part of the value created without the labor of the capitalist can be appropriated legally by the capitalist, in other words, without violating the corresponding right to the exchange of commodities (Deleuze and Guattari 1987: 568).

Likewise, by housing a worker, the owner of the property makes possible the creation of value through work, and thus contributes to the production of the value it then extracts through rental demands on the worker's income. This type of violence thus '*contributes to the creation of that which it is directed against*' (Deleuze and Guattari 1987: 447), and as such seems natural.

19.2.3 Money

The third 'head' of the apparatus of capture is currency. Here also, there is the direct comparison of the objects exchanged and, at a higher level, the monopolistic appropriation of the means of comparison. Direct comparison occurs at the level of the incorporation of specific goods and services into the global circulation of money, which means that the price of these goods and services can be evaluated 'objectively' in relation to the whole of what is already exchanged on the market. With respect to monopolistic appropriation, two different perspectives must be taken: on the one hand, taxation, and on the other, finance. '*It is taxation that monetarizes the economy; it is taxation that creates money, and it necessarily creates it in motion, in circulation, with turnover, and also in a correspondence with services and goods in the current of that circulation*' (Deleuze and Guattari 1987: 443). But the issuance of money is also due to the actions of banks who '*ensure the equivalence, conversion, and circulation of goods-money throughout the economy*' (Deleuze and Guattari 1987: 568).

What brings together taxation and the issuance of money is the production of financial value, and in the case of housing, this is visible in the production of property values. The value of a property determines how much tax is generated when it is transacted, inherited or held; mortgage loans are a major source of new money creation. The production of property values is thus intertwined with state power and its regulatory apparatus, and is thus facilitated at the individual level by proximity to decision makers and at the collective level by particular policies that socialize risk, incentivize particular types of investment through tax advantages and modify the regulatory apparatus to the advantage of particular groups. The statement 'I know how to get more from my property' encapsulates the importance of special knowledge, of being in the secret, to maximize the production of property values in a given context.

19.3 Haussmann's Planning Practice

The argument of this chapter is that Haussmann's planning practice was an attempt to disrupt the capture of Parisian property, predicated on the three modalities described above: control, extraction and production. To understand this disruption, it is first necessary to briefly describe the state of planning legislation when he became Prefect of the Seine, and to explain the changes he introduced.

The 1807 law, passed under Napoleon's First Empire, gave planning authorities efficient tools to modify the urban fabric. For example, towns could prepare plans with the desired 'alignment' (street layout), for both existing and news streets (Sutcliffe 1981), and thus render illegal layouts that did not conform to these plans. The 1807 law was thus meant to foster the incremental widening of streets by refusing to allow buildings to be erected or repaired if they overstepped the decreed alignment. But it also placed the powers of expropriation in the hands of the executive, not the judiciary: it was the administration who decided on the works to be undertaken in the name of public utility, who pronounced the required expropriations and who established the compensations that should be distributed.

In practice, however, the 1807 law never received full application in terms of both its street widening and street creation (and expropriation) clauses. It quickly came under attack in the legislature and the courts, with property owners finding legal ways to resist demands to have their properties pulled back to the property alignment and to ensure that expropriations were separated out from the concept of 'alignement'. Expropriations became subject to stringent requirements and long administrative procedures that could be threatened by the refusal of a single property owner, rather than treated as fixed lines on a map that property owners needed to yield to. Over and beyond its defeats in the courts and legislature, Roncayolo shows that the 1807 law was never applied very rigorously during the July Monarchy (1830–1848):

> 'The figures of the property owner and of the tax payer (often the same person) inspire the utmost respect. Thus, one refrains from scrupulously implementing the measures of the 1807 law on alignments and city extensions, notably with respect to the drawing up of a general plan. One limits oneself, under the fastidious gaze of the architects of the Civil Building's Council, to ensuring the smooth join at the limits of existing and new urban tissue' (Roncayolo, in Agulhon 1983: 97).

It is clear that Napeolonic planning and its manifestation in the 1807 law had quickly been defeated by property owners and the courts.

It is only in 1850—after the cholera epidemic of 1849—that the focus returned to strong planning laws. The Second Republic (with Napoleon's nephew as president) pushed through an efficient compulsory acquisition instrument, which made it possible for the municipality to acquire properties that were in an unsanitary condition. Expropriation legislation was further strengthened by the 1852 decree, passed by Napoleon III (after his 1852 coup) only a year before the start of Haussmann's tenure as Prefect of the Seine. Its most important article is the following:

> In all expropriation projects aimed at widening, straightening or creating Parisian streets, the Administration will have the possibility of including the totality of affected buildings in its plans, in those cases where it considers that remaining sections are not of a size or shape that would allow for salubrious constructions to be erected on them (Faure 2004: 443).

This made it possible to expropriate not only the required buildings themselves but also the strips of land that bordered the future path of the boulevard. This made it possible for the city to capture real estate that would increase in value immensely after the boulevard was opened, and that could then be sold at a high margin. The huge expenditures needed to expropriate, demolish, clear, level, pave, etc., could then be recouped by selling the land bordering the new boulevards.

This clause of the 1852 decree thus strengthened the hand of public powers over planning but the lengthy administrative procedures that had been accumulated through the trials and errors of the expropriation legislation since 1807 still remained. In addition, for all the changes introduced to the expropriation legislation by both the 1850 and 1852 laws, they did not strengthen the 1807 law as concerns the issue of alignments, understood here as the need to draw up general plans and limit repairs and new constructions that did not conform with that general plan. By themselves, these new laws did not challenge the prevailing planning power. But they did set the context for Haussmann to creatively fashion his planning practice. What will be presented here on Haussmann's planning practice is drawn from the analysis of letters of complaint sent by property owners to the Minister of the Interior (Haussmann's direct superior) between 1853 and 1859 (for more information see Paccoud, 2016, 2019). Two main types of administrative decisions were the focus of most of their complaints: refusals of any repairs, minor constructions or property development because the building in question was not on the proper street alignment (street widening) or because a new street was to be opened up over the property (street creation).

In the case of street widening procedures, Haussmann made sure that alignments were properly enforced, no matter how long they had been tolerated. He thus systematically opposed attempts by property owners to consolidate their properties that were not on the proper alignment. The question of whether repairs or renovations would strengthen a property that sat outside of the alignment was important because it determined the effectiveness of the gradual approach to urban change street widening represented. The stricter the municipal authority was on the kinds of repairs that should be considered as reinforcing a property, the more quickly the state of the properties would deteriorate and the more quickly the property owners would have to destroy or pull back their properties. On the other hand, if (as during the July Monarchy) property owners were allowed to reinforce their properties or mask the extent of their state of disrepair, they could prolong the existence of a property outside of the proper alignment. In street widening cases, Haussmann thus pushed for the strict application of a law that had become dead letter.

As concerns street creations, as mentioned above the 1852 law gave the executive the power to decree expropriations but the separation of the concept of expropriation from that of alignment after the 1807 law meant that the lengthy administrative procedure of the 1841 law remained. There were a number of administrative steps that needed to be taken between the moment when the public became aware of expropriation plans for their neighborhood and the moment at which the expropriation of their particular property would be enacted. In order to limit speculative opportunities during this time, Haussmann forbade property owners in the area under expropriation from making any changes to their properties, even if they were on the proper alignment. This was the case from simple projects decided with the Emperor to expropriations just about to be officially decreed: once the plan had been decided, it was illegal to build or rebuild on the wrong alignment. In effect, Haussmann was bracketing out over 40 years of jurisprudence and was applying the 1807 law as it was meant to be used, and more, by including not just plans officially decreed but also any project that had been decided by his administration. The next section will show how this planning practice constituted an important disruption to the private capture of urban development.

19.4 Disrupting the Apparatus of Capture

19.4.1 Against Control

As stated above, the operation of the apparatus of capture depends on property owners being able to exercise full control over their properties. This in essence means defending the sanctification of property rights. The strength of the right to property is its past: one of the main victories of the French Revolution, it is a symbol of the destruction of absolutism: 'the declaration of the rights of man of 1791 was the culmination of a battle of many centuries to defend the dignity of the individual against claims that the public good should always have priority over him' (Zeldin 1981: 30). Haussmann's practice, and especially his use of the 1807 law to forbid owners from repairing or building if their property was on the planned path of a new street, was thus attacked by property owners as a return to the arbitrariness of absolutist rule, a subverted reference to the Revolution to serve those claiming to act in the public interest. This elevation of the discourse of property owners to the level of property rights served to obscure the rationale invoked by Haussmann to limit property rights, namely, that it is to be understood as the contribution of property owners toward solving the problems of the city. The sanctification of private property finds a clear expression in this property owner's attack on Haussmann's extension of construction freezes to street creation cases:

> Upholding private property, its inviolability, and the respect of the laws that protect it are general interests of an order greater than those that for example call for the establishment of new avenues of communication and private property would no longer be really inviolable, the laws which consecrate and establish the conditions under which expropriations can be undertaken would be nothing more than a dead letter, the day where simple administrative projects would suffice to despoil property owners of the rights and advantages of property itself [...] Mr Prefect assumes the power to say: you will not build, because the administration has a project to open new streets that demand the sacrifice of private property! It is impossible to think without dread about the consequences that would inevitably follow from a triumph, even if incomplete, of doctrines such as these, material consequences first, followed by much worse moral consequences[1] (Case 234, 1856).

The protection of private property rights in the face of Haussmann's planning practice is couched in terms of 'control' by another property owner, who is appealing to the Minister of the Interior to overturn Haussmann's decision:

---

[1] *All translations of the archival material into English are the author's.*
«Le maintien de la propriété privée, son inviolabilité, et le respect des lois qui la protègent, sont des intérêts généraux d'un ordre supérieur à ceux qui appellent par exemple l'établissement de voies nouvelles de communication, et la propriété privée ne serait plus réellement inviolable, les lois qui la consacrent et fixent les conditions de l'expropriation ne seraient plus qu'une lettre morte, le jour où il suffirait de simples projets administratifs pour dépouiller les propriétaires des droits et des avantages constitutifs de la PP elle-même [...] Monsieur le Préfet s'arroge le pouvoir de lui dire: tu ne bâtiras pas, parce que l'administration a le projet d'ouvrir des voies nouvelles qui exigeraient le sacrifice de la propriété! Il est impossible de songer sans effroi aux conséquences qui suivraient inévitablement le triomphe, même incomplet, de semblables doctrines, conséquences matérielles d'abord, conséquences morales bien autrement graves ensuite.»

> Worn-out, I come once again, Mr Minister, to appeal to Your Excellency to urge you to accelerate the sale of my property which is of no use to me anymore since it is no longer under my control or to have the authorization to build delivered to me and to establish as valid the compensation I request. I am counting on your compliancy, Mr Minister, to give me an answer and especially to do me justice; as it is completely unacceptable to be a property owner but not to be able to make use of my property[2] (Case 107, 1854).

Haussmann's planning practice was seen as such a threat because it jeopardized the ability of property owners to manage the level of absolute rent in accordance with their interests. The opposition between Haussmann and property owners on the interpretation of the 1807 'alignement' legislation thus centers on who has control over the development of the urban fabric. Property owners had the weight of tradition on their side: because the principle they invoked was so firmly entrenched in the post-Revolutionary era, any attempt to intrude into the private realm was easily attackable through legal and administrative channels. The excerpt from a property owner's letter to the Ministry of the Interior in 1857 is a clear example of the way in which Haussmann's practice was distorted to highlight only his violation of the rights of property, and thus mask what he intended to do with his newfound control over the development of the city:

> While I wait [for expropriation], I must retain the free disposition of what is mine; a threat of expropriation is not equivalent to an expropriation. Mr Prefect of the Seine is the guardian both of the financial interests of the City of Paris and of its general interests as concerns the improvement of the street network. You will not allow that he abuse of his competencies over the street network to serve, while violating the right to property, what he believes to be the financial interests of the city, and that he sacrifices larger interests to smaller ones[3] (Case 238, 1857).

For Haussmann, there is no distinction between the financial interests of the city and the improvements to the street network: both are integral to his practice since the limits he puts on the right to property are only a means to make property owners contribute to the overall improvements. By driving a wedge between financial and general interests, this property owner is trying to obscure Haussmann's call for all to contribute by framing his attacks on the right to property as serving only financial gains, completely divorced from the overall improvements to the city. Haussmann's cupidity can then be opposed to the sacrosanct right of property.

---

[2] «Fatigué je viens de nouveau, Monsieur le Ministre, me pourvoir auprès de Votre Excellence pour vous prier d'accélérer la vente de ma propriété qui ne me sert plus à rien puisque je n'en suis plus le maître ou de me faire autoriser à bâtir et à fixer valable l'indemnité que je demande. Je compte sur votre obligeance, Monsieur le Ministre, pour me faire une réponse et surtout me rendre Justice; car propriétaire je ne puis disposer de ma propriété ce qui est intolérable.»

[3] «En attendant je dois conserver la libre disposition de ma chose; menace d'expropriation n'équivaut pas à expropriation. Monsieur le Préfet de la Seine est gardien tout à la fois des intérêts financiers de la Ville de Paris et de ses intérêts généraux en ce qui touche l'amélioration de la voie publique. Vous ne permettrez pas qu'il abuse de ses attributions en matière de Grande Voirie pour servir, en violant le droit de propriété ce qu'il croit être l'intérêt pécuniaire de la ville, et qu'il sacrifie les grands intérêts aux petits.»

19.4.2 Against Extraction

In addition to taking back control over the development of the urban fabric, Haussmann sought to ensure that building codes were respected. This in practice led him to scrutinize the housing conditions of the Parisian population, and as such place limits on the extractive practices of property owners. For example, he wrote the following to the Minister of the Interior in 1859 concerning the actions of a Baroness who was also a multiple property owner:

> 'All the repairs that have been prescribed aim to consolidate the constructions elevated by Mme Bonardi outside of all generally accepted construction practices. The improvements made to the buildings since the condemnation cannot suffice to remedy such a unfavourable situation: in vain does Mme Bonardi appeal to the inconvenience the works will cause her and her tenants, this is a question of a very secondary order in the face of which the regulations cannot falter. The administration must to the contrary be tough towards speculators who under the pretext of building worker housing raise for the lowest possible price, in densely populated neighborhoods, poor constructions that won't survive longer than a tenancy and which jeopardize the lives of tenants'[4] (Case 84, 1859).

This property owner refused to implement the eight precise repairs and improvements Haussmann's architects and engineers had identified as essential to safeguard the integrity of the two buildings visited and to protect the 61 worker households (for a total of 200 people) living there. Instead, through her lawyer, she argues that no damage is noticeable in the main building even though it had supported the weight of the 37 cabinet makers who worked there as well as of their workstations and tools for the last two years. She thus concludes that time has given proof of the solidity of her building, and that if she were forced to conduct the prescribed repairs she would have to evict the worker households and raise future rents.

Haussmann points out in another letter to the Minister of the Interior, this time for a case in 1854, that a property owner housing 80 worker families is using this fact to attempt to escape his administration's order for the property to be pulled back to the decreed alignment. Haussmann's preoccupation was with the state of the property and the improvement to the street network the realignment would allow, as shown in the following excerpt:

> Besides, the administration cannot preoccupy itself with the cause that ruins this pillar; it cannot make a distinction that the law has not made. It follows from principle that a building subject to an order to be pulled back, whatever the cause of its ruin, can be rebuilt only on the alignment decreed by law; and if Mr Masson rebuilds a pressure point as essential as a pillar, it would result from this a clear and direct fortification of the façade. As far as the

[4] «Tous les travaux prescrits tendent à la consolidation de constructions faites par Mme Bonardi en dehors de toutes les règles de l'art; les améliorations exécutées depuis la condamnation ne peuvent suffire pour remédier à un état des choses défavorable: en vain Mme Bonardi s'appuie t'elle sur le trouble que les travaux peuvent causer à elle et à ses locataires, c'est une question très secondaire devant laquelle les règlements ne peuvent faiblir. L'administration doit au contraire se montrer sévère à l'égard de spéculateurs qui, sous le prétexte de faire des logements d'ouvriers, élèvent au plus bas prix possible, dans les quartiers populeux, de mauvaises constructions dont l'existence, limitée d'avance à la durée d'un bail, compromet la vie des locataires.»

> dismissal of tenants and the inactivity of the factory, Mr Masson mentioned them only for the purpose at hand; indeed the state of the building surely made him expect that he would be forced to rebuild it very soon[5] (Case 104, 1854).

The same argument is used by the owner of a large apartment building in the 8th Arrondissement in which 20 worker households lived. She wrote to the Minister of the Interior in 1856 because she was refused the authorization she had requested to repair her facade on grounds that her property had to be pulled back 25 cm from the street. She argued that such a small gain to the width of the street was useless, as well as the following:

> As this house is occupied as of July by twenty small households, getting rid of these apartments at a time when small lodgings are this rare, especially in this neighbourhood, is to go directly against the government's stated goal of encouraging new constructions of this kind[6] (Case 63, 1856).

For Haussmann, this is a property owner who is driven by private interest only and the house is in such a state of disrepair that it cannot be allowed to persist any longer. He recommends the Minister not to heed the property owner's request, given that the house

> 'has arrived at the end of its time; now, its reconstruction being mandatory, whatever the extent of the retrenchment, the administration cannot authorise, even temporarily, repairs that to be actually effective, would not only consist in a partial renovation but also in the rebuilding of the main columns, piers, etc. It would establish a precedent all the more regrettable that in almost all neighbourhoods the grounds Mrs de Montgomery put forward could be evoked[7]' (Case 63, 1856).

What Haussmann was disrupting here by refusing to tolerate rental housing in advanced states of disrepair was the impunity property owners thought they could enjoy when renting out poor quality lodgings to the working class.

---

[5] «D'ailleurs, l'administration ne peut se préoccuper de la cause qui a détérioré cette pile; elle ne peut faire une distinction que la loi n'a pas faite. Il est de principe qu'une construction soumise à retranchement, quelle que soit la cause de sa ruine ne peut être reconstruite que sur l'alignement arrêté légalement; et si le Sieur Masson reconstruit un point d'appui aussi essentiel qu'une pile, il en résulterait une réconfortation évidente et directe du mur de face. Quant au renvoi des locataires et au chômage de l'usine, le Sieur Masson n'en a parlé que pour les besoins de la cause; car l'état de la maison devait lui faire pressentir qu'il serait obligé de la reconstruire dans un délai rapproché.»

[6] «Cette maison étant occupée au mois de Juillet par 20 petits ménages, supprimer ces locations dans un moment où les petits logements sont si rares surtout dans ce quartier, c'est aller directement contre le but que se propose le gouvernement en encourageant les constructions nouvelles de cette nature»

[7] «Elle est donc arrivée à la limite de sa durée; or, sa reconstruction étant obligatoire, quelle que soit la qualité du retranchement, l'administration ne peut autoriser même temporairement une réparation qui d'ailleurs pour être efficace, ne consisterait pas seulement dans un ravalement partiel mais encore dans la reconstruction des piles, trumeaux, etc. Ce serait établir un précédent d'autant plus fâcheux que dans presque tous les quartiers, on pourrait évoquer les motifs mis en avant par Madame de Montgomery»

19.4.3 Against Production

The transformation of Paris under Haussmann gave rise to numerous speculative opportunities, as portrayed in Zola's 1871 novel *La Curée [The Kill]*. The novel follows the rise of Aristide Saccard, who, as a municipal employee was able to draw on his knowledge of future developments to rapidly accumulate wealth. The novel shows the importance of secret knowledge and political connections in the production of property values: he starts out by purchasing houses on the path of future boulevards and uses friends in the expropriation committee to receive highly inflated financial compensations.

The available archival material shows that Haussmann's administration was keenly aware of the opportunities the public works gave rise to for those with property and the right political connections. For Haussmann, the planning administration needed to be able to limit the actions of property owners on the path of new streets in order to ensure that the costs of the public works remained manageable—without such limits, speculation would ensue, as he explained to the Minister of the Interior in an 1857 letter:

> The administration should, without taking any account of its projects and without making a single comment, provide all construction permits requested by individuals within the limits of the old alignments, even as an administrative procedure is under way to adopt new plans. Until the decree declaring expropriations of public utility, everyone should be completely free to take all possible actions to block its subsequent execution; the administration alone would have its hands tied. In this system, the inevitable administrative delays would be a time reserved for private interests, always very apt and very active, to speculate on the projects submitted to the public enquiries, and commit all possible frauds against general and municipal interests. It would be the moment, or never, to invoke the maxim: summum jus, summa injuria[8] (Case 205, 1857).

Two cases in the archives provide illustrations of the production of property values Haussmann was worried the public works could lead to. In both cases, Haussmann denied the property owners' request to build on their land because it was on the path of a future boulevard, and they wrote to the Minister of the Interior to have this decision overturned. The fact that property owners sought to overturn the decisions of the planning administration by appealing to the Minister of the Interior (Haussmann's direct superior in the administrative hierarchy) shows that they were used to drawing on political connections to bypass regulations. In the first case, Haussmann justifies

---

[8] «L'administration devrait, sans tenir aucun compte de ces projets et sans faire la moindre observation, donner toutes les autorisation de construire demandées par les particuliers dans la limite des anciens alignements, alors même qu'une instruction serait ouverte en vue de l'adoption de plans nouveaux. Jusqu'à l'apparition de décret déclaratif de l'utilité publique de ceux-ci, chacun devrait être complètement libre de faire toutes les combinaisons possibles pour en entraver l'exécution ultérieure; l'administration seule aurait les mains liées. Dans ce système, les délais inévitables de l'instruction seraient un temps réservé aux intérêts privés toujours fort habiles et fort actifs, pour spéculer sur les projets soumis aux enquêtes, et ourdir toutes les fraudes imaginables à l'encontre des intérêts généraux et municipaux. Ce serait le cas, ou jamais, d'invoquer la maxime summum jus, summa injuria.»

his course of action to the Minister of Interior in the face of accusations by the property owner that he is arbitrarily restricting property rights:

> These are not vague projects that will be carried out far in the future, but projects of a certain and current nature which do not allow the municipal administration to tolerate building activity which could make the undertaking more difficult and more costly. Besides, is it not doing individuals a favour to forbid them from putting up new buildings that would almost immediately be subjected through a public utility decree to servitudes restricting their use? Mr. Bergeront, called to my offices, was informed of the situation of his property. At first, he seemed to understand that the municipal administration could not grant him the permission he had requested; but soon, he came back with a less conciliatory spirit and, hoping to take advantage of the opportunity that seemed offered to him to force the administration to acquire his property on the spot, he was not afraid to ask to be compensated, not only for the current value of this building, but also for the potential profits that his new constructions would, according to his calculations, bring him. The property is rented out, and well rented out, according to its market value; it is therefore fully productive [...] The refusal I notified Mr. Bergeront of the authorisation to erect new buildings on the rue de Longchamps does not cause him any noticeable harm. Finally, there is a real analogy here between his situation and that of a property owner subject to a normal alignment procedure who can neither repair nor improve his building, but who continues to make the most out of it in the state in which it finds itself and who is cannot take advantage of speculations he is forbidden to undertake to claim compensation[9] (Case 234, 1857).

In the second case, the property owner, a count, seemed to know all the administrative steps required in an expropriation, and used this knowledge to maximize his financial gains from the process. For example, as soon as he knew that his property was destined to be expropriated, he approached the municipal administration to have them acquire it directly. When this failed, he still proceeded with his initial building plans, in the process adding value to his property. This is Haussmann's perspective on the events:

> Mr de Luçay did indeed offer to sell his property to the City, but his deliberately exaggerated proposals prevented any agreement. Since then, Mr. de Luçay has erected a building [...]

[9] «Il ne s'agit donc pas de projets vagues et d'une exécution lointaine, mais de projets ayant un caractère de certitude et d'actualité qui ne permet pas à l'administration municipale de tolérer des constructions qui pourraient en rendre l'entreprise plus difficile et plus onéreuse. N'est-ce pas d'ailleurs rendre service aux particuliers eux-mêmes que de leur interdire l'édification de bâtiments neufs qui se trouveraient presque immédiatement frappés des servitudes de grande voirie par l'effet d'un décret d'utilité publique? Mr Bergeront, appelé dans mes bureaux a été informé des conditions dans lesquelles se trouvait placée sa propriété. Il a paru d'abord comprendre que l'Administration Municipale ne pouvait lui délivrer la permission qu'il avait demandé; mais bientôt, il revint animé d'un esprit moins conciliant et, espérant alors profiter de l'occasion qui lui semblait offerte de contraindre l'administration d'acquérir sur le champs sa propriété, il ne craignit pas de demander à être indemnisé, non seulement de la valeur actuelle de cet immeuble, mais encore des bénéfices éventuels que ses constructions nouvelles devaient, suivant ses calculs, lui procurer. La propriété est louée, et bien louée, d'après sa valeur vénale; elle est donc en plein produit. Il existe même des baux d'une durée assez longue. Le refus que j'ai fait au Sieur Bergeront de la permission d'élever des constructions neuves sur la rue de Longchamps ne lui cause pas de préjudice appréciable. Enfin, il y a une véritable analogie entre sa position et celle d'un propriétaire frappe d'alignement par mesure ordinaire de voirie qui ne peut ni réparer ni améliorer son immeuble, mais qui continue à en jouir dans l'état où il se trouve et qui n'est admis à se prévaloir des spéculations qu'il ne peut pas y faire, pour réclamer une indemnité.»

without prior declaration or production of plans [...]. So this owner, knowingly and without waiting for the outcome of the complaint he made to Your Excellency, has given his property considerable added value, by erecting buildings that are already slated for demolition. These practices, which frequently occur in the path of planned roads, warrant severe repression. Suffice it to say that I strongly urge Your Excellency to reject the complaint by Mr de Luçay, whose clear intention is to take advantage of his property's location to hold the city to ransom. Adopting this property owner's system would make the carrying out of the projects of public utility decided for Paris more and more costly, a carrying out which is becoming increasingly necessary, however, since population increase in this city has resulted in insufficient living space and justifies the municipal administration's efforts to assimilate to the city through clever piercings the open land at its circumference[10] (Case 52, 1857).

Rather than spurring speculation on, Haussmann worked in his daily practice to limit the opportunities the public works provided to those with right the know-how and political connections.

19.5 Haussmann and Haussmannization

It is now possible to return to the notion of haussmannization in light of Haussmann's planning practice in the first half of the public works. This practice aimed to disrupt three facets of the capture of housing by wealthy property owners: control (by limiting property rights), extraction (by enforcing building standards) and production (by limiting speculation). In doing so, Haussmann came under intense attack by property owners and their lawyers. His interpretation of planning law was eventually defeated by court cases brought to the Council of State by property owners by the end of 1850s (Benevolo 1967; Gaillard 2000; Harvey 2006). These defeats had serious implications for the way in which Haussmann conducted the public works in the 1860s and set the frame for the use of this planning law in other French cities which underwent similar transformations (Lacave 1980; Darin 1988). Haussmann pushed on with the public works in the 1860s, without the backing of planning law, and these became a source of speculative profit for property owners and developers.

[10] «Mr de Luçay m'a en effet proposé de vendre sa propriété à la ville, mais ses propositions exagérées sans doute à dessein, ont empêché tout arrangement. Depuis, Mr de Luçay a élevé sans déclaration ni production préalable de plans, une construction compose d'un rez-de-chaussée et d'un étage en pan de bois doublant par derrière le bâtiment existant sur la rue de la Pépinière et formant retour en aile sur la cour. Ainsi ce propriétaire, en connaissance de cause, et sans attendre le résultat de la réclamation qu'il a adressé à votre excellence, a donné à sa propriété une plus-value considérable, un faisant des constructions condamnées d'avance à la démolition. Ces faits, qui se produisent fréquemment sur le parcours de voies de communication projetées, méritent une répression sévère. C'est assez dire que j'insiste vivement auprès de Votre Excellence pour lui faire rejeter la réclamation de Monsieur de Luçay dont l'intention évidente est de profiter de la situation de son immeuble pour rançonner la ville. L'adoption du système de ce propriétaire aurait pour résultat de rendre de plus en plus onéreux l'exécution des projets d'utilité publique adoptés pour Paris, exécution qui devient cependant de plus en plus nécessaire, depuis que l'accroissement de la population de cette ville y rend insuffisant l'espace habitable et justifie les efforts de l'administration municipale pour y assimiler par des percements intelligents les terrains encore vagues à la circonférence.»

Elsewhere, I have argued that Haussmann's early planning practice was of Saint-Simonian inspiration (Paccoud 2019). Saint-Simon was a utopian socialist whose influence permeated French society in the nineteenth century and who called for decision-making to be in the hands of the productive class, i.e., all those working to produce things useful to society: factory owners and workers, those involved in agriculture or trade, as well as artists and scientists. In a similar vein, Haussmann's planning practice was predicated on the refusal that property owners benefit from the public works just by virtue of being property owners. In his mind, property owners should contribute to the city's transformation by relinquishing the speculative opportunities opened up by the process—and if they were unwilling to do so, strong limits should be placed on private activity. In the context of extreme wealth inequalities, this planning practice can be seen as an attempt to elevate the general interest above that of property owners. It is thus difficult to reconcile this early planning practice with two ideas at the heart of the notion of 'haussmannization': that Haussmann was an omnipotent planner and that the aim of the public works was to systematically rid Paris of its poor neighborhoods.

The assumption that Haussmann had full control over the city and could push through, relatively unhindered, Napoleon's III ambitions for Paris is best summarized by Chapman & Chapman (1957: 251) for who

> 'Haussmann [...] was fortunate in that for seventeen years he was able to exercise his considerable talents to the full',

but it can found in a number of other accounts. This is also the case for Engels –

> "By 'Haussmann' I mean the practice which has now become general of making breaches in the working class quarters of our big towns"

– and Benjamin:

> 'As for the phantasmagoria of civilization itself, it found its champion in Haussmann and its manifest expression in his transformations of Paris (2002: 14–15)'.

This personification of the public works under the omnipotent figure of Haussmann comes to be one of the defining features of the concept of haussmannization, which has invoked in urban contexts as different as Dakar (Vernière 1977), Brussels (Van Loo 1994), London (Schubert & Sutcliffe 1996), Cairo (Volait 2003) and Managua (Rodgers 2012).

This personification of the public works is linked to Haussmann's particular intent in rolling out the public works: proto-gentrification (Lefebvre 2000; Merrifield 2014), a process through which the center of the city was to be reconquered by the upper classes (Ceaux 1975). Drawing on Henri Lefebvre, Andy Merrifield summarizes this process thus:

> Haussmann patented not only what we'd now call the gentrified city, with its commodification of space; he also pioneered a new class practice, bankrolled by the state: the deportation of the working class to the periphery, a divide-and-rule policy through the urbanization itself, gutting the city according to a rational economic and political plan (Merrifield 2005: 700).

This interpretation of the public works brushes aside recent work that has shown that the destruction of working-class areas and the displacement of their inhabitants from the center of the city was a gradual—with roots in processes pre-dating the public works (Bonvalet and Tugault 1984)—and spatially uneven process (Bourillon 1994, 1999; Faure 2004).

This critique of the concept of haussmannization is no way intended to downplay some of the concerns with the public works. For example, as Ann-Louise Shapiro (1982: 487) has pointed out, it is clear is that 'there was no public attempt to rehouse those displaced by demolitions, and private industry failed to take up the task, preferring rather to chase the windfall profits of the luxury housing market.' His planning practice in general did not provide much voice to those affected by the demolitions, and traces of the way in which this population experienced the public works are scarce (as the public works were seen as an initiative emanating from the Emperor himself, it may be that any negative press about them was censored at source). The problem with the concept of haussmannization is that it comes to overcode interpretations of the public works, obscuring how Haussmann attempted to limit property owner capture of housing markets.

Even though Haussmann was defeated in his planning practice, his actions showed that regulation could be achieved through sheer political will. The first decade of the public works thus represents a particular temporality, a singular period in which political will was able to supersede private interests. Seeing Haussmann as a figure engaged in a political conflict with property owners provides a middle ground between the purported omnipotence found in the concept of haussmannization and accounts of the public works, such as Harvey's, that downplay Haussmann's agency:

> The towering figure of Haussmann dominates the state apparatus of Paris throughout the Second Empire. To say that he merely rode out the storm of social forces unleashed through the rapid accumulation of capital is by no means to diminish his stature, because he rode out the storm with consummate artistry and orchestrated its turbulent power with remarkable skill and vision for some sixteen years. It was, however, a storm he neither created nor tamed, but a deep turbulence in the evolution of French economy, politics and culture, that in the end threw him as mercilessly to the dogs as he threw medieval Paris to the démolisseurs (demolishers) (Harvey 2006: 152).

The difficulties Haussmann faced in defending his planning practice illustrates the fundamental difficulty of achieving autonomous regulation, which requires the political will to supersede private interests. Benevolo concludes his history of the European city with a different approach to regulation:

> 'The quality of urban space depends upon a fluctuating balance between spontaneity and regularity, and upon a combination of public control and private initiative that can either succeed or fail, stimulate creative change or paralyse it' (Benevolo 1993: 217).

What our study of Haussmann's planning practice has shown is that such a balance is by definition unachievable, public control and private initiative are two forces drawn by their fundamental principles in opposite directions. Setting a balance between public control and private initiative is a political decision, not a technical one—and it is thus a highly conflictual process. Regulation cannot be called for in abstraction of political considerations, it must be actively fought for in the face of the opposition of private interests. The case of Haussmann seems to have informed Benevolo's account in a way that differs from the one presented here. This is the lesson Benevolo takes from Haussmann's defeat in the Council of State:

> This outcome exacted an enormous price in terms of economic and spatial balance; from that date property speculation exerted an excessive influence on urban transformations. The remedy, however, was not the permanent public management of the entire urban terrain, a solution incompatible with the European tradition and one which led to disastrous results in socialist Eastern Europe. What was needed instead was the re-establishment of an appropriate degree of competition between public and private interests. It is a problem of degree, like many other of the problems associated with the European tradition. This new combination of public and private interests proposed in the first decades of the twentieth century operates along chronological as opposed to spatial lines; it calls for public intervention at those moments when the urban fabric undergoes change, while giving free rein to private initiative in the intervening periods (Benevolo 1993: 217-218).

The call for public intervention during periods of change seems to be a direct consequence of his careful study of Haussmann's public works. But we have seen that there is an important distinction between public intervention and public control: while public intervention took place over the whole period of Haussmann's tenure in the form of the public works (street and park creations, infrastructure projects, erection of monuments, etc.), public control in the sense of the regulation of private activity was lost with Haussmann's defeats in the Council of State.

There is thus a clear difference between the existence of regulations (in legislation, as manifested by regulatory tools or by an investment in individuals and institutions) and the political will that is needed to carry out regulatory functions. This is because regulation exists in opposition to socioeconomic necessity: it is by definition a curtailment of private activity and thus a brake on the circulation of capital. Regulations may exist, but that does not guarantee in any way that they are being used as envisioned. And those who decide to use them to their full power will soon hear lawyers knocking at their door. Regulation is evanescent because it relies on political will, which is easily broken in the face of socioeconomic necessity.

Viewed as an episode in which regulation was successful at bracketing out structural considerations, Haussmann's planning practice was a practical answer to a problem still with us today: how do we reclaim urban space from power and capital? Haussmann's particular answer was appropriate to the time and place in which he faced this problem, and thus drew on Saint-Simonian ideas. Other answers, in different places and times, will develop other answers drawing on other ideational configurations. Viewing the Haussmann event in this way allows a different kind of comparison from that offered by the concept of haussmannization. While this concept offers the promise of taking the experience of Paris and using it to understand what occurs in another city, there is a risk of drawing comparisons across experiences that are not comparable: a process is used to fixate the sense an event expresses, but there is no guarantee this sense will be similar across scholars. The comparison proposed here is more limited, focused on isolating within the transformation of Paris the opposition between a practice foregrounding the collective interest and powerful private interests. The comparative gesture is thus: can this opposition between the regulation of private activity for the good of the collective and the inalienable freedom of individuals to act in abstraction of this collective be found elsewhere?

19.6 Conclusion

This chapter has argued that it is important to add to the list of interpretations of the Haussmann event the fact that he was a planner who sought to regulate private activity over urban development to adapt the city to industrialization. I have shown that his tenure as Prefect of the Seine could be split into two distinct periods: between 1853 and 1859, Haussmann drew on Saint-Simonian ideas to force through property owner capture of planning; after 1859, losses in the courts handed back planning power to property owners. Some of the strong critiques attached to the process of 'haussmannization'—proto-gentrification, the consolidation of urban capitalism—thus correspond to the consequences of the return of property owner influence over planning in the context of extreme wealth inequalities. The demise of Haussmann's early planning practice shows that housing and urban development are important sites of capture—disrupting the mechanisms that underlie this capture requires regulatory interventions, and thus political will.

Confronting housing as an apparatus of captures requires putting in place measures that limit property owner control, extraction and production. For Haussmann, taking back control over planning meant curtailing property rights as soon as the first administrative steps had been taken to modify the urban fabric. To limit property owner extraction, he was strict in the application of planning regulations, foregrounding the safety of occupants over the productivity of investments in housing. Finally, he worked to ensure that property owners could not use the public works as another opportunity for the production of housing value. Though it might seem heavy-handed today, his initial approach was effective and enabled rapid, comprehensive and cost-efficient changes to Paris' street network—what Choay (1996) has called an urbanism of regularization, based on the interconnection of comprehensive and layered systems that fulfilled the twin objectives of circulation and hygiene.

The Haussmann event, as presented here, thus recenters the theoretical debate on the tension at the heart of the capitalist city: as an object of capture, how can housing (and the city more generally) be made accessible to all who need it, at a cost they can afford? While planning is the set of tools that is supposed to help manage this tension, it can easily fall prey to the influence of powerful interest groups or to the growth imperative (Paccoud et al. 2022). My argument in this chapter is that the lesson of Haussmann's early planning practice is political: even though his interpretation of planning law was eventually defeated in the courts, his actions show that it is possible to confront property owner capture over urban development and live to tell the tale.

References

Benevolo L (1967) The origins of modern town planning. Routledge & K. Paul, London
Benevolo L (1993) The European city. Blackwell Publishers, Oxford
Benjamin W (2002) In: Eiland H, Mclaughlin K (eds) The arcades project. Harvard University Press, Cambridge
Bonvalet C, Tugault Y (1984) Les Racines du Dépeuplement de Paris. Population (French Edn) 39(3):463–481
Bourillon F (1994) La Rénovation du Quartier Saint-Victor sous le Second Empire. Recherches Contemporaines 2:79–112
Bourillon, F. (1996) 'Rénovation 'Haussmannienne' et Ségrégation Urbaine'. In: Fourcaut A (ed) La Ville Divisée: les Ségrégations Urbaines en Question, France XVIIIe-XXe. Créaphis, Grane, pp 91–104
Bourillon F (1999) 'Changer la Ville: La Question Urbaine au Milieu du 19e siècle'. Vingtième Siècle. Revue d'Histoire 64, Numéro Spécial: Villes en Crise? 11–23
Ceaux J (1975) Rénovation Urbaine et Stratégie de Classe: Rappel de Quelques Aspects de l'Haussmannisation. Espaces Et Sociétés 13–14:19–31
Chapman JM, Chapman B (1957) The life and times of Baron Haussmann: Paris in the second empire. Weidenfeld & Nicolson, London
Choay F (1996) La Règle et le Modèle: sur la Théorie de l'Architecture et de l'Urbanisme. Editions du Seuil, Paris
Darin M (1988) 'Les Grandes Percées Urbaines du XIXe Siècle: Quatre Villes de Province', Annales. Histoire, Sciences Sociales 43(2):477–505
Deleuze G, Guattari F (1987) A thousand plateaus: capitalism and schizophrenia. In: Massumi B (trans). University of Minnesota Press, Minneapolis
Engels FR [1872] (1942) The housing question. Lawrence and Wishart, London
Faure A (2004) Spéculation et Société: les Grands Travaux à Paris au 19e Siècle. Histoire, Economie Et Société, 3:434–448
Gaillard J (2000) Paris, la Ville: 1852–1870. L'Harmattan, Paris
Harvey D (2006) Paris, capital of modernity. Routledge, London
Lacave M (1980) 'Stratégies d'Expropriation et Haussmannisation: l'Exemple de Montpellier', Annales. Économies, Sociétés, Civilisations 35(5):1011–1025
Lefebvre H [1972] (2000) Espace et Politique: Le Droit à la Ville II. Edition Economica, Paris
Loyer F (1987) Paris XIXe Siècle: L'Immeuble et la Rue. Hazan, Paris
Marx K (1933) The civil war in France [1871] (1933). Lawrence and Wishart, London
Merrifield A (2005) Metropolitan birth pangs: reflections on Lefebvre's the urban revolution. Int J Urban Reg Res 29(3):693–702
Merrifield A (2014) The new urban question. Pluto Press, London
Paccoud A (2016) Planning law, power, and practice: Haussmann in Paris (1853–1870). Plan Perspect 31(3):341–361
Paccoud A (2019) Badiou, Haussmann and Saint-Simon: Opening spaces for the state and planning between 'post-politics' and urban insurgencies. Plan Theory 18(3):339–358
Paccoud A, Hesse M, Becker T, Górczyńska M (2022) Land and the housing affordability crisis: landowner and developer strategies in Luxembourg's facilitative planning context. Hous Stud 37(10):1782–1799
Paccoud A (2023) Haussmann and the rebuilding of Paris, 1853–70. In: Mayne A (ed) The Oxford handbook on the modern history of slums. Oxford University Press, Oxford, pp 266–280
Piketty T, Postel-Vinay G, Rosenthal JL (2006) Wealth concentration in a developing economy: Paris and France, 1807–1994. American Economic Review 96(1):236–256
Pinheiro EP (2002) Europa, França e Bahia: Difusão e Adaptação de Modelos Urbanos. Universidade Federal de Bahia, Bahia
Rodgers D (2012) Haussmannization in the tropics: abject urbanism and infrastructural violence in Nicaragua. Ethnography 13(4):413–438

Roncayolo M (1983) La Production de la Ville. In: Agulhon M (ed) La Ville de l'Age Industriel: le Cycle Haussmannien. Le Seuil, Paris, pp 91–125
Schubert D, Sutcliffe A (1996) 'The 'haussmannisation' of London?: the planning and construction of kings way-Aldwych, 1889–1935'. Plann Perspect 11(2):115–144
Scott J (1982) Seeing like a state: how certain schemes to improve the human condition have failed. Yale University Press, New Haven, CT
Shapiro AL (1982) Housing reform in Paris: social space and social control. Fr Hist Stud 12(4):486–507
Smith N (1996) The new urban frontier: gentrification and the Revanchist City. Routledge, London
Sutcliffe A (1981) Towards the planned city: Germany, Britain, the United States, and France, 1780–1914. St. Martin's Press, New York
Van Loo A (1994) L'Haussmannisation de Bruxelles: la Construction des Boulevards du Centre, 1865–1880. Revue De L'art 106(106):39–49
Vernière M (1977) Les Oubliés de l'"Haussmannisation" Dakaroise. Crise du Logement Populaire et Exploitation Rationnelle des Locataires. L'espace Géographique: Régions, Environnement, Aménagement 6(1):5–23
Volait M (2003) Making Cairo modern (1870–1950): multiple models for a 'European style' urbanism, in: Nasr J, Volait M (eds) Urbanism. Imported or Exported, Chichester: Wiley.
Zeldin T (1981) Anxiety and hypocrisy. Oxford University Press, Oxford
Zola E [1871] (1999) La Curée. Gallimard, Paris

Part VIII
The Modern City

Chapter 20
Modern City: A Breaking Point in the History of Cities

Inés del Pino Martínez

Abstract The present article describes the modern city as a "breaking point" in the way of living in cities and their future growth. It signifies the end of a historical period marked by the First and Second World Wars, which immediately forced to think of solutions that had industry as an ally of change. The center of transformation was the product of eleven meetings held between 1928 and 1959 by a group of European intellectuals who founded the International Congress of Modern Architecture (CIAM), developed in periodic and itinerant meetings in ten countries. A model for the modern city was discussed, and concepts were created based on a language different from that of the past. The Athens Charter is the reference document for modern urban planning. However, each city will apply the principles with variations depending on the place. In that sense, Latin America was a laboratory of experimentation and Bogota is a case in point. The "CIAM projects" were never built, but they led to reflections on the city, developed the concepts of architecture and the modern city, thought from specialists for whom economy, efficiency and technology will be the priority in the emergence of modern urbanism.

Keywords Cities · CIAM · Le Corbusier · José Luis Sert · Urban planning · Modern city

20.1 Introduction

In the twentieth century, several crises heralded the end of one era and the beginning of another. The period of the break leaves in view some transformations that took place gradually. It is worth to mention the urban solution to liberate the city of Paris in the nineteenth century from the Haussmann plan, which was the beginning of a profound change in the city and in architecture that would continue in the first half of the twentieth century with the First and Second World Wars (1914–1954). As a result, large urban areas were destroyed, exposing their weaknesses behind the facades of

I. del Pino Martínez (✉)
Pontifical Catholic University of Ecuador, Quito, Ecuador
e-mail: idelpinom@puce.edu.ec

F. Carrión Mena and S. Rodríguez Alvarez (eds.), *From the City as a Project to the City Project*, The Urban Book Series, https://doi.org/10.1007/978-3-031-94226-6_20

the buildings in 1930 (Barsac 2020). In this sense, Marina Waisman (1995) points out that more important than continuities are the breaks or crises in history, and from these turning points, it is possible to clearly understand the flow of history, the moments in which society's ways of thinking, interests and possible future scenarios emerge. In this sense, the insertion of modern urbanism in the midst of two wars that spanned the first half of the twentieth century is a turning point that concludes with the determination of another social and political order that is visualized in the city.

The modern city proposes to re-found the existing city on its vestiges, through the planning of functional cities divided by zones, one for each human activity: living, working, recreating and circulating, with which the validity of the heterogeneous city enters into crisis as a truth sustained for centuries. The discourse of the new city, which will also be applied in architecture: "sun, space and greenery", principles that most people share, but whose application was not accepted in Europe. On the other hand, urban land acquires a differentiated cost according to the conditions of location in the territory, which leads to socio-spatial segregation. In addition, in the past, the structuring of the city was the responsibility of an enlightened elite and influential people close to power. The modern city is planned by an elite of specialists and technicians: urban planners, architects, engineers, and finally sociologists and historians, who are the mediators of reason, in other words, the management of the city and politics are strongly related.

At the center of the debate on urban transformation and architecture will be a group of European visionaries who between 1928 and 1959 created a language for the new order in urbanism, but which extends to other disciplines. This group, initially made up of 29 members, almost all born in the last decade of the nineteenth century, gathered to create alternatives to the lack of housing and to rebuild the cities destroyed by the war. They organized the International Congress of Modern Architecture (CIAM), which will be repeated every year, and then, because of the war, it will be more irregular. Each time it will be held in a different country, first to identify the problems of the destroyed cities and then to find solutions through urban and architectural projects in alliance with industry. Of this group, the most visible figure is the Swiss-French architect Charles-Édouard Jeanneret-Gris, known as Le Corbusier, who will lead the ideas of urban transformation, of a compact city with the separation of human activities, large green spaces and industrial centers on the periphery.

Le Corbusier made efforts to test his proposal in practice, however, it was not accepted by the mayors and politicians of the time. When the model was transferred to Latin America, in association with other professionals such as Paul Lester Wiener and José Luis Sert, regulatory plans were carried out that did not become reality either, however, his ideas were spread around the world, adapted to local circumstances and were the benchmark for the modern city according to the particular circumstances of each one. The city conceived by the CIAM group was criticized in its last meetings for having an individual and technocratic conception, when the city is supposed to be a collective product. It is possible that the introduction of the social and cultural component shaped new visions of the contemporary city.

20.2 The International Congresses on Modern Architecture and Their Reflections

The first meeting of CIAM was held in the castle of La Sarraz, Switzerland (1927), and the second one in Frankfurt, Germany (1928), in both meetings a common problem was noted, the significant loss of housing during the war throughout Europe, which led to the need to establish a link between the design of residential buildings and industry. The parameters were: design of high-rise apartment housing units, in horizontal property, which would allow, on the one hand, to optimize the final cost of housing, and on the other hand, the provision of centralized basic services for economic reasons. Economy in construction, product standardization, efficiency, rationality and functionality, and costs is tied to produce an efficient and profitable product.

The discussion on affordable housing under new parameters is transferred to the design and planning of the city, reiterating at the urban scale the link between the city and industry. The city must be efficient, planned, with an agile, systematic and orderly mobility. High-rise housing has a massive character, and this condition makes it visible in the urban stain of the future with an impact that affects not only its presence in the landscape but also in urban engineering. Discussions addressed the transformation of the city, architecture and industrial design, under similar aesthetic references, with the validity of new truths. In 1930, the Soviet cities were a reference for the West in terms of the organization of this type of cities and it aroused interest to listen to the solutions and projects carried out.

The third meeting was held in Brussels in 1930, the emphasis was directed toward urban and regional planning, and collective housing as a solution to reestablish the demand of that time, with the focus on economy and technology applied in construction, and the generation of a new logic based on reason that will characterize modern architecture. The experience of city planning in the Soviet Union in affinity with the separation of functions approaches was considered with interest. A diagnosis of 30 Soviet industrial cities was presented whose operation proposed separate production, housing and storage areas, reducing costs and transportation, since they were connected to railroad lines that systematically carried products from the production points to the collection and marketing points. This meant a qualitative change in the traditional agricultural production method since its reordering incorporated ideas related to the capitalist world; as expected, this idea declined in the socialist countries since they had no empathy with the policies of equity in the economic management between the countryside and the city.

Of the 11 meetings, the fourth, held in Athens in 1933, is perhaps the most representative in terms of the ideas of architecture and the modern city. It brought together 100 people aboard a cruise ship that traveled between Marseilles—Athens—Marseilles. This ship anticipates the ideas of the "machine for living" in analogy with naval architecture where nothing is superfluous, nothing is missing, everything responds to a function, space and place, and to a logic of living in a space that demands a different way of life, in which the technique accompanies a layout and operation that must be efficient for the voyage through an unknown environment such as the sea. In this artifact, the idea of space optimization is clearly identified, as well as its functionality and ornamentation, necessary for a specific function.

In this meeting, the "Athens Charter" was produced, a document that proposes the paradigm of the modern city, the charter reiterates the ideas agreed upon in previous meetings in the logic of efficiency and functionality in urban design and architecture, and the separation of human activities in "zones", a proposal with predominance of technique, reason, systematic processes, in order to increase the productivity of the worker in general through a circulation system separated according to the type of vehicle: the first, destined to heavy transport, a second, to collective transport, the third to automobiles and personal means of mobilization, and the fourth, located on the surface, would be destined to pedestrians. The purpose of urban planning is to establish a rational and functional city, designed by specialists who look at the city and its optimal functioning from a laboratory. For this meeting, 33 urban plans for European cities were presented with statistical studies and growth projections in the diagnosis, the plans propose the application of the postulates of separation of functions and their projection into the future. The Athens Charter of 1933 was adjusted and published in 1943.

However, this city was not in the logic of the common citizen of that time, accustomed to circulate in the heterogeneous city where human activities are interspersed in the same space. The craftsman had his workshop in his house, all the supplies were in the neighborhood where he was known and knew the neighbors, the neighborhood was a space of reciprocity and identity among its members. The ideas of the modern city designed by CIAM represent not only a change in the way of building houses and the city, but also a change in the way of life that had been in force up to that time. In the modern logic, the city prioritizes individuality, both in the way of inhabiting architecture and in the way of experiencing urban space.

The fifth CIAM meeting was held in Paris in 1937. An outstanding presentation was by José Luis Sert (1902–1983), who added to the discussion a theme inherited from the era of the Industrial Revolution: industrial cities and the problems derived from their functioning, for which he proposed to expand planning on at least three scales: the regional as an enveloping space of the city and the city itself. The topography and geography of the places destined for agricultural production, as well as economic and social factors.

Between the fifth and sixth meetings in 1947, held in Bridgwater, United Kingdom, there is a ten-year gap due to the coincidence with the events of the Second World War (1939–1945), the social and political turbulence, and the exile of some active CIAM members, which made it difficult to organize meetings. Thus, dealing with the reconstruction of cities was the concern that required immediate action. The group had a positive attitude toward the loss, considering it an opportunity to transform the city into a modern one. In this sense, the publication of the book *Ville Radieuse*, was an urban proposal for a million inhabitants devised by Le Corbusier, which shortly before the fifth meeting, aims to efficiently solve the housing demand. The central ideas are: to reduce the urban area by means of high-rise residential buildings; ample green spaces for contemplation, enjoyment and air purification; functional apartments in minimum dimensions optimizing the space of the rooms. For the proponents of CIAM, individual freedom is in the living space, and the outdoor green space is the place for collective life and recreation.

CIAM VII was held in Bergamo, Italy, in 1949. It is arguably the first post-war meeting. Urban plans were presented for discussion in collaboration with José Luis Sert and Paul Lester Wiener. The urban projects test applications in intermediate scale cities. The urban proposals confront different populations, topographies and natural environments: Tumaco is a coastal city on the Pacific, in southern Colombia, in a territory of abundant vegetation, tropical climate, vulnerability to a series of natural events such as floods and earthquakes; while Chimbote is located on the Pacific coast, in north-central Peru, in direct contact with the desert.

Lima and Bogota were capital cities that at the time promised growth in a short time for economic and political reasons, so they were projected for 1,500,000 inhabitants. The proposals are somewhat different from CIAM's initial approach because they integrate other elements in the civic center: a church, entertainment spaces and recover the colonial square, as well as relevant historic buildings, however, the historic area is transformed. The censuses do not have information for the same years as the project, which allowed the elaboration of approximate growth projections. Another contribution to CIAM VII was the presentation of a method to show the systematization of the data that will be used for the urban proposal called "CIAM grid". This method was designed by Le Corbusier and the French group ASCORAL in 1947 (Bright 2011).

The 1951 CIAM VIII redirected its focus with different themes such as the strengthening of the "hearts of the city" proposed by José Luis Sert, understood as historic centers or consolidated spaces of the city, where daily activity and urban density are concentrated. The organizer of this meeting was Jacqueline Tyrwhitt (1905–1983), South African urban planner and journalist. Ernesto Roggers and José Luis Sert participated in this meeting, the latter addressed the need to reverse the generalized tendency of the population to move to the periphery of cities, with the aim of: "bringing man closer to man and making people exchange ideas and be able to discuss them freely". To this end, he proposes a network of city hearts as active centers that bring together the dispersed population; he allocates the city hearts exclusively to pedestrians and the insertion of specialists from sociology in urban planning (Mumford 2007, 112).

The proposal of the "heart of the city" and the configuration of a network of "hearts" that connect consolidated urban centers shows between the lines that the radical separation of human activities is applied in the planning of ex nihilo cities, as in the case of Brasilia, but not in cities with cultural pre-existence where destruction or displacement means the loss of the symbolic value of urban spaces, and in response, the tendency to reconstitute the space of the city arises, giving continuity to the history and spaces of identity. It seems that the criteria that gave strength to CIAM's discourse in the first meetings were distorted, which led to a distancing among the members and finally to its dissolution.

CIAM IX was held in France, in the town of Aix en Provence, in 1953, which showed the separation of the team. Three groups were identified at this meeting: the first dealt with the excessive growth of some cities near industrial centers in North Africa and proposed the improvement of living conditions on the scale achieved by these cities, a proposal synthesized with the expression "habitat for the majorities". The second is identified with the disagreement of Le Corbusier and Walter Gropius about African cities. The third, led by young people, rejects the pre-war CIAM positions and reforms the ideas of the Athens Charter (Mumford 2007, 114).

CIAM X was held in Dubrovnik (former Yugoslavia) in 1956. The division of the group continued with the rejection of the division of functions proposed by CIAM in its first meetings, which was one of the innovations in the conception of urban planning. The marked differences made the end of the congresses visible. At this meeting, the Team X group assumed the restructuring of CIAM, but in reality, it was its death, which was declared at CIAM XI, held in Otterlo, Netherlands in 1959.

To close the title of CIAM's tour, it should be noted that the first five meetings were preceded by the work of the International Committee for the Resolution of Problems of Contemporary Architecture (CIRPAC), where the basic ideas were discussed in advance, for which there was an executive body that channeled the proposals to be debated in the plenary sessions, as well as the organization of the congresses themselves (Mumford 2007). Subsequently, there is no similar activity in the studies published on the subject.

20.3 CIAM Moves to America

The discourse disseminated by Le Corbusier was forceful as it quickly positioned itself in the world, with followers and opponents; everyone wanted to be modern, understanding the term as total innovation, with emphasis on light, space, and greenery, particularities that differentiate the modern city from the past. The discussion incorporated new issues in places that were not affected by the war: poverty, insufficient basic services, migration, segregation, among others. Other challenges are related to topography, geography, climate, culture, specific political, and economic interests that force to qualify the original proposal.

Le Corbusier's twelve trips to the United States and to several South American countries such as Brazil, Argentina, Uruguay, Paraguay, Peru, and Colombia attracted interest in the proposal of urban planning and modern architecture (Segre 2006). In these countries, the architect gave lectures to academics, met with politicians and influential technicians in the cities interested in learning about his projects, also met with French families and invited some young professionals to do internships in his workshop in Paris (CEDODAL 2024).

In that sense, the relationship with Latin America and the collaboration of architects from nine countries is present in the "black book" that contains the attendance list of the staff that worked in the studio on *rue de Sèvres* in Paris between 1932 and 1965, among them: Rogelio Salmona and Germán Samper from Colombia, and Augusto Tobito from Venezuela, who participated as resident architects in the construction of Chandigarh. Other collaborators of the office were: Reinaldo Valencia and Alberto Peña from Colombia. Emilio Duhart worked in the studio, and later, when he returned to Chile, he designed the Economic Commission for Latin America (ECLAC) building. Other Chilean disciples were Roberto Matta, Roberto Dávila Carson, Guillermo Julián de la Fuente. Uruguayan architects include Justino Serralta, Carlos Clemont, Carlos Gómez. From Argentina: Teodoro González, Vicente Medel, Pedro de la Mora from Mexico. Juan Kurchan and Jorge Ferrari are on staff, as well as Ernst Robert de Carvalho and Adolf Franz Heep from Brazil; Efraín Pérez Chenis from Puerto Rico and Roberto Wakeham from Peru (Quintana 2018).

The participation of young people in Le Corbusier's workshop constituted an international meeting and a school of specialization for a group of architects who in the future would be the bearers of modern thought in their countries. Thus, it could be said that the influence of the principles of modern architecture and urbanism has several forms of insertion in the world. The first, the one spread by the members of CIAM and Le Corbusier as its representative and activist. The second comes from his direct collaborators, who through their experience in the *rue de Sèvres* workshop at some point were authors of projects that followed the principles of modern architecture and the modern city with their own contributions. Another actor in the dissemination of the principles of modern urbanism are the universities and faculties in which architecture and urbanism are taught; engineering and construction techniques with which the modern movement materializes its ideas. In this way, the model expands to propose different solutions and interpretations based on a matrix that emerged in the CIAM.

So what does modern urbanism propose? It proposes to control the overcrowding of the population and the spontaneous growth of the same, it seeks to be a healthy city with the support of urban engineering and industry; organized and compact; efficient, with green space where collective living takes place. In the densification of the urban land, that means of the property, housing goes from being individual to collective in horizontal property buildings, its presence in the urban stain turns it into a leading urban component. In this conception, all the dwellings will have light and sun all year round, it is the individual shelter, with its complement, access to a green space. This idea alludes to the new image of the city and progress, and to a new monumentality (Le Corbusier 1964).

The city is expected to be efficient, and to this end, human activities are concentrated in "zones": the first one is administrative: public and private offices, financial centers, commerce. The second zone is collective housing, segregated according to the sector of the city and the type of population. By concentrating housing, it is possible to optimize the circulation of the different types of transportation, the buildings are elevated from the ground, which makes it possible for pedestrians to circulate underneath. The third zone is constituted by the green city, intended for sports and as an urban lung. The vehicular circulation is hierarchized in seven types of roads (the 7 V rule), considered as a blood and respiratory system, applied in Chandigarh. The system starts with the national highways and ends with a secondary road that surrounds the green zone of the cities, linking schools and sports centers. The bicycle route (V8) was added later.

In practice, the application of the urban model was embraced by the discourse: a city in interaction with nature: sun, space, and greenery. However, few projects were contracted, due to their high cost, to political decisions, to being a radical proposal that ignores local collective practices, and to ignoring the "spirit of place" that creates recognition and identity of a place (Norberg-Schulz 1979).

Le Corbusier disseminated the concepts of the city and modern architecture discussed at CIAM during his stays in Latin America and the United States. On each occasion he took with him an urban proposal for the city that received him and at least one architectural project that one of his contacts had requested. The cooperation between his office in Europe or the United States and teams of local technicians was the mechanism applied to carry out the Regulatory Plan of Bogota. It was a rare international experience at that time, however, this interaction allowed him to somehow evaluate the model of the universal modern city outlined in the Charter of Athens, while the local counterpart made variations that will give the urban project its roots, however, the proposal will show dualities and mismatches between the groups involved in the planning of the city.

The vision of the modern city promoted by Le Corbusier in his visits to Latin America had a convincing discourse that was immediately assumed by influential politicians and technicians who anticipate what is to come; the theory was immediately taught in universities and faculties of architecture; thus, the diffusion multiplied and was inserted among the new generation of professionals, officials and local and national planning offices. Since the transformation is radical, the discourse was applied in some cases without further reflection and without a vision of the future, individually and not in terms of community coexistence; affecting the city that hosts the project, the built heritage, public spaces and places of citizen concentration.

On the other hand, under the tutelage of Le Corbusier and his team of collaborators in Latin America, José Luis Sert and Paul Lester Wiener tested plans for cities with populations over one million inhabitants (Buenos Aires), and plans for intermediate and smaller cities. The first results were reported at CIAM VII: Lima with 595,000 inhabitants in 1940 (PNUMA, s.f.) and Chimbote, in northern Peru, 15,000 inhabitants in 1948 (Macrotrends 2024). The Colombian city of Tumaco had just 12,000 inhabitants according to the 1938 census (Gutiérrez 1948), Medellín, 358,187 inhabitants in 1951 (Ramírez 2011), and Bogotá 648,324 inhabitants in 1951 (Departamento Administrativo Nacional de Estadística 1951).

A variant discussed in the last CIAM was "the heart of the city", an idea led by José Luis Sert (1902–1983), a Catalan architect who collaborated with Le Corbusier in the last stage of the group, who made this idea his own to put it into practice. The proposal has as a reference the diagnoses made for the plans in several Latin American cities that did not live through the war; the proposal is to activate the consolidated historical urban nuclei that coincide with spaces that correspond to the foundational origin or to spaces of collective recognition in the long duration, in this case, the conformation of a "network of city hearts" makes sense.

Table 20.1 lists the capital cities of some South American countries visited by Le Corbusier, their size and complexity: Bogota, Rio de Janeiro, Montevideo and Buenos Aires, to contrast the population at the time of the study and the relationship

Table 20.1 Population of some Latin American cities between 1930 and 1962

City	Year	Population
Bogota	1930	240.356
	1953	765.360
Rio de Janeiro	1920	2717.244
	1960	6709.891
Lima	1940	595.000
	1961	1846.000
Caracas	1936	3364.347
	1961	7523.999
Montevideo	1930	655.389
	1958	989.616
Buenos Aires	1947	2981.043
	1960	2966.634

Sources CEPAL (2023), Archivo de Bogotá. Alcaldía Mayor (2019), Yepez and Marero (2021), Contreras (1994), IFEA (2015), Ministerio de Economía (1950), Fernández et al. (2024)

with their growth over a period of 25 to 30 years. The statistics have limitations since in some countries the first population census was applied in the middle of the century. The table shows that Le Corbusier chose a group of cities with the largest population and size to apply in his projects. Sert, on the other hand, took a group of intermediate and smaller cities with different objectives: international and local development, control of future growth.

In the cities with the largest populations, there is a growth that can triple in 40 years as a result of vegetative growth and immigration, due to the impulse given by the construction of public works and industrialization. It is worth thinking that the application of urban plans at the same time as the excessive population growth did not allow for an immediate adjustment of the figures, therefore, the application of the plan became obsolete.

20.4 Bogota: Two Approaches in the Same City

Urban plans in Latin America found their own solutions that in some way modified the European references, since this continent did not experience war or massive destruction. However, the twentieth century is the one of the demolitions of colonial architecture and layouts, as can be seen today in Bogota, Caracas and Lima, as examples of Andean countries in the Pacific region, whose representatives and politicians were determined since the 1920s to show progress in the city and give a renewed image of the capital cities, i.e., the interest to innovate was present in parallel with the CIAM meetings in Europe, which confirms an atmosphere of change in several places simultaneously.

The ideas of the modern city arrived in Latin America at the same time as they were being discussed in Europe. After the Athens Charter, Le Corbusier himself was in charge of disseminating and perfecting his ideas in order to obtain contracts that would make possible the application of a universal city. In 1929, he made a trip to Buenos Aires invited by the Society of Friends of Art to give a series of lectures. In the same year, he went to Sao Paulo, Rio de Janeiro, and Montevideo where he also gave lectures. The discourse of the modern city immediately gained followers in this part of the continent.

As background, Bogota is characterized by having tested at least nine urban intervention projects between 1923 and 1961; most of the proposals emphasize the improvement of vehicular circulation, the widening of roads and the incorporation of urban growth areas (Table 20.2). This concern is related to the excessive and widespread growth of the city, which in 1938 had 325,658 inhabitants and in 1950 had 638,562, i.e., it doubled its population in 12 years; this circumstance is not particular to Bogota but to several Latin American cities, as given in Table 20.1. The first interventions in Bogota in the 1930s were aligned with the trend of the "garden city" and the nineteenth century city in order to achieve fluid vehicular circulation, expand public space and connect the different neighborhoods of the city, for which the main ravines were systematically filled in Niño and Reina (2014).

The difficulty in implementing a large-scale project to regulate the growth of the city and the pressure to solve immediate problems such as the provision of municipal services constitutes a duality to be solved by the municipalities. One solution was to undertake major works in stages and with a low budget, with different scopes depending on economic circumstances and political interests; this was feasible on the main avenues. It could be said that the projects executed under this modality took longer, but they were completed. In this way, service coverage was achieved and important areas of the city were united to improve social mobility and urban growth. The interest in transforming the city is visible through cartography, urban plans, the prolongation and widening of roads, such as Carrera Décima, in a sustained process, in which each advance meant for the elites to fulfill the aspiration of showing a modern city and the image of "progress" (Niño and Reina 2014).

Le Corbusier's arrival in Bogota in 1947 was an event, not only because he was received by the mayor of the city, but also because of the idea that he was the person who could solve the problems of the city at that time. One of his first activities was to give two lectures in which he presented the principles of architecture and modern urbanism as a whole; urbanism as a science and a social computer. He brought out the principles and laws essential to life: sun, space, and greenery. His proposal created wide expectations among politicians and technicians. In 1949, he signed the contract for the Master Plan, also known as the Pilot Plan for the city, which was drawn up by Le Corbusier and legalized with decree 185 in 1951. This document contains the objectives and arguments that justify the elaboration of the plan, which establishes the limits of the city and its surroundings (Metropolitan Plan). The Regulatory Plan establishes the zones of human activity, building standards, type

Table 20.2 Proposed plans and projects for Bogota and status of implementation

Name of the plan	Data	Responsible	Observations	Application
Bogota Future (1923–1925)	Emphasis on road organization	Engineer Ricardo Olano and Enrique Uribe Ramírez, head of Public Works of Cundinamarca	Abstract and geometric plan, with a continuous grid that runs from the San Cristobal river in the south to 112th street in the north, with diagonal avenues, monuments and equipment in the squares, parks, and public spaces	It was not applied, neither did it have a function as the city's computer
Bogota's Development Plan (1930–1931)	First urban planning document of the city according to Juan Carlos del Castillo	Harland Bartholomew	Expansion of public services	Proposal
Road Plan (1936)	It proposes neighborhoods and urban intervention, based on the ideas of the nineteenth century European city of the *City Beautiful*	Karl Brunner, head of Bogotá's Urban Planning Department between 1934 and 1939, and later as a consultant to the city	Network of main roads, parks and tree-lined avenues, *Garden City* guidelines Caracas avenue appears Streets 1,6,13,22,26,34 goes diagonally and joins National Park	Proposal, implementation
Soto Bateman Plan (1944)	The city is divided into zones and uses	Mayor Jorge Soto del Corral and his Secretary of Public Works, Alfredo Bateman	Road widening, including Carrera Décima to decongest downtown, proposes two diagonal roads that reach near Plaza Bolívar	Construction of Carrera Décima

(continued)

Table 20.2 (continued)

Name of the plan	Data	Responsible	Observations	Application
Agreement 92 (1944)	Most of the Soto—Bateman proposal is retained		Define roads that will be opened or widened Only the diagonal Antonio Nariño avenue was proposed Opening of Francisco Soto avenue	It was applied
Agreement 70 (1944)	Regulates the valorization tax		Helps to concretize roads	It was applied
Plan of the Colombian Society of Architects (1945)	Aspects of vehicular circulation and transit	Colombian Society of Architects	Rationalist urbanism Continuity of Carreras 4, 7, 13 and Caracas Proposed Las Americas avenue Proposes a railroad line	Proposal
Proa Magazine Plan (1946)	Rationalist ideas, First study of the complex The ideas of the plan of the Society of Architects are collected Traditional reticle, with a single diagonal, which breaks the reticle	Proa magazine	It points out the ways to build the next 12 years The central area is looked at in detail Expansion of Carrera Cuarta and 17 Maintains Carrera Séptima Proposes Cundimarca avenue, San Francisco avenue	Awaiting technical comments and interested parties
Le Corbusier Pilot Plan (1951)	Demolish the entire downtown from Sexta street to 26th street and from Carrera Tercera to Caracas street	Le Corbusier	Industrial, massive, uniform and egalitarian urbanism, for automobiles Important roads: V-1: Cundinamarca and Centenario avenues. V-2: Libertador avenue, Caracas and Séptima	The plan was completed and submitted. It was not implemented

(continued)

Table 20.2 (continued)

Name of the plan	Data	Responsible	Observations	Application
District Pilot Plan (1957)	City established since 1954 as a Special District		The municipalities of Usaquén, Suba, Engativá, Fontibón, Bosa and Usme are integrated Classification of roads Concentric roads such as Boyacá avenue Emerges the Avenida de los Cerros, on the National Park	Not implemented
District Plan (1961)	Review of the previous Plan Road structure that the city currently has		System of radial roads and peripheral rings Regional connections Hierarchical roads from north to south and from east to west, peripheral roads	It was applied

Source Niño and Reina (2014)

of buildings, roads; and a plan for the Civic Center that contains the design of the administrative, government and financial area, as well as the public space.

His collaborators were José Luis Sert (1902–1983) and Paul Lester Wiener (1895–1967), members of CIAM, who emigrated to America due to inconveniences during the wars in Europe. Since 1945, they had been working on urban plans from their office: "Town Planning Associates (TPA)" in New York. Paul Schulz also collaborated with them in supporting projects of strategic importance for the United States in South American cities (Fundación arquitectura y ciudad 2021). In this context, they developed plans for other Colombian cities that are currently documented in specialized magazines: Medellín (Schnitter 2004; Ospina de los Ríos 2021), Cali (Espinosa 2006), Tumaco (Goossens 2014); in Peru: Lima (Ludeña 2002), Chimbote (Facho and Asto 2016); among others.

The study of Bogota was carried out between 1947 and 1953, a complex period in the political arena in Colombia marked by the assassination of Jorge Eliecer Gaitan, candidate for the presidency of Colombia, in one of the most violent episodes in history: the "Bogotazo" in 1946, followed by short-lived governments. Nevertheless, the technical team continued with the contract. The first stage was to develop a Pilot

Plan that had a local counterpart called the Bogota Regulatory Plan Office (OPRB) formed in 1947 by a technical team from the municipality that provided the required technical information and conducted preliminary studies for the development of the initial sketches of the Pilot Plan for the first stage of work.

The studies were carried out in coordination between the Bogota, New York and Paris offices. In the second stage, 1949, Le Corbusier made a new trip to Bogota to discuss the ideas elaborated in a preliminary sketch, make technical decisions in conjunction with the Bogota office and propose the stages of future work. The third stage was the development of the Pilot Plan elaborated by Sert and Wiener in coordination with Le Corbusier; the document made adjustments requested by the OPRB and incorporated the need for connection with the world through an airport. The fourth stage was undertaken by Sert and Wiener with detailed plans of the Civic Center, the development of appropriate legislation for its implementation.

The final delivery of the Plan for Bogota in 1950 first outlines the general policies and actions in the city and its radius of influence in the Metropolitan Area, thereby changing the scale of Bogota. Then, the Regulatory Plan proposes the zoning of the city and its future projection. According to this document, the city maintains its location in a north-south direction, following the hillsides to the east; and a green space barrier to contain urban growth to the west, that is, the savannah. It divides the city into zones, densifying the city by means of high-rise housing buildings with different characteristics depending on the sector of the city. Part of the Regulatory Plan is the development of the Civic Center, located in the historic area of Bogota, which concentrates government architecture, financial buildings, and commerce. Sert and Wiener's proposal modified Le Corbusier's proposal, which caused problems. The handover included the legislation and financing of the plan.

Le Corbusier's idea for the Civic Center was to create a new, open, illuminated administrative area, in which the citizen could walk short distances between buildings and green spaces without the need to use a vehicle. The proposal is radical, the civic center concentrates the functions of the national and local government. The ministries are concentrated in a 15-story building, as is the municipal building. Apparently, this caused their interest to decline and they turned their efforts to Chandigard, but the plan was nevertheless submitted by the group. Apparently the final project did not include some plans with the master's ideas.

Sert's Civic Center proposal moves away from the initial idea and recovers the square, the street and the 100 × 100 m block, and in this unit, the dwellings of up to four stories introduce greenery into the internal courtyards. The planners also create an intermediate or buffer area between the Civic Center and the surrounding areas. Monumental buildings are planned for government buildings and financial institutions. In other words, the center is not radically transformed, but the public space close to the collective memory is maintained. This idea is in line with what Sert proposed since the meeting in Hoddesdon (England) as the "heart of the city", and the plan for Bogota is the opportunity to develop it. The plans of the proposal allow to understand that the blocks have in their interior pedestrian walkways that communicate with those of other blocks, where the commercial and entertainment activity maintains the life of this urban space. In the civic center the spaces of the city are reconciled with the vegetation and the existing extensions at that time, that is to say, it is a more conservative proposal. For Sert and Wiener, the civic center is the fifth urban function, in addition to the four human activities proposed by CIAM.

The delivery of the project coincided with the change from the conservative government of President Roberto Urdaneta Arbeláez (1951–1953) to the military government of Gustavo Rojas Pinilla (1953–1957), who took power after a coup d'état. In this context, the political interests and priorities of the Colombian city changed and the plan was not executed. Some drawings and plans generated by the Master Plan (1950), some plans worked on by the OPDB, and others derived from the project by Le Corbusier, Sert and Wiener are compiled in the "Historical Atlas of Bogota" (Cuéllar and Mejía 2007) and suggest that the period from 1947 to 1953 was one of great intensity of collective work, discussion, and rupture to achieve the theoretical ideas for a modern international city with a city project that considers the place and somehow maintains its local identity.

Criticism of the plan was permanent and of a different nature, the first, that the plan did not represent the feelings of the population, since it was carried out in the United States. Although in reality, the coordination was international, between the offices of the municipality in Bogota, the consultants in New York, and Le Corbusier's workshop in Paris. The second inconformity was from the real estate developers of the clandestine urbanizations located in the periphery of the city, and which were inventoried in the 1950 plan of Bogota made by the engineer Martínez (2007; Hernández 2017).

20.5 As a Closing

The historical facts of the first half of the twentieth century show that in less than fifty years the war in Europe put in crisis the way of life of the entire population of that continent with singular effects in the world. The destruction and loss marked a before and after in several aspects: technological innovation, changes in the aesthetics of things and the idea of healthy cities with the integration of the vehicle as a system that dynamizes the city.

The restitution of the state had as a proposal to reestablish at the same time the cities and architecture with a radical transformation with the support of technology and reduced economic resources. Throughout this process, reason and efficiency were imposed to achieve the expected results. This profound change can be called a point of rupture in history and culture; it involved radical positions such as thinking about the re-foundation of the city and architecture, in terms of economy, efficiency and systematization of processes, without considering the pre-existing values of the culture.

Despite the loss and the crisis, the CIAM discussion was an urban laboratory that might not have been possible under other conditions. The solutions and circumstances varied according to the place and the political decisions taken by the governments. The experience in Latin America had other characteristics, there were no wars, but there were natural or social and political events unrelated to the wars in Europe. The reception of the discussions at CIAM arrived almost immediately in Latin America where there was the necessary openness to receive innovative ideas for their cities. The arrival of Le Corbusier accelerated an expectation of change toward modernity, which had already arrived in some cities, but at a slow pace since the beginning of the century.

Thus, the ideas of the luminous, open, green, austere, and healthy modern city had to accommodate to other variables such as cultural pre-existences, different natural environments, and discontinuous historical processes; marked social differences that are understood under different logics according to the regions of the world. This circumstance forced Sert and Wiener to rethink the modern city devised by CIAM on American soil, resulting in a city with its own characteristics, and for this reason, the transformation was not radical. The plan was finalized, but not implemented, however, the intellectual environment was left with the possibilities that could be experienced after the discussion of the plan with possibilities of application in the future. It is worth asking what would Bogota have been like if the final version of the plan of Le Corbusier, Sert, Wiener and the OPRB was applied in practice?

On the other hand, the temporal breaking point was deep, since it involved the creation of a new language to denominate architectural and city objects, given that the concept of "style" applied to architecture as heir to the history of art disappeared. In this context, it is the first time that a historical period generates objects simultaneously with the gestation of language. In the modern period, which identifies and qualifies the qualities of objects, the architectural orders, the traditional square and the open space disappear; housing is reconfigured, the absence of ornament is present in modern aesthetics and the emphasis is on the quality of space: lighting, sunlight and hygiene, large garden or forest spaces. With this, the need arises to invent terms to name the architecture and the city represented in the drawings of the projects.

References

Archivo de Bogotá. Alcaldía Mayor (2019) Archivos a fondo. https://archivobogota.secretariageneral.gov.co

Barsac J (Director) (2020) Le Corbusier [Film]. Cine Service Technique; Antenne 2, la S.E.P.T., Mision Cable, Channel Four, RTSR

Bright P (2011) Comunicación y enseñanza de la arquitectura. Revistas.utadeo.edu.co, 83–87

CEDODAL (2024, enero 10) Le Corbusier en el río de la Plata, 1929. [Video]. YouTube. https://www.youtube.com/watch?v=b1Ct8IS7ixw

CEPAL (2023) CEPALSTADT: Bases de datos, publicaciones y estadística [Informe regional]. CEPAL-ONU

Contreras C (1994). Sobre los orígenes de la explosión demográfica en el Perú: 1876–1940 [Investigación]. IEP Ediciones

Cuéllar M, Mejía G (2007) Atlas histórico de Bogotá: Cartografía 1791–2007. Cargraphics

Departamento Administrativo Nacional de Estadística (1951) Censo de población de Colombia-1951. DANE

Espinosa L (2006, 30 de octubre) El Plan Piloto de Cali. Dialnet. https://dialnet.unirioja.es/descarga/articulo/4015058.pdf
Facho A, Asto R (2016, 8 de junio) Chimbote, entre la utopía y la informalidad: Parte 2: El Plan Regulador. Habitar. https://habitar-arq.blogspot.com/2016/06/chimbote-entre-la-utopia-y-la_8.html
Fernández T, Wilkin A, Biramontes T (2024, 18 de enero). ¿Localidades informales, excedentarias o segregadas? La heterogeneidad urbana en Uruguay según el censo de 1963. Territorios 46:1–28
Fundación Arquitectura y Ciudad (2021, 30 de mayo) Visitas memorables. https://fundaayc.com/2021/05/30/visitas-memorables-17/
Goossens M (2014, 14 de julio) Jorge Gaitán Cortés y la introducción del urbanismo moderno en Colombia. Redalyc. https://www.redalyc.org/pdf/3416/341633874017.pdf
Gutiérrez H (1948) Informe geológico sobre la isla del Morro. Bahía de Tumaco. Departamento de Nariño. Servicio Geológico Nacional (SGC-Colombia)
Hernández C (2017) De Le Corbusier a Sert en el Plan Director de Bogotá (1947–1953). https://repositorio.unal.edu.co/handle/unal/29255
IFEA (2015) Las barriadas de Lima. IFEA
Le Corbusier (1964) Los tres establecimientos humanos. Poseidón
Ludeña W (2002) Lima: Poder, centro y centralidad. Revista Eure 28(83):45–65
Macrotrends (2024, 15 de julio) Chimbote population. https://www.macrotrends.net/global-metrics/cities/22070/chimbote/population
Martínez J (2007) Algunas parcelaciones clandestinas en los alrededores de Bogotá. In: Mejía M, Cuellar G (eds) Atlas histórico de Bogotá. Atlas Histórico de Bogotá, pp 108–109
Ministerio de Economía (1950) Primer censo de población del Ecuador 1950: Volumen único. Ministerio de Economía. Primer_Censo_de_Pob_del_Ecuador_1950_Volumen_Unico.pdf
Mumford E (2007) El discurso del CIAM sobre el urbanismo, 1928–1960. Bitácora 11:96–115
Niño C, Reina S (2014) La carrera de la modernidad: Construcción de la carrera décima en Bogotá (1945–1960). Segunda Edición
Norberg-Schulz C (1979) Genius loci: Towards a phenomenology of architecture. Rizzoli
Ospina de los Ríos N (2021) Todo en uno: Un conjunto urbano en el centro de Medellín. Repositorio Universidad Nacional de Colombia. https://repositorio.unal.edu.co/handle/unal/80629?show=full
PNUMA (s.f.) Evolución de la población de Lima y Callao. http://www.pnuma.org/deat1/PDF's/GEO%20Ciudades/GEO%20Lima%20y%20Callao/61-70.pdf
Quintana I (2018) Hijos de la rue de Sèvres: Colaboradores latinoamericanos de Le Corbusier en París (1932–1965). Universidad de los Andes
Ramírez S (2011) Cuando Antioquia se volvió Medellín, 1905–1950. Anuario colombiano de historia, 37, 67–89. Revistas UNAL
Real Academia de la Historia (s.f.) José Luis Sert López. https://dbe.rah.es/biografias/8201/jose-luis-sert-lopez
Schnitter P (2004) Wiener and Sert's pilot plan for Medellín: contract and presentation. http://www-etsav.upc.es/personals/iphs2004/pdf/194_p.pdf
Segre R (2006, agosto) Cuerpo, naturaleza y abstracción. Café de las Ciudades. https://cafedelasciudades.com.ar/articulos/le-corbusier-los-viajes-al-nuevo-mundo/
Urbanopolis (s.f.) Plan Voisin y la Ville Radieuse de Le Corbusier
Waisman M (1995) La arquitectura descentrada. Escala
Yepez B, Marero G (2021) El siglo XX en Venezuela: Una revolución demográfica [Ensayo]. PRODAVINCI

Chapter 21
From IAPI to Ceilândia: Occupation and Dislodgement in the Construction of Brasília

Maribel Aliaga Fuentes, Benny Schvarsberg, Ana Elisa Carnaúba, and Júlia Bianchi

Abstract Across the memories of the construction of Brasília, federal capital and modern heritage, the process of urban space development has commonly been portrayed by dualities: the planned and disorder, the modern and the archaic, erudite and popular or order and disorder. These contrasts, when juxtaposed with everyday experience, reveal that these dichotomies permeate a fundamental conception; the need for shelter and the right to land, especially for the dispossessed, poor, people of color, indigenous people and quilombolas. This article proposes to retell this story, based on the juxtaposition of documents, clippings from periodicals of the time, images and narratives.

Keywords Brasília · Pilot plan · Migration · Informal settlements · Social exclusion

This article is part of the research "Occupations and Evictions" by @amarelinhaobservatorio to which the researchers are linked. The work aims to investigate the capital through other approaches, and in this text, we have the contribution of professor Dr. Benny Schvarsberg.

M. A. Fuentes (✉) · B. Schvarsberg · A. E. Carnaúba · J. Bianchi
University of Brasilia, Brasilia, Brazil
e-mail: marialiaga@unb.br

B. Schvarsberg
e-mail: benny@unb.br

A. E. Carnaúba
e-mail: ana.carnauba@aluno.unb.br

J. Bianchi
e-mail: julia.bianchi@aluno.unb.br

F. Carrión Mena and S. Rodríguez Alvarez (eds.), *From the City as a Project to the City Project*, The Urban Book Series, https://doi.org/10.1007/978-3-031-94226-6_21

21.1 Introduction

From IAPI to Ceilândia, the title of this work uses narratives and historical documents to retell the stories of migration, occupation and evictions; important elements, however, erased from the capital's historiography. The aim here is to show how an apparently informal process is controlled and governed by the State, maintaining a large population of manual workers on the city's margins.

This work is the result of some investigations about Brasília, both from the point of view of urban planning and also from the professor's research. Dr. Benny Schvarsberg, such as the research on women and cities, carried out by Observatório Amarelinha, more recently the research "Occupations and evictions" which proposes to investigate the capital through new approaches, that go beyond traditional historiographical issues that make the vulnerable invisible and, mainly, erase female existences in the construction of the capital.

The investigations aim to understand the origins of exclusion and territorial division based on imaginary borders that repeat the country's urbanization process. Starting with migratory differentiation, while engineers and technicians are called pioneers, manual construction workers are candangos, and the Plano Piloto and the modernity of the new capital do not belong to them. These are left with margins, improvisation, occupations and later evictions.

Our subject of study is the construction of Brasília, looked at and rewritten now with new characters and narratives. We intend to show that the modern city carries within itself the colonial and slaver genesis, in which only some have rights to the city, while the others, those who built it, only have the red dust of the cerrado left. A situation that continues to this day in the Brazilian capital, which has the highest GDP in the country. As a general objective, we seek to understand the initial occupations and evictions offered to the population as an opportunity to own land. How the health and environmental argument is used in official discourse to justify the mass displacement of workers and, at the same time, allows the middle class to occupy the same area that was called environmental protection. To rewrite part of this history, we worked with historical documents, newspaper and magazine clippings, public and family images, as well as oral reports. All this material is made available to us like a kaleidoscope that through the common thread of occupation and eviction is superimposed and juxtaposed forming the text presented here. By understanding that Brasília is still a relatively new study context, and with material not always accessible to other languages, we made an effort to seek references for other and new research than ours and, as far as possible, we provided access to other researchers.

21.2 The Construction

In the 1950s, Brazil was a predominantly archaic and conservative country, attempting to break away from its agrarian past and embrace modernity. One of the developmental strategies was to bring the capital to the center of the country, flattening the Cerrado,[1] in pursuit of this modernity. Implementation, earthmoving and infrastructure efforts began with the inauguration of President Juscelino Kubitschek in 1957, and was soon called for an international public competition[2] for the city's project design. There was an international jury, and 27 proposals submitted, by architects from Brazil and abroad, and the winning project was by architect and urban planner Lúcio Costa, which blended bossa with modernity through grand avenues, rationality and sectorized division of spaces—this proposal was called Report for Plano Piloto.[3]

Brasília, the capital of hope, as it was known during its construction, relied on a strong discourse, both to persuade the intellectual elite, especially from Rio de Janeiro, proud of its status as national capital, and to attract the working class, which would come from all corners of this vast and poor country. For the former, there were benefits and incentives; for the latter, there was the fallacy of belonging. This utopian discourse contrasted with reality, making it clear that there were limits within the Plano Piloto, a control over the profile of the residents who would settle there. On a territorial scale, few housing policies were aimed at the working class. Given the demand for labor to build the city, there was uncontrolled migration, for which the project had no response. The modern way of living implemented in Brasília, once it did not incorporate the workers, proved to be selective and segregating, repeating the situation in which the country still lived in and overflowed, with the misery of the vast majority of the population.

In the quest for housing, informal occupations began to appear on the margins of construction sites, remaining there after the completion of the buildings. These occupations were characterized by precariousness and unhealthy conditions, but also by community and cultural bonds, forming truc communities. The scenario of these occupations, classified as favelas, did little to reflect the image of modern discourse. Soon, the State, although the majority holder of public lands, began to implement housing policies based on eviction action and removals, referred to in official reports as "transfers". In this context, there were significant investments in migration control policies, the lack of housing supply, as well as campaigns to eradicate informal

[1] Cerrado is the brazilian ecoregion present in the region. It refers to a vast tropical savanna f, characterized by a diverse landscape of grasslands, shrubs, and scattered trees. The Cerrado is renowned for its rich biodiversity, containing a wide array of flora and fauna species, many of which are endemic to the region.

[2] Lucio Costa's winning project flirts with this mysticism, with the idea of this profound Brazil that is modernizing. His proposal, a sketch summarized in two large simple axes that manifest the beginning of a new chapter of history. More about the projects can be found at Brasília Invisíveis: https://artsandculture.google.com/story/6wWBc1VBYOpuKw?hl=pt-BR.

[3] The full text of the winning proposal by Lucio Costa, the Report for Plano Piloto, can be access at: https://www.ipatrimonio.org/brasilia-plano-piloto/.

occupations, reallocating construction workers and their families, and all those who had no place in the planned space, to peripheral regions outside the Plano Piloto."

21.2.1 Meta 31: A Summary of Modernity

Candidate for president Juscelino Kubitschek took part in a rally in the city of Jataí, Goiás, on April 4, 1955. Among the people, a member of the public interrupted him with a question: since the candidate talked so much about complying with the Constitution, would he comply with the article which envisaged the move of the capital to the Central Plateau? It was Toniquinho from Farmácia, who was then 28 years old. Juscelino hesitated for a moment, but made the commitment.[4]

In the mid-1950s, the then presidential candidate Juscelino Kubitschek—JK, during a rally in the city of Jataí, took advantage of the question made by Toniquinho da Farmácia, to declare that he would fulfill the constitution and move the capital to the center of the country. The idea of reallocating the capital of Brazil was not new; it had already been present on the political horizon since the late nineteenth century, with the mapping expeditions of the central plateau carried out by the Cruls Mission.[5] However, it was the economic transformations and the cultural impact of the post-World War II era that were fundamental in boosting the country forward, as it was still stuck in its colonial past, and was looking for the path to modernity. This transformation finds one of its greatest symbols in the construction of Brasília.

JK, the elected president, assumed office in 1956 and launched the Plano de Metas program,[6] which aimed to improve Brazil's infrastructure, with the intention of encouraging the country's transformation through industrialization. Upon taking office on January 31, 1956, he incorporated into his discourse the slogan '50 in 5', i.e., fifty years of progress in five years of government, seeking to detach from the outdated and underdeveloped past that until then permeated state policy. There were 30 development goals, to which the 31st goal, known as the "meta-synthesis", was added, envisioning the construction of Brasília and the transfer of the federal capital (Fig. 21.1).

To accomplish this feat, JK called in a team of experts to implement the city's infrastructure. The team included important figures such as Oscar Niemeyer, who proposed holding an international ideas competition. Many entered the competition, and 27 teams submitted their proposals, with the winning project design by architect

---

[4] Travel rally by then-candidate Juscelino Kubitschek to publicize his Target Plan: http://memorialdademocracia.com.br/card/construcao-de-brasilia.

[5] Expedition carried out to evaluate the Central Plateau of Brazil, led by Belgian engineer Louis Ferdinand Cruls. These expeditions aimed to find the right location for the new capital of Brazil, to find out more about the mission read: From the Cruls quadrangle to the historical and cultural heritage of humanity: https://www.senado.gov.br/noticias/especiais/brasilia50anos/not02.asp#:~:text=A%20chamada%20Miss%C3%A3o%20Cruls%20demarcou,the%20same%20nominated%20powder%20Varnhagen.

[6] Goal Plan: https://jk.cpdoc.fgv.br/fatos-eventos/plano-de-metas.

Fig. 21.1 Brasília and its centrality-distance from Brasília to each state capital. Embroidered reinterpretation of the map that appears in James Holston's book, The Modernist City. *Source* Ana Elisa Carnaúba-embroidered on raw cotton fabric

and urban planner Lucio Costa. The suddenness of the event and the modernist ideology itself erased the memories of the pre-existing territory, an erasure justified by the birth of a new city. "Lúcio Costa presents the problem of founding a capital as if there were no history. He does not justify the Plano Piloto as the result of a consideration of Brazilian historical conditions or a history of architectural ideas. In fact, he dehistoricizes the problem, presenting it in terms of a foundation myth, made by divine inspiration" (Holston et al. 1993, p. 71).

With the Plano Piloto for Brasília (Fig. 21.2), Lucio Costa established himself as an urban planner, and for the main projects the new city, a name requested and already known by JK during his term as mayor of Belo Horizonte, was Oscar Niemeyer, who had carried out important works in the Pampulha complex between 1940 and 1945. With Niemeyer's work already changing the requirements of architecture in

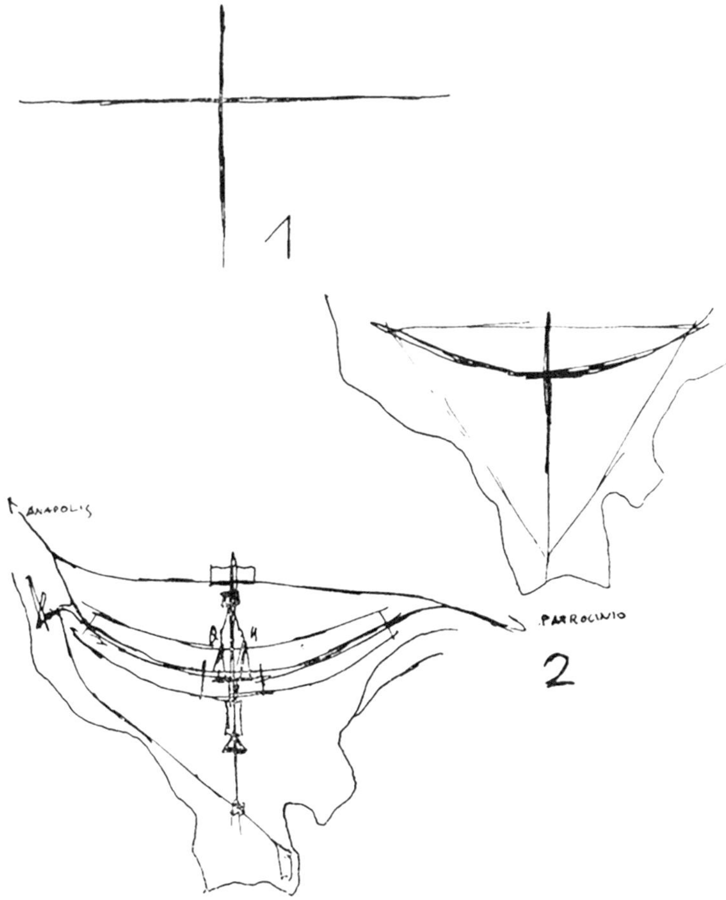

Fig. 21.2 Lucio Costa's sketch for Brasília. *Source* Brasília Pilot Plan Report (2018, p. 29)

the country, nothing was more promising than to highlight him at the center of the territory that promised the newest Brazil. This dynamic of modern architecture that was taking place in the country "would, therefore, be proof of the technical development achieved by Brazil, which in the context of developmental nationalism, could no longer be identified as an agricultural country, a supplier of raw materials" (Espada and Gautherot 2016, p. 31).

21.2.2 The Promised Land

The movement toward the construction of a new capital drew both internal and external attention, with advertisements spreading the news across the country. The

Fig. 21.3 Workers in the construction of Brasília. *Source* Collection of the Carnaúba Family

discourse surrounding the construction and its celebrated call enchanted a significant portion of the vulnerable population in need of new beginnings and better living conditions. The unknown appeared to be attractive and abundant, nurturing hope and encouraging thousands of people to move their roots to lands that were still empty but already delineated by layouts that promised a new meaning to the vigor and bond with the belief of a new country (Fig. 21.3). As a result, a new migration process begins; this displacement early in the construction phase centralizes the diaspora toward the middle of the Brazilian territory.

The first workers arriving assume an important role and are exalted as heroes of the nation for their strength and faith in a project that required much devotion and dedication from those who arrived seeking to establish a new life filled with new opportunities. In this first moment of Brasília's history, the worker redefines the designation of the "candango",[7] which previously had a negative connotation, and

[7] In the book "Os Filhos dos Candangos" by Edson Beú, there is an excerpt that explains that construction workers, construction workers, were known as ''candangos". At the time there was this distinction, "pioneer" was an honorific title, while the word "candango" had a derogatory meaning. Trunk explains that in its "original meanings: bad, ordinary, villain, extracted from Kimbundukangundu, according to the dictionary Aurelio. It was used by inhabitants in southwestern Angola to attack the colonizers of that continent. When being trafficked to the northeastern sugarcane region, the Africans maintained the pejorative meaning to refer to the Portuguese masters and, later, to the Brazilians themselves. Over time, however, the depreciation target was reversed. The word became synonymous with cafuzo, mestizo, mameluco and black. Later, its meaning was expanded, starting to generically cover the poor populations of the interior of the country, especially itinerant

begins to receive notable recognition. The "candangos" are recognized by the nation as those who left their native states behind to fight and build a new country. There is also a desire within them to contribute, through their work, to history and materialize their effort so that it would be embodied in the country's mythical narrative.

21.3 IAPI: Between the Formal and Informal City

At first, men arrived alone and settled in accommodation provided by construction companies and also in informal settlements surrounding the Plano Piloto. They came from different places, but primarily from the north and northeast of the country, territories punished by drought and underdevelopment. Beú highlights that in 1959, a year before the inauguration, "already there were 56,953 inhabitants in the city under construction. It is known, from the testimony of the candangos themselves, that the flow of northeastern migrants was influenced by the severe drought that struck the region at the time" (Beú 2013, p. 48).

At the beginning, workers could not bring their families; in other words, the rules were clear and established that there was no availability to make these people stay, a city that was already born without any housing planning for those who worked in its construction and arrived here in search of opportunities. It is worth remembering that Lucio, in his Memorial Report about the city, states that the highest officials would live alongside the simplest workers in nearby areas, only differentiating materials, finishes and construction standards.

With the abundance of work, human displacement to the capital took on greater proportions, as Brasília was consolidating, and the families of workers and more builders arrived. There was no planning to receive them, and occupations within and around the Plano Piloto grew. As a result, territorial conflicts of informal occupations also grew, worrying those who wanted to keep Lucio Costa's project immaculate and protect the Plano Piloto. Therefore, there was a constant practice of removals that occurred while the city was still under construction. On one hand, Lucio Costa's Report for Plano Piloto, despite being succinct, described how differences in standard from one block to another would be neutralized by the proposed urban plan, to which everyone had a right, differing only according to the needs of each individual or each family. His inclusive speech regarding the right to the city proposed for the central zone contrasted to what was imagined for the surrounding areas. As he stated, "in this sense, it must be prevented from enclaving favelas both in the urban and rural periphery. It is up to the Urbanization Company to promote decent and economical accommodations for the entire population within the proposed scheme" (Costa 1991). It was clear that in the Plano Piloto, there was a limit, control and profile of residents who would occupy the new capital of Brazil. If in the Plano Piloto everything was planned, on the other hand, there was no preparation for the migratory outburst that

workers, with low professional qualifications, a predominant situation among those who arrived at the construction site of the future capital".

Brasília began to experience, revealing the situation of poverty in which the country still lived and overflowed.

Informal settlements also grew around the Plano Piloto as Brasília was being built. Among these informal centers that marked the city's history we have the Vila do IAPI (Instituto de Aposentadoria e Pensões dos Industriários), located behind the hospital of Brasília, an area that now corresponds to the Museu Vivo da Memória Candanga. The territory was situated alongside the Cidade Livre, now called Núcleo Bandeirante and Candangolândia. In addition to its proximity to the end of Asa Sul, it was also adjacent to Brasília Airport (Fig. 21.4).

Similarly to IAPI, many other informal settlements were consolidated during this short period of time, the housing deficit in the new capital would extend for a long time and, up to this day, directly impacts the peripheral areas of the city. The camp was also used by the public authorities to shelter the homeless who were within the limits of the Plano Piloto. This population was taken to the IAPI camp, where they were registered and assigned to occupy a plot of land, despite being informal, the camp

Fig. 21.4 IAPI, aerial view. *Source* Secretariat of Social Communication—Public Archive of the Federal District

had streets and lots divided by NOVACAP itself,[8] as Epstein describes, this action by public authorities aimed to have an understanding of the arriving population.

The truck containing the household effects and the boards and roof tiles from the shack would proceed at high speed to the Social Security Invasion, where a Social Service Foundation office, housed in a small shack, was the first stop. Here the foundation's representative made a record of the name, number of dependents, identity documents, and the like, of the head of the household. Any civil document—identity card, birth certificate, elector's registration card, social security employment record or veteran's identification—was accepted. Many squatters had no documents at all (Epstein 1943, p. 121).

IAPI quickly consolidated with other camps in its neighborhood. Contiguous to the IAPI were the Vila Tenório, Vila Esperança, Vila Sayão, Placa da Mercedes, Morro do Querosene and Curral das Éguas. The shacks were built of wood, a temporary material leftover from the construction, and what is understood is that there was no clear and distinct division between the other occupations, only delineations with street names. Epstein will describe the location of IAPI and state that there was a main road known as Avenida Principal, which was a wide street adjacent to the hospital access road. In addition, it had public buildings for community use, revealing it to be a village that possessed a certain autonomy. The author, describes:

The landmarks along the length of the Main Street are mostly the larger buildings used for some public purpose. There are three schools, one near the gully known as the Interplanetary School in impartial tribute to the sputniks and satellites, the two others—located in the middle of the street, which widens to accommodate them-provided by the Education Secretariat. All are of wood. Further above them is a church of the Pentecostal Assemblies of God, with a Potemkin-village or movie-set facade facing down the hill, proving to be false only when one passes alongside the church to see that the building itself is not much higher than any other. Behind the church are the small post maintained by the Social Service Foundation, which was staffed during the early part of our stay, and a police post which was usually empty or provided a place for NOVACAP work crews charged with reassembling shacks moved to the invasion to carry on conversations when (as frequently happened) they were not working (Epstein 1943, p. 108) (Fig. 21.5).

The shared memories of people who arrived with similar purposes and found themselves facing the same difficulties of living in still precarious conditions, generated an energy of communion, in terms of community, of neighborhood. In these dynamics it was possible to relearn the new provisional and unforeseen way of living, far from the bases and structures they knew they had before they began to risk their hopes in a city that was beginning to be built.

Therefore, the daily dynamic between the community exactly reflects the spirit of unity in which the informal occupations lived. Migration was becoming increasingly

[8] The Companhia Urbanizadora da Nova Capital do Brasil (Novacap) was created by law, on September 19, 1956, by the then president of Brazil, Juscelino Kubitschek de Oliveira. The sole purpose was to manage and coordinate the construction of the new Capital of Brazil. To learn more about the company and the construction of Brasília, visit: https://www.arpdf.df.gov.br/revista-brasilia/.

Fig. 21.5 Removal of shacks from IAPI toward Ceilândia. *Source* Public Archive of the Federal District

larger, an, at a given moment, the neighborhood began to grow and accommodated itself to the ways of life that precariousness and lack of public policies were offering. Of all the ambiguities and mysticisms that Brasília was shaped, the issue of housing is the most evident, since its entire consolidation process was due to the strong migration of workers[9] and families who arrived to work and occupied the peripheral areas of the city.

21.3.1 The City Is One

The builders of Brasília and their followers thought that, by planning a city, they would be eliminating the traditional conflicts of our cities. The logic of real estate

[9] BEÚ, narrates that in the beginning "workers of simple origin, the construction workers, were elevated to the status of "national heroes", as symbols of the strength, faith and capacity for achievement of a people, a situation in which the "pioneers' 'would try include themselves, also calling themselves "candangos" (Beu 2013. p. 32).

capital and socio-spatial segregation are structuring elements of this conflict, whether in the planned city or not; because of this, planning techniques must start from conflict and not its denial (Paviani 1991, p. 81).

On April 14, 1970, the first article was reported in Correio Braziliense about the attempt to erase the IAPI, with the following headline: "IAPI invasion will disappear" (Fig. 21.6). The article tells how the eradication process that was already taking place in other parts of the territory was turning its attention to the favela. The public authorities were developing a plan to change the satellite cities and this plan would be sent to governor Hélio Prates da Silveira. The Social Service Secretariat would be in charge of carrying out a survey of the favela population that lived there. This eradication process began to be constantly publicized in the media, with the aim of making the population aware that measures were being presented and plans were being approved, so that the situation the Village was going through and all its consequences would come to an end. But this speech obviously brings even more conflicts, contradictory feelings such as hope and despair among the population.

"The city is one" was the catchphrase used in the campaign to eradicate favelas. In addition to being constantly publicized by newspapers as an incentive for the population's support, there is also publicity in schools, so that children can sing and spread the withdrawal process faster. On July 15, 1970, an editorial in Correio discusses the slogan "The city is one"[10] (Fig. 21.7) and reports how Brasília is seen as a cold city, but that everyone is mistaken, it actually cares so much about the residents and those who are there that it even suffers from this campaign in a "purely humanitarian sense, to support the favela residents". The propaganda needed support and donations, so several events were held to raise funds for this project, including building materials, clothes, school supplies, cash donations and everything that could be of help to the families who would be eradicated from their temporary homes.

IAPI became the largest favela in the Federal District and, in addition to its proximity to the Plano Piloto, it revealed the greatest of all contradictions in Brasília, planning and the lack thereof. Over time, there was a reversal of reverence for the workers, who went from being portrayed as candangos to being considered invaders, undesirable of that space. Therefore, a series of removal measures took place and, with the social problems in IAPI, it became the focus of a quick solution so that it does not spill over into the margins of the planned city. These tensions and withdrawals happened constantly, while other distance-based plans are being approved for these people. One of the main justifications for the removal of people from Vila do IAPI is the discourse of the sanitary ring-environmental and water protection that Brasília had in its plan (Fig. 21.8). With this, "the valued land expels the favela dweller, hence their condition of 'half-citizen', of 'incomplete citizen', to whom they evaded full rights and freedom" (Paviani 1991, p. 220).

To achieve its objectives, the eradication campaign reinforces stereotypes, thus the IAPI is constantly reported in local news, revealing it to be a dangerous, dirty, and

[10] This public eradication campaign and propaganda tools are portrayed in the film "A cidade é uma só" (The City Is One) by Adirley Queirós: A cidade é uma só [Internet]. 2013 [cited February 16, 2024]. Available at: https://www.vitrinefilmes.com.br/filme/a-cidade-e-uma-so/.

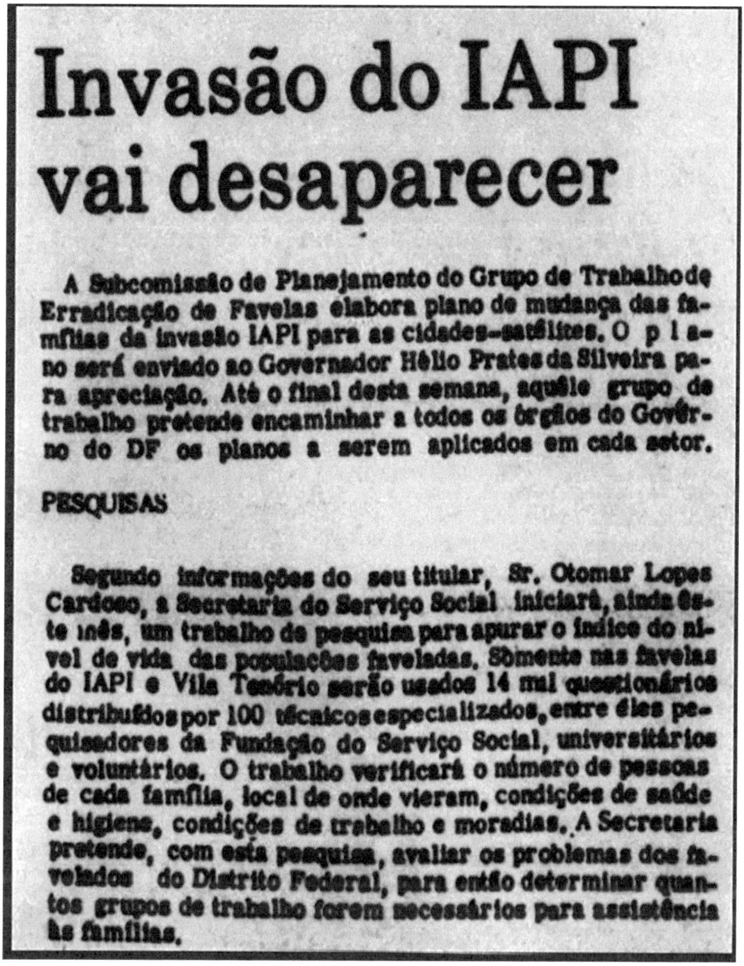

Invasão do IAPI vai desaparecer

A Subcomissão de Planejamento do Grupo de Trabalho de Erradicação de Favelas elabora plano de mudança das famílias da invasão IAPI para as cidades-satélites. O plano será enviado ao Governador Hélio Prates da Silveira para apreciação. Até o final desta semana, aquêle grupo de trabalho pretende encaminhar a todos os órgãos do Govêrno do DF os planos a serem aplicados em cada setor.

PESQUISAS

Segundo informações do seu titular, Sr. Otomar Lopes Cardoso, a Secretaria do Serviço Social iniciará, ainda êste mês, um trabalho de pesquisa para apurar o índice do nível de vida das populações faveladas. Sòmente nas favelas do IAPI e Vila Tenório serão usados 14 mil questionários distribuídos por 100 técnicos especializados, entre êles pesquisadores da Fundação do Serviço Social, universitários e voluntários. O trabalho verificará o número de pessoas de cada família, local de onde vieram, condições de saúde e higiene, condições de trabalho e moradias. A Secretaria pretende, com esta pesquisa, avaliar os problemas dos favelados do Distrito Federal, para então determinar quantos grupos de trabalho forem necessários para assistência às famílias.

Fig. 21.6 The IAPI invasion will disappear. *Source* Correio Braziliense, April 14, 1970

precarious place. The perspective that these articles bring is always about a place of fear, deemed unsuitable for a new city. To exemplify this, shortly after the beginning of the first withdrawals, in March 1971, there is a report in the Correio Braziliense on June 13 of the same year that discusses the difference between the IAPI and Ceilândia. The opening lines of the text say:

Promiscuity. Mud. Crimes. Dirty children. Barefoot children. Children without school. No health post. No recreation. Lack of security. Of water. Of light. Of sewage. Of hospitals. Of living space. Misery. In a few strokes, here is a portrait of the so-called "invasions", especially the largest of them, the IAPI Invasion. Not only for tourists, the favelas of the Federal District are a surprise. There are many people who live here and do not know the full extent of the misery of the invasions. For a large

EDITORIAL

A cidade é uma só

Brasília vem sendo acusada, no correr dos tempos, de ser uma cidade fria. Criou-se o mito da falta de calor humano. A lenda ganhou foros de verdade e é pesquisada sèriamente por estudiosos, como é prova a pergunta feita ao Governador por um estagiário da Escola Superior de Guerra. De tanto se falar na algidez de Brasília até parece que seus Palácios não são de mármore mas feitos de blocos de gêlo.

No entanto, Brasília não é nenhuma ilha polar, refratária ao calor. Recebe-o diretamente do sol, que no Planalto tem uma luminosidade mais ativa e fecundante e irradia-o através de sua gente que, por ser uma amálgama de brasilidade, é calorosamente humana.

A campanha da erradicação das invasões, que se constituiu em ponto de honra do Govêrno do Distrito Federal e em meta primeira da Secretaria dos Serviços Sociais poderá servir de ponto de partida para os estudiosos preocupados com a falta de calor humano de Brasília, levando-os, do caminho da sociologia aos do folclore e à separação da realidade da lenda.

A eliminação das invasões, se tem como causas necessidades administrativas, possui, concomitantemente, bases profundamente humanas. À defesa do anel sanitário, à preservação das águas de consumo e de recreação, à obediência à ordenação do planejamento, junta-se a necessidade de se promover o homem à sua condição, mormente numa cidade que surgiu para redimí-lo através da irradiação do progresso em tôdas as suas formas.

Nesta cidade acusada de falta de calor humano, a campanha oficial de erradicação de invasões deu origem a uma outra campanha, de sentido puramente humanitário, de amparo ao favelado. As condições subumanas em que vive a grande maioria dos invasores, quando divulgadas e conhecidas, começou a formar "aquela corrente" de que fala o hino da Copa, apenas com a diferença de que não é de entusiasmo cívico mas de entusiástica solidariedade.

Comandando essa campanha, a Sra. Vera Prates da Silveira integrou-se na comunidade brasiliense de um modo todo peculiar.

Tôda a pessoa que muda seu domicílio sente, no princípio, uma certa hostilidade do meio-ambiente. A falta das tradicionais amizades, e a natural dificuldade de formar novos amigos; a certeza de encontrar alguém em determinado lugar que passa a inexistir; a necessidade de aquisição de novos hábitos e a de enjeitar os antigos, tornam o período de adaptação tedioso, aqui em Brasília ou em qualquer outra cidade.

Para a espôsa de um Governador, o caminho natural seria a convivência com as pessoas de sua categoria social. Por certo, ela as frequenta. Mas a efetiva integração na comunidade brasiliense só a alcançou no momento em que resolveu enfrentar, corajosamente, a miséria que assola milhares de lares (se barraco é lar!) e dezenas de milhares de sêres humanos, procurando levar-lhes, não o auxílio passageiro da esmola, mas a base que sirva para a reconstrução de suas vidas em moldes dignos e estáveis.

Nessa campanha se deve engajar a cidade, pois "a cidade é uma só". Uma só na monumentalidade de seus Palácios, mas também uma só na vergonha de suas invasões; uma só na inigualável beleza arquitetônica e na funcionalidade de seu urbanismo, mas também uma só no primarismo quase pré-histórico de suas favelas; uma só no confôrto de suas superquadras, mas também uma só na miséria de seus barracos.

Com sua iniciativa, a Sra. Vera Prates da Silveira retesou as cordas humanas de Brasília, nela se integrando pela porta do sentimento humano. A resposta da cidade também só pode ser uma só, demonstrando que não lhe falta calor humano. A voz da Primeira Dama não será eco a clamar no deserto.

Fig. 21.7 The city is one. *Source* Correio Braziliense, July 15, 1970

Fig. 21.8 Territory is a right (Please do not show here the Brazil we don't want to see). *Source* Ana Elisa Carnaúba—embroidered on raw cotton fabric

part of the population, the IAPI Invasion is nothing more than an ugly landscape at the entrance to the most beautiful Capital in the world, a pile of shacks on the margins of the Plano Piloto, something more or less uncomfortable to explain about.[11]

Despite the official discourse proposing eviction as the only solution, on July 25, 1970, the headline "Unemployment is low in the IAPI invasion" in the Correio Braziliense brings an article revealing the results of a survey conducted by the Secretariat of Social Services, which showed that the majority of the population in the favela were employed workers, with a very low number of unemployed individuals. These data were collected during the campaign process to estimate the existing population that would be removed from the favela. There was still uncertainty about where these people from the invasion would be taken. Initially, some of them were transferred to Taguatinga, a satellite city that had existed since 1958, despite Lucio Costa acknowledging in the Pilot Plan Report that satellite cities would be a solution for urban expansion "after" the occupation of the Pilot Plan. However, they emerged even before or simultaneously with its initial implementation.

21.3.2 Ceilândia

The transfer from IAPI to Ceilândia began on March 27, 1971. According to Correio, the first families who arrived were welcomed with a Thanksgiving Service, alluding to the first event in Brasília, which was the inaugural mass. It was as if both events

[11] IAPI x CEILÂNDIA. Como a dignidade humana é construída. Correio Braziliense. June 13, 1971.

sought blessings for the beginning of a city. This was also a way to prevent major protests from those who were arriving and finding themselves in a place with no infrastructure to receive them. Regarding these removal stages, Correio Braziliense reports:

The scheme for removing families initially foresees an average of 20 shacks per day, which will be gradually increased until reaching 90 shacks per day. Families are notified in advance and a team of social workers studies social cases, while an ambulance from the Hospital Foundation remains at the removal work in case of any emergency, along with a team from the Public Security Secretariat and the Military Police.[12]

In the historical context of this removal process, in addition to the supervision of the Public Security Secretariat and the Military Police to avoid protests and consequently rebellions, there is also the scenario of the Military Dictatorship. The entire brutal process of removing these residents was covered up by false assistance that these people would receive in the new territory. One day before the change, Correio Braziliense announced that the transfer would be made following a detailed scheme from the Public Services Secretariat and concludes with:

The Invasion Eradication Campaign-CEI will help families in the new location, both with regard to material aid, as well as social guidance, so that they can consolidate as quickly as possible in your new place of residence.[13]

The population was transported in trucks, carrying all the belongings that remained after the demolition of their old shacks. When the residents arrived in Ceilândia, they were supposed to rebuild their homes, and in a way, their lives, in a space where there was nothing, only small marks on the ground delimiting their new lots. Dissatisfaction was widespread as Ceilândia lacked any urban infrastructure. Without water, sewage, and sanitation, they realized that the promises made were far from being fulfilled. For those women who used to wash their clothes near IAPI, dealing with the situation of being in a new city where there was no water was challenging. The distance from the Plano Piloto generated even more precariousness, as many residents worked in the city center and had to deal with transportation costs to get to work.

(…) the change of families caused a significant reduction in their income, due to the distance of the new nucleus in relation to the employment center (at the time, the Pilot Plan) and mainly due to the cost of transport, which practically prevented the participation of minor children and women in the composition of family income, which is basically supported by the meager earnings of the head of the family, generally perceived as construction workers. The vast majority of women were prevented from carrying out their activities, such as washerwomen (one of the professions most practiced by people from the old Vila), who found themselves unable to work, as Ceilândia, in the early days, did not had water (Gouvêa 1995, p. 92) (Fig. 21.9).

This extensive removal process will last approximately eight months and, when the last shack leaves IAPI, the Correio of November 9, 1971 brings the article saying that: "Satisfied, its owners, about ten trucks left the site, which will now be transformed

[12] Correio Braziliense of March 27, 1971—Removal of the invasion begins today.

[13] Correio Braziliense of November 9, 1971—IAPI invasion ends: until last shack.

Fig. 21.9 Arrival in Ceilândia. *Source* Public Archive of the Federal District

into a large forest." In the corner of the newspaper you can see an image of Governor Prates da Silveira with a hoe digging the soil and planting a forest tree. This gesture is an attempt to give new meaning and consequently erase the history of this territory.

21.4 After the City

The removal process of vulnerable populations, supported by the official discourse of defending a sanitary cord, is actually an ethnic, social and racial cleansing, where some have more right to the city than others. More than sixty years later, the territory of IAPI, which was intended to be a green belt and a sanitary ring around the Plano Piloto, now consists of irregular condominium developments for the middle class.

The former occupants of IAPI found only the red dust of the territory in Ceilândia. They understood very early on that the city is one only for those who remained in the Plano Piloto. The displaced were relocated 30 km from Plano Piloto, in the middle of nowhere with no urban infrastructure, no water, no electricity. But they were promised a subdivided plot to convince them that they would have their property there. Ceilândia took a long time to establish itself as a city, and today it is one of the largest cities within greater Brasília, with about 400 thousand inhabitants, including within its territory the largest favela in the country, Sol Nascente (Fig. 21.10).

Returning to memories and headlines about the deliberate disappearance of a favela highlights how unwanted bodies in certain territories are assaulted and symbolically erased from the historiography of the city, with a selection of memory shaping heritage. In a way, the population living in informal settlements subverted their ways of living, adapting to the precariousness, requiring new adjustments in their way of housing. It was not synonymous with comfort, let alone choice. As Brasília grows in

Fig. 21.10 Satellite maps of IAPI in 1964, 1975, 2007 and 2021. *Source* Geoportal, Seduh/GDF

its discourse and as a cradle of modernity, it has an aversion to any account that recalls the past but ignores the very current Brazil that overflowed through construction sites and city outskirts.

After the city, there is more city, another unplanned city that renews and expands continuously. Ironically and irreverently to the architect's design, Brasília does not end with the construction of the Brazilian capital.

References

Beú E (2013) The children of the candangos: Brasília from the perspective of the periphery. Brasília, DF: Editora UnB, University of Brasília, 143 p

Costa L (1991) Relatório do Plano Piloto de Brasília [1957]. In: RELATÓRIO do Plano Piloto de Brasília. GDF, Brasília, p 38

Democracy Memorial [Internet] [cited February 16, 2024]. Democracy Memorial—"I will change the capital", promises Juscelino. http://memorialdademocracia.com.br/card/jk-percorre-o-brasil-duas-vezes

Epstein DG (1943) The city sprouted: the rise of Brasília [Internet]. University of California Press, California. https://www.jstor.org/stable/26924964

Espada H, Gautherot M (2016) Monumentality and shadow: the civic center of Brasília by Marcel Gautherot, 1st ed. Annablume, São Paulo, SP, Brazil, 173 p

Federal AP do D. Revista Brasília [Internet]. Public Archive of the Federal District [cited February 16, 2024]. https://www.arpdf.df.gov.br/revista-brasilia/

Google Arts & Culture [Internet] [cited February 16, 2024]. Invisible Brasílias: meet the finalists of the Plano Piloto competition. https://artsandculture.google.com/story/brasílias-invisíveis-conheça-os-finalistas-do-concurso-do-plano-piloto/6wWBc1VBYOpuKw
Gouvêa LADC (1995) Brasília: the capital of segregation and social control: an assessment of government action in the area of housing. 1a. ed. Annablume, São Paulo, SP, Brazil, 154 p
Holston J, Coelho M, Holston J (1993) The modernist city: a critique of Brasília and its utopia. Companhia das Letras, São Paulo, 362 p
http://memorialdademocracia.com.br/card/construcao-de-brasilia. Construction of Brasília [Internet] [cited September 30, 2019]. http://memorialdademocracia.com.br/card/construcao-de-brasilia
IPHAN, organizer (2014) Brasília pilot plan report [Internet], 3rd ed. Secretariat of State and Culture of the Federal District, Brasília. https://www.ipatrimonio.org/brasilia-plano-piloto/
JK CPDOC [Internet] (2013) Goals plan [cited February 16, 2024]. https://jk.cpdoc.fgv.br/fatos-eventos/plano-de-metas
Kubitschek J (2000) Why I built Brasília [Internet]. Federal Senate (Brazil 500 Years Collection). https://www2.senado.leg.br/bdsf/handle/id/1039
Paviani A (Organizer) (1991) The conquest of the city: popular movements in Brasília. Editora UnB, Brasília, DF, 266 p
Public Archive of the Federal District (Federal District, Brazil) (1991) Companhia do Desenvolvimento do Planalto Central, Distrito Federal (Brazil), organizers. Brasília Pilot Plan Report. Brasília: The Government, 75 p
SOUSA NHBin (1996) Taguatinga: a candanga story. In: Paviani A, Schvarsberg B (organizers) Brasília, housing and exclusion. Ed. UnB, Brasília, DF (Brasília Collection)

Newspaper Articles

IAPI invasion ends: until the last shack. Correio Braziliense [Internet]. March 27, 1971 [cited February 20, 2024]. https://memoria.bn.br/DocReader/DocReader.aspx?bib=028274_02&pesq=%22IAPI%22&pasta=ano%20197&hf=memoria.bn.br&pagfis=16479
IAPI x CEYLAND, how human dignity is built. Correio Braziliense [Internet]. June 13, 1971 [cited February 20, 2024]. https://memoria.bn.br/docreader/DocReader.aspx?bib=028274_02&pesq=%22IAPI%22&pagfis=12050
Removal of the invasion begins today. Correio Braziliense [Internet]. March 27, 1971 [cited February 20, 2024]. https://memoria.bn.br/DocReader/DocReader.aspx?bib=028274_02&pesq=%22IAPI%22&pasta=again%20197&hf=memoria.bn.br&pagfis=9866

Part IX
The Socialist City

Chapter 22
The Soviet and East European City During State Socialism: An Overview

Yulia Shaffer, Žaklina Grgić, and Sonia Hirt

Abstract State socialism, first established in Russia following the Bolshevik Revolution of 1917, emerged as a formidable rival to capitalism for over 70 years. The state socialist system was characterized by distinct ideological principles, an alternative economic structure, centralized political power, and a unique approach to civic life. During the height of state socialism in the USSR and Eastern Europe, urban planning was reimagined and implemented in a manner that reflected the regime's key goals and values. Although this political and economic system has long since receded in this part of the world, the cities it shaped continue to stand as enduring legacies of that era. In this article, we offer a comprehensive general overview of urban planning and urbanism under state socialism in Eastern Europe in general and the Soviet Union in particular, examining the socialist city during that era both as an evolving process and as a deliberate project of the regime.

Keywords State socialism · Eastern Europe · Soviet Union · Sotsgorod · Microrayon

22.1 Introduction

State socialism was established first in Russia after the Bolshevik Revolution of 1917 and came to dominate much of the globe after World War II in the footsteps of the Soviet victory over Nazi Germany. Definitions of "state socialism" vary, but there is a broad consensus that the term implies a distinct socio-political order where the public sector (the state) dominates large segments of economic, political, and civic life (e.g.,

Y. Shaffer (✉) · Ž. Grgić · S. Hirt
University of Georgia, Athens, GA, USA
e-mail: yulia.fadina@uga.edu

Ž. Grgić
e-mail: zgrgic@uga.edu

S. Hirt
e-mail: sonia.hirt@uga.edu

F. Carrión Mena and S. Rodríguez Alvarez (eds.), *From the City as a Project to the City Project*, The Urban Book Series, https://doi.org/10.1007/978-3-031-94226-6_22

Kornai 2021). At some point, the world's most populated countries, China and India, were considered socialist too, as did many other nations in Africa, Asia, and Latin America. It is generally agreed upon that the state socialist period in Europe and the former USSR ended in 1989–1992 and that these societies have moved toward different versions of neoliberal capitalism since then. China is still formally socialist, although its economy has long been a mixture of socialism and capitalism. Several small and mid-sized states, such as Cuba and Venezuela, are considered socialist to this very day. Still, the once-mighty Soviet-led socialist system has long ceased to be a powerful geopolitical alternative to capitalism as it was during the Cold War.

Any discussion of socialism and socialist cities must consider the significant variation of socialism that existed in the twentieth century. Soviet socialism was quite different from Chinese socialism. Socialist Yugoslavia was vastly different from socialist Albania, Bulgaria, or Romania, despite their geographic proximity. All socialist countries shared some basic common features: the public sector in socialist societies yielded extraordinary powers exceeding those in capitalist countries; a single party, the communist party, ran all politics, although on occasion, communist parties co-existed with others that served as nominal opposition; the national economies were centrally planned; and governments sought to reduce, if not totally eliminate, the class conflicts that characterize the capitalist world (Kornai 2021). But within these broad parameters, there was plenty of distinction. There was more political freedom in some countries than in others. The USSR fully dominated some, whereas others, like Albania and Romania, acquired some geopolitical independence. Yugoslavia, which first broke with the USSR in 1948, became one of the founding members and leaders of the Non-Aligned Movement in 1961. In contrast, neighboring Bulgaria remained a reliable Soviet satellite. Private actors could own land, property, and businesses in the more liberal states, such as Hungary, Poland, and especially Yugoslavia, but much less so in the USSR. Some orthodox socialist states, like Bulgaria, had remarkably high homeownership rates; others, like Hungary, although more liberal, had much lower homeownership rates. In this essay, we cannot address the complex variety of socialisms; hence, we will try to cover some of the shared general characteristics of socialist cities in Europe and the former USSR (especially its European part), leaving other parts of the world to other authors.

The spatial product of state socialism—the "socialist city"—is a perfect example of a city conceived as a "project." Socialist regimes espoused an environmentally deterministic philosophy; they believed that the physical environment is a decisive factor in shaping social relations and individuals' characters (Kopp 1970). Few other twentieth-century political regimes invested so much energy and effort into transforming urban forms as did socialist governments. By designing and building massive public spaces—parks, boulevards, plazas, civic buildings, cultural centers— socialist governments hoped to teach their citizenry the value of the public realm and the common good. They hoped that the "socialist city" would cultivate an ideal, new "socialist man" and "socialist woman," who would be selflessly dedicated to serving their beloved countries and countrymen (Hirt 2012). It is an understatement to say that the grand socialist utopia did not succeed.

Within the ongoing discourse on the relevance of the term "socialist city," or "sotsgorod" as the Russian term was, two basic viewpoints still duel. Scholars aligned with the "ecological school" of thought assert that the socialist city represented but an iteration of modern-day industry-driven urbanism. In this interpretation, industrialization was seen as the key independent variable, whereas urbanization was seen as the dependent variable (Szelenyi 1996). Because the key independent variable, industrialization, was shared by socialist and capitalist regimes, the dependent variable, the city, was not fundamentally different in contemporaneous countries, whether socialist or capitalist. Furthermore, some aspects of socialist urbanism appeared to have existed only on paper. For instance, socialist authorities were invariably proud that their large cities exhibited fewer signs of poverty, marginality, and income- and ethnicity-based segregation. However, socialist cities were never truly "classless," and neither was socialist society. Hence, authors like Smith (1996) posit that the socialist–capitalist contrasts in urbanism may have been more conceptual than tangible.

The alternative interpretation of the socialist city follows from the ideas of the "historical school" of thought. This school has two wings: neo-Marxist and neo-Weberian (Szelenyi 1996): the former emphasizes the importance of the political economy, and the latter, the importance of institutions and ideologies. One way or the other, the two wings of the historical school maintain that whether because of the uniqueness of the socialist political economy or because of the uniqueness of the socialist guiding philosophy (these two can be deemed independent variables), the dependent variable, the city, was principally different in socialist versus capitalist settings (e.g., Hirt 2013; Zarecor 2018).

Regardless of the debate, there is consensus that state socialism did make a notable difference in urbanism and that the socialist city of Eastern Europe, and the USSR was at least somewhat distinct from the prevailing American and West European models. This is the line of argument that this essay will seek to defend. The essay is organized as follows. We first discuss the processes that led to the production of urban forms under socialist auspices. Second, we highlight the results of these processes—the physical features of the socialist city. We end with some observations on the current lives of formerly socialist cities.

22.2 The Socialist City as Process

22.2.1 State Socialism as a Governance System Alternative to Capitalism

So, what was state socialism, and how did it impact urban development over time? State socialism is different from other forms of modern-day governance, such as classical liberalism, libertarian socialism, and state capitalism. Under state socialism, as noted already, the state controlled nearly all aspects of economic and political life, hence the common description of socialist societies as "totalitarian states." The state owned most of the means of production and guided nearly all planning and building of cities. In the former Soviet Union and most East European nations of that period, concepts like civil society, democracy, or private property had little meaning (Andrusz 1984; Zarecor 2018). The most obvious exception was Yugoslavia, where the economy operated under a less centralized model referred to as the Yugoslav "self-management model" (Estrin 2010). In contrast to state socialism, in all other governance systems mentioned above, the government has a much more limited involvement in the economy, leaving it primarily to private forces (Chomsky 2005). The socialist ideology was based on the teachings of Karl Marx and Friedrich Engels and further developed by Vladimir Lenin, the leader of the Russian Communist Party and the Bolsheviks. After the October Revolution of 1917, Lenin's Bolsheviks abolished the Imperial Russian government and took power, propelling one of the grandest political experiments of all time.

The early Soviet period started with establishing state ownership of urban land, major real estate, and the means of production, in a process known as nationalization. This enormous undertaking pertained to all large-scale industries (e.g., coal, metallurgical, oil, chemical, engineering, textile, food, etc.), and all private banks, railways, foreign trade, and the merchant fleet (see A Commission of the C.C. of the C.P.S.U. (B.) 1939). More specifically, in 1921, Lenin put forward his theory-heavy New Economic Policy (NEP) which, however, did not envision total nationalization. Following Lenin's death, when Joseph Stalin became General Secretary of the Soviet Communist Party in 1924, he phased out Lenin's NEP and introduced the first statewide Five-Year Economic Plan. This plan called for further nationalization, heavy industrialization, and the collectivization of all private peasant farms and all agricultural production (Andrusz 1984; Meyerovich et al. 2019). In Stalin's belief, the collectivization of farms was necessary because "industrial development could only proceed by replacing small-scale agricultural commodity production with large-scale production and the direct appropriation of surplus product by the state" (Harrison 1980, p. 58).

With urban land, real estate, and large means of production in the hands of the state apparatus, the USSR attempted to organize the entire economic life of their massive country through central planning. Soviet planning was top-to-bottom and very comprehensive. It was meant to ensure the rational allocation of national resources and distribute them efficiently and equitably, so that workers' exploitation and class conflict are eliminated and all people—workers, farmers, and intellectuals—would have similar access to what they need (Hausladen 1987).

Nationalization policies were adopted in all socialist countries as they formed the Eastern Bloc after World War II. However, the extent of nationalization policies differed significantly by country. Whereas in all countries some types of private property were permitted (the socialist term for this was "personal property" and it was understood that such property might include apartments, dachas, cars, and other material items for household consumption), in the most orthodox states like Bulgaria, the limitations on private property were strict. Yet, Poland never fully nationalized agricultural land and agricultural enterprises and Hungary was famous for its "goulash communism"—a version of state socialism where small-scale private ownership and initiative co-existed within the broad socialist framework (e.g., Bartha et al. 2023).The nationalization campaigns, which were often carried out violently, were seen as necessary to dismantle the exploitive, unjust capitalistic order. Yet over time socialism developed its own exploitations, injustices, and inequalities. And so, while authoritarian, the socialist states never became fully equalitarian.

22.2.2 *Political Ideology as a Driving Force in Planning*

Ideology stands out as one of the most significant differentiating factors affecting the development process in socialist versus capitalist cities (Borchert 1981; Crowley and Reid 2002). The design and planning of cities, just like art, architecture, and literature, were powerful instruments through which the socialist state sought to impose its values and will (e.g., Jenkins 2012). A modest private realm (as in the relatively small apartments that made the post-WWII mass-housing estates) and a generous public realm (as in magnificent parks and plazas) were perceived as catalysts for promoting socialist collectivism over the petty but destructive private individualism that socialist ideologues believed capitalist countries were regretfully dominated by (Andrusz 1984; Smith 1996).

Since cities were the seats of industry and therefore homes of the proletariat, they were considered settings of the highest form of social life. There, in cities, a socialist consciousness was supposed to grow, the evils of competitive capitalism were to be eliminated, and an inherent unity of people was supposed to develop (Fisher 1962; French and Hamilton 1979). The belief that the urban environment is capable of changing individual character and transforming society was certainly not unique to socialist societies (Hausladen 1987) but appears to have been more dominant there than anywhere else in the world (Kopp 1970; Clark 1985; Crowley and Reid 2002; Engel 2022). As Czepczynski (2008) points out, socialist cities were socialist not just because they were developed under that regime, but because socialist ideology was rooted in all levels and scales of the urban development process.

22.2.3 *Primacy of National Economic Planning*

Another distinguishing feature of the socialist system was the supreme role given to national economic planning. As noted earlier, socialist leaders believed that central planning and management were key to achieving efficient production and equitable distribution of resources. To achieve this goal, the main instrument was the 5-year national economic plan, mentioned earlier, dating back to 1928 in the Soviet Union and the late 1940s in the rest of the socialist world. These plans covered all sectors of national industrial development and were subsequently taken to all other layers of government (republican, provincial, municipal) to be implemented and expanded upon in detail to fit the particular local circumstances. However, economic plans were frequently hindered by inconsistencies, contradictions, false forecasts, and false reports (Zarecor 2018). Initially, central economic planning may have helped the Soviet Union, especially after the Revolution and the Civil War and, also, after the destruction endured by World War II, but these plans were considered highly ineffective in the late socialist period. They were obstacles to innovation and may have greatly contributed to the economic stagnation and the eventual demise of the entire European state socialist system (Roberts and LaFolette 1990; Kiss 2024).

At the municipal level, the main planning tool was the 30-year-long master plan or general (comprehensive) plan (GenPlan in Russian). This type of plan was also in wide use in capitalist countries through the twentieth century, but there were several principal differences between the master plans developed in socialist versus capitalist settings. First, the socialist urban master plan was but the local, spatial translation of the national economic goals. Since urban land belonged to the state, land for urban development was administratively allocated, rather than sold to private parties as in free-market economies (Bertaud and Renaud 1997). The state had the power to determine the location, the pace, and the form of urban growth (Hausladen 1987; French and Hamilton 1979). The planning system followed a strict, vertically hierarchical order, with all major goals coming from the national capital and dictating the terms of local and regional development (Hess and Metspalu 2019).

Once a municipality received its national directives for economic growth, it would then engage in calculating its future population, future factories, future housing, future parks, etc., so that these national directives could be met. Room for local rethinking of national objectives was limited. Planning was mostly a discourse between state economists and local architects. Public participation in formulating the plan's goals and objectives was largely absent. And, because the socialist state-owned urban land and production means, it acted as both the planner and the developer of urban spaces (Enyedi 1996; Hirt 2005). The greatest exception to this top-down economic model was Yugoslavia, where the constituent republics had a substantial level of autonomy in decision-making and the central governing premise, institutionalized since the 1950s, was that the various industries were not owned and managed by the central republican apparatus but by the workers employed in the factories themselves. Local authorities in Yugoslavia also had a greater say in urban master planning (Estrin 2010).

22.2.4 *The Soviet City: A Simplified History*

In this section, we provide greater detail on the main historic phases of planning the socialist city. To narrow down this broad topic, we will focus on cities in the Soviet Union: the country that invented state socialism as a governance system and stayed socialist a bit longer than its East European neighbors. Three main periods can be distinguished: the Leninist-Stalinist era (1918–1953), the Khruschev-Brejnev era

(1953–1986), and the brief Gorbachev's Perestroika era that concluded the Soviet experiment (1987–1991) (Engel 2022).

The years after the October Revolution and the Russian Civil War saw brave theoretical explorations into the ideal socialist city (Engel 2022). There were two opposing schools of thought– the urbanists and the dis-urbanists (Gobova 2024). Urbanists advocated for dense, small, self- sufficient urban settlements near production sites, while dis-urbanists promoted the idea of decentralization, advocating for residential settlements to be placed far from industries. This concept resembled Ebenezer Howard's Garden Cities idea, which gained popularity during that time in Europe, the United States, and elsewhere due to growing public dissatisfaction with the effects of industrialization on cities (Howard 2003; Gobova 2024). During this phase, there was little construction due to the scarcity of resources and the novelty of the ideas. Still, much utopian and visionary planning was undertaken (Engel 2022). Some Western star-architects, most famously Le Corbusier, found the Soviet Union an especially welcoming place for monumental grand designs which, fortunately or not, remained mostly on paper (Mumford 2009). This is how Soviet architects and urbanists from the 1930s articulated the overarching goals of planning for socialist urbanism: *(1) to correct the ills inherited from the era of capitalism; (2) to develop a [wholly] new pattern for the city which will indicate clearly the inherent unity of the people, the classlessness of the society*" (Fisher 1962, p. 252).

The Stalinist period, which lasted until Stalin's death in 1953, was characterized by large-scale industrialization of existing cities, and by the planning and building of many new industrial towns (we will revisit these at the end of this section). Following the massive devastation of Soviet cities by the Nazi armies, Stalin faced the urgent imperative of revitalizing and expanding the industrial base. This was hardly the time to invent wholly new types of cities, so the utopian streak of Soviet planning gradually faded. Industry, as well as monumental ceremonial projects celebrating the state's power, became the top priority. Resources for adequate housing remained limited (French and Hamilton 1979). A key aspect of this planning approach involved the deliberate spatial separation of industrial and residential zones, often accomplished by implementing green belts or buffers while maintaining proximity to minimize commute times and distances to places of work (Milyutin 1974; Collins and Alex 2021). Service amenities were strategically distributed among neighborhoods to fulfill daily needs, with more specialized services concentrated in the city center. This hierarchical organization, predicated on the principle of providing all citizens with equitable access to essential goods and services while minimizing travel time and effort, whether by foot or public transportation, reflected the socialist vision for a just society (French and Hamilton 1979). This period was also marked by ambitious master plans, the beautification of streets, the construction of grand public spaces, and the establishment of architectural landmarks, frequently crafted in the Greco-Roman style—a style that came to be known as Stalin's neo-classicism or, perhaps strangely, "socialist realism" (Gobova 2024). In summary, Ford (2013) contends that plan-making during that era exhibited four main elements: (1) a cohesive urban layout, (2) imposing structures, (3) efficient and versatile urban transportation, and (4) organized neighborhoods. This approach was driven both by the state's desire to display its power and by its commitment to a socialist ideology that emphasized equality (Adams 2013).

The second phase spans the most stable and prosperous decades of Soviet history—the rules of Nikita Khrushchev (1954–1963) and Leonid Brejnev (1964–1982). This was the first Soviet period during which housing renovation and construction became a top priority. In the 1950s and 60s, new housing came mostly as urban infill in medium-sized apartment buildings that came to be known as "Khrushevki." Later on, with the application of industrialized methods of construction, new housing was built on green-fields in the urban peripheries in the form of prefabricated high-density housing blocks—the signature Soviet "microrayon" or the core neighborhood unit of the socialist city (Logan 2020; Engel 2022). During the Khrushchev era, the architectural paradigm shifted from monumental neo-classicism to minimalist modernism (Logan 2020; Engel 2022). Khrushchev's ideological stance was that architectural embellishments were extravagant, wasteful, and unfit for socialist societies (van der Straeten and Petrova 2022). During the Brejnev era, Soviet modernism devolved into an even more austere and Spartan version of its formerly creative self, giving the Soviet microrayons a rather depressing look (Lizon 1996). Priority was given to fast and inexpensive construction methods. These methods ensured that citizens of the mircorayons had similar, standardized living spaces, conveying the official message of socialist equalitarianism (Fisher 1962). Regrettably, they typically produced environments of subpar quality and esthetics, at least compared to older, traditional neighborhoods.

By the 1980s, almost half of the urban population of socialist countries resided in microrayons (e.g., Mezentsev et al. 2024). With all their esthetic deficiencies, the microrayons came close to achieving some of the central promises of socialist ideologues. The state, through its many subsidiaries, distributed the apartments so that households of very different social and ethnic backgrounds would routinely end up living in direct proximity. The Soviet city was, hence, less socially segregated than its Western counterparts (Engel 2022). Although the individual apartment units were modest by Western standards, the green spaces between the large, standardized apartment buildings were generously sized to reflect the primacy of the socialist public realm. They were also well equipped to provide the diverse citizens of the microrayons with sufficient opportunities to play, rest, and socialize (Mezentsev et al. 2024). Public services were plentiful and always included kindergartens and schools in strategic locations. This had much to do with the fact that Soviet women—unlike women in capitalist societies—were expected to be an important part of the workforce. This is not to say that Soviet women achieved equal political power to men. Yet, they certainly participated in the economy more fully than women of the Western world during the same time period (e.g., Lapidus 1978).

Throughout the socialist period but especially through the 1930s, 50s, 60s, and 70s, the USSR undertook a massive campaign of constructing new towns over their colossal territory (Bertaud and Renaud 1997; Kotkin 1995). Whereas all existing cities were supposed to be reconstructed along the principles favored by socialist ideologues, the existing fabric of these cities sometimes stood stubbornly in the way of radical transformation (unless it was bombed out in World War II). Hence, perhaps the purest form of the socialist city—the true "sotsgorod," a term pioneered by Soviet architect Nikolay Milytin in 1930—was neither reconstructed Moscow nor Leningrad but rather, precisely these new towns. The most common type was a town centered around a strategically important industry. Some examples: Magnitogorsk in the Ural Mountains, renowned as Magnetic Mountain for its iron ore resources, established in 1929; Khimki, established in 1939, as a major Soviet aero-defense center; Dubna, incorporated in 1956, a science research center, and Tolyatti, which is an old city renamed after the Italian communist Palmiro Togliatti in 1964 and reconstructed to become the Soviet "Detroit"—an automobile industry hub that was envisioned and implemented in cooperation with Italy's Fiat. The building of these towns from almost scratch allowed them to be portrayed as ostensibly clear from the "ills" of capitalist urbanism (Hausladen 1987).

The late Brejnev years marked the onset of Soviet economic stagnation, which was coupled with major geopolitical blunders, such as the Afghanistan war. From about 1975 on, the once-feared USSR entered a downward spiral. The heroic efforts of the last Soviet leader Mikhail Gorbachev (1985–1991) to reform the entire system—the Russian term is "perestroika"—did not succeed, culminating in the dismantling of the Eastern Bloc in 1989 and the disintegration of the Soviet Union in 1991. Due to the challenging political and economic circumstances, the avant-garde spirit of Soviet urbanism faded altogether. Major new urban projects were rarely undertaken.

In the next section, the authors zoom in on some of the spatial characteristics of the "typical" socialist city that resulted from the distinctive socialist process of spatial production. We have organized the discussion into two subsections: one considering the relationship between urban centers and peripheries, and the second, the relationship between the public and the private realms.

22.3 The Socialist City as Product

22.3.1 Distinctive Overall Form: Centers and Peripheries

One of the best-known contrasts between socialist and capitalist cities of the second half of the twentieth century was their overall spatial articulation. In short, capitalist cities tend to expand horizontally, covering vast regions surrounding the existing built fabric—a process that is derogatorily known as "urban sprawl" and exists in its most advanced form in the United States. In capitalist conditions, this sprawl occurs as a result of the combination of the efforts of private and public actors. In the American case, for instance, suburban peripheries have been growing for some one hundred and fifty years. Initial suburbanization came in the form of private single-family homes linked by railroads, street cars, and later, private roads. Since the mid-twentieth century, the American federal government invested massively in highways and other infrastructure that further propelled suburban growth.

There was no clear equivalent to this process in socialist conditions. The state owned the land both in the centers and the peripheries of cities. With no land value, there were no incentives for increasing the population density in historic downtowns. Often, socialist cities experienced the opposite: population density rose with distance from the city center, first in the rings of the pre-war urban neighborhoods and even more so, in the socialist microrayons (Bertaud and Renaud 1997). Population density reached a sharp zero at the outer limit of the microrayons, whose last tall buildings served as the clear edge of the city, almost as if they were a wall around it. Beyond them were greenbelts, forests, and agricultural land separating the urbanized areas from nearby towns and villages. Where single-family houses or lower-rise structures were constructed in the periphery of cities, often within or around old villages, these were typically small vacation homes (dacha in the USSR, vikendica in the former Yugoslavia, villa in several other Slavic-language-speaking countries). Western-style suburbanization was perceived as both inefficient and unduly individualistic—i.e., unfit for socialist societies. Socialist planners were proud to have avoided it. This is how Soviet planners explained this idea back in the 1930s: "*From the capitalist era, we have inherited cities in which we see a striking difference in the amenities of the center and the outskirts.... We strive to eliminate the distinction between the city and the countryside. Buildings in the socialist cities should be surrounded by greenery, with streets opening into grand squares*" (Ladyzhensky 1936, p. 11). The microrayons of the post-war period, which were the socialist answer to capitalist suburbs, were supposed to become those wonderful neighborhoods "*surrounded by greenery, with streets opening into grand squares*" (Ladyzhensky 1936, p. 11). The reality, however, fell short of the grand promise (Fig. 22.1).

Another distinct feature of the socialist city was that socialist downtowns looked and functioned differently. Unlike capitalist historic cores, socialist ones were not under pressure to serve as business or tourism centers. Surely, some like Prague were always beautiful and attracted both foreign and domestic visitors. Moscow's inner-city center had some of the most distinguished cultural hubs in the world (Smith 1996). East Berlin stood out, at least partially, as a consumer "paradise" among other socialist capitals (Standley 2013). But, overall, as land in socialist cities did not have commercial value, socialist planners were able to load city centers with a rich plethora of civic, cultural, government, and other ceremonial functions organized in gigantic plazas and other public spaces to the neglect of housing and commercial services. One could posit that the historic urban centers—as legacies of the old bourgeois order—were to an extent purposefully neglected (Mariotti and Leetmaa 2024) to overcome the inequalities between the centers and the once-poor outskirts. Yet the socialist downtown—not just in Moscow but universally, as we discuss further in the next section—was the proud home of fine theaters, museums, cinemas, galleries, and universities, all in service of the cultured citizen of the socialist state (Hirt 2013).

Fig. 22.1 Residential areas built during socialist times aimed to provide plentiful public green space—an example from Weissensee (Weißensee) Neighborhood in Berlin, Germany, formerly in East Berlin *Source* Žaklina Grgić (2020)

22.3.2 A City for All: Public Monumentality and Private Restraint

As we have emphasized already, the generosity of public spacc was one of the most distinctive aspects of the socialist city. Monumental squares were dedicated to the celebration of the state's power and triumphs. Instead of commerce, socialist downtowns housed almost exclusively magnificent administrative buildings, beautiful cultural centers, and grand plazas. Religious and sacral buildings and monuments, which might have had an important role in the city's history until the socialist era, lost significance, as socialist societies were proudly secular (Kelly 2016). Far from being capitalist-style central business districts, socialist city centers were "spaces of power, the venue for parades and official celebrations, demonstrating the greatness of the socialist ideology" (Mezentsev et al. 2024, p. 153). The size and number of monuments, usually dedicated to war heroes and political leaders, were certainly more impressive than what could be found in contemporaneous capitalist urban settings. Similarly, cultural and educational institutions usually came in colossal and highly ornamental styles, such as the wedding-cake-style structures of the famous Seven Sisters of Moscow (Zubovich 2021). The Seven Sisters is a set of skyscrapers built in the Stalinist style, which was replicated in ceremonial buildings in other cities, such as the Latvian Academy of Sciences in Riga and the Palace of Culture in Warsaw (Fig. 22.2). Where city centers were destroyed in World War II, as was the case of Warsaw, they were meticulously restored to reflect both national and socialist pride (Minkenberg 2014).

The splendor of public space extended under the street surface, too. Special attention to extravagant design was given, for example, in creating the metro stations in Moscow and Kyiv. These were envisioned in the 1930s as palaces of the people and included fancy murals, sculptures, and chandeliers (Grava 1977). Comparing

Fig. 22.2 Latvian Academy of Sciences, Riga, Latvia built in the Stalinist style. *Source* Žaklina Grgić (2019)

Moscow's or Kyiv's metro stations with those of major US cities like Boston or New York makes the latter look like dingy tunnels and the former like gorgeous art museums—and this was as true during socialism as it is today.

Socialist dedication to grandeur and luxury did not carry into the construction of ordinary, private spaces, especially housing units. The right to housing was guaranteed to every person in socialist constitutions. To a large extent, the austerity of residential design in the microrayons was a result of this commitment. Housing for millions and millions of new urban residents moving from the countryside to take part in socialist industrialization could only be provided, if it was relatively inexpensive. Initial planning for housing was time-consuming and resulted in working people waiting for years to obtain a modern apartment in one of the microrayons. Often, at least in the earlier socialist periods, residents lived in overcrowded conditions. For instance, in 1930, the Russian city of Prokopyevsk had only 1.7 m^2 of residential space per person (Häfner 2014). Living conditions were improved somewhat through time: in the 1950s, the average living space was still less than 5m^2 per person, in mid-1970s, it was around 8 m^2 per person. Moscow's conditions were better with around 10m^2 per person (Morton 1980). While the working class lived in modest conditions, despite voicing their concerns more actively over time (Morton 1987), the elite groups resided in spacious and attractive apartments, often in prime locations (Roberts and LaFollette 1990). Some of the Seven Sisters in Moscow had residential wings where the acclaimed artists, scientists, and politicians of Soviet society thrived.

Further, although housing in the microrayons may have looked too humble from the viewpoint of Western middle classes, it did not necessarily look that way to the citizens of socialist cities who were coming from either dilapidated old apartments in city centers or from villages with substandard peasant huts. A sunny apartment with a modern kitchen and bathroom in a microrayon seemed, to many, a superior alternative. Still, the relative modesty of private space was not only a matter of economic necessity but also, a purposeful ideological choice (Crowley and Reid 2002). In some extreme cases, private spaces were reduced to the bare minimum. For example, the famous Romanita building in Chisinau, Moldova, was envisioned in a way that the apartments didn't have private kitchens but rather a common one on each floor (Rusu 2015). However, such experiments were too radical to carry on for very long: even socialist families liked having their own kitchens.

By the 1980s, microrayons had become the largest residential areas of capital cities and other major centers. They housed approximately 60% of households in Sofia, 77% in Bratislava, and 82% in Bucharest (Hirt 2013). The public spaces in these microrayons—the greenery between the residential buildings—although not as glamorous as the central ceremonial spaces, were well-sized and well-maintained. They were also well connected by public transport. Private car usage was minimal and discouraged. The most essential services—from schools to shops— were reachable either on foot or by mass transit. Schools, groceries, and polyclinics were all planned to be accessible within 500 m of each residence (Pallot and Shaw 1981). This was the socialist method of planning cities "for all."

22.4 Coda

The adventures of socialist urbanism seem to have ended some thirty years ago, at least in Europe. Or did they? The socialist city idea is not bound by a specific time or place but by the process of how it was made (Zarecor 2018). Outside of Europe and the former USSR, socialist principles of urban development are still in operation as evidenced by the massive industrialization and urbanization projects now undertaken in China, where to this very day state ideologues speak of overcoming class conflict and the contrast between urban prosperity and rural poverty. Further, the socialist spatial legacy cannot be fully deconstructed, even by the most enthusiastic of neoliberal advocates today. Surely, the statues of Marx and Lenin are gone, but the ceremonial city centers and the massive housing estates that socialism built are now an integral and likely permanent part of the fabric of East European cities (e.g., Light and Young 2010). Aside from Germany, where the cities of the former East truly shrunk after the fall of the Berlin Wall, no other country has attempted to demolish socialist-era microrayons on a massive scale. On the contrary, some of the better-designed ones such as Novi Beograd (New Belgrade) and Novi Zagreb (New Zagreb) are today deemed as these cities' most desirable neighborhoods.

Yet some aspects of socialist urbanism are gone for good. With the withdrawal of the public sector from urban development, the opening of urban peripheries to private development, and the rise of car ownership, formerly socialist cities have been sprawling as intensely as Western cities. Forests and green spaces have been vanishing to be replaced by gated compounds, malls, and surface parking lots (Boentje and Blinnikov 2007; Hirt 2013). And while color and commerce have returned to downtowns, public spaces have yielded to private ownership and use, and all pretense of socio-spatial equality has been abandoned. Perhaps the post-socialist urban transition should have been better thought out. But there will be no second chances for the former Second World.

In this essay, we provided a brief historical review of the socialist-era Soviet and East European city, the city of the former Second World. We emphasized the much greater role of the building compared to that in capitalist countries and the importance of socialist egalitarian ideology. We delineated some of the distinct physical features of the socialist city, such as the importance of the microrayon as the chief neighborhood unit that socialist governments built in the post-World War years around the historic cores of large cities. Although the microrayon was designed in accordance with modernist urban design principles that were also used in capitalist countries, it was distinct from its Western counterparts both in terms of the scale of its application (i.e., microrayons comprised a much greater share of the fabric of socialist cities compared to capitalist cities) and in its land-use specifics (e.g., microrayons included a greater share of public services). We also discussed that the socialist city, reflecting socialist ideology, had more generous public spaces but less luxurious private spaces compared to what we typically find in capitalist settings. Our arguments were based on a thorough and critical review of the literature on the topic, while providing only a few examples of landmark projects in select cities. This approach is the main limitation of this essay, as we leave the in-depth empirical analysis of specific socialist cities and projects within them to other authors in this volume.

References

A Commission of the C.C. of the C.P.S.U. (B.) (ed) (1939) History of the Communist Party of the Soviet Union (Bolsheviks). Francis White Publishers

Adams JS (2013) Monumentality in urban design: the case of Russia. Eurasian Geogr Econ 49(3):280–303

Andrusz GD (1984) Housing and urban development in the USSR/Gregory D. State University of New York Press, Andrusz

Bartha E, Tamás K, Mezei B (2023) From socialism to neoliberalism: lessons from Eastern Europe and Hungary. In: State socialism in Eastern Europe: history, theory, anti-capitalist alternatives. Springer International Publishing, Cham

Bertaud A, Renaud B (1997) Socialist cities without land markets. J Urban Econ 41(1):137–151

Boentje JP, Blinnikov MS (2007) Post-Soviet forest fragmentation and loss in the green belt around Moscow, Russia (1991–2001): a remote sensing perspective. Landsc Urban Plan 82(4):208–221

Borchert JR (1981) The socialist city: spatial structure and urban policy by R. A. French and F. E. Ian Hamilton. Geogr Rev 71(4):472–474

Chomsky N (2005) Government in the future. Seven Stories Press

Clark K (1985) The Soviet novel: history as ritual. Indiana University Press

Collins GR, Alex W (2021) Introduction. In: Sotsgorod: the problem of building socialist cities. https://mitp-arch.mitpress.mit.edu/pub/o8smvxr0

Crowley D, Reid ES (ed) (2002) Socialist spaces: sites of everyday life in the Eastern Bloc, 1st ed. Berg Publishers

Czepczynski M (2008) Cultural landscapes of post-socialist cities. Representation of power and needs. Ashgate

Engel B (2022) The concept of the socialist city plans and patterns of Soviet urbanism. International Planning History Society Proceedings, TU Delft

Enyedi G (1996) Urbanisation under socialism. In: Harloe M, Szelenyi I, Andrusz G (eds) Cities after socialism: urban and regional change and conflict in post-socialist societies. Wiley-Blackwell, pp 100–118

Estrin S (2010) Self-management: economic theory and Yugoslav practice. Cambridge University Press

Fisher JC (1962) Planning the city of socialist man. J Am Inst Plann 28(4):251–265

Ford LR (2013) World cities and global change: observations on monumentality in urban design. Eurasian Geogr Econ 49(3):237–262

French RA, Hamilton FEI (1979) The socialist city: spatial structure and urban policy, 1st ed. Wiley

Gobova N (2024) New ecological planning and spatial assessment of production sites in socialist industrial Yekaterinburg (formerly Sverdlovsk) in the 1960s–80s. In: Mariotti J, Leetma K (eds) Urban planning during socialism. A view from the periphery. Routledge, pp 209–226

Grava S (1977) Politics and design of the Moscow metro. Ekistics 43(256):174–178

Häfner L (2014) Julia Franziska Landau, Wir bauen den großen Kuzbass. Bergarbeiteralltag im Stalinismus 1921–1941. (Quellen und Studien zur Geschichte des östlichen Europa, Bd. 80.) Stuttgart, Steiner 2012. Historische Zeitschrift 299(1):244–246

Harrison M (1980) Why did NEP fail? Econ Plan 16(2):57–67

Hausladen G (1987) Planning the development of the socialist city: the case of Dubna New Town. Geoforum 18(1):103–115

Hess DB, Metspalu P (2019) Architectural transcendence in Soviet-era housing: evidence from socialist residential districts in Tallinn, Estonia. In: Hess DB, Tammaru T (eds) Housing estates in the Baltic countries. The legacy of central planning in Estonia, Latvia, and Lithuania. Springer, pp 139–160
Howard E (2003) Garden cities of tomorrow. Organ Environ 16(1):98–107
Hirt S (2005) Planning the post-communist city: experiences from Sofia. Int Plan Stud 10(3–4):219–224
Hirt S (2012) Iron curtains: gates, suburbs, and privatization of space in the post-socialist City. Wiley-Blackwell
Hirt S (2013) Whatever happened to the (post)socialist city? Cities 32(Supplement 1):29–38
Jenkins J (2012) Building the revolution: soviet art and architecture. Des Issues 28(4):86–92
Kelly C (2016) Socialist churches: radical secularization and the preservation of the past in Petrograd and Leningrad, 1918–1988. Northern Illinois University Press
Kiss D (2024) From reverse colonial trade to antiurbanism. Frustrated urban renewal in Budapest, 1950–1990. In: Mariotti J, Leetma K (eds) Urban planning during socialism. A view from the periphery. Routledge, pp 79–98
Kopp A (1970) Town and revolution: Soviet architecture and city planning, 1917–1935. Translated by Thomas E. Burton. George Braziller
Kotkin S (1995) Magnetic mountain: Stalinism as a civilization. University of California Press
Kornai J (2021) The Socialist system: The political economy of communism. Princeton University Press
Ladyzhensky AM (1936) Reconstruction of USSR cities. Science and Life 9:11–15
Lapidus G (1978) Women in Soviet society: equality, development, and social change. University of California Press
Light D, Young C (2010) Reconfiguring socialist urban landscapes: the 'left-over' spaces of state-socialism in Bucharest. Hum Geogr 4(1):5
Lizon P (1996) East Central Europe: the unhappy heritage of communist mass housing. J Arch Educ 50(2):104–114
Logan S (2020) In the spaces between buildings: the case of South City, Prague. In: Drummond LBW, Young D (eds) Socialist and post-socialist urbanism: critical reflections from a global perspective. University of Toronto Press
Mariotti J, Leetmaa K (2024) Urban planning during socialism: views from the periphery. Routledge Research in Historical Geography Series. Routledge
Meyerovich MG, Men'kovskii VI, Kerebcov IL (ed) (2019) "Социалистический город": Идея и её воплощение в Советском Союзе 1920–1930-х годов ("Socialist City": Idea and its embodiment in the Soviet Union in 1920–1930ss), ИЯЛИ Коми НЦ УрО РАН
Mezentsev K, Provotar N, Gnatiuk O (2024) Courtyards, parks and squares of power in Ukrainian cities—planning and reality of everyday life under socialism. In Mariotti J, Leetma K (eds) Urban planning during socialism. A view from the periphery. Routledge, pp 153–175
Milyutin NA (1974) Sotsgorod: the problem of building socialist cities (Trans. Arthur Sprague). MIT Press
Minkenberg M (ed) (2014) Power and architecture: the construction of capitals and the politics of space, 1st ed. Berghahn Books
Morton HW (1980) Who gets what, when and how? Housing in the Soviet Union. Sov Stud 32(2):235–259
Morton HW (1987) Housing quality and housing classes in the Soviet Union. In: Herlemann H (ed) Quality of life in the Soviet Union. Westview Press, pp 95–115
Mumford E (2009) CIAM and the communist bloc, 1928–59. J Archit 14(2):237–254
Pallot J, Shaw DJB (1981) Planning in the Soviet Union. University of Georgia Press
Roberts PC, LaFollette K (1990) Meltdown: inside the Soviet economy. Cato Institute
Rusu D (2015) International scientific committee on 20th century heritage: socialist modernist heritage in the Republic of Moldova. Heritage at Risk, pp 147–151

Smith DM (1996) The socialist city. In: Harloe M, Szelenyi I, Andrusz G (eds) Cities after socialism: urban and regional change and conflict in post-socialist societies. Wiley-Blackwell, pp 70–99
Standley M (2013) Here beats the heart of the young socialist state': 1970s' East Berlin as socialist bloc tourist destination. J Arch 18(5):683–698
Szelenyi I (1996) Cities under socialism and after. In: Harloe M, Szelenyi I, Andrusz G (eds) Cities after socialism: urban and regional change and conflict in post-socialist societies. Wiley-Blackwell, pp 286–317
van der Straeten J, Petrova M (2022) The Soviet city as a landscape in the making: planning, building, and appropriating Samarkand, c. 1960s–1980s. Cent Asian Surv 41(2):297–321
Zarecor KE (2018) What was so socialist about the socialist city? Second world urbanity in Europe. J Urban Hist 44(1):95–117
Zubovich K (2021) Moscow monumental: Soviet skyscrapers and urban life in Stalin's capital. Princeton University Press

Chapter 23
Moscow, the Soviet City: The Impact of the Soviet Urban Plan in the City of Moscow Today

Rosa Cristina Laureiro

Abstract The Soviet era deeply influenced Moscow's architectural landscape reflecting the evolving political ideologies of its time. Following the 1917 Bolshevik Revolution, Constructivism emerged as a radical departure from pre-revolutionary styles, emphasizing functionalism and avant-garde esthetics. Architects like Konstantin Melnikov and Moisei Ginzburg pushed innovative designs that sought to embody socialist ideals, resulting in iconic structures such as the Rusakov Workers' Club and the Narcomfin Building. These projects underscored the Soviet vision of communal living and social equality. With Stalin's rise, Constructivism gave way to monumentalism, aligning architecture with the state's propaganda goals. The 1935 Master Plan for Moscow introduced wide avenues, river embankments, and the iconic skyscrapers known as the "Seven Sisters." The unrealized Palace of Soviets symbolized this era's aspirations. Post-Stalin, Khrushchev's reforms prioritized mass housing, resulting in the widespread construction of "Khrushchyovkas," utilitarian apartment blocks that addressed the city's housing crisis. Through successive political changes, from Brezhnev's stagnation to the collapse of the Soviet Union, Moscow's architecture mirrored the state's shifting priorities. Today, the legacy of Soviet architecture remains a defining element of the city's identity, blending revolutionary ideals with authoritarian monumentalism, alongside contemporary modern buildings that are constantly reshaping the skyline with innovative designs and global influences.

Keywords Moscow · Architecture · Soviet housing · Urbanism

23.1 Introduction

The image of the city of Moscow has been shaped throughout history over the political vision of its leaders, who have left an indelible mark on its urban landscape with drastic social changes and revolutionary renewals. From the times of the tsars to the

R. C. Laureiro (✉)
Quito, Ecuador
e-mail: rosacl@mac.com

F. Carrión Mena and S. Rodríguez Alvarez (eds.), *From the City as a Project to the City Project*, The Urban Book Series, https://doi.org/10.1007/978-3-031-94226-6_23

Soviet era and beyond, each period has left its imprint on the architecture, urban planning, and culture of the city. Leaders have used Moscow as a blank canvas to embody their political and social ideals, constructing imposing monuments, entire neighborhoods, and symbols of power that have defined the city's identity over the centuries.

Moscow's ascent as a pivotal political and cultural hub predates the Bolshevik revolution of 1917 by centuries. Tracing its origins to the twelfth century, Moscow began as a fortified outpost strategically positioned along the extensive trade routes traversing the vast expanse of the Russian landscape. Gradually, it evolved into the capital of the Grand Duchy of Moscow, with the iconic Kremlin emerging as the epicenter of its burgeoning influence. The Kremlin, characterized by its formidable red brick wall and ensemble of towers and cathedrals, including flamboyant onion domes, served as a visual testament to Moscow's architectural glory. The display of colors and intricate patterns adorning the renowned Red Square represented the fusion of Byzantine, Renaissance, and Baroque influences, encapsulating Moscow's rich of cultural heritage.

By the turn of the twentieth century, the city had undergone significant shifts, driven notably by the abolition of serfdom in 1861. This event spurred a considerable migration of rural inhabitants to the city, providing a vital labor force essential for reconstructing Moscow after the devastation wrought by the Napoleonic War and the subsequent Moscow Fire. Industrialization, urbanization, and a burgeoning population further propelled the city's transformation. Architecturally, Moscow's skyline began to reflect its newfound economic and cultural prominence, with a blend of neoclassical, eclectic, and art nouveau styles leaving lasting impressions.

However, beneath the surface of prosperity, discontent brewed. Social inequality, political repression, and the looming threat of revolution simmered within Russia. These tensions, coupled with economic disparities, foreshadowed the seismic changes to come. Despite outward signs of growth and development, underlying societal fissures hinted at the unrest and upheaval that would soon reshape Moscow's landscape.

23.2 Constructivism's Dawn (1917)

The emergence of Constructivism in Russia, after the 1917 revolution, marked a radical shift in architectural paradigms. The Bolsheviks, filled with a utopian vision of societal reorganization and political fervor, sought to dismantle the remnants of tsarist rule, and forge a new urban landscape reflective of their revolutionary ideals in Moscow, the capital of Soviet Russia, once again since 1918.

The modernist movement, which endured for less than a decade until 1929, represented an absolute departure from traditional architectural conventions, marked by a deliberate rejection of decorative elements. Originating during a period of post-revolutionary tumult characterized by the Civil War between Bolsheviks and the

Whites, the emerging Soviet state, in its formative years, lacked the capacity to exert comprehensive control over architectural projects.

Against this backdrop of uncertainty, the emergence of Constructivism served as a bold assertion of autonomy, as vanguard architects seized the opportunity to cultivate the style independently, encouraged by the belief that innovative designs pioneered in the capital would be tolerated. Spanning various art forms, including painting, sculpture, graphic design, and architecture, the Constructivist movement encapsulated a holistic reimagining of artistic expression in the service of ideological renewal. In 1920, VkhTEMAS was created by decree from Lenin, a higher education project with close ties to Bauhaus, to prepare master artists of the highest qualifications for industry.

At the core, the architectural aspect of Constructivism was the profound connection between architectural form and function, with an unwavering emphasis on functionality over ornamental embellishments. It was characterized by bold experimentation, avant-garde esthetics, and a fervent embrace of socialist principles. Buildings were designed to serve specific purposes, and their designs were driven by the functional requirements of the spaces they housed.

Architects such as Konstantin Melnikov, Vladimir Tatlin, and Moisei Ginzburg advocated innovative designs that sought to transcend traditional notions of form and function, and fervently promoted prioritizing function, calling buildings "machines for living. There was much influence from Le Corbusier, who was already an important presence within the architects in the city, much before he traveled in 1928 to Moscow to participate in the contest for the City of the Future.

The emergence of Constructivism was an act of defiance against traditional norms, as the Soviet state, in its formative years, lacked the energy to exert full control over architectural endeavors. Vanguard architects seized the opportunity to develop the style independently, this movement extended beyond Moscow to the provinces, under the assumption that the innovative designs pioneered in the capital were permissible.

23.3 Trade Union Initiatives

The 1920s witnessed an infusion of energy through projects commissioned by trade unions. These initiatives were fueled by the ethos of constructing Clubs of Culture, designed specifically for workers in the suburbs at a large scale, with common areas for 1000 or more people. These clubs were envisioned not merely as recreational spaces but as dynamic meeting places, featuring theaters, art rooms, sports facilities, and more.

Konstantin Melnikov's Rusakov Workers' Club, on Stromynka, 6, erected in 1929 in Moscow, exemplifies these ideas. The building's design is characterized by bold geometric shapes, a cylindrical structure with a series of cantilevered volumes radiating from its core. The use of circles, cylinders, and rectangles creates a dynamic and visually striking composition. It houses an auditorium and assembly hall within a visually striking and functionally efficient space. Characterized by a grid of rectangular windows and the use of exposed materials like cement and glass, the design prioritizes transparency, honesty in construction, and a focus on the essential elements. Melnikov's innovative approach to form and function in the Rusakov Workers' Club stands as an iconic representation of early Soviet architectural ideals, reflecting the movement's commitment to a socially conscious and industrially informed built environment.

The Kauchuk Factory Club on Plushchikha, 64: Commissioned by Chemical industry trade unions in 1927, showcased a circular main building which is the theatrical hall embraced by two buildings for various activities and a sports center. The design incorporated two staircases from the street leading to the second level, emphasizing a multi-functional approach.

Ilya Golosov was another staple in this movement, who later evolved from Constructivism to early Stalinist designs. In 1927, he was commissioned the construction of The Zuyev Workers' Club on Lesnaya 18, completed in 1929. Purposefully constructed to accommodate various facilities for workers of the Moscow tram, the building showcases an innovative glazing treatment at its corner, resulting in a remarkably photogenic feature. The interior comprises a sequence of club rooms and open foyers leading to an impressive 850-seat auditorium, blending innovative design with practicality.

23.4 Communal Living Experiments and Unique Projects

During this era, communal houses served as experimental ventures aimed at establishing a novel lifestyle centered around collective living. Addressing the housing needs of the masses became a focal point for the Association of Contemporary Architects (OSA), led by Moisey Ginzburg. Their objective extended beyond mere provision of shelter in major urban centers; it aimed to redefine housing within the context of a society transitioning toward full socialization.

This initiative unfolded through a meticulous three-stage design process spanning five years and observed by the international community. The inaugural phase, presented formally at the 1926 Comradely Competition, concentrated on conceptualizing communal living spaces as a response to the housing crisis in Moscow, the subsequent stages involved rigorous scientific, technical, and methodological investigation of constructing systems and materials. Six such experimental communal houses were erected around Moscow, for workers to be close to factories and industrial enterprises and were characterized by compact individual apartments accompanied by expansive communal areas.

Notably, one standout project, Narcomfin Housing Building, gained global acclaim as a pioneering model of Soviet avant-garde housing. It was designed by Moisey Ginsburg and Ignaty Milinis in 1928, for high-ranking workers of the Ministry of Finance, at Novinsky Boulevard 25.

The Narcomfin Building's innovative approach involved constructing blocks with a blend of reinforced concrete frame and cinder blocks, and reeds as insulation at the end of the concrete beams, a cost-effective strategy contributing to a groundbreaking spatial arrangement. The six-story structure with long horizontal windows featured corridors solely on the second and fifth floors. To access their apartments, residents utilized additional staircases, constructed in dual levels, and stacked vertically. This ingenious design not only optimized financial resources but also added a distinctive dimension to the project. It was aimed to motivate communal living, offering elements like social areas to share, and to motivate an equality of roles between both genders by reorganizing everyday activities.

After numerous years of continuous interventions, the architectural firm led by Ginsburg's grandson has successfully restored the building, transforming it into an esteemed architectural monument.

The Konstantin Melnikov House (1927–29), the most internationally recognized architectural icon of Moscow's avant-garde of this period, corresponds to a meticulous architectural work aimed to represent familial principles. Securing a parcel of land adjacent to Arbat 10, Melnikov committed to a 15-year mortgage to materialize his visionary residential project during a period when private property and land ownership were receding from Soviet society.

Through avant-garde architectural innovations, Melnikov secured approval for the construction of what would be recognized as an "experimental house." While primarily designed for familial occupancy, the structure distinguished itself by incorporating progressive wall tension-spreading technology, resulting in a commendable 20% reduction in brick consumption.

Its cylindrical towers, adorned with a captivating arrangement of hexagonal windows, epitomize a harmonious blend of artistic expression and functional design. Melnikov's innovative use of geometric patterns extends to the interior, where modular living spaces showcase adaptability to climate and light, and a visionary approach to domestic life.

The walls were built with 200 hexagonal open cells, facilitating an equitable distribution of tension. Subsequently, these cells were selectively bricked up, culminating in the retention of 60 hexagonal windows. The inter-floor partitions were meticulously fashioned using wooden boards laid upon ribs and grooves. In a time where industrialization was the main discussion in all aspects of production, this home was built almost entirely, manually.

Currently, the Melinkov House, following prolonged disputes among its heirs, is in the possession of the government, and a restoration initiative has started.

Throughout this decade, the nationalization of land and the abolition of private property emerged as a pragmatic response to the urgent need for public housing. Families previously housed in cramped barracks were relocated to expropriated apartment buildings and noble houses constructed during the eighteenth and nineteenth century. The process of communalization involved converting existing apartments into communal homes (known as "domma-kommuny") by collectivizing the use of their facilities. Typically, this entailed assigning individual families to single rooms within the house, while transforming a designated kitchen and entrance hall into shared spaces for cooking and socializing, respectively. According to Bliznakov

(1993), it is estimated that by the conclusion of 1921, approximately 865 of these communal apartments had been registered in Moscow.

> How is the housing question to be settled, then? (...) There is already a sufficient quantity of houses in the big cities to remedy immediately all real 'housing shortage', provided they are used judiciously. This can naturally only occur through the expropriation of the present owners and by quartering in their houses homeless workers or workers overcrowded in their present houses. (Engels 1970, pp. 30–31)

However, the demand for housing continued to outstrip supply, forcing innovative solutions to address the housing crisis. Communal housing emerged as a crucial intervention, offering a collective approach to accommodation that sought to maximize available space and resources.

23.5 Stalinist Era (1927)

However, the euphoria of Constructivism was short-lived, giving way to the authoritarian dictates of Stalinist orthodoxy following Joseph Stalin's rise to power in the late 1920s. In the eager pursuit of projecting power and control, the established Soviet government dealt with contrasting aspirations and contradictions as it endeavored to transform Moscow into a modern metropolis, shedding the traces of its provincial past. The requirement to showcase the flourishing Soviet state spurred a concerted effort to produce a new urban identity reflective of socialist ideals.

By the onset of the 1930s, with the consolidation of state and the need to show the world the grandiose capital of the new Soviet super power, monumentalism emerged as an imperative objective for posterity, represented by the ambitious Master Plan for the Reconstruction of the City of Moscow in 1935, ordered by Stalin.

This comprehensive urban design aimed to overhaul Moscow's cityscape through a series of multifaceted initiatives. These efforts included expanding infrastructure, such as widening roads, which featured 16 main avenues designed to accommodate military parades. Additionally, river embankments were constructed to mitigate flooding risks, while the implementation of a fantastic metro system revolutionized transportation within the city. Inaugurated in 1935, this magnificent and opulent stations served as a showcase of the achievements of Soviet society through the integration of lavish materials and meticulous craftsmanship.

The plan also incorporated a green belt, spanning 10 km around Moscow, to preserve natural landscapes amid urban expansion. Furthermore, the urban development strategy involved expanding housing options to accommodate a projected increase in population. With a current population of 4 million residents, the plan aimed to provide housing for an additional 5 million people that would work on the fast-growing industries and factories being built rapidly.

This ambitious endeavor sought to address the pressing need for adequate housing while accommodating anticipated population growth. Additionally, in 1932, the construction of the Moscow Volga River canal was undertaken to resolve the growing city's water supply challenges. The emergence of labor camps, or Gulags, contributed to providing an endless source of manpower for the construction of immense projects.

23.6 Palace of Soviets

Central to this grand vision appeared the idea of the iconic Palace of Soviets, envisioned as the pinnacle of architectural expression and propaganda, symbolizing the triumph of socialism.

The competitions for the construction of the Palace of Soviets, a political convention center, lasted almost a decade and sparked enthusiastic debates within architectural circles, regarding the trajectory of socialist urbanism. Alexei Tolstoy's 1932 article in Izvestia Gazeta, articulated the quest for a new architectural paradigm rooted in monumentality; he said that the Palace of Soviets was the medium of how the proletariat, having taken a hold of history, must put forth its own form in architecture.

This was an era also marked by widespread demolitions, as historic neighborhoods and landmarks fell victim to the relentless march of progress. Urban planning became synonymous with social engineering, as the regime sought to reshape Moscow in its own image, imposing rigid hierarchies of space and function upon the urban fabric. The grand cathedral Christ the Savior was to be replaced by the Palace of Soviets. More than 160 projects were presented by international and local architects, including Walter Gropius and Le Corbusier. The winning design was crafted by a team led by Boris Iofan, a prominent architect renowned for his neoclassical sensibilities.

Their proposal envisioned a colossal structure crowned by a towering statue of Lenin, soaring to a height of 415 m, intended to eclipse the Eiffel Tower in Paris and the Empire State Building in New York. The architectural ensemble was to incorporate elements of neoclassical grandeur, with monumental colonnades and a vast central hall capable of accommodating tens of thousands of people.

However, the grand vision of the Palace of Soviets was never fully realized. Construction commenced in 1937 but was halted by the outbreak of World War II in 1941, and the subsequent focus on military priorities. The partially completed structure served as the site for the world's largest open-air swimming pool, known as the Moskva Pool. The debate and discussion on the completion of the project lasted through the Soviet era as a metaphor of the rise and fall of the USSR.

23.7 The Urban Plan of Moscow

The Urban Plan of Moscow in 1935 marked a pivotal moment in the city's architectural history, laying the groundwork for the transformation of its skyline with the introduction of the vysotky, or Moscow skyscrapers. The winning design, spearheaded by a team of architects led by Alexey Shchusev and Igor Rozhin, envisioned a bold departure from traditional urban planning principles, embracing a vision of verticality and modernity.

Significant to the plan was the concept of radial development, which sought to decentralize the city's core while promoting the construction of high-rise buildings along key transportation arteries. This approach aimed to alleviate congestion in the city center while fostering a sense of dynamic urban growth. The vysotky, characterized by their soaring heights and distinctive architectural styles, were designed as Government buildings, housing projects for privilege party members, hotels near train stations, and the vast Moscow University. Symbols of Soviet propaganda such as the sickle and hammer, stars, flags, wheat leaves, and sculptures of workers begin to adorn the new buildings around the city, even the facades of the previous constructivist monuments.

Construction of the Seven Sisters as they are known today commenced in the late 1930s and the last one was finished in 1953, reshaping Moscow's physical landscape. Despite the challenges posed by war and political upheaval, the vysotky endure as symbols of Moscow's architectural heritage giving the city its distinct image.

While senior party officials had been accustomed to the luxuries of hotels and former aristocratic palaces that had been expropriated, the construction of the Government House or House of the Embankment in 1928 heralded a shift toward more communal living arrangements. Designed by Boris Iofan, the building where he lived until his death initially served as a transitional structure offering more than five hundred traditional family apartments with modern services as hot running water and telephones, private bathrooms, and small kitchens, alongside communal facilities. Upon its completion in the spring of 1931, the Government House boasted an array of amenities including a bar-restaurant, theater, library, recreational rooms, sports facilities, and extensive communal services such as banking, laundry, post office, childcare, healthcare, and shopping facilities.

These apartments were assigned to individuals contributing to the advancement of the new communist agenda, party officials, intellectuals, scientists, etc., symbolizing the broader societal transition toward communal living and shared amenities. Amid Stalin's purges, the building faced scrutiny, leading to the imprisonment and execution of many residents in Gulags nationwide. Numerous novels recount the trials of its renowned inhabitants in this social experiment turned open prison under constant surveillance for Stalin's favor, excellently described in Slezkin's (2017) novel Government House "…It was the rear of the vanguard, the private world of public figures, the place where revolutionaries lived and the revolution died."

Today, the massive building stands as a testament to modern living, and the apartments are among the best located and sought for in the city.

During the 1930s to the 1950s, Moscow witnessed the proliferation of housing projects commonly known as Stalinkas. These residential complexes were conceived as a response to the pressing need for housing in the rapidly expanding city. While officially denying the existence of social classes in the socialist society, the architectural typology of Stalinist buildings reflected a clear hierarchy. Residential blocks were often divided into two sectors: smaller cinderblock structures were erected for the working class, while larger brick buildings adorned with ornate facades were reserved for the Communist party elite.

Communist party housing, distinguished by either white or red brick facades, with neoclassical columns and décor, grandiose halls and luxurious interiors with superior insulation, high ceilings and spacious rooms, meant to reflect the regime's commitment to providing dignified accommodation for the proletariat. Constructed using robust materials such as brick and concrete, Stalinkas were designed to withstand the harsh climatic conditions of Moscow and endure for generations. At these four floor buildings, units were meant for single family use, with private kitchen and bathrooms but communal spaces for socializing. Despite their utilitarian purpose, these housing projects incorporated elements of esthetic grandeur, evoking a sense of permanence and stability. Until the collapse of the Soviet Union, these 'Stalinkas' stood as the epitome of prestige in the country's real estate market. Access to these coveted apartments was limited, reserved for esteemed figures such as military leaders, renowned scientists, professors, artists, and members of the Communist party.

In stark contrast, residential units allocated for the working class presented a different reality. Some of the façade decorations were still used, but many of these buildings failed to meet basic standards, earning the designation of "temporary structures." These buildings were constructed with lower quality bricks, the interior finishes were very poor, using discarded construction materials for the internal walls, floors, and ceilings. Living conditions were often inadequate, with spaces shrinking to a fraction of the officially required minimum. The discrepancy between the opulent accommodations of the elite and the cramped living conditions of the working class underscored the inherent inequalities within Soviet society.

Moreover, Stalinkas were often complemented by communal facilities such as schools, shops, and recreational areas, fostering a sense of community and social cohesion among residents.

While examining the historical significance of Stalinkas and their role in shaping the urban fabric of Moscow during a transformative period in its history, it becomes evident that these projects, which were eagerly sought after by the population, proved insufficient to meet the demand for housing. As a result, in the two or three-room apartments, up to three families were compelled to cohabit, leading to overcrowding, and heightened social tensions. Furthermore, as financial resources from the government after the war decreased, construction efforts began to compromise on the quality of materials employed.

23.8 Khrushchev's Thaw

Khrushchev's rise to power marked a departure from the repressive regime of his predecessor, with his leadership characterized by a degree of de-Stalinization and a push toward a more open and reform-oriented approach. As soon as one month after Stalin's death, Gulags were releasing their prisoners across the Soviet Union. The 1960's in the USSR overflowed with optimism, marked by the race for the conquest of space with USA, the historic launch of Sputnik and Yuri Gagarin's pioneering outside the orbit, signaling expanded opportunities for women and youth.

After Stalin's death, his favored architectural style quickly fell out of favor. The transition had already begun toward the end of his era but addressing critical issues like the severe housing shortage, lacked political traction.

Khrushchev initiated profound changes in architecture and urban planning shortly after Stalin's era ended in 1953. In 1954, one year before becoming the leader of the Soviet Union, Khrushchev discussed the challenges of prefabricated housing during what became known as the Builders Conference. The speech title was:'On the extensive introduction of industrial methods, improving the quality and reducing the cost of construction'. During the long speech, he also condemned the construction of Moscow's skyscrapers as excessive and wasteful and publicly criticized architects, employing strong ridicule in his political rhetoric, a trait that became part of his regular public speeches.

The architectural landscape shifted from monumental neoclassical displays to a mass housing campaign, featuring the implementation of ambitious architectural projects aimed at modernizing the city and improving the living conditions of its inhabitants. Khrushchev placed a strong emphasis on the construction of mass housing developments, later known as Khrushchevka, to address the chronic housing shortage in the capital.

Khrushchev finally canceled the original Palace of Soviets project forever and launched a new competition for its design. The new location, near Moscow University rather than the city center, was intended to commemorate Lenin. Khrushchev advocated for a more modest structure, departing drastically from the previous emphasis on monumentalism.

23.9 Kruchevkas

The Khrushchev-era housing developments, known as Khrushyovkas, constructed from 1954 onward, including during the Brezhnev era, may not have boasted high-quality standards, but they epitomized sheer standardization. The implementation of panel housing technology in the Soviet Union originated from French architect Raymond Camus's pioneering work and his patented technology in 1949. A decade later, the USSR acquired its inaugural production line for panel manufacturing from France, followed by licensing agreements with Camus's company for the mass production of concrete components. Soviet engineer Vitaly Lagutenko further refined this technology to suit local conditions, resulting in the development of the K-7 series of Soviet five-story buildings.

This method relied on concrete panel plants and an expedited assembly schedule that symbolizes the quintessential Khrushchev-era housing unit. A staggering 64,000 units were erected in Moscow alone from 1961 to 1968, marking just the beginning of a widespread construction effort. The speed of construction resulted in low quality buildings, with low ceilings, thin walls, and poor insulation.

The series of construction models evolved to more than 10 typologies that shifted to 9 or 12-story buildings with elevators by the late 1960s. The last 5-story Khrushyovka was completed in Moscow in 1971. Across the rest of the USSR, the construction of Khrushyovkas continued until the collapse of communism. Today, millions of such units remain occupied, far surpassing their intended design lifespan.

Access to these homes was determined by job status, with workers in unions, factories, and industries relying on the plant director for allocation. While the wait could last years, many families, eager for private space after living in communal apartments (komunalkas), found it worthwhile. Requests for future housing often began with the birth of children, leading to multi-generational living within the same family.

Despite facing construction challenges and design flaws, the buildings from this era served as a viable housing solution for numerous families. They provided access to a wide array of communal amenities including supermarkets, laundries, schools, medical facilities, sports venues, public transportation, and green spaces, all contributing to the societal advancement of the time. There exists a palpable nostalgia for the sense of community that thrived in these developments, fueling efforts to revitalize entire neighborhoods.

However, the lack of proper maintenance and inferior construction have resulted in the deterioration of these buildings, originally designed to endure for up to two decades. Consequently, there are plans underway to demolish large sections of these structures and replace them with new, high-rise buildings backed by private investment.

In 1961, the first public building of this period was constructed, The Palace of Congress located within the Kremlin fortress. Part of the historic Armory Chamber was demolished to make room for a theater and convention center with a capacity for 6000 people. In modernist style, Architects M.V. Posokhin, A.A. Mndoyants, G.V., designed a concrete building with large vertical glass openings to reflect the domes of the Kremlin ensemble's cathedrals, prioritizing functionality and structure.

The same architects designed four massive buildings for official use and hotels on New Arbat Street, commonly known as the Open Books due to their shape. They face the South side of the avenue, with an emphasis of horizontal openings in the facades 26 stories high. This succession of massive vertical planes were utilized as large luminous screens for propaganda, and reflected flags, stars, or the initials USSR. Today, they continue to serve as screens, with generous commercial areas at street level, as in their construction in 1964–1968.

23.10 The Era of Stagnation

Leonid Brezhnev rose to power following the ousting of Nikita Khrushchev in 1964. Brezhnev's ascent marked a return to a more conservative and stable leadership style after the tumultuous years of Khrushchev's rule. Known for his cautious approach to domestic and foreign policy, Brezhnev sought to consolidate power within the

Soviet hierarchy and maintain stability within the Communist Party. His leadership was characterized by a focus on maintaining the status quo and preserving the Soviet system, rather than implementing bold reforms. Under Brezhnev's rule, the Soviet Union experienced a period characterized by economic stagnation, political repression, and a decline in social and cultural freedoms.

In 1968, the Soviet Union led a military intervention of Warsaw Pact troops into Czechoslovakia, aiming to suppress reformist movements in Prague. While this intervention temporarily stifled reform momentum within Czechoslovakia, it inadvertently strained the unity of the communist bloc and awakened a stronger oppression in the Soviet society.

The need for housing, a chronic symptom among the Muscovite population, continued to drive the rapid construction of massive blocks. The typologies from the Khrushchev era evolved in height, incorporating freight elevators to increase from 5 stories to 9, 12, and even 14. But essentially, urban developments remained the same, resulting in a landscape dominated by standardized, utilitarian buildings with little room for architectural innovation or esthetic expression.

Following the economic boom due to the high price of oil starting in 1970, an intense period of public building construction began. The modernist style of that time was adorned with classical elements, returning to traditional façade ideas. The distinctive style of Socialist Realism that had dominated Soviet architecture in earlier decades gradually declined. Instead, there was a move toward more nondescript and functionalist architecture. The emphasis was on practicality and meeting the basic needs of the population rather than creating inspiring or ideologically charged buildings.

Massive constructions such as the Rossiya Hotel, by architect Dmitry Chechulin, appeared. An enormous project in front of the Kremlin on the site of a complex of demolished religious buildings. It had more than 3,000 rooms to accommodate delegations meeting at the Congress Palace. It was heavily criticized for having no relation to its surroundings and blocking the view of the Kremlin, but it was very popular as it held the world record for hotel capacity for a time. It was demolished in 2006, giving way to a park with multiple cultural spaces designed by architects Diller Scofidio + Renfro (New York) in 2012.

In 1981, the original 1930's project of the Government House in the neoclassical style by architect Dmitry Chechulin was revitalized and built. Originally the project was to house the office of the national airlines, Aeroflot, but it morphed into the oval Government Building, which is still a national landmark.

As the Soviet Union approached its fall, the architectural landscape became very diverse, reflecting the change of ideologies and the liberation of the state pressure in the planification of buildings.

Brutalist architecture in Moscow of the 1970's often served institutional or governmental functions, with buildings housing administrative offices, cultural institutions, and educational facilities. The imposing presence of these structures conveyed a sense of authority and permanence, reflecting the Soviet regime's desire to assert control and project strength.

Despite its severe appearance, brutalist architecture also had a certain poetic quality, with its stark forms and textured surfaces creating dynamic interplays of light and shadow.

The Tass News Agency building, Architects Yegerev and Shaykhet, 1977; The Children Musical Theater by Architect Natalia Sats, 1971; The Russian Academy of Sciences, Arquitects Platanov and Zakharov, 1973; are some examples of this Brutalist diverse representation in Moscow.

By the end of the decade, Architect Andrei Meyerson designed a series of residential blocks in the northwest of Moscow, all of which left their mark in the history of Soviet architecture as it was the first building in the URSS, constructed entirely in reinforced concrete. The design was originally intended for a hotel during the 1980 Olympic Games, which was to be located near a public park (the spaces between the "legs" would have provided a direct link to the surrounding nature). After plans changed, however, the block was built to house workers of a nearby airplane factory. The "Centipede"—as this house is affectionately known among locals—was not the only, but certainly the most successful homage in the Russian capital to Le Corbusier's Unité d'Habitation. Excerpt from Nikolai Vassiliev's article in: SOS Brutalism: A Global Survey. Catalog DAM + Wüstenrot Foundation, Zurich (Park Books) 2017.

For the celebration of the Summer Olympic Games of 1980, a series of constructions were planned with a budget of 9 billion dollars, possibly the most extensive architectural boom after World War II. The sports buildings, stadiums, courts, pools, sports museums, and housing buildings for over 5000 athletes and attendees were a display of Soviet modernism that remains relevant. Although some have undergone changes in use and renovations to keep them operational, they are still in use and well-maintained. There has been a delay in officially designating them as historical buildings, but they are already part of the Soviet modernist heritage. The housing buildings were immediately occupied by city residents after the games, given the ever-present need for housing.

23.11 Conclusion

Moscow has seen significant changes as a city since the post-Soviet era, with advancements in areas like economic development, infrastructure, and cultural diversity, alongside ongoing challenges related to the environment, sustainability, and social inequality. The city's continuous urban expansion in peripheral regions, reminiscent of Soviet projects but adapted for contemporary needs, signifies its evolution.

Under the leadership of Sobianin since 2010, with support from Putin, Moscow has maintained consistency in projects, solidifying its place as a center for state political messaging. The city's economic growth is evident in its modern skyline and status as a global financial hub, driven by foreign investments and government initiatives. Infrastructure enhancements demonstrate a forward-thinking urban planning approach, though sustainability concerns persist.

Culturally, Moscow thrives as a hub of artistic and traditional expressions, although social inequality remains a significant issue in its diverse landscape. The development of urban areas showcases a fusion of tradition and innovation, shaped by deliberate urban planning strategies and enormous private investments. Moscow's historical importance and political prominence underscore its role in shaping Russia's narrative on the international stage, especially now that the future is uncertain amid the ongoing conflict with Ukraine.

References

Bliznakov M (1993) Soviet housing during the experimental years, 1918 to 1933. In: Brumfield W, Ruble B (eds) Russian housing in the modern age: design and social history. Woodrow Wilson Center Press, pp 85–148

Engles F (1970) The housing question. Moscow

Slezkine Y (2017) The house of government: a saga of the Russian revolution. Princeton University Press

Chapter 24
From the Industrial to the Post-industrial City. The Reunification of Berlin: Searching for the 'Centre' the New City Project

Paola Alfaro d'Alençon

Abstract Berlin's division into East and West during the Cold War resulted in contrasting urban development and socio-political trajectories. Through qualitative research, including archival analysis and case studies, this paper aims to provide insights into the dynamics of urban development processes in both division and reunification of the city. Along with the focus on the urban transportation space and its concepts in Berlin it explores resulting spatial transformations, particularly the proliferation of city projects post-unification with the political decision of unification, driven by social reforms but also by the imperative to integrate two formerly divided cities to one functioning urban fabric. It addresses: (1) Urban development concepts transitioning from the Cold War era to post-Fordist urbanism; (2) key influences on urban planning; and, (3) disputes in urban development and the emergence of a new spatial identity stemming from unification. Consequently, urban planning emerged as catalysts for new options and paradigms. In doing so, the idea is presented of the reunification of Berlin as a city project, where junctures were lost, because they failed to articulate themselves to the general process or because their forms of production were not inscribed in the 'state-of-the-art' logics of the predominant political order.

Keywords Berlin's division and reunification projects · Urban transportation concepts · Cold War urbanism · Post-Fordist urbanism

24.1 Introduction

The division and reunification of Berlin, provide a compelling case study of how not only demarcated physical boundaries where shaped but also engendered contrasting approaches to urban development. During the division, West Berlin embraced capitalist ideologies, whereas East Berlin adhered to socialist principles, thereby giving

P. Alfaro d'Alençon (✉)
Frankfurt University of Applied Sciences, Frankfurt, Germany
e-mail: paola.dAlencon@fb1.fra-uas.de

F. Carrión Mena and S. Rodríguez Alvarez (eds.), *From the City as a Project to the City Project*, The Urban Book Series, https://doi.org/10.1007/978-3-031-94226-6_24

rise to disparate urban landscapes and development paradigms. The reunification of Berlin in 1989 brought forth a plethora of challenges and opportunities by the imperative to integrate two formerly divided cities into one functioning urban fabric. This process was driven by social reforms promoting cohesion but also by instruments to foster economic growth across the newly unified city. The reunification also catalysed the emergence of a new spatial identity in Berlin, characterised by the synthesis of diverse cultural, architectural, and historical elements from East and West. These shifts underscore the iterative process of urban development, in which diverse stakeholders actively engage to shaping the city's evolving trajectory.

Against this background, this paper examines Berlin's urban development related with the transportation system—particularly the Stadtbahn, the inner-city rail network—as deeply intertwined with urban projects and urbanisation dynamics. Since the industrial revolution, transportation systems have been integral to the city development, as well as city division, unification to its ongoing transformations into the present day. This relationship is particularly tied related to the dichotomy between capital-identity, representative urbanism, and the functionalist metropolis. The paper employs both historical research and contemporary empirical analyses to reveal this relationship. In doing so, it explores key spatial transformations, particularly the proliferation of city projects before and post-1989 following the political decision of unification and is structured in two main parts:

(1) Presenting urban development concepts transitioning from the Cold War era to post-Fordist urbanism; (2) An exploration of post-unification urban development and the emergence of new spatial identities; Each section includes subchapters that explores major influences and debates surrounding urban planning, transportation infrastructure developments and specific urban projects.

24.1.1 City Development and the Stadtbahn

In 1871, architect August Orth introduced the concept of the Stadtbahn, an inner-city railway network intended to complement Berlin's existing freight transport infrastructure. Orth envisioned the project, known as the Berliner Centralbahn as one for suburban use with a clear metropolitan transportation bias (Orth 1871). This represented a novel concept: Prioritising the transportation of people over goods. Drawing inspiration from established mass transportation systems in metropolises such as New York and London, Orth examined their role in shaping urban development (Die Berliner Stadteisenbahn, 1891). He emphasised the necessity of improved transportation links to connect Berlin's rapidly growing industrial base with key urban centres, commercial hubs, and industrial zones. The Stadtbahn played a pivotal role in enhancing access to Berlin's high-dense city areas and nerves with important business corridor between Hamburg and Munich. In consequence the new rail network spurred with its intra- and inter-city linkage, as a catalyst Berlin's industrial development (Siewert 1978). Industrial growth took place along the Spree River, and with the expansion of factories to the than city's outskirts, such as Siemensstadt in

Charlottenburg and AEG in Wedding. Following the development of the project's end station Charlottenburg, inhabitant and building activities in adjacent areas to the station increased. The area surrounding the Stadtbahn's final station in Charlottenburg witnessed a surge in residential and commercial development, culminating in the eventual merger of Charlottenburg with Berlin, transforming it into a key cultural and economic centre (Alfaro d'Alençon 2013). In fact, the Stadtbahn was both as a provider of rapid access to services for different urban activities, and a developer, fulfilling the constant need to extend the railway system to achieve the new demands of the extending city.

However, transportation dynamics had already begun again in the immediate post-war period. This is exemplified by plans as Hans Scharoun's collective plan (1946), which contrasted with Walter Moest's Zehlendorf Plan (1946/47), and based on Hermann Jansen's 1910 conceived urban skeleton plan. As this paper will further demonstrate, the Stadtbahn was massively altered during the Cold War. The division of Berlin meant the general boycott of the GDR-operated transportation infrastructure in West Berlin, and the city's urban and transportation planning policies diverged along ideological lines. In West Berlin, the focus shifted ultimately to the U-Bahn (subway system), while in East Berlin, the Stadtbahn remained central to urban development projects. This division of transport infrastructure reflected broader geopolitical tensions and ideological divides, influencing urban projects such as the centre development in East Berlin and the creation of large housing estates in the east and in the west on the periphery of the city (Alfaro d'Alençon 2013).

24.1.2 From 'City as Project' to 'City Project'

In the post unification period, Berlin's transportation networks underwent significant transformations. Beginning in the late 1990s, the city's rail system became the focal point of ambitious redevelopment projects. These projects, exemplified by the 'Planwerk Innenstadt', a major post-unification development plan formulated along 'Urban Renaissance', sought to revalue rail space as an integral component of urban regeneration. The station, viaducts and rail corridors became central to the broader goal of revitalizing inner-city areas, improving infrastructure and attracting investment.

The particular place, its specific potential to bring out urban qualities became in the 1990s a significant issue for scholars. A common thread was the provision of attractive urban spaces with ambitious architecture and urban design. The specific spatial development and facilities were intended to serve as anchor points that may influence their surroundings or act as a catalyst for urban regeneration. In this context, place-specific characteristics carried a unique potential for economic development and became increasingly significant for the description of the post-Fordist city with growing tourist-orientation (Porter 1995) and creative industries (Florida 2005). Such redevelopment strategies reflect the rise of 'urban entrepreneurialism', where cities actively produce and market distinct identities to attract investment, tourism, and creative industries, 'the active production of places with special qualities assumes a pivotal role in spatial competition between localities, regions, and nations' (Harvey 1989:295). This paradigm shift in urban development is characterised by a high degree of image and symbol-rich city and location strategies as urban space for festivalisation, eventisation, aestheticisation as well as flagship developments. These tendencies are also related to mega-events, such as the 2006 opening of Berlin's main-station after the unification, Europe's largest railway hub, which was timed to coincide with the World Cup and featured participation by the chancellor, positioning it as a public event. However, these trajectories are interrelated to provide the hosting cities with new urban or real estate redevelopment that typically accompanies such events, as noted by Camerin and Longato (2021:126).

Hence, in doing so this paper emphasises the idea of 'From the City as a Project to the City Project' as the evolution of Berlin's urban landscape and identity in Cold War and post-unification exemplifying, as a transition from the traditional notion of the city as a passive recipient of development projects to a more participatory approach embodied by diverse actors in particular from the predominant political and social order that shape the concept of the 'City Project'. And in particular, against this background, the reunification of Berlin is presented in this paper as a city project, where junctures were lost, either due to failure to articulate themselves to the broader process or because their production forms were not inscribed or aligned with the 'state-of- the-art' logics of the predominant political order.

Calling the city a theatre of social action Mumford laid out already the fundamental proposition about urban life, the potential of urban planning and citizenship from a relational perspective (Mumford 1961). Social needs, their interactions and places to display them remain primary for Mumford, while the physical organisation of the city and economics remain supporting issues and are not reasons alone for the development. They serve to enrich the social drama. If urban space were to lose this sense of dramatic dialogue, it certainly would not survive. Urban development is a process, a 'genesis' (Lefebvre 1991:163), where former configurations (spatial and/ or social) sometimes compliment or even resemble one another. With the process character of urban development, each new addition inherits and reorganises that which came before it. Previous forms might be converted through new approaches; space becomes diverse and ultimately hybridised (Lefebvre 1991). Since an existing space may outlive its original purpose, which has determined its form, function, and structure, it may in a sense, become vacant and susceptible to being converted or put to use for a purpose quite different from its initial one. The built environment is continuously being transformed. Consequently, against this background urban planning emerged as catalysts for novel options and paradigms.

24.2 Forcing the Division of Berlin Through the Train: Urban Development Concepts During the Cold War

The London Protocol of 1944 formalised the division of Berlin into four Allied occupation zones. This agreement marked the beginning of the city's division, which became further entrenched after the introduction of currency reform in 1948. A few days later, the Soviet Union initiated the more than a year-long blockade of West Berlin, effectively making the city the epicentre of Cold War tensions. The declaration of Berlin as the capital of the German Democratic Republic (GDR) in the East in 1949 and the constitution for West Berlin under the control of the Western Allies sealed the political division. This was followed by the construction of the wall, which spatially separated the city hermetically. With the division of Berlin's urban territory into East and West Berlin, the inner-city transportation network Stadtbahn began to function as two separate, independent entities (see Fig. 24.1). Two key factors one spatial and one political contributed to this division: The physical separation of the city and the political control over the railway system, both in a context of growing tensions between the Eastern and the Western worlds (Lüpke 1983).

24.2.1 Physical Division: A Cut Through the City

The physical division of Berlin, and in particular the way the city was divided from north to south was most evident in the cut through the middle of the Stadtbahn system at Friedrichstrasse station, which became a focal point in the city's division. Along with Albert Speer's East–West Axis and the Spree River, this part of the train system represented one of the most important East–West connection in the city geography. From August 13th 1961 until reunification on October 3rd 1989, the station served as a border crossing point between East and West Berlin. The station was split into two distinct sections, reflecting the separation between the two Berlin's. The station square became a highly controlled zone, home to the 'Tränenpalast' (Palace of Tears), the border point for visitors entering East Berlin territory. For West Berlin, the division meant a dislocation from the former historical centre. Prior to the city's partition, the Stadtbahn network had been central to Berlin's urban development, and the station at Friedrichstrasse had been a key transit point linking the city's districts. Another element that illustrated the physical division was the East–West State border surrounding West Berlin. The border cut off network access to the city hinterland, so that the Stadtbahn could no longer offer its services of regional network interconnection, particularly for West Berlin, which was entirely encircled by the GDR.

Fig. 24.1 Division of Berlin with the train. The development of the S-Bahn network from 1945 to 1989 (*Source* Cabrini, in: Alfaro d'Alençon 2013). This figures illustrate the evolution of Berlin's railway network from 1945 to 1989, highlighting the impact of the city's political division on its transportation infrastructure. New lines are marked in red, with their completion dates indicated, showcasing the role of the Stadtbahn in shaping the city's urban landscape during the Cold War: **In 1945**, the net structure shows its historical radial structure, with lines extending from the city outward to the suburbs, reflecting integrated transportation system that exist before the division of Berlin. **In 1950**, following the division of Berlin, the radial lines connecting West Berlin were disconnected at the border. A new ring structure was established within East Berlin to serve the eastern sectors in GDR territory. **1961, post-wall construction**: After the construction of the Berlin Wall, the outer ring of railway line was cut off from the rest of the city by the wall. The inner radial lines, continued to operate across border, with trains passing through the Friedrichstrasse *station*, located in East Berlin without stopping, thus providing direct connections between East and West Berlin. After the construction of the wall, new separate transit points for passengers and freight traffic were established. These routes functioned independently, reinforcing the physical and political division between the halves of the city

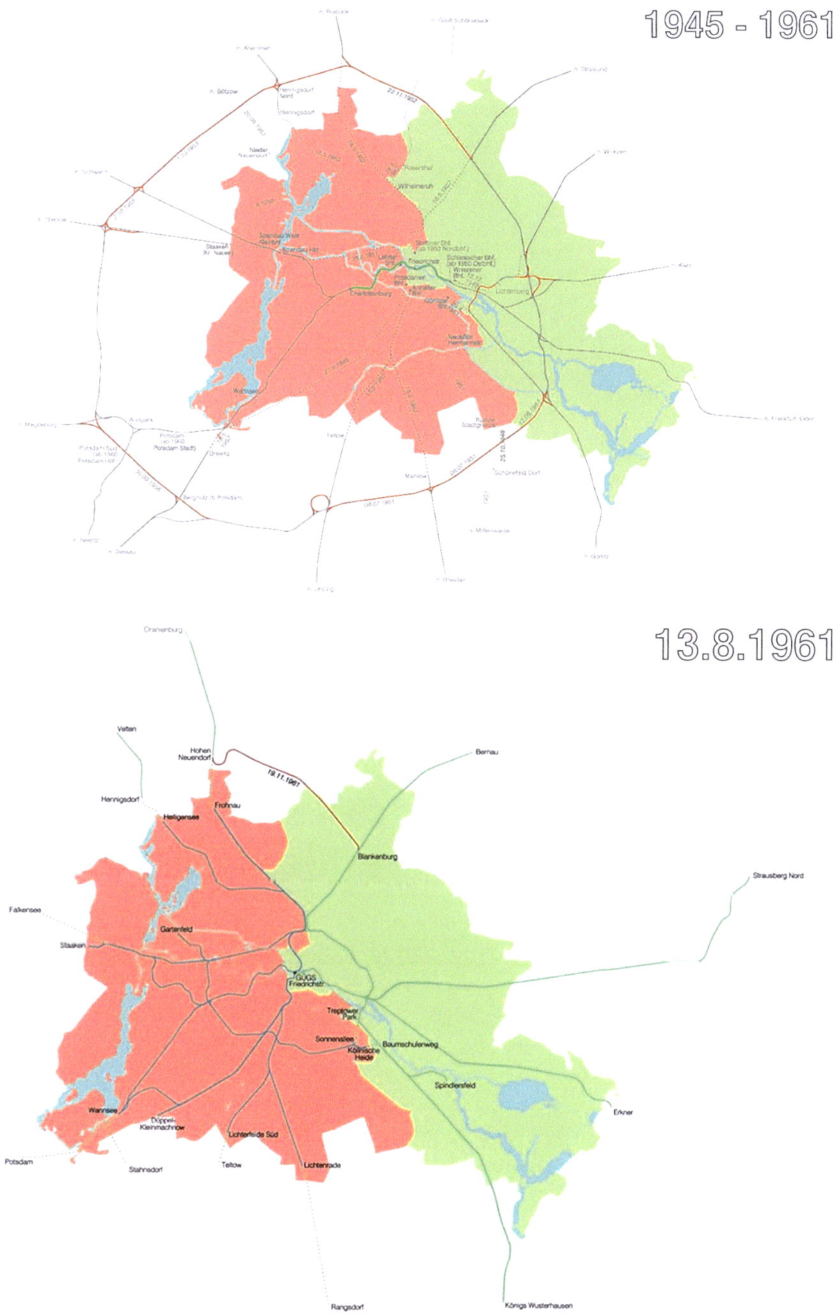

Fig. 24.1 (continued)

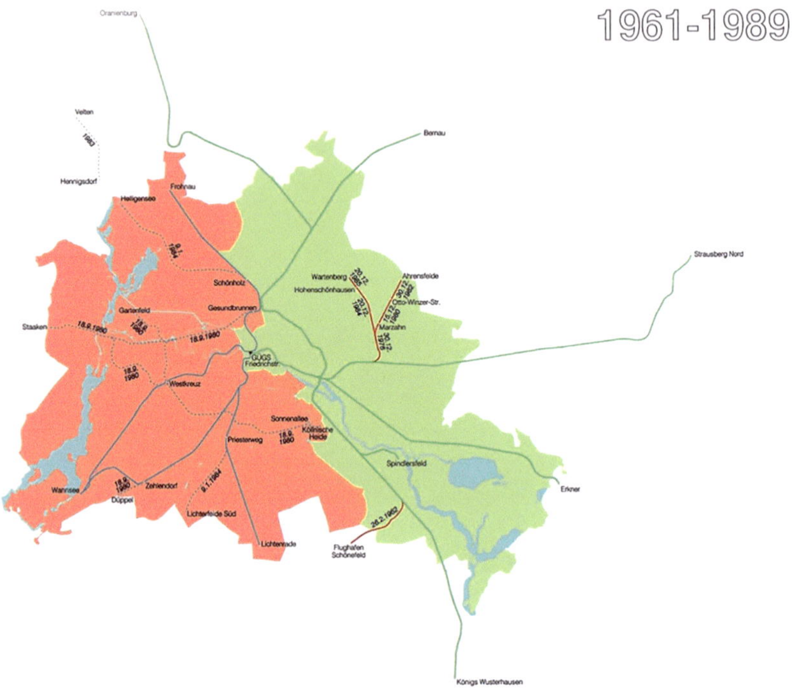

Fig. 24.1 (continued)

24.2.2 Political Division: A Paradox of Control

The political factors behind the division of the Stadtbahn system were equally significant. After the division, the control of rail operations in both East and West Berlin was held by the Deutsche Reichsbahn in the GDR, while the Deutsche Bundesbahn (DB) was established for the Federal Republic of Germany. Because of this paradoxical situation—meant that the DR retained control of the network, the system was boycotted in West Berlin. The situation became even more complicated after the Berlin Wall was erected in 1961. The government opposed to the GDR's control of the rail network, effectively boycotted the Stadtbahn system. The system became a symbol of East–West division, and its operations in West Berlin were gradually diminished. This boycott had both organisational and physical consequences. The Stadtbahn system was increasingly marginalised, and the division of the transportation network altered not only the movement of people within the city but also its overall urban planning and development. By 1961, the Stadtbahn had become not just a transportation system but a political tool, emblematic of the broader ideological divide between East and West. The infrastructure itself—once an integrated element of Berlin's urban identity—was now a physical barrier that reinforced the city's separation.

24.2.3 The Urban Concepts in the Divided City

The 'Berlin Hauptstadt' (Berlin Capital City) competition, held in West Berlin in 1957, was intended as the final attempt to produce a public master plan including the urban territory of both East and West Berlin. In 1959, the 'Berlin Hauptstadt der DDR' (Berlin Capital City of the GDR) competition, launched in East Berlin, represented a clear move towards independent decision-making. Both proposals offer significant insights into the political circumstances of the time and its influence on urban planning prospects the divided city. The Stadtbahn, featured prominently in in both competition plans, as it largely comprised as transportation system of the competition territory.[1]

The 'Berlin Hauptstadt' competition aimed to reconstruct Berlin as a unified city, although it was spearheaded by the West Berlin administration. The area designated for development was primarily located in the eastern part of Berlin—the historical district of the city. Academics and municipal authorities designed the framework for the competition, with a focus on new typologies. A possible new beginning was to be achieved through concepts of radical modernisation and urbanisation, and envisioned Berlin as a model for future urban planning. The key objective was to provide a solid basis on which the notion of a capital city could be built for a

[1] See Competition outline Hauptstadt Berlin. Planungsgrundlagen für den städtebaulichen Ideenwettbewerb. Bundesminister für Wohnungsbau u. Senator für Bau- und Wohnungswesen (1957, pp. 56–80).

new state—one that could shape its capital as a canter of administration, culture, and economy. The concept of a 'capital' was central to the competition, and the proposals were presented as: *'The concept of 'capital' signifies action of politically creative determination and of shaping order. The complexion of the new capital will however not be that of nationalist governmental power. Rather, it will be dominated by the ideas of democracy and international collaboration on equal footing. Here, the aspirations of the individual must find expression as much as the order of the community".*[2]

However, the GDR government considered the competition a political provocation, mainly because the call had been launched solely by West Germany, presenting Berlin as a unified city. The project remained unrealised, and only two years later, another competition was launched by the GDR with the equally provocative title 'Berlin, Capital of the GDR', with a similar development area to that of the West Berlin competition. In 1960, the first prize was awarded to architects Kaiser, Gericke and Swiss. The proposal envisioned the rebuilding of this area with office towers, new commercial centres and a central building at the crossroads of the two axes between Alexanderplatz and Marx-Engels-Platz. However, during the early stages of construction in 1967, economic challenges and technical difficulties led to the abandonment of this plan. The project was replaced with developments in nearby areas, including the Friedrichsforum, the television tower, and the Palast der Republik (Palace of the Republic), a centre for cultural and political events.

Despite of being addressed by two distinct planning bodies—one based in West and the other in East—the central historical area of Berlin, situated in the Mitte district, represented the action space for both. Both circles embraced the concept of a multifunctional, polycentric city, reflecting the broader trends of the time. In this concepts, former railway lines and newly constructed highways were transformed into major transportation corridors, linking production and residential areas throughout the city. In addition, peripheral areas were extending the city for everyday activities with production, industrial zones, commercial spaces and housing areas spanning from single-units to mass-residential areas (Dant 2006:103). Giedion describes planning concepts as shifting, fluid with planners attempting to reorganise space using 'three-dimensional networks and large-scale structures' (Giedion 1976:508). Networks organised space from the macro to micro level, with roads and paths (group form), and neighbourhoods with working areas and large-scale road frameworks (mega structure). Precedence was given to systems of movement and the efficiency of infrastructure over patterns of defined public space. This approach aligns with the broader functionalist urbanism that dominated post-war Berlin planning, underscoring the relevance of these urban examples in shaping the city's post-1945 development.

[2] See Competition outline Hauptstadt Berlin. Planungsgrundlagen für den städtebaulichen Ideenwettbewerb. Bundesminister für Wohnungsbau u. Senator für Bau- und Wohnungswesen (1957:12).

24.2.4 Competition, Construction and Counter-Construction: The Contrasting Development of Housing Areas in East and West Berlin

To analyse the development of housing districts in both post-war East and West Berlin serve as a good basis for understanding urban planning approaches and building standards, since they were a key concern in both parts of the city. These projects, although emerging from competing and conflicting political ideologies, were developed simultaneously and reflect the contrasting urban visions of the two administrations. Berlin stands out globally as a city where the geopolitical tensions between East and West have been indelibly and directly shaped architecture and urban planning. The clash of divergent ideologies sparked a series of competitions for urban design and architectural projects, most notably before the construction of the Berlin Wall in 1961. A striking feature of this rivalry was the stark contrast in the size and character of housing districts on either side, with significant disparities in the scale of development (Flierl 2013). The rivalry produced two distinct urban typologies, emblematic of the political and social ideologies that defined each side.

In East Berlin, the iconic Stalinallee project exemplifies the Eastern approach. The grand boulevard designed by architect Henselmann as part of the 'Collective Plan' (1946) was intended as a symbol of socialist architecture. It featured monumental 'Residential Palaces', broad boulevards, and large squares with gateway and tower structures. The design drew inspiration from linear city concepts and from modernist architects such as Hilbersheimer and Le Corbusier (Flierl 2013). Henselmann envisioned modern forms of housing, including row houses, detached residences, and standalone structures (see Figs. 24.1, 24.2, 24.3 and 24.4). In contrast, at the Western sector the Hansaviertel, developed as part of the Interbau project in 1957, the international building exhibition overtly embraced the International Style in accordance with the principles outlined in the CIAM-Charter of Athens (Figs. 24.5, 24.6 and 24.7). This project was a collaborative effort involving prominent architects such as Aalto, Le Corbusier, Eiermann, Gropius, Jacobsen, Mies van der Rohe, and Taut. The Hansaviertel was conceived as a public project, focusing on functionalism and modernist principles.

The Eastern model, characterised by Socialist Realism, revitalised regional architectural traditions, embodying the idea of 'socialist in content, national in form'. In contrast, the Western model, with its embrace of the International Style and open urban landscape concept 'Stadtlandschaft' (city-landscape) sought an open, flexible urban form. The architectural legacy of post-war Berlin, particularly during the 1950s and early 1960s, is defined by juxtaposition of ideologies. Situated along the central East–West axis, structures on both sides of the Berlin Wall embodied distinct global post-war architectural and urban design movements. Both the Eastern and Western governments recognised the power of architecture and urban design as tools for shaping their public image and promoting their respective ideologies. With the Cold War rivalry intensifying, both sides of Berlin were transformed into 'showcases' of

Fig. 24.2 Construction of Stalinallee in East Berlin. *Source* https://www.mdr.de/geschichte/ddr/alltag/stalinallee-berlin-karl-marx-allee-sowjetunion-102.html. The iconic Stalinallee showcases the Eastern approach, featuring grand boulevards with 'Residential Palaces' and squares with towers and gateways. Designed by Hermann Henselmann and built by the state, the 1946 'Collective Plan' embraced the concept of a linear city. Henselmann envisioned modern housing forms, including row houses and standalone buildings

Fig. 24.3 A vintage art poster titled NATIONALES AUFBAUPROGRAMM BERLIN. *Source* https://www.stalinbauten.de/stalinbauten/

Fig. 24.4 View from Berliner Fernsehturm (television tower) on Stalinallee. *Source* https://commons.wikimedia.org/wiki/File:Bundesarchiv_Bild_183-G0408-0092-001,_Berlin,_Blick_vom_Fernsehturm,_Nacht.jpg

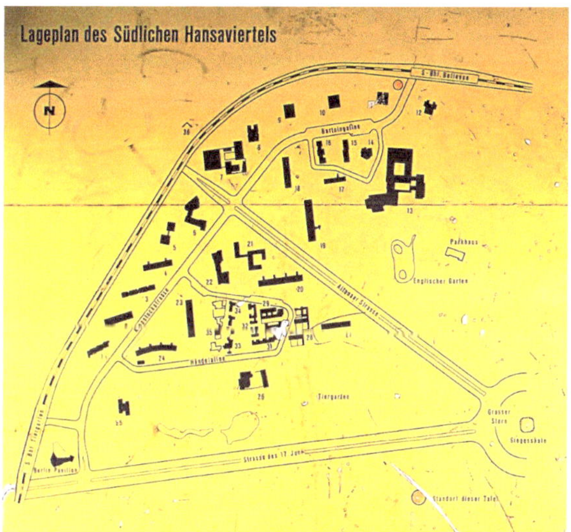

Fig. 24.5 Site map of Hansaviertel showing the architects of the single buildings. *Source* https://commons.wikimedia.org/wiki/File:Map_of_S%C3%BCdliches_Hansaviertel,_Berlin.jpg In the Western sector, the Hansaviertel as part of the Interbau 1957 embraced the International Style, following the CIAM-Charter of Athens. The Western model adopts the International Style and the open urban concept of 'Stadtlandschaft'. Post-war Berlin's architecture of the 1950s and early 1960s stands out for its cultural and political contrasts. On either side of the Brandenburg Gate, these buildings exemplify two major post-war architectural and urban design movements

Fig. 24.6 Interbau Plans—This image shows an aerial view of the plans for interbau. *Source* https://commons.m.wikimedia.org/wiki/File:Interbau_Plans.jpg

Fig. 24.7 Train lines in East and West Berlin and the new housing settlements in West and East Berlin

their political systems, each side using extensive external resources. The duplication of urban projects, cultural institutions and architectural ensembles in both sides document the competition between the two regimes and the temporary autonomy of East and West Berlin (Oswalt 2000).

24.2.5 *The S-Bahn in East and the U-Bahn in West for New Residential Districts in Berlin*

In the 1960s and 1970s, substantial suburban housing developments were undertaken in both East and West Berlin. Although these developments were similar in nature, the key distinction lay in their integration with the existing transportation networks. In the East, the S-Bahn remained the primary mode of transit, while in West Berlin; emphasis was placed on the U-Bahn (subway), which underwent significant expansion during this period (see Fig. 24.7).

In East Berlin, the S-Bahn was central to both residential and industrial development (Fig. 24.10). By the late 1950s, large-scale housing projects were planned with direct connections to the S-Bahn, underscoring its role as the main transportation link (Fig. 24.11). The first example was the new Stalinallee housing complex completed in 1959, developed near the newly planned S-Bahn station Jannowitzbrücke, which was intended to integrate with a new commercial and cultural axis, including important facilities, like the cinema Kino International, the restaurant Moscow, and the Hotel Berolina. Already by 1962 Schönefeld, the main GDR airport had been connected to the extended S-Bahn network. This significant network enlargement spurred the development of new residential areas in the 1980s. The line and its stations opened before the housing projects were finished and were used by the construction workers. The Marzahn district (Fig. 24.13), a major housing area in the northeast, was connected to the S-Bahn by 1982. From 1984–85, a new line along the outer ring opened to connect housing areas as Hohenschönhausen and Hellersdorf.

In contrast, in West Berlin preference was given to the increased use of the Untergrundbahn (U-Bahn), the subway network, which was expanded considerably in the 1960s (Lüpke 1983:42). Transportation network, commercial activities and the localisation logic in West Berlin were configured without including the railway system. The divergence between the railway and the development of important areas in the city became evident in the development of new housing districts, such as Hansaviertel (Fig. 24.8) or Gropiusstadt (Fig. 24.9) and production areas such as Siemensstadt. In 1970, new industrial areas located in proximity of the U-Bahn provided working places for approximately one of every four West Berliners (Lüpke 1983:42). A new public transportation plan was proposed in the form of the so-called '200 km Plan'. This foresaw the mass redevelopment of the subway network via extensions, renewals of several subway lines and even the building of completely new ones.

The newly built subway lines generally had fewer interchanges with the already installed S-Bahn network. For example, new U-Bahn lines were constructed in

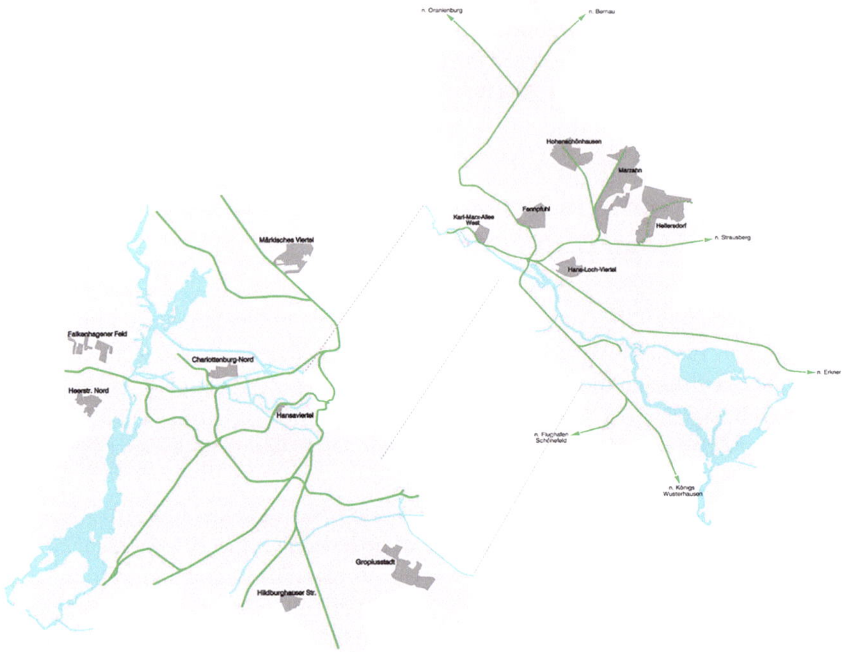

Fig. 24.8 Train lines in West and East Berlin and new housing settlements in West and East Berlin (*Source* Cabrini, in: Alfaro d´Alençon 2013). The relationship between train lines and housing developments in East and West Berlin presents notable contrasts. In East Berlin, there was a strong spatial connection between the Stadtbahn lines and the newly developed mass housing areas. Following the model of the Marzahn housing settlement, most new developments were planned in close proximity to S-Bahn stations. In contrast, the West Berlin Stadtbahn lines did not exhibit the same degree of integration with high-density housing. Instead, the new settlements in West Berlin were primarily oriented around the subway system

direct competition with the existing S-Bahn routes, despite the proximity of the two networks. Stations on the U-Bahn were frequently located close to S-Bahn stations, yet often carried different names, reinforcing the separation between the two systems (e.g. Wilmersdorfer Strasse (U-Bahn) vs. Charlottenburg (S-Bahn). In East Berlin, the integration of transportation and urban space continued to evolve, with the opening of a new station building at Ostbahnhof (East Station) in 1987. This station was renamed Hauptbahnhof (Central Station). As was the case of Alexanderplatz, this station developed into a public space well incorporated within the surrounding neighbourhood (Fig. 24.12).

Fig. 24.9 Train lines in West and East Berlin and new housing settlements: *The Interbau Hansaviertel, 1957*. (*Source* Cabrini, in: Alfaro d´Alençon 2013). This quarter was developed as part of the International Building Exhibition and was intended to serve as a new model for architecture and urban planning. Given its exemplary status, it is interesting to observe that the nearby Tiergarten S-Bahn station and the associated viaducts played an insignificant role in the overall design. As is typical in large-scale housing developments of this time, a new centre featuring commercial and cultural amenities was planned in proximity to urban transportation. However, the subway line was built a few years later, meaning that the centre was planned not around the existing *S-Bahn*, but rather in anticipation of the future subway line

24.3 Searching for the Centre with the Train: The 'Critical Reconstruction', Selective Memory, Identity and Urban Development for Berlin

> *The "Plan" does not invent a new city, rather it rediscovers lost arteries of Berlin's inner-city. Rooted in the building tradition of the European city, necessary reconstruction orientates itself to the organisation of the city's streets, public parks and squares as well as its blocks, respecting the existing stock and avoiding demolition.*[3]

The city exemplified fragments and historical and ideological ruptures, evident in its urban grid and build environment (Oswalt 2000). Its reunification brought profound political, economic, social, and cultural challenges. Reunification brought profound political, economic, social, and cultural challenges. Bombing during the war, followed by the Cold War's political divide, left significant tracts of inner-city

[3] Www.stadtentwicklung.berlin.de/planen/planwerke/de/planwerk_innenstadt/einleitung/index.shtml (accessed August 12th 2009).

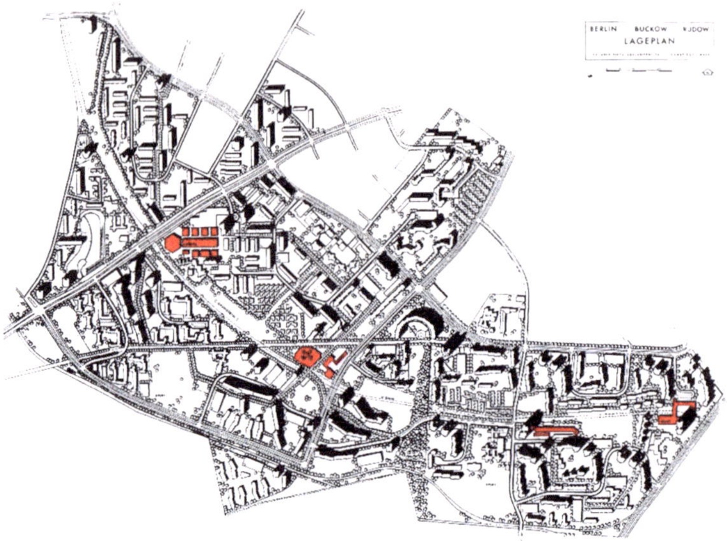

Fig. 24.10 Train lines in West and East Berlin and new housing settlements: The Gropiusstadt, 1965 (*Source* Cabrini, in: Alfaro d´Alençon (2013)). In 1962 the construction of the quarter began in the periphery of West Berlin. The connectivity to the *S-Bahn* was not considered. The new subway line U7 was developed to address the area's transit needs. The four sections of the new centres within the housing development were positioned around the new subway stations

land available for redevelopment. The reunified city, however, faced not only to cope with now redundant capitalist and communist planning policies, but also to reconcile the matter of a civic realm with doubled functions such as industrial areas, cultural venues, political centres and (transportation) infrastructure systems (Fig. 24.14).

The integration and merging of its historical layers—along with the preservation of significant building structures—became central themes in the debates surrounding Berlin's architectural and urban development. The influential factors were the return of the government to Berlin as the new capital city for Germany, the gradual withdrawal of government subsidies from industries, and the influx of young people moving to the city. With expectations that Berlin's population would double, a new dense urban core was envisioned. The pressing issues of urban development, identity formation, and spatial organisation were inherent to the debate and the Planwerk Innenstadt (Inner City Plan) developed in the 1990s. The plan was the first comprehensive urban development proposal for a reunified Berlin, devised by the City Senate and in collaboration with diverse architects, administrations and urban planners. The plan outlined a master vision for the 30 squares kilometres the city, particularly for the inner-city areas, encompassing both City West and City East. The proposal aimed to place new urban programmes in the neutral 'terrain vague' between East and West, creating a new urban nucleus—distinct from the existing city fabric (Oswalt 2000:86). The concept emphasised bridging the divides of the

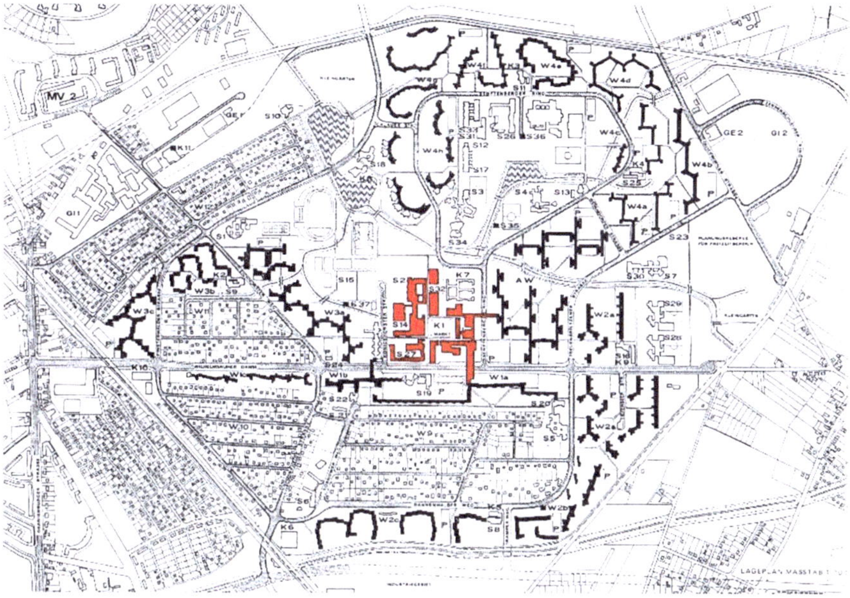

Fig. 24.11 Train lines in West and East Berlin and new housing settlements: The Märkisches Viertel (*Source* Cabrini, in: Alfaro d´Alençon 2013). The area was bordered by the northern ring line and the Berlin Wall to the east. Despite the proximity of the Wittenau S-Bahn station, the planning was carried out without incorporating this existing transport link. The master plan envisioned the construction of a new subway station as part of an extension of the current U8 line. However, this extension was later realised, leaving the mass housing area with limited connectivity to Berlin's city centre, contributing to its relative isolation from the urban core

former East and West Berlin, while also fostering a cohesive urban landscape yet historically resonant urban landscape.

Efforts to redefine, reconstruct, and repurpose Berlin's historical centre following reunification encountered multiple obstacles. The large-scale dismantling of the former GDR's central functions left a functional void, rendering the area's spatial structure, significance, and economic value increasingly uncertain. This functional gap was further exacerbated by the physical vacancies in the city centre. Additionally, much of the societal knowledge regarding the history and unique characteristics of the urban fabric in central Berlin had largely faded. Finally, it is essential to highlight Berlin's subcentres, which embody one of the most significant achievements in the city's urban planning: The polycentric structure of the metropolitan area (Bodenschatz 1995). Numerous town halls from the imperial era still stand today of this tradition and persisted throughout the twentieth century, particularly in the western parts of the city, where subcentres experienced significant development during the Weimar Republic and post-war periods with large department stores and cultural buildings.

The challenge, however, was not just restoring a historical identity but also repositioning the city on a national and global scale. The *Planwerk Innenstadt* reflects these

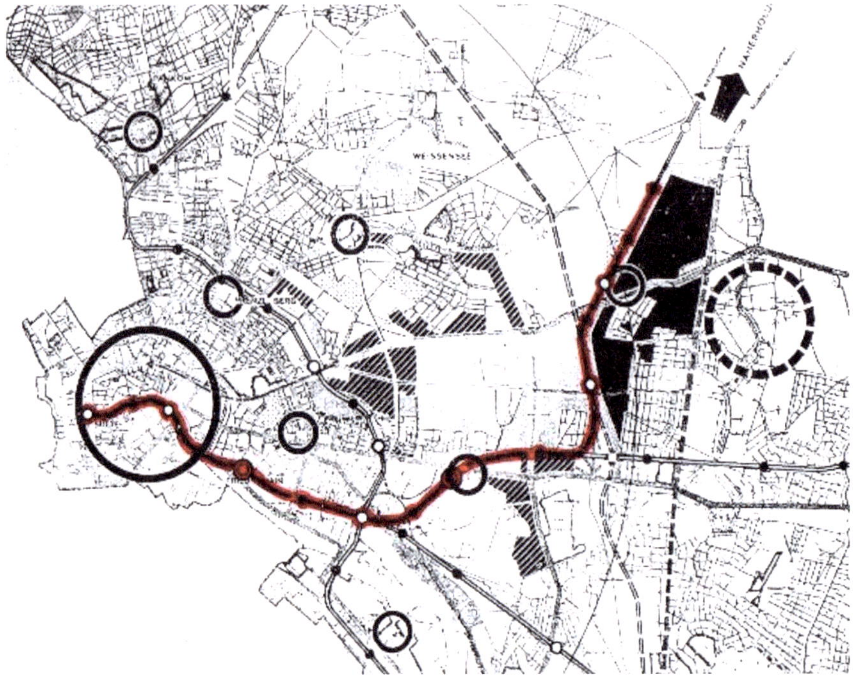

Fig. 24.12 Train lines in West and East Berlin and new housing settlements: The Hans-Loch Quartier, 1972 (*Source* Cabrini, in: Alfaro d´Alençon 2013). Located on the outskirts of East Berlin, the Hans-Loch Quartier in the 1970s underwent significant planning around existing train infrastructure. The Hans-Loch S-Bahn station, originally located further from the new residential area, was relocated closer to the development to improve accessibility. The repositioned station became the central hub for the community, surrounded by essential amenities such as schools, shops, and sports facilities. This strategic relocation exemplifies how East Berlin's planners leveraged the S-Bahn network to integrate transportation with community services, fostering a self-sufficient urban environment

challenges, addressing issues such as: (1) Demographic structures and changes; (2) urban typologies; (3) transportation networks, and sustainable development strategies. Central to the plan was the revival of the city's mixed-use zones, combining labour, retail, transportation, entertainment, and residential functions. The idea was to strengthen Berlin's polycentric structure by reinforcing existing urban centres and integrating transportation infrastructure, such as the Stadtbahn, into the new developments. The overarching goal was to avoid the creation of homogenous, monofunctional urban spaces by tailoring development to the unique conditions of each district.[4] Given Berlin's traditional European urban model, characterised by a clear

---

[4] Project group: Senatsverwaltung für Stadtentwicklung, Umweltschutz und Technologie: Plan development: Peter Jürgens, Holger Orb, William Rauch, Frank Studzinski, AG Arch.-Büro Albers u. Hoffmann-Axthelm: Prof. Bernd Albers, TobiasNöfer, Dr. Dieter Hoffmann-Axthelm, AG Arch.-Büro Ortner u. Dr. Neumeyer: Prof. Manfred Ortner, Markus Penell, Prof. Dr. Fritz Neumeyer Büro für Städtebau Machleidt and Partner: Hildebrand Machleidt, Stefanie Kirchner,

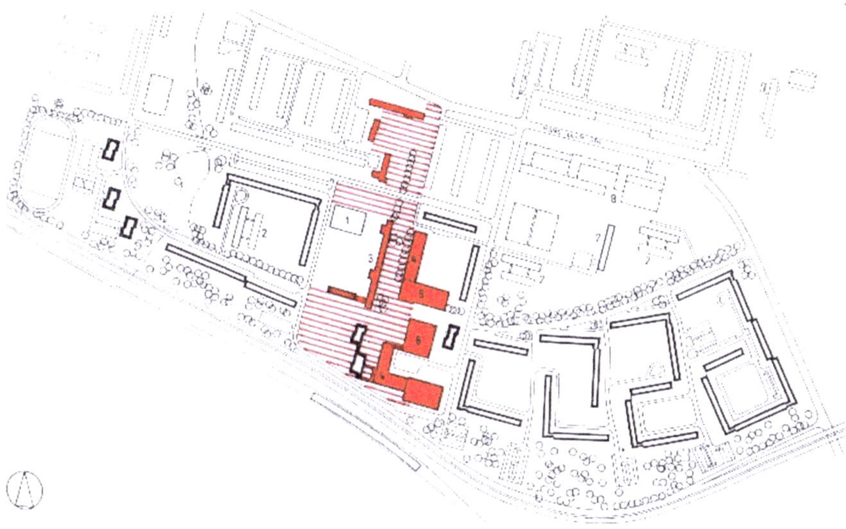

Fig. 24.13 Train lines in West and East Berlin and new housing settlements: The Alexanderplatz, 1974 (*Source* Cabrini, in: Alfaro d´Alençon 2013). The quarter was part of the redevelopment of Alexanderplatz in East Berlin. The East German government was tasked with integrating the existing Stadtbahn into its new urban planning proposals. The development showcased an example of housing and urban projects and its spatial relationship to the transportation infrastructure, with particular attention given to how these residential areas were connected to the broader urban transit network. This close integration of housing and transportation was intended to promote accessibility and efficient mobility within the rapidly urbanising landscape of East Berlin

division between public and private realms, the challenge was to envision public spaces that supported both transportation functions and the quality of urban life. To ensure the implementation of the planning proposal, economic feasibility studies were applied to the areas of Planwerk Innenstadt.

24.3.1 The Critical Reconstruction and *Planwerk Innenstadt*

The new Berlin may be the world epicentre of this brand of dialectic redesign. The rebuilding of the city—not the immediate post-war slash-and-burn urbanism of misguided bureaucrats and traffic engineers on both sides of the Wall all seeking

GRI: Bodo Fuhrmann, Christiane Dumm, Participating Administration: Bezirksämter Mitte, Kreuzberg, Schöneberg, Tiergarten, Charlottenburg, Friedrichshain Senatsverwaltungen: Bauen, Wohnen und Verkehr, Schule, Jugend und Sport, Finanzen Urban Studies and analyses of demographic development: SenStadtUmTech: Heidrun Nagel, Dietrich Flicke, City Forum/CityProject: Philipp Meuser.

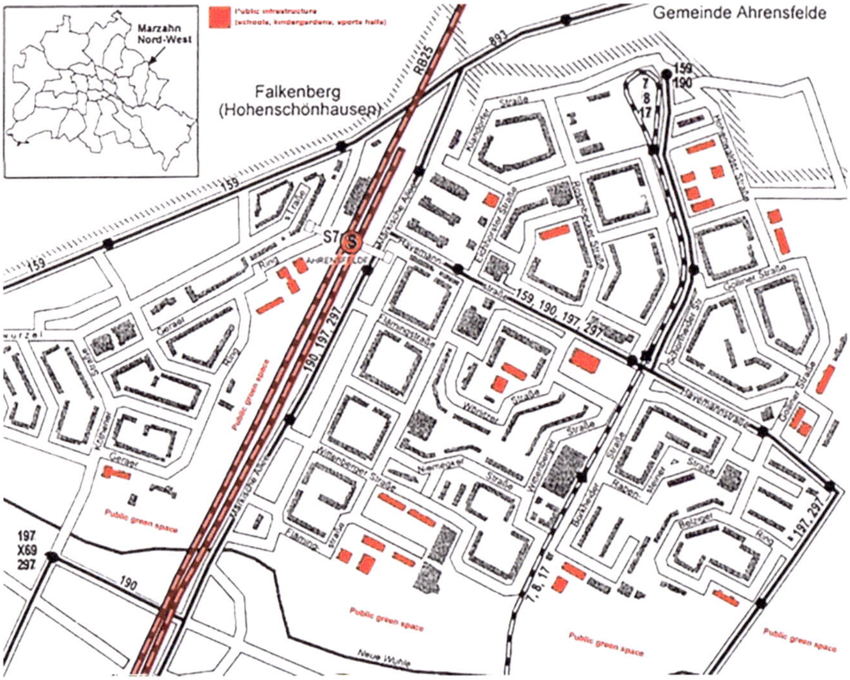

Fig. 24.14 Train lines in West and East Berlin and new housing settlements: Marzahn, 1979 (*Source* Cabrini, in: Alfaro d´Alençon 2013). In 1979, the construction of the large housing area Marzahn began. The plan foresaw a close relationship between the development of housing, industrial areas, public space and the new train station Marzahn. This station also worked as a centre equipped with residential towers and pedestrian areas. This example became the Marzahn model and was copied in the 1980s in other peripheral housing areas like Hohenschönhausen, in the outer ring, and Hellersdorf, north of the Ostbahnhof

to exorcise the ghosts of Germany's past, but the post-Wall Critical Reconstruction, whose interpretative redesign formulas were originally tested in West Berlin's International Building Exhibition, or IBA, in the mid 1980s—exemplifies, on a large scale, the stealth conservatism. (Schmaling 2005:24).

The post-Wall rebuilding of Berlin exemplified a specific form of urban redesign, often referred to as Critical Reconstruction. This approach was meant to address the tension between preserving the city's historical urban fabric and incorporating modern developments, creating a common historical and spatial configuration, and assisting the future development of the inner-city realm. As Schmaling (2005) observes, Critical Reconstruction involved a dialectical, interpretive approach to rebuilding the city, aiming to repair the damage inflicted by Albert Speer's mega scale and decades of post-war functionalist planning. The strategy applied the Städtebauliches Leitbild (urban planning concept) of the Compact City[5] accompanied by the

[5] In the discourse on sustainable urban development and as anti-movement to urban sprawl in the early 1990s in Europe the proposal arose of a Compact City (e.g. Ascher). The compact city concept

traditional Berlin city block structure and the 22 m eaves height in accordance with the city's building code, as to weave with different urban structures for the pattern of a uniform master plan. The planning process was highly participatory, involving public workshops where scholars, architects, and citizens could voice their concerns and critiques over a five-year development period.

However, Critical Reconstruction faced significant criticism ensued promptly: Firstly, with the expectations to become a capital city on a par with the physical grandeur of its European counterparts, Critical Reconstruction was transformed into a rigid set of architectural imperatives under Berlin's post-Wall city planning director Stimmann, who explicitly focused on Berlin's elusive genius loci 'Das steinerne Berlin' (The Stony Berlin) and to rebuild a city unified by Prussian classical order and by streets lined with their building codes (Jaeger 2006). Secondly, academics in the architectural and urban planning realm, as to mention the prominent once as Rem Koolhaas or Mathias Ungers criticised a form of selective memory, denying the urbanistic heritage of the former German Democratic Republic (GDR) and imposed a narrative that ignored the city's socialist past (Huyssen 1994). To exemplify the urban design advocated by Critical Reconstruction, Kleihues and his followers adeptly utilised nostalgic imagery from black-and-white postcards portraying scenes of bourgeois comfort and the lively streets of Berlin during the Roaring Twenties. It's worth noting that this imagery overlooked the presence of Nazi times, and the socialist urban development of East Berlin (Schmaling 2005). This selective focus led to further debates, particularly as critics argued that the city's future should not be based solely on nostalgic images of its past by prioritizing the architecture of the Gründerzeit. Berlin's late nineteenth-century expansion period, the city's post-war urban developments and the unique character of East Berlin were largely overlooked.

24.3.2 Political Demolition vs Historical Reconstruction

Against this background, the deconstruction of Communist-era Berlin, including the demolition of the Palast der Republik (Palace of the Republic), was a defining aspect of the post-Wall urban transformation (Figs. 24.15 and 24.16). The building a symbol of political power in the German Democratic Republic (GDR), stand where the former Berliner Schloss (Berlin City Palace). The building, built in 1443 on the Schloßplatz, was heavily damaged during the Second World War and demolished in 1950. The Berlin City Palace was than reconstructed with the Humboldt Forum within the 2010th year, while the Palace of the Republic itself was demolished in 2006. Till today scholars dedicates a focus to the site, thereby reflecting on existence and history, as well as addressing the debated question of why the Palace of the Republic was demolished and replaced by the Humboldt Forum—designed to

aims for high-density mixed-use and dense urban form. Urban activities are located within micro circulation radius to ensure better access to services and facilities based on its accessibility from the public transport network, walking, and cycling.

resemble the baroque palace and incorporating parts of the original concrete foundation of the palace preserved underground (Humboldtforum.org 2024). Discourses demonstrated that the demolition of the Palace ultimately constituted a political project. Contrary to the widely circulated legend among opponents of the demolition, this demolition was democratically legitimised through several votes with overwhelming majorities, unlike hardly any other construction project in the history of the Federal Republic of Germany. The reasons for rejecting the building were diverse: For some—by no means all—opponents of the GDR, the construction primarily symbolised the authoritarianism of the Socialist Unity Party. For certain traditionalists in the Berlin urban landscape, the aesthetic modernism of the building was abhorrent. Some nationalists desired the return of the palace facades to demonstrate the continuity of the second Federal Republic with the first German unification of 1870. Prussia enthusiasts among planning and architectural circles saw an opportunity to improve Prussia's image (Bernau 2022). Hence, the vision was set on a view that the GDR's urban planning contribution was incongruous with the city's prewar, modernist past and the post-modern urbanism of the future with former urban structures. With priority given to the 'Gründerzeit'—the period of significant urban expansion in Berlin from 1870 to 1910 the city experienced rapid industrialisation, population growth, and architectural development—rather than the developments of the 1970s (Huyssen 1994).

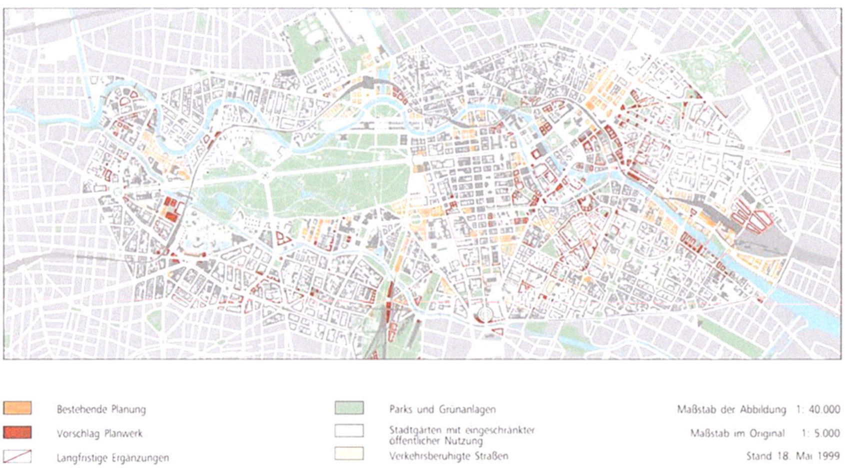

Fig. 24.15 Planwerk Innenstadt for Berlin. *Source Senatsverwaltung für Stadtentwicklung, Bauen und Wohnen, Berlin.* The Planwerk Innenstadt initiative cantered its inner-city redevelopment strategy on promoting urban density and reconstructing the historical morphological patterns of Berlin, with specific objectives that encompassed: **a** Analysing demographic structures and tracking population changes over time; **b** assessing the utilisation and typologies within urban spaces to better align with contemporary needs; **c** developing an integrated urban traffic network, with particular attention to the roles of the Stadtbahn, central stations, and broader transportation systems as core elements of Berlin's renewed urban concept; and **d** formulating planning strategies aimed at achieving sustainable development across the various facets of the proposal

Fig. 24.16 Left: Destruction of the Lenin Monument in 1991. Right: Palast der Republik in GDR times. *Source* Vera Rüttimann. https://static99.evangelisch.de/get/ccd/JEkYeuEib4RptHoqjdTAeMMu00082181/i-311. *Source* https://commons.wikimedia.org/wiki/File:19860503400NR_Berlin_Palast_der_Republik_Marx-Engels-Platz.jpg

24.3.3 The Transportation-Planning Concept 'Pilzkonzept' for Berlin

The *Pilzkonzept* (Mushroom Concept), formulated in 1990, arose from a broader discussion on the reorientation of Berlin's railway system to meet the city's evolving transportation needs (Fig. 24.17). The central aim of this proposal was to foster intermodality by enhancing urban connectivity and providing seamless access to long-distance networks. This concept was integrated into Berlin's larger urban planning strategy, particularly the Planwerk Innenstadt, which sought to restructure the city's infrastructure after reunification (Kuhlmann 2006: 106). The Pilzkonzept emphasised the development of a decentralised network of transportation hubs rather than focusing solely on the creation of a single central station. The initial plans advocated for multiple long-distance junctions distributed throughout Berlin's polycentric structure, facilitating urban mobility while respecting the city's spatial organisation. One of the main outcomes was the establishment of the Hauptbahnhof (central station), which became a key intermodal hub designed to link Berlin with major German cities and other European capitals.[6]

24.3.4 The Renaissance of the Stations' with Projekte 21

Following the establishment of Planwerk Innenstadt and the Pilzkonzept, the early 1990s saw the launch of the 'Renaissance of the Stations' initiative, with the 'Projekte 21' programme spearheaded by the state, Deutsche Bahn AG, and a network of architects and urban planners. The primary objective was the revitalisation and

[6] See www.s-bahn-berlin.de/fahrplanundnetz/sbahnhofanzeige.php?ID=119 (accessed March 5th 2024).

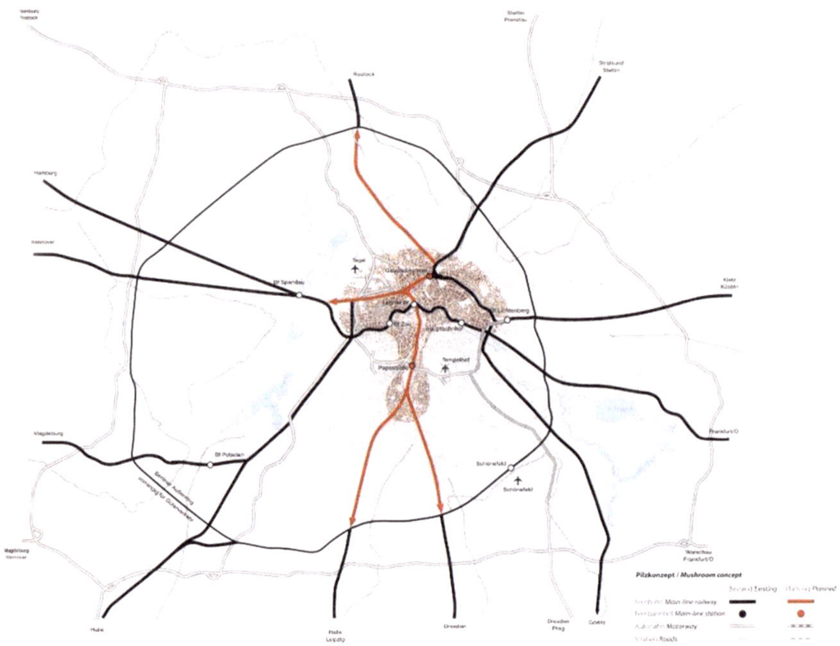

Fig. 24.17 *Pilzkonzept* (*Source* Kuhlmann 2006: 156). The intention of the concept was to connect suburban, trans-regional and national networks through defined points of inter-modality, catering to dense urban interconnectivity and providing direct access to long-distance networks. In detail, the proposal was the development of the north ring of the *S-Bahn* for international traffic and the development of the north–south track as a tunnel, connecting the northern section of the inner ring to the suburban line. With the proposal of a central station in the former space of the Berlin wall between East and West

upgrading of Berlin's inner-city railway stations, focusing on the integration of transportation infrastructure with urban redevelopment. This initiative was critical to the transformation of the area surrounding Berlin's Hauptbahnhof, which was historically a peripheral railway site. As Berlin's urban centre densified, the surrounding railway spaces were increasingly viewed as valuable urban assets. The 'Projekte 21' programme sought to redevelop these spaces for mixed uses, blending transportation, commercial, and cultural activities. The renovation process emphasised additionally a dialogue between historical and modern architectural elements, ensuring that the stations' iconic features, such as facades and glass roofs, were preserved while incorporating new design elements to suit contemporary needs.

24.3.4.1 Expanding Functionality: Railway Stations as Urban Centres

In line with global trends of adaptive reuse, many of Berlin's redeveloped stations aimed to extend their functionality beyond mere transportation hubs. Former track areas were repurposed for cultural and commercial activities, transforming stations into focal points of urban life. The 'Projekte 21' initiative was not merely about restoring these historical structures; it also involved reprogramming them to serve as symbols of Berlin's post-industrial identity, akin to other successful transformations like the Tate Modern in London or the Hamburger Bahnhof in Berlin, designed by the architect Kleihues.[7] The initiative was accompanied by a massive rebuilding of railway stations in the late nineties with its aim was to meet transportation, urban and commercial demands in an integrative planning development. The functionality of the stations selected for renovation was extended, adjoining areas (Dürr 1997).

According to the authors of 'The Renaissance of the Stations', they would create important urban areas and improve the city's quality of life (Steguweit 1997). In the framework reference was made to the historical image of stations and their squares during their zenith in the 1920s. The railway station was an urban highlight, a first-class address for a city's commercial and social space and as such, it formed the surrounding neighbourhood, the Bahnhofsquartier. The associated exhibition to the projects 'The Renaissance of the Stations', on the role of the railway in Germany was held in Berlin in 1997. The primary motive of the exhibition was to promote the increased reconstruction of railway areas and stations. One of the most important players of this initiative were the Deutsche Bahn. Its new management structure made profitability and economic viability the key issues. DB's intention to go public resulted in an interest in profit-oriented businesses as well as a revaluation of the stations and railway owned land and properties. As far as the railway authorities intended to provide as much transportation coverage as possible for the benefit of the population, the Deutsche Bahn AG developed marketing strategies and positioned themselves in the transport market as 'the brand with high level of potential recognition' (Steguweit 1997).

24.3.4.2 The Central Station in the Sparsely Populated Area

The completion of the Hauptbahnhof (Central station), officially opened on May 26, 2006, marked the culmination of Berlin's railway station renaissance. The ceremony, coinciding with the 2006 FIFA World Cup, was widely covered in the media, drawing attention to both the station's architectural significance and its potential as a key urban landmark. However, despite the media fanfare, the location of the central station in a relatively sparsely populated area raised concerns about its future viability. While the station architectural and transportation functions were widely praised, its integration into Berlin's urban fabric remained a topic of debate. Critics questioned the lack of residential and commercial development in the surrounding district, suggesting that the station might not attract the volume of passengers anticipated (Jäger 2006).

[7] Paul Kleihues's writings and article titled "Städtebau ist Erinnerung " (Urban Design is Memory), featured in "Internationale Bauaustellung Berlin 1984/87," In: Die Neubaugebiete. Berlin: Bauausstellung Berlin GmbH, 1993], pages 14–29 on "Critical Reconstruction" mirror many of Rossi's earlier points and also integrate concepts from Colin Rowe and Leon Krier, with whom Kleihues collaborated in the early 1970s.

Some viewed the station as a missed opportunity to create a more vibrant urban area, given its location outside the existing city core. However, the station's success in attracting retail tenants and its role as a transportation hub indicated that the commercial potential of the surrounding area might eventually be realised.

24.4 Conclusions: Urban Projects as a Laboratory for Learning About the Process of Urban Space Production

It is evident from Berlin's urban development, from division to reunification that urban space emerges from hybridisation, shaped by different social, political, and economic forces, their conflicts and negotiations, spatial ruptures, and subsequent reconfigurations. Urban planning in Berlin has functioned not only as a technical and architectural process but as a powerful catalyst for broader social and political transformations. Planners, developers, investors, and residents—produce a dynamic urban landscape marked by diverse ideologies, shifting political agendas, and changing needs.

24.4.1 Urban Space as Political and Social Catalyst

The complex legacy of Berlin's division brought to contrasting urban development trajectories in East and West Berlin and highlights the profound impact of political systems on the shaping of urban space. In conclusion, this paper underscores the transformative impact of urban planning initiative, the significance of the 'City Project' in shaping critical dynamics, particularly regarding political processes. Berlin's division created not only physical barriers but also divergent socio-political visions for urban life. In this context, urban design became a tool for constructing a new civic identity. This reshaping of forms stands as a physical indicator of the constant social impact on the city space. In this sense, urban space is not a static backdrop but a site of negotiation between the past and future, where Cold War era ideologies and urban concepts interact to form new urban paradigms. As Koolhaas suggests, cities thrive of opposites rather than through an idealised order. Berlin's urban evolution exemplifies this complexity and spontaneity inherent in urban development processes, emphasising the need for a nuanced understanding of cities as constantly evolving entities.

In the process of reunification, urban planning became an arena for negotiation between competing visions of future and history, presenting Cold War era and post-Fordist urban concepts that would influence the city's development into the twenty-first century. However, the process was not a simple synthesis of two cities but rather an ongoing negotiation between historical memory, political ideologies, and emerging globalised urban trends. These urban interventions not only redefined the city's physical form but also its social fabric, ultimately contributing to a new spatial narrative for Berlin. Despite the transformative impact of reunification, this paper also highlights missed opportunities for deeper integration. The failure to address persistent spatial and cultural divisions post-1989 raises important questions about how urban design can mediate between political ideologies and the lived reality of its citizens. Urban interventions in Berlin often fell short of creating a truly cohesive urban space that bridged the symbolic and physical divides between East and West. Thus, the reunification of Berlin should not be seen merely as a process of spatial integration, but as a site of ongoing contestation where competing visions of urban identity, history, and modernity continue to play out.

24.4.2 Transportation Projects an Ingredient for the City as a Project

From this paper it can be concluded, that transportation projects are an important component in the creation of city projects. As this paper presents, transportation spaces are not merely functional elements but are deeply tied to the city's collective memory and identity. The territorial expansion of the urban transportation space seen by the Stadtbahn, sets a standard for the idea of interaction of infrastructure lines with their spatial environment. The railway line with its arches partially morphologically reconfigured the inner-city of Berlin by breaking through the Hobrecht plan with its strict division of orchestrated public area and private inner block space. Lefebvre's concept of urban space as a product of social interaction resonates here, as Berlin's transportation systems—particularly the Stadtbahn—reflect the dynamic and contested processes of urban development.

In the divided city, transportation spaces were not just functional but became symbolic sites where connections and boundaries were constantly redefined. The post-1989 proliferation of new city projects highlights the role of transportation infrastructure in both the physical and symbolic integration. However, the transformation of these spaces was not merely a technical exercise in redeveloping infrastructure; it was an ongoing negotiation of identities, values, and visions for the city's future. The research shows that transportation spaces, especially rail systems like the Stadtbahn, have become sites of both mass transit and mass consumption, transforming into commercial hubs that reflect broader shifts in urban life. These spaces, once seen as 'non-places', have evolved into urban areas. However, there are no rules or historical automatism that can be applied whereby the refurbishment of a station automatically results in the improvement of the station square or the surrounding area. Moreover, the interaction between the transport infrastructure and its immediate surroundings—particularly the city—has led to the creation of diverse urban environments. These spaces, once perceived as peripheral or utilitarian, have become active sites of cultural production, social interaction, and identity formation. In this sense, the city's transportation systems are not just conduits for movement but active components of the urban fabric, shaping the city's spatial and social identity in ways that go beyond mere function.

24.4.3 Concluding Remarks: The Evolving City Project

Ultimately, the urban transformation of Berlin, from its division to reunification, exemplifies the idea of the 'City Project' as an ongoing process of negotiation and adaptation. Urban projects in Berlin—whether transportation infrastructure or larger city planning initiatives—are not simply the result of technical decisions but are shaped by political forces, social aspirations, and historical legacies. The city's ongoing evolution, driven by diverse actors and interests, offers valuable lessons about the complexities of urban space production. These insights challenge traditional models of urban planning and highlight the need for a more nuanced understanding of how cities evolve, adapt, and reconcile competing visions of the future.

Acknowledgements The authors would like to sincerely thank Julia Tasic for her valuable support as a research assistant at DFG KOPRO Int. Her contributions, particularly in preparing and organizing the visual materials, greatly facilitated the development of this publication.

References

Alfaro-d'Alençon P (2013) The production of urban space through transport infrastructure: the case of the Stadtbahn in Berlin. Fakultät VI Planen Bauen Umwelt, Technische Universität, Berlin

Bernau N (2022) Gab es keine Alternativen? Gedanken zum Abriss des Palastes der Republik in Berlin, dem das Humboldt Forum ein Themenwochenende widmet. In: Baunetz. https://www.baunetz.de/meldungen/MeldungenGedanken_zum_Abriss_des_Palastes_der_Republik_in_Berlin_dem_das_Humboldt_Forum_ein_Themenwochenende_widmet_7917055.html

Bodenschatz H et al (1995), Berlin - Auf der Suche nach dem verlorenen Zentrum, Junius Verlag, Hamburg (e-book en https://harald.bodenschatz.berlin/)

Camerin F, Longato D (2021) Urban impacts of Spain 1982 and Italy 1990 FIFA World Cup: a comparative analysis with more recent sports mega-events. Urban Res Practice 16(1):109–126. https://doi.org/10.1080/17535069.2021.1986126

Dant M (2006) In Jayne M (ed) Cities and consumption. Routledge, Abingdon UK

Dürr H (1997) Renaissance der Bahnhöfe: Die Stadt im 21. Jahrhundert. In: BDA, DB AG, DAZ & Meinhard von Gerkan (Edit.). Braunschweig/Wiesbaden, Vieweg Verlag

Flierl T (2013) Karl-Marx-Allee and Interbau 1957. Berlin postwar heritage between confrontation and co-evolution. https://doi.org/10.11588/ih.2013.4.20080

Florida R (2005) Cities and the creative class. Routledge, New York

Giedion S (1976) Raum Zeit Architektur. Birkhäuser Verlag, Basel, Boston, Berlin

Harvey D (1989) The condition of postmodernity. Blackwell, Oxford

Harvey D (2000) Possible urban world. In: Lotus International 110, Milan

Hegemann W (1930) Das steinerne Berlin. Geschichte der größten Mietskasernenstadt der Welt, Frankfurt a. M, Vieweg- Verlag
Huyssen A (1994) Monuments and memory in a postmodern age. In: Young JE (ed) The art of memory: holocaust memorials in history. Prestel Verlag, München
Jaeger F (2006) Der Baumeister des steinernen Berlins. Berlin. https://www.tagesspiegel.de/berlin/der-baumeister-des-steinernen-berlin-1312411.html
Kleihues P (1984) Städtebau ist Erinnerung, Internationale Bauausstellung Berlin 1984/87. In: Die Neubaugebiete. Bauausstellung Berlin GmbH, 1993, Berlin, pp 14–29
Koolhaas R (2015) There's been very little rethinking of what cities can be. https://www.fastcompany.com/3044008/rem-koolhaas-theres-been-very-little-rethinking-of-what-cities-can-be
Krier R (1989) Über architektonische Komposition. Klett-Cotta, Stuttgart
Kuhlmann B (2006) Bahnknoten Berlin: Die Entwicklung des Berliner Eisenbahnnetzes seit 1838. Verlag GVE, Berlin
Lefebvre H (1991) The production of space. Blackwell Publishing, Malden, USA
Mumford L (1961). The city in history. Harcourt, Brace & World, Inc., Harvest Books, New York
Neumeyer F (1990) Realität und Disziplin. In: ARCH+99
Orth A (1871) Berliner Centralbahn: Eisenbahnproject zur Verbindung der Berliner Bahnhöfe nach der innern Stadt. Verlag von Ernst & Korn, Berlin
Oswalt P (2000) Berlin, Stadt ohne Form: Strategien einer anderen Architektur. Prestel Verlag, München
Peters D (2009) The renaissance of inner-city rail station areas: a key element in contemporary urban restructuring dynamics. Critical Planning. 16(1):162–185
Peters G (1995) Kleine Berliner Baugeschichte – Von der Stadtgründung bis zur Bundeshauptstadt. Stapp Verlag, Berlin
Porter M (1995) The competitive advantage of the inner city. Harvard Business Review
Schmaling S (2005) Regeneration: design as dialogue, building as transformation. Harvard Design Magazin 23
Siewert HH (1978) Die Bedeutung der Stadtbahn für die Berliner Stadtentwicklung im 19. Technical University Dissertation. Jahrhundert, Hannover
Schulz M (2002) Wohnen in Marzahn Nord-West: Stärken und Potentiale eines Wohngebietes, Geographisches Institut, Humboldt Universität zu Berlin. Heft 65
Steguweit C (1997) Renaissance der Bahnhöfe: Die Stadt im 21. Jahrhundert, Braunschweig/Wiesbaden, Vieweg Verlag, In: BDA, DB AG, DAZ & Meinhard von Gerkan (Edit.)
von Lüpke D (1983) S-Bahn in Berlin (West) Konzepte zu ihrer Integration und Modernisierung. Regional Planning Institute at the Technical University of Berlin, Berlin

Chapter 25
Havana: Past, Present and Future

Gina Rey

Abstract This work contains a reflection on the emergence and development of the City of Havana so that its origins, subsequent development and the process of transformations that arose with the triumph of the Revolution and the proclamation of its socialist character can be known, and how it impacts a large city in a small country that also had to prioritize the development of the rest of the country, a situation that demanded large economic resources to make it possible. It attempts to present the social and economic achievements obtained from the application of the principles of socialism and the limitations and mistakes that occurred in the complex task of moving toward a better society for all.

Keywords Havana · Urban development · Heritage · Architecture · Socialist city

25.1 Introduction

Havana is not exempt from some of the urban problems of Latin American cities today. Fortunately, the city was not subjected to the indiscriminate demolitions to which many cities in the world have been subjected, with the consequent loss of its cultural heritage and its own identity.

As the capital of a small country with limited resources for its development, in spite of its recognized social advances, it has been affected for more than 60 years by the blockade and the current global economic situation. At present, in spite of the achievements in health, education and culture, the city presents a high degree of deterioration of its built-up area and its urban services infrastructure, circumstances in which the investment programs undertaken have not been able to reverse in the necessary dimension. As a result, these programs have not been able to satisfy to the necessary extent the difficulties of the daily life of the people of Havana, who continue to feel the effects of the crisis in terms of food, public transportation, energy, water supply and sanitation, and housing, among others. Maintaining jobs and social

G. Rey (✉)
Havana, Cuba
e-mail: ginarey@cubarte.cult.cu

F. Carrión Mena and S. Rodríguez Alvarez (eds.), *From the City as a Project to the City Project*, The Urban Book Series, https://doi.org/10.1007/978-3-031-94226-6_25

security, ensuring access to health and education, improving food, housing, and basic services for all, and developing the economy while preserving the environment are the main challenges the country faces today.

The needs of the city and its complexity demand substantial investments, which is why it is necessary to introduce a new model of economic and urban management with innovative instruments to optimize the use of available resources and provide a more effective response to social needs in order to continue advancing toward a fairer and more advanced society and city.

25.2 Antecedents of the Urban Development of Havana

Havana's privileged geographic position and the natural qualities of its large protected bay made it the meeting place for the ships that made up the fleet system, in order to protect them from attacks by corsairs and pirates during their voyage from the New World colonies to the Spanish Metropolis. The designation by the colonial government in 1561 of the port of Havana as a meeting place for the fleets brought about a substantial change in the dynamics of the city's growth, which began to be visible in the sixteenth century and reached its peak during the seventeenth century.

The gathering of the fleets in the port of Havana and their permanence during long stays determined the stay of a floating population of about 20,000 inhabitants, which exceeded the resident population at that time, and which was made up of the ships' crews, soldiers and passengers from Spain and the New World, who stayed in the houses of the settlers and in the convents.

Travelers arriving in the New World, including artists, writers, scientists, merchants, explorers and adventurers, influenced the culture and the cosmopolitan character of the population of the nascent colony and contributed with their books, research and drawings to the knowledge about Cuba and life in the nascent town. The relevance that the city achieved from its origins, earned it the title of city in 1592, and those of Key of the New World and Antemural of the West Indies in 1634 by Royal Decree of the crown of Spain.

The regulations of the founded settlement were governed by the Laws of the Indies, until 1574, when the Municipal Ordinances for the Villa de La Habana of Alonso de Caceres were issued (Fernández Núñez 2002), which regulated the urban layout, buildings and life in general in the new human settlements. From the urbanistic point of view, the construction of the walls and the fortresses that constituted the defensive system of the city were undoubtedly the most relevant events that took place in the young population.

During the seventeenth century the process of growth of the original settlement outside the walls occurred spontaneously along the existing roads that departed from the walled enclosure, radially, in different directions and communicated with the surrounding territory (Sorhegui 2007). The city walls created a marked social stratification between the walled enclosure where the wealthiest families lived and the suburbs outside the walls where poor families and free slaves settled. This growth was influenced by the existence of the Real Arsenal, where the ships of the Spanish Crown were manufactured, and where the primitive industries were located.

The increase of the population brought as a consequence that the urbanization outside the walls acquired an official character, and for this reason the original roads were paved and became causeways, which formed the main roads of the city, and along them were located the highest quality buildings and the most important urban functions. The first neighborhoods outside the walls were also founded following the layout of the roads that connected the Villa with other settlements on the periphery, such as Guanabacoa, Santa María del Rosario and Santiago de Las Vegas, which by the end of the century had also increased their population.

Havana's nineteenth century was characterized by a remarkable urban development in which the urban structure was consolidated and technological advances were made that placed Havana at the same level as the great European and North American cities. Due to the influence of the Industrial Revolution, there was an increase in the number of industries concentrated around the port, which was modernized, and at the end of the century, the railroad, the construction of the Albear Aqueduct, telephone and telegraph services, public lighting and street paving were inaugurated.

In the first half of the nineteenth century, important works were also carried out during the government of Captain General Miguel Tacón, which contributed to urban development due to the quality of the buildings constructed, such as the Tacón Theater (Chateloin 1989), today the Alicia Alonso Grand Theater of Havana, the creation of important public spaces such as the Paseo de Carlos III and the improvement of the roadways. The railroad system inaugurated in 1837.

The area outside the city walls expanded to the Belascoaín Causeway and in the second half of the century the Urbanization Plan of New Havana, promoted by Antonio María de la Torre, was undertaken in order to order the irregular growth of the area of Havana outside the walls, today known as Centro Habana, an urbanization that first extended to the Belascoaín Causeway and in the early twentieth century to the Infanta Causeway.

The emergence of two new urbanizations in that century generated important changes in the urban structure and in the use of urban codes. The first of these, El Cerro, was developed at the beginning of the century along the roadway of the same name, following an open urbanization model of isolated residences surrounded by gardens; this was the place where the Creoles settled and created institutions that contributed to the gestation of the Cuban nationality.

In the second half of the century, the urbanization of Vedado, promoted by the Count of Pozos Dulces, was inspired by the advanced ideas of modern urbanism and assumed the model of the garden city; many of the officers of the Liberator Army resided there when the war of independence ended. This remarkable development is produced by the boom of the sugar industry, with the location of a large number of factories in the western part of the country, and the tobacco industry, which was a determining factor in the economic power acquired by the Creole class.

As a result of urban expansion in various directions, at the end of the century the city consolidated as a metropolitan and polycentric area, as its urbanization spread along the roads and other thoroughfares, forming a conurbation with the historical nuclei that had originally been founded as independent settlements.

The twentieth century was marked at the beginning by the American intervention and the birth of the mediatized Republic. In this period, urban infrastructures were developed with the establishment of the electric tramway system, the construction of the sewage system and the organization of solid waste collection, which significantly improved the hygienic and sanitary situation of the city. The construction of Havana's Malecon was also of great urban transcendence as it opened the city to the sea in an extensive maritime front.

The increase in sugar prices as a consequence of the two world wars produced large profits that were invested in the real estate development of Havana, a process that gave rise to urban developments known as repartos, distinguished by their architectural and urban quality, such as Miramar, the old Country Club and the Biltmore for the upper class, and others for the middle class such as Santos Suárez and La Víbora (Rodríguez 1998).

As a result of the boom in real estate development, the urban area expands and social stratification is accentuated by the increase in precarious conditions in the plots of land and the old central neighborhoods and in the precarious neighborhoods of the periphery.

In the fifties, there was a great boom in tourism and with it a modernizing zeal that resulted in the construction of hotels to attract American tourism, and major infrastructure works such as the construction of the Bay Tunnel that responded to the increase of the automobile motivating the urban development of the east of the city. New urban developments were built following the influence of the American suburban urban model, designed to encourage the use of the private automobile. This process resulted in a major expansion of the city, which doubled its urban area and led to its consolidation as a metropolitan area.

25.3 Havana of the Socialist Revolution

The triumph of the Revolution in 1959 found a capital with disproportions and contrasts very marked in relation to the rest of the country, which presented a state of abandonment and impoverishment of its population in rural areas, hence the development option assumed was to concentrate the greatest resources in order to reverse this situation, prioritizing the social and economic development of the most backward territories. As a result of this policy, the high migratory movements to Havana that had been occurring up to that time diminished, together with a significant reduction in the birth rate as a result of the active participation of women in social life, which led to a considerable reduction in Havana's growth rates, unlike the predominant tendencies in Latin American cities.

That macrocephalic Havana of the fifties, which hid its poverty behind a mask of opulence, became the object of study of an incipient urban planning that arose from the advanced ideas of a group of young Cuban students and professionals, who, together with the struggle against the regime and the established social order, embraced architecture and urbanism, a discipline that by its very essence entails a public service and social architecture, as a paradigm of humanistic training and a deep sense of ethics and high social commitment, embraced architecture and urban planning, a discipline that by its very essence entails a public service and social architecture, as a paradigm of humanistic training and deep ethical sense and high social commitment, which represented the fight against commercialism in architecture and the defense of social urbanism.

With the triumph of the Revolution in January 1959, and the declaration of the socialist character of the Cuban project in 1963, new ideas and projects emerged in which utopia could be touched with the hand. Relevant facts were the nationalization of large foreign and Cuban companies and the socialization of social services such as public health, education and culture and the development of social housing.

This was a very intense stage in which the rhythms of socioeconomic transformations had a strong territorial impact and demanded accelerated responses from planners and urban planners. Havana, with 1 million 500,000 inhabitants, concentrated 20% of the country's population, 90% of port activity, 60% of industry and more than 90% of hospital beds. These great territorial differences determined the political decision to change the situation, for which massive programs of housing, schools, hospitals, industries and road works were undertaken, which allowed to bring development to the rest of the country and reduce the existing imbalances. As a result of the transformations, migration to the City of Havana was considerably reduced (GDIC 1998).

As a large city in a small country with limited economic resources, together with the imposition of an iron economic blockade by the United States, were the causes that prevented the country from having the necessary economic resources for its development. However, thanks to the policies implemented after the triumph of the Revolution, Havana has been able to preserve its historic center, a World Heritage Site, the valuable architectural heritage existing in other urban areas and preserve its urban image by not being subjected to indiscriminate demolitions as a result of land speculation that other cities have suffered with the loss of their own identity.

In the 1960s, housing complexes of great social importance were conceived and built to eradicate slums. A very significant example was the construction of the Camilo Cienfuegos City and other housing developments carried out by the nascent National Institute of Savings and Housing (INAV) in which an outstanding architectural and urban quality was achieved, designed by Cuban architects and urban planners on very favorable land near the coastal strip. In addition, large educational facilities were built, such as the José Antonio Echeverría University City and the Art Schools, and advanced scientific institutions such as the National Center for Scientific Research (CNIC).

From 1970 onwards, collaboration with socialist countries was established, particularly with the Soviet Union, which brought about significant changes in the country, both in the economic and social fields. This is why large-panel construction systems of Soviet technology were assimilated, based on the repetition of building models that made massive construction possible with the purpose of satisfying the social demand for housing for the most disadvantaged groups. The new neighborhoods were located in the periphery because of the facilities they offered to the construction process, which produced an extensive growth, and the free land in the already urbanized area was not taken advantage of.

Of great importance in the search for housing solutions was the creation in the 1970s of the Microbrigade Movement, created by organized labor collectives for the construction of housing for their workers, in which they contributed their labor force and the state provided the land, equipment, construction materials and technical assistance. A good part of the areas were under construction at the time of the triumph of the Revolution and were therefore urbanized or in the process of urbanization.

However, there was greater creative freedom in the programs of social works, in which groups of great architectural quality were developed, such as the Art Schools, the Jose Antonio Echeverría University City, and the refunctionalization of existing buildings for new functions as university centers, the project for a Clinical Surgical Hospital in a building originally built for a Bank and other buildings of private social clubs that were dedicated to schools, sports facilities and social recreational centers. This set of actions meant a reappropriation of the city by its inhabitants.

During this stage, the first City Master Plan of the revolutionary period was carried out, which contributed significantly to the definition of housing development zones, new industrial areas, road structure and green areas, sports, recreational areas on the beaches, and agricultural areas such as the Havana Cordon. The creation of the Physical Planning activity allowed the realization of plans in the main cities of the country, which had a positive influence on the development of the territories in a harmonious way, and therefore reduced migration to Havana (Baroni 2003).

In the 1970s, projects of great environmental relevance were developed, these were the large urban parks on the outskirts of the city: Lenin Park, the National Botanical Garden and the National Zoo, which gave access to the population to enjoy new forms of recreation healthier and instructive in an environment of great environmental and landscape quality. Through Fidel's visionary idea, the Lenin Vocational School was inaugurated to guide the training of the country's future scientists. Programs of great social importance were also carried out, such as the housing development zones of Alamar, San Agustín and Abel Santamaría, among others, and the programs of facilities for polyclinics, family doctor's offices, schools and children's circles, which were executed by the innovative movement of the microbrigades of the work centers and the social ones in the neighborhoods.

In the 1980s, two urban developments were developed: Las Arboledas, which follows the "garden city" model, and Villa Panamericana, inspired by the urban codes of the traditional city. These projects of notable urbanistic quality were very encouraging because they represented a qualitative advance in relation to the previous housing developments, which nevertheless constituted isolated examples although they managed to overcome the prevalence of prefabricated housing developments.

During this period, important health institutions were created, among them the Hermanos Ameijeiras Hospital, institutes dedicated to various medical specialties, social programs for health and education works such as polyclinics and children's circles, the Gato Aqueduct system, the EXPOCUBA fairgrounds, and the development of the Havana Metropolitan Park began.

In 1984, the Executive Committee of the Council of Ministers approved Havana's Master Plan until the year 2000, which recognized the need for a plan to order its development and serve as a guide for the location of investments to be made in the city. In the 1990s, as a result of the economic crisis experienced by the country during the special period, many of the projects proposed by the plan could not materialize, and problems worsened. However, during this period, the development of the Western Scientific Pole was undertaken, an important project linked to the biopharmaceutical industry, and urban agriculture was developed. The Great Metropolitan Park of Havana was also created and its development began.

The Group for the Integral Development of the Capital (GDIC), created in 1987, was significant in promoting an urbanistic will with a more integral conception of Havana's development, and at the same time disseminating innovative instruments and technological advances to address the solution of urban problems. The results of the diagnosis of the city's problems and potentialities coincided with those of the institutions and the inhabitants in the fundamental problems: Housing, transportation, water supply and solid waste collection, and the preservation of emblematic places such as the Malecón and the Almendares River (Oliveras and Barbón 2007).

At the same time, an experience was initiated at the community level, the Integral Neighborhood Transformation Workshops, which in its scope of action addresses the physical and social dimensions in a comprehensive manner, and creates self-management capacities to promote greater participation of community actors in initiatives aimed at improving life in those communities with the most precarious living conditions (Rey 2013). This experience was recognized by UN-Habitat when it was selected among the Best Practices at the Habitat II World Conference in 1996. Although its development has been limited by various factors linked to its own management and the availability of economic resources, it has been a sustained experience for more than 30 years, covering 20 Havana neighborhoods.

At the inauguration of the Group on May 27, 1988, the leader of the Revolution, Fidel Castro, said: "The declaration of Old Havana as a World Cultural Heritage Site helped us become aware of the need to preserve the values of the city in other areas such as Centro Habana, Regla, La Víbora, Luyanó, because each one has its own characteristics"."

The Development Strategy for Havana prepared in the early 1990s combined two levels: Global reflection on the projection of the city and direct action on social reality and the rescue of culture.

25.4 Havana Today: Challenges and Potentialities

Havana is the capital of a country that has achieved significant social progress. The revolutionary triumph prevented the city from being subjected to the indiscriminate demolitions that many cities in the world have suffered, with the consequent loss of its cultural heritage and its own identity, although it is not exempt from the urban problems of Latin American cities today.

Due to its condition as a large city, capital of a small and blockaded country, with limited economic resources for its development, Havana currently presents a problem of great complexity that is concentrated in the high degree of accumulated deterioration of its built-up area and urban infrastructure, which demand large financial resources, circumstances that the investment programs undertaken have not been able to reverse in the necessary dimension and require a change in the way of managing the city, to increase the efficiency of the urban development processes.

Havana is currently facing a demographic challenge caused by the reduction of its population, in which the main causes are identified as the accelerated process of population aging, the significant reduction of the birth rate and the external migration of young people (Oficina Nacional de Estadística 2018). This trend creates a complex scenario for which a set of actions have been implemented as part of a program aimed, among others, at stimulating the birth rate to gradually overcome the current situation and achieve a positive evolution in the future.

Centralized urban management has not stimulated community participation, preventing better use of the resources available in the territory. Scientific advances and innovative capacity to provide effective solutions to urban problems and development demands have not been incorporated into practice to the extent necessary.

Reversing the high degree of accumulated deterioration of the city's building stock, particularly housing and its infrastructure, entails large-scale rehabilitation investments that require advanced technologies and large economic resources. Housing programs have been dominated by a sectoral and centralized approach and, as a result, have lacked a comprehensive vision that considers the multiplicity of social, economic, cultural, and legal-normative factors involved in both formal and informal production, as well as their interrelationships (Mesías and Suárez 2002).

The city, beyond its deterioration, has preserved the beauty of the urban environments of different periods, attractions that distinguish it internationally, because fortunately it was not subjected to the indiscriminate demolitions that other cities have suffered, with the consequent loss of much of its cultural heritage and its own identity (Dirección Provincial de Planificación Física 2018).

Although the quality of the city's architecture is internationally recognized and admired, paradoxically, in the new architectural and urban planning projects, the loss of these attributes has been the object of criticism, which has been concentrated in the constructive interventions in the urban fabric that have caused a sensitive affectation of the quality of the urban environment.

The most far-reaching and innovative urban project that has had nationally and internationally recognized achievements have been the recovery of the Historic

Center of Old Havana (Dirección Provincial de Planificación Física 2018). This experience was extended with successful results to other historic centers in the country for the recovery of their cultural heritage, their historical memory and the improvement of their image, which has also encouraged tourist activity with a favorable social impact, although these experiences have not been replicated in other heritage areas of the city such as Centro Habana, Guanabacoa, Cerro and 10 de Octubre, where comprehensive programs are needed to rehabilitate and recover their valuable built heritage and improve the quality of life of their inhabitants.

25.5 Strategic Visions of Havana's Urban Development

A city of great cultural wealth, with a population of high cultural level, natural resources, social development, open, cosmopolitan by its own origin, Havana as the capital of Cuba has the ideal conditions to assume advanced forms of urban management and citizen participation, with the strengthening of the collective conscience and social cohesion cemented in Cuban society.

Havana possesses strengths that allow it to move toward sustainable urban development in its environmental, economic, social and cultural dimensions, based on its potential for development, its social model, state land ownership, the urban plan that orders its development and the capacity to implement advanced government management. As a heritage city, it has a high degree of conservation and authenticity of its architectural and urban heritage and treasures an extraordinary historical memory that must be preserved. Its areas of heritage value extend beyond the Historic Center of Old Havana, declared World Heritage, and include other areas of heritage value such as the Historic Center of Guanabacoa, the Plaza de la Revolución, El Vedado, La Rampa, the Art Schools and the Camilo Cienfuegos City.

The strategic visions for the future of Havana must contemplate an ecological approach in all projects, the link with nature, through measures that promote the resurgence of the symbolic city-port relationship, the relevance of its urban landscape composed of the sea and the coastal edge, the large parks, the natural landscapes, its vegetation and its rivers.

The fundamental objectives of Havana's Urban Development Plan are the preservation and recovery of its built heritage, the improvement of infrastructure, transportation, housing, health and education services, and the computerization of urban services to improve the quality of life and increase social equity and inclusion. The achievement of these noble objectives must be supported by an economic development that takes advantage of its cultural and environmental values with great attractions for tourism, cultural industries and creative enterprises, universities and scientific research.

The plan identifies three major urban areas: The central, the intermediate and the periphery, which have different urban characteristics, different population densities and degrees of land occupation, and in which coexist different qualities of urbanization and housing stock. In the central area, the oldest, the high degree of deterioration contrasts paradoxically with the high values of its cultural heritage, and therefore the priority should be directed to the comprehensive rehabilitation of its housing stock to improve its habitability conditions, which have as economic sustenance the resources generated in the valorization of its cultural, tangible and intangible heritage, cultural industries, design and information technology.

In the intermediate area, programs are required for the conservation and rehabilitation of the built-up area, the improvement and completion of services and infrastructure, and the use of available areas within the urbanized area for new construction to favor the compacting of the city and contain territorial expansion.

In the periphery of the city, where the most depressed neighborhoods and precarious informal settlements are concentrated, the strategy should be aimed at compensating for existing imbalances, based on programs to improve housing, infrastructure, social services, and accessibility to urban transportation, which will favor better physical and social integration with the city. The demand for jobs can be satisfied through the development of technology parks linked to universities and research centers and urban agriculture, complemented by local development projects that promote the use of existing cultural and natural resources.

The city has been productive since its origin and must maintain this condition. To this end, the urban plan contemplates the relocation of polluting industries, the technological modernization of existing industries, and the development of new high-tech industrial facilities capable of creating productive chains that take advantage of their opportunities.

25.6 For an Advanced Urban Management

The needs of the city and its complexity require substantial investments, which is why it is necessary to introduce an advanced urban management model with innovative instruments to optimize the use of available resources and provide a more effective response to social needs by strengthening the city government's capacity to manage the urban plan, reconcile investments and define its priorities in concert with economic stakeholders (García Pleyán 2014).

The city demands a management model that combines centralization with decentralization of those functions that require it, and the development of projects with local governments as protagonists, through the articulation of economic and social actors, state actors, cooperatives, entrepreneurs and communities, encouraging through various means their social responsibility and commitment to contribute to their economic and social development. The successful management model of the Historic Center of Old Havana, due to its results, should be a reference for the future urban management of the city.

As a result of the effects of the special period in the nineties of the last century, the City Master Plan was affected in its role as a guiding instrument for city planning and development, which has had a negative impact on the quality of life and the urban image. The authority of the urban plan is reinforced in the country's Economic and

Social Policy Guidelines, the Programmatic Platform of the Economic and Social Development Model to the year 2030, in which a political will is expressed to return to territorial and urban planning its role of foresight and projection of development, and means a support to the governing institution of physical planning and local governments (Carrión 2006).

The recent approval by the Council of Ministers of Havana's General Plan for Urban and Territorial Planning until 2030, creates a legal basis to guide a better planning of its future development and to promote the territorial conciliation of the investment plans of the different sectors, a crucial aspect to achieve an advanced urban management that meets the requirements of the city's development.

The management of the Urban Management Plan requires an instrumentation that translates its guidelines and proposals into decisions for the responsible administrative bodies and that is assumed as a guide for urban planning actions, in such a way as to overcome the sectorial and short-term vision that prevails in urban management. In order to achieve this objective, it is strategically important to establish its compatibility with economic investment plans that guarantee the integrality of the projects and actions to be carried out in the city. A legal framework for land management must also be implemented to promote the rational use of this valuable resource.

25.7 The Havana Can Be the Future

Havana is internationally admired for its cultural heritage, artistic culture and urban image. That urban image that makes it unique does not need to be transformed, but it does need to be conserved and enriched, preserving its valuable architectural heritage, reversing the deterioration of its building stock and significantly improving the precarious conditions existing in central and peripheral urban areas.

The principle of conservation does not deny the development of new projects if they are harmonious with the environment, with emphasis on the height of the buildings is one of the aspects that produce a greater rupture in the urban image. Fortunately, the city has unique attractions and does not need to erect skyscrapers or create a 'downtown' area that would threaten the image of its significant urban heritage sites, as has occurred in other cities that have become settings for highly speculative real estate development, resulting in the loss of both cultural heritage and local identity. Its favorable situation, as the urban land is almost entirely state-owned, means that it is not subject to the pressures of the real estate market, so that this valuable resource is at the disposal of social and economic needs and should therefore be used rationally in urban development.

The recovery of historic centers and urban heritage areas requires greater awareness of their cultural values and greater appropriation by government entities and local communities to ensure that they become major attractions for the enjoyment of the population, tourism, job creation and the generation of resources to finance the improvement of the living conditions of their inhabitants (Carrión 2006).

The recognition of culture as a development factor is becoming increasingly widespread. Havana can be a creative city that enhances the valorization of cultural heritage, artistic culture and design as factors of economic development, because they represent a great opportunity for the development of cultural and creative economies that have a promising future in Havana, as shown by the growing trend of the recording industry, publishing houses, international events, festivals and cultural enterprises, which are joined by design, computerization, communication and audiovisuals.

The feasibility of developing the local economy in the urban environment has been demonstrated by the growing number of small-scale economic enterprises that have emerged in recent years, which contribute a strong dynamic to the processes of economic revitalization and have an impact on the improvement of the economy of the city and its inhabitants. At the same time, they are forming a dense fabric of associative forms, based on the synergies created between cultural heritage, traditional trades, design and new information and communication technologies, gastronomy, lodging and a wide variety of services.

The revitalization of life in the neighborhoods must combine the rehabilitation of the housing fund with the promotion of local development projects that take advantage of existing resources. These processes must have the innovative capacity to promote local economic development based on the opportunities offered by existing endogenous resources such as cultural heritage, traditional economies, associative and other forms capable of generating resources to improve the precarious living conditions concentrated in the central areas and in the informal neighborhoods of the periphery; these should be managed by local governments, with the integration of state actors, cooperatives, entrepreneurs and communities, encouraging through various means their social responsibility and commitment to their locality, for social and economic improvement.

The introduction of advanced instruments, supported by information technology and innovation, is a strategy for optimizing the use of available resources and providing a more effective response to social needs, including e-government, the citizen portal and the computerization of urban services, which are in the process of being implemented. Other feasible initiatives include citizen consultation on urban plans and projects and the creation of urban observatories, which have had positive results in several Latin American cities. The implementation of citizen education processes starting at school is essential to foster roots, a sense of belonging and the creation of a collective conscience regarding the behavior, care and conservation of the city.

Urban management must strengthen the role of municipalities in channeling local initiatives, promoting citizen participation and the role of neighborhoods as urban structures that, because of their strong identity and social cohesion, are the repositories of Havana's rich intangible heritage. The changes demanded in urban management include the strengthening of the role of municipal governments as promoters of projects and investments of local interest and of popular councils to encourage the participation of social and economic actors in their demarcation.

The implementation of a mass rapid transit network by bus is an imperative, together with the organization of traffic by means of exclusive roads and lanes and an increase in the number of routes to cover the entire territory. The completion of the city's road structure will make it possible to achieve interconnection and accessibility to all parts of the city. A very important aspect is the improvement of pedestrian mobility to ensure the safety of a growing elderly population.

Havana can be a green city, hence the importance of continuing to move forward with the water network rehabilitation program to reduce water losses in the aqueduct system, improve the quality of service and, as a priority, promote water saving and reuse in various ways, including low-consumption technologies.

The sanitation of the rivers and the bay to reduce pollution have response in the execution of the investments in the sewage system that includes the treatment plants and the rehabilitation and construction of the networks. For the improvement of the environment and urban hygiene, it is imperative to create an integral system for the management of urban solid waste with the use of appropriate technologies for its recycling and economic use for various purposes.

As part of the concept of an ecological city, the structuring of a Green System is conceived, integrated by the large parks built in the seventies and eighties: Lenin Park, the National Botanical and Zoological Gardens and the Metropolitan Park of Havana, which has been gradually implemented. The proposal of the Green System of the City's Master Plan contemplates the solution of the great deficits of urban trees existing in the most densely populated areas. To this end, it proposes the restitution of urban trees in the streets, the creation of public spaces such as parks, micro-parks, park-streets with trees and regulated circulation, which will have a very positive impact on environmental improvement.

25.8 Major Projects to Boost Havana's Economic Development

Havana needs projects capable of translating the ideas outlined above into concrete achievements and generating economic resources to finance social problems. The existing city, as a high expression of the culture and cultural identity of a people, is the starting point for projecting the future. An understanding of its forms and its capacity for modification and transformation will lead to the formulation of new urban intervention projects.

What is the image that Havana should project as a city and as the capital of Cuba, expressive of its society, its cosmopolitan character and its universal vocation, within the framework of existing problems and available resources, how to implement an urban management that makes it possible to improve the functioning of the city and its image, raise the urban culture and effectively implement investment plans? Taking into account the preservation of its cultural heritage, what are the projects and actions to be carried out in the city to contribute to its economic sustainability, what would be the ways to achieve a greater dissemination of the city's values to institutions and inhabitants in order to raise awareness and their participation in the city's project?

The city has the capacity to develop urban projects that take advantage of its great opportunities, that serve as elements of economic dynamism at different scales, with the integration of a diversity of urban actors, and the implementation of innovative management models (Rey 2007).

The complexity of projects in the city requires an urban intervention that has the city government as the leader and coordinator of a plurality of actors made up of the public, cooperative and private sectors, associative entities and other non-governmental organizations, and those that may require foreign or mixed financing due to their large economic magnitude. An integrated insertion of the actors is the best guarantee that the urban project will be able to produce the dynamizing effects in the economic and social spheres to reach the materialization of its objectives.

The great project of the city for the twenty-first century is that of the Bay of Havana, due to its valuable historical-cultural heritage as a historic port, its natural and environmental values and its privileged location, and its synergic potential as a dynamic element of urban development in all its diversity. The recovery of the cultural heritage of the historic port and its surroundings, including the fortresses, the industrial and port heritage and the urban centers of Regla and Casablanca, create a highly attractive scenario for aquatic activities and the location of a diversity of innovative urban functions such as creative industries, communication technologies, exhibition centers and major cultural shows.

Another major project to be developed during the rest of the century is the urban rehabilitation and revitalization of Centro Habana, the old Havana outside the city walls that arose as a result of the expansion of the historic center during the eighteenth to twentieth centuries, a period in which a different urban fabric was generated from that of the historic center, with an architecture of great richness and diversity. In this area important cultural processes in music, theater, plastic arts, productive activity and manifestations of popular cultures that have transcended to the present day and constitute valuable components of the national culture were generated. It is considered the part of the city most representative of the spirit of Havana, in which the life of its people in the street becomes an image of great vitality and attractiveness.

The Centro Habana project envisages the revitalization of the traditional streets and the recovery of their architectural heritage, their commercial life and the insertion of cultural activities along their entire length. In this way, these vital arteries, which connect the city center with the periphery, would be rescued and make up a very unique urban image with their endless succession of columned portals that inspired Alejo Carpentier when he called Havana "the city of columns." The implementation of a management model that takes the Historic Center of Old Havana as a reference point will make it possible to allocate the economic resources obtained to support the rehabilitation of the housing fund and the improvement of the quality of life of the resident population (Leal 2006).

The city-university-research relationship is an essential aspect of the dynamization of the urban fabric, for the production of innovations and people with the capacity to promote projects that have as a platform the urban university campus to develop scientific-technological parks and business incubators. A strategic project is the creation of a development corridor that includes the overlapping of the gravitational areas of two universities, the University of Technology and the University of Computer Science, which create an innovative environment for technological development. This corridor connects with the area where the facilities dedicated to research-production in the field of biopharmaceutical industry are concentrated.

The continuation of the city's Green System is an environmental imperative, and as part of it, the Havana Metropolitan Park is of great ecological, recreational and scenic importance, due to its privileged location surrounded by densely populated areas in the central area of the city.

25.9 Final Thoughts

Havana needs projects capable of translating the ideas outlined above into concrete achievements and generating economic resources to finance social problems. The existing city, as a high expression of the culture and cultural identity of a people, is the starting point for projecting the future. An understanding of its forms and its capacity for modification and transformation will lead to the formulation of new urban intervention projects.

Havana has the conditions to become an international benchmark in terms of quality of life and sustainable development, so that the most advanced ideas in the field of urban planning can be applied on a global scale. For this purpose, it has the main resources: Its society, its culture and the scientific potential for innovation.

At the current juncture, the challenge is to take advantage of all the potentialities of an advanced society such as Cuba's to take on the challenge of Havana's development, with an integrated model of urban management that enhances cooperation between the different economic and social actors and citizen participation in urban development. Cuba has set out to face this challenge and move toward sustainable development in accordance with the 2030 Development Goals and the New Urban Agenda of UN-Habitat, which postulate among their objectives the achievement of sustainable cities and communities.

The premises for establishing the master lines of Havana's future development include the preservation of its valuable architectural and urban heritage, relevant both for its magnitude and its degree of conservation, its internationally admired urban image and its natural values due to the presence of the sea, the bay, the rivers and its vegetation. The preservation of heritage values acquires great significance in new urban interventions in the context of these valuable areas that require a harmonious integration between the existing city and the new architectural and urban developments.

Havana must preserve its character as a historic city, and in that spirit its historic port, sites and areas of historic-cultural value and the integrity of its historic fabric loaded with tangible and intangible values must be preserved, recognizing as opportunities for development the synergies produced by the interaction of a great diversity of resources and actors in a limited territory.

However, the strengths of Cuban society create the conditions to aspire to an ecological, creative, inclusive and innovative city that generates innovative and wealth-producing means, integrating culture, economy and society in an interactive system.

This exceptional city that is celebrating the 500th anniversary of its founding opens itself to the future with the political will to preserve its values and social achievements, to preserve the extraordinary heritage of its past, its cultural heritage and identity, and the aspiration to achieve sustainable development with a growing quality of life, providing spiritual and material well-being to its inhabitants. At the present juncture, the challenge is to take advantage of all the city's potentialities, to materialize the political will to make Havana a city of a socialist, sustainable and socially inclusive country, where the most advanced ideas in the field of social development and urbanism can be put into practice.

References

Baroni S (2003) Hacia una cultura del territorio. GDIC

Carrión F (2006) Gestión o gobierno de los centros históricos. In Manejo y gestión de los centros históricos. Ediciones Boloña

Chateloin F (1989) La Habana de Tacón. Letras Cubanas

Dirección Provincial de Planificación Física (2018) Plan General de Ordenamiento Territorial y Urbano de La Habana 2013–2030

Fernández Núñez JM (2002) Ciudades y Regulaciones Urbanas. Grupo para el desarrollo Integral de la Capital

García Pleyán C (2014) La Habana, ¿Una ecuación posible? Revista Temas

GDIC (1998) Estrategia. GDIC

Leal E (2006) La cultura como eje del desarrollo de los centros históricos. In Manejo y gestión de los centros históricos. Ediciones Boloña

Mesías R, Suárez A (2002) Los Centros vivos, La Habana, Lima, México, Montevideo: alternativas de hábitat en los centros antiguos de las ciudades de América Latina. Ciencia y Tecnología para el Desarrollo

Oficina Nacional de Estadística (2018) Encuesta Nacional de Migraciones (2016–2017). Oficina Nacional de Estadística

Oliveras R, Barbón MR (2007) Hacer ciudad y hacer barrios: Los Talleres de transformación integral del barrio. In: Hearn A (ed) Cultura, tradición y comunidad: perspectivas sobre la participación y el desarrollo en Cuba. Imagen Contemporánea

Rey G (2007) Los grandes retos de la ciudad. En: Ponce G (ed) La Habana. De Colonia a Metrópoli. Agencia Española de Cooperación Internacional

Rey G (2013) Del barrio a la ciudad: una experiencia de planeamiento participativo en barrios de La Habana. In: Bolívar T, Guerrero M (ed) Ciudades de la gente. Casas de infinitas privaciones. Abya Yala – CLACSO

Rodríguez EL (1998) La Habana: Arquitectura del siglo XX. Blume

Sorhegui A (2007) La Habana en el mediterráneo americano. Imagen Contemporánea

Part X
Epilogue

Chapter 26
Epilogue. The Importance of the City as Anticipation

Jesús M. González-Pérez

Abstract Cities have concentrated the technological and scientific advancements of various societies throughout history and, in many instances, are even an expression of these advancements themselves. In one way or another, all cultures have applied some form of regulation to urban forms or the design of new cities. The grid and the orthogonal plan are present in all stages and cultures. Until the nineteenth century, the principles of what is known as pre-scientific urbanism were responsible for organisation and/or planning. Urban theories relied on urban planning treatises and the design of city models, often utopian and projected under the principles of ideal cities. Many cities had an author's signature. Scientific urbanism resulting from urban growth due to the Industrial Revolution introduced not only new techniques and the emergence of complex urban laws, but also a succession of global movements and influences. The twenty-first century is a world of cities. Urbanisation is widespread and urban society dominates. The challenges for urban planning are extraordinary and, undoubtedly, new proposals and actions are required. However, it is essential to read and understand experiences from the past. Only in this way can we ensure good future choices.

Keywords Planned city · Orthogonal plan · Urban utopias · Ideal cities · Urban theories · Urbexplosion

26.1 Introduction

Every urban grouping is essentially a historical creation. According to Aldo Rossi (1982, 222), "cities are the text of history." The city itself is a repository of history yet, at the same time, history can be understood as the study of the very foundation of urban events (Rossi 1982). In this sense, many of the cities analysed in this book are, in themselves, ideas of the city, having influenced and transcended the history of urbanism beyond their physical form and the remnants we may preserve.

J. M. González-Pérez (✉)
University of the Balearic Islands, Palma, Spain
e-mail: jesus.gonzalez@uib.es

F. Carrión Mena and S. Rodríguez Alvarez (eds.), *From the City as a Project to the City Project*, The Urban Book Series, https://doi.org/10.1007/978-3-031-94226-6_26

The first civilizations and cities emerged and developed at different times from 3500 BCE in seven different places across the world: Sumer in southern Mesopotamia (modern Iraq), Egypt, the Harappan culture in the Indus Valley, China along the Yellow River, the Toltec and Olmec cultures in the Valley of Mexico, the Maya in Guatemala and Honduras, and the Inca on the coast and in the mountains of Peru. The first three are known as "dead" cultures from which the Greek, Roman, and Western Christian civilizations of Europe evolved. The three American cultures are similarly considered "dead," having been destroyed by Spanish conquerors in the sixteenth century (Aztec and Inca) or even before the conquest (Maya). Chinese culture is the exception, as it has lasted without interruption from the end of the third millennium BCE to the present day (Araújo de Oliveira 2022). The main goal of this final chapter is not to provide new research but to present a series of concluding reflections as a closure to the previous 25 chapters, focusing on the weight of the history of planning and the birth of urbanism that have led, along with many other factors, to a predominantly urban contemporary society in a world of cities.

26.2 The Topodynamic Corridor and Urbexplosion

Although contested (White 2012), Luc-Normand Tellier (2009) has developed an interesting theory about the history of the world seen through the spatial dynamics of urban and economic systems, and two key concepts: the topodynamic corridor and the urbexplosion. The topodynamic corridor refers to continental or global-scale corridors where urbanisation and economic development cluster. The three main corridors are the Grand Corridor, extending from Sumer northwards across the Mediterranean to Great Britain, and eastwards through southern and southeastern Asia to Japan; the Asian Corridor, traversing India, Indonesia and China; and the American Corridor, stretching from New York westwards to Los Angeles and Tokyo, and eastwards through London and Berlin to Moscow. Within each of these corridors, urban development is the result of a succession of urbexplosions which, from spaces of economic concentration, drive the progressive growth of a typically transnational urban system (Fig. 26.1).

Throughout history, one urban explosion replaces another. Typically, the time sequence is also spatial so that, for example, the first urban explosion centred in Sumer was replaced by the Hellenistic system centred to the west, in the Eastern Mediterranean, which was in turn followed by the Roman system even further west. According to Tellier (2009), this represents a space-time trajectory or topodynamic inertia essential for understanding history.

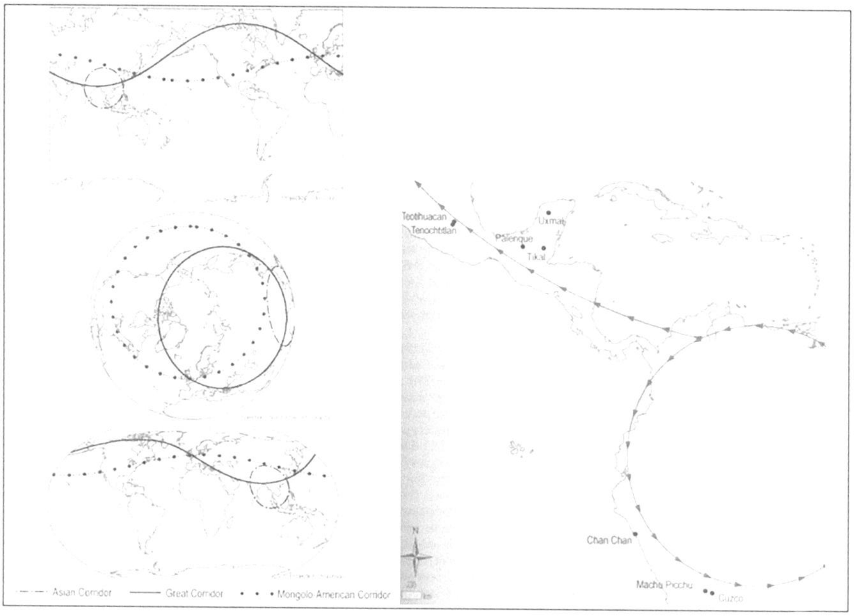

Fig. 26.1 Topodynamic corridors (left). Two pre-Columbian corridors (right). *Source* Tellier (2009)

26.3 Spontaneous City Versus Planned City

Since the origins of civilization, cities have been classified into two types: planned and spontaneous. Planned cities are commonly associated with ideal city models, to the extent that the most radical examples of exhaustive urban planning primarily come from the design of ideal cities. The desire to achieve a perfect physical and social environment drove thinkers and urban planners to design perfect and optimistic images of what the ideal city would be, with a relevant strong religious or ideological faith in each specific instance. Ideal images have mostly appeared during periods of social change, when the disintegration of a previous economic order facilitated the emergence of new cultural experimentation; a new urban model attempting to be applicable to different territories under various circumstances (Rosenau 1986). Although an organic expansion of the city would seem to represent the opposite end of the scale, the urban form of spontaneous cities also adheres to a set of basic principles and the total absence of rules is not a common thread in the history of urbanism. The relative repetition of formal schemes and the concern for adapting new growth to a pre-existing core or adhering to a set of norms when choosing a location for settlement, distributing religious and grand buildings, or constructing defensive walls could also be classified as conscious planning (Lois-González et al. 2012) (Fig. 26.2).

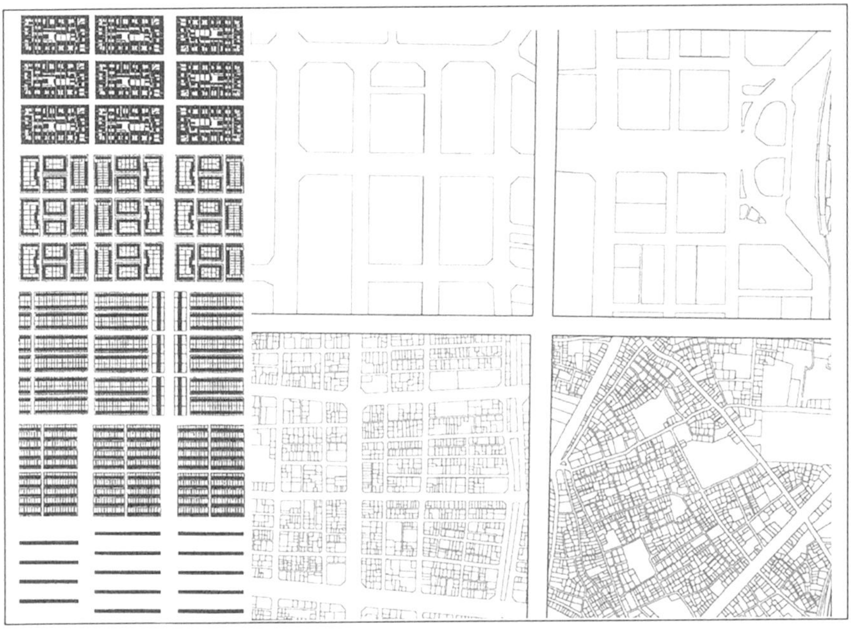

Fig. 26.2 Formes urbaines: de l'îlot a la barre (left). Right: streets, street blocks and plots at the same scale in Tokyo: Shinjuku (Nishi), Chiyoda (Marunouchi), Chuo (Ningyocho) and Sumida (Mukoujima). *Source* Panerai et al. (2004) and Satoh (2003)

26.4 The Grid and the Orthogonal Plan

The history of urbanism owes much of its prominence to the application of grid schemes in city planning. The orthogonal plan is characterised by its grid of streets intersecting at right angles (Capel 2002). These streets intersect transversally and longitudinally until, through the repetition of prototypes, they create chessboard, checkerboard or grid patterns. The grid comprises the intersection of two series of parallel lines, implying priority for a public movement system which, according to Galantay (1977), significantly contributes to maximising efficiency. The grid plan is the simplest to design and implement, with a layout that facilitates the three phases of the urbanisation process, sometimes through a single unit operation: parcelling, urbanisation and construction. The street and road acquire special significance. Buildings cease to be the structuring elements of the urban plan and the road network spearheads the entire urbanisation process in the new planned, orthogonal city. Buildings are forced to adapt to the urban form designed by the rationalised road network.

Signs of this approach are found in different civilizations that had no apparent contact: the Harappan culture in India (2154–1750 BCE), the Assyrians in ancient Mesopotamia (eighth century BCE), the Shang dynasty in China (1765–1123 BCE), the extensions of Tenochtitlan in Mexico, etc. Although adapting previous models, the scheme designed by Hippodamus and publicised by Aristotle remains relevant after 2000 years.

The orthogonal plan was the urban model for the first planned cities, but spread widest at the height of Ancient Greece. From this time on, the relationship between grid and planned city seems fully established. The model was used for the development of Roman urbanism, new mediaeval cities, the ideal cities of the Renaissance, urbanisation in the Inca Empire (fifteenth century), urbanism during American colonisation (sixteenth–nineteenth centuries), internal reforms of European cities, Spanish population expansions (nineteenth century), British new towns and a significant part of the monotonous and characteristic peri-urban developments in contemporary sprawl. While it is true that nowadays there are many more forms of streamlining urban plans, the grid remains prominent.

The simple and easy-to-apply drawing type is possibly the driving force for the grid's rapid spread over space and time. However, it is not the only reason. Authors such as Galantay (1977) and Capel (2002) highlight other advantages: it is the most convenient system for supervising and dividing land, and one of the easiest to execute using a rope and a ruler; it is a model of easy adaptation; it is useful for planning new settlements, as well as for the expansion of a pre-existing core; it facilitates recording, census preparation and tax collection; it simplifies buying and selling plots and properties, and with it, land speculation; and it expresses the superiority of conquering peoples and supports acculturation among the conquered. The grid was used to demonstrate the rationality of civilised life against barbaric and indigenous anarchy.

In opposition to the irregular and unplanned layout of imperial metropolises, one recognisable aspect throughout the history of urbanism is that the urban form of new colonial settlements developed from regular and geometric schemes, fundamentally through orthogonal plans (Lois-González et al. 2012). The Greek, Roman, Spanish, British, French and German empires applied planned urban schemes in conquered territories, but were unable to move away from organic growth in most home metropolises. Examples include the Greek colonies in southern Italy and Sicily; the Roman military colonies developed from the castra (fortified legionary camps); the new Spanish cities in America resulting from the so-called Laws of the Indies; the new cities in USA, including those with French origins such as Washington or New Orleans, etc.

These ordered geometric forms were valued and adopted by urbanists in industrial cities. Most modern urbanism authors view the use of the orthogonal plan positively. Just as the Renaissance had done, nineteenth-century utopians and Soviet constructivists magnified the use of geometric schemes. The progressive urbanism of architects like Le Corbusier found geometric design to be the most suitable way to achieve order, rationality and functionality. The progressive city rejected all artistic legacy from the past with the aim of conforming to the laws of "natural" geometry (Choay 1983). Although less influential, the opposing position is held by the neo-romantic authors of the late nineteenth century, who expressed a particular aversion to orthogonal plans. Against the large, abstract, open spaces with almost infinite perspective, Camillo Sitte advocated for a space cut out against the continuity of a backdrop of buildings. An ardent defender of landscape recovery and the life in ancient (mediaeval) cities, he supported a space adapted to the sinuosity of the terrain, prevailing winds or the comfort of users. In his nostalgic city of limited and close perspectives, opposition to any principle of symmetry is a constant. Le Corbusier challenged the street to the extreme. Sitte considered it a fundamental organ for a closed and intimate space. Straight forms versus curves, geometry versus irregularity, wide spaces against closed and intimate places, standardised forms versus typological diversity contrast progressive principles of the Athens Charter adherent with those of culturalists like Sitte. The Swiss architect considered the Viennese urbanist the embodiment of a retrograde vocation of the past: "The religion of the donkey path has just been created. The movement has started from Germany as a consequence of a work by Camillo Sitte" (Choay 1983).

26.5 The Birth of Urbanism

In this book, we have seen how urbanism as a reflection of the city emerges and evolves in parallel with history. However, as the development of an individual urban project referring to a specific site, urbanism is closely tied to architecture and is primarily understood as an art form. From the Renaissance until the second half of the twentieth century, political and religious power extensively utilised urban projects, leading to ambitious constructions, from Versailles or Aranjuez to Washington and Canberra (Zoido et al. 2013). During the Enlightenment, social considerations replaced religious ones and we entered a secular stage inaugurating our contemporary period.

Nevertheless, the origin of modern or "scientific" urbanism as a discipline with its own theoretical corpus is recent, dating back to the mid-nineteenth century when urban problems associated with industrialisation were insurmountable for eighteenth-century urban planning techniques. Urban layout proposals were not immediate and initially put forward as measures for internal reform; later proposals were for large-scale decongestion projects, even involving the creation of new cities. A key figure in understanding the discipline's scientific institutionalisation was Catalan engineer Ildefonso Cerdà. His "Teoría General de la Urbanización" (1867) is considered the first systematic work in the history of urbanism and the first on the theory of the city since the Renaissance era.

As the twentieth century progressed, urbanism increased its role as a regulated public practice. This evolution was likely spurred by the abandonment of utopianism and the gradual assumption of everyday administrative responsibilities (Lois-González et al. 2012). Authors such as Gravagnuolo (1998) lament this situation and criticise the intention to establish the foundations for the new discipline by promoting interactions with other related and presumably more "scientific" disciplines such as economics, law or statistics. The result was a discipline laden with abstract devices from nomothetic sciences. According to the author, the two-dimensionality of zoning or the algebraic jargon of standards is causing urbanism to forget the discipline's ultimate goal: designing urban forms (Gravagnuolo 1998).

26.6 Utopian Thought and Ideal Cities

The origins and foundations of ancient cities are often associated with fantastic tales and legends, at time with a historical basis and, at others, completely invented. The blending of real and mythological characters is so common that figures such as Theseus, Romulus or Aeneas are considered fundamental to the evolution of the history of Western civilization. Similarly, urbanism as a reflection on the city stems from visionary attitudes towards sacred places, often associated with games of symmetry and geometric figurations, and at other times linked to esoteric or hermeneutic interpretations, and the imagination of ideal societies that develop within a specific urban framework.

Throughout history, utopias have always been imbued with urban forms, and the main currents of urbanism have been influenced by some form of utopian thought. A long tradition of theorists have attempted to describe their model of the ideal city, and their proposals were charged with utopianism, ideology, and philosophy, to varying degrees of interest in applying their general projects to the real city. Most developed their ideas from studying urban art and, from the mid-nineteenth century, urban theory. Examples include Vitruvius, Filarete, Alberti, Owen, Morris, Garnier, Gropius, Cerdà, Le Corbusier, Sitte, Howard, Unwin, Wright, Geddes, Mumford, Lynch and Rossi. (Fig. 26.3).

However, other non-urbanist authors have approached the study of urban problems and designing an ideal urban framework from other even more expressive and artistic disciplines, such as philosophy, literature and cinema. The first two areas include figures as diverse as Victor Hugo, Saint Augustine, Campanella, Georg Simmel, Oswald Spengler, Martin Heidegger, Pyotr Kropotkin, Charles Dickens, Friedrich Engels, Karl Marx, Jules Verne, Italo Calvino, Hermann Hesse and Thomas More. The unique perception of urban reality and the creativity of certain filmmakers regarding different ways of creating cities have made cinema a fundamental tool for researching and reinterpreting the urban world (Costa 2003). On the one hand, cinema as an art form has the ability to clearly express contemporary urban phenomena or processes: the functionalist city and critique of the modern movement (*Playtime*, Jacques Tati), the art of architecture (*The Belly of an Architect*, Peter Greenaway), the diffuse city (*American Beauty*, Sam Mendes), the armoured city and New Urbanism (*The Truman Show*, Peter Weir and *Edward Scissorhands*, Tim Burton), etc. There are even films where the city becomes a main protagonist in the plot: *Metropolis*, *Playtime*, *Blade Runner* or *Brazil* (Niney 1994). In any event, these visions from critical perspectives do not inspire optimism. On the other hand, cinema has uniquely drawn inspiration from visionary urbanism to contextualise or feature scenes and

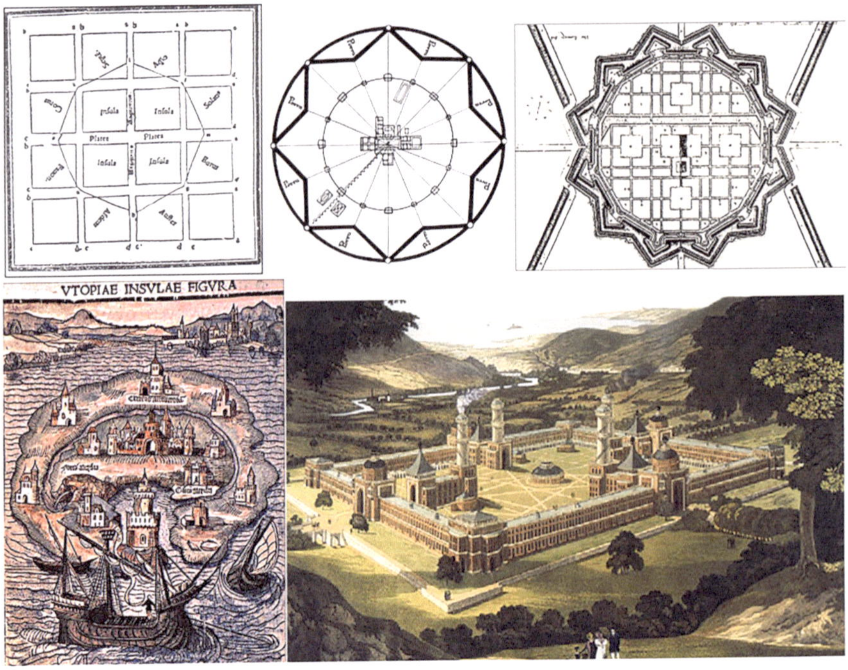

Fig. 26.3 Ideal cities. Vitruvius (15 BCE), Antonio Averlino "Filarete" (1457–1464), Vicente Scamozi (1615), Tomás Moro (1516), Robert Owen (1838). *Source* Rosenau (1986), Chueca (1998) (top). Cover of the 1516 edition of Utopia published in Leuven and Engraving of New Harmony by F. Bate, London, 1838 (bottom)

plots which, as true premonitions of the futuristic city, have become excellent examples of the modernist (e.g. *Metropolis*, Fritz Lang) and postmodern movements (e.g. *Blade Runner*, Ridley Scott) (Lois-González et al. 2012).

26.7 The Planned City in History: A Synthesis

The chapters of this book with their historical analyses, case studies, urban theories and models have shown how urbanism anticipates urban problems only in certain instances and how, for the most part, it is understood as a solution and response to the constant challenges facing the city.

For centuries, the city has been the product of urban theory, and architectural and urban treatises that have outlived their time and acquired global, even timeless, influence. For centuries, cities have borne the mark of their creators and, in one way or another, our cities are heirs to treatises or proposals by Hippodamus in Ancient Greece; Vitruvius in the Roman era; Alberti, Filarete or Scamozzi in the Renaissance; or Cerdà, Le Corbusier and Rossi in the industrial era.

We are aware of the city in the Ancient World primarily through writings found in works by classical authors and discoveries in prehistoric and protohistoric archaeology. The crisis of the thirteenth century began as a crisis of the city. Although the abundance of proposals, texts and urban practices from Antiquity declined during the Middle Ages, the period experienced prolific construction of new cities. The first signs of the modern city were seen in the fifteenth century and its urban models extended until the late eighteenth century. Despite the persistence of mediaeval characteristics and the resurgence of styles from the classical Greco-Roman world, the preservation and dissemination of urban treatises by well-known authors launched a new transitional phase towards contemporary urbanism. In comparison with the backdrop of the general anonymity of mediaeval artists, the Renaissance is characterised by strong individualism. The ideal plan is an aspiration present in artists of the calibre of Leonardo, Michelangelo, Raphael or Dürer. The monumental nature of many current European cities is largely thanks to Baroque works (Lois-González et al. 2012). The industrial city and the birth of urbanism represent a profound change: urbanism becomes scientific with successive movements on a global scale, while the urbanisation process becomes widespread and cities expand in both size and number. More than ever, urbanism must respond to the challenges of the industrial city (Table 26.1).

26.8 A World of Cities: Conclusion

The urban explosion resulting from the industrial and post-industrial eras has multiplied both the urban population and urbanised spaces. We are faced with a society that is both fully and predominantly urban. The classical attributes used to define the urban (population, activities, morphology, social product, way of life, etc.) extend the concept of the city and incorporate intermediate spaces with complex boundaries, which are sometimes addressed with the confusing use of neologisms referring to both processes and spaces: alongside the more traditional terms peri-urban, rururban or suburban, many others have emerged, such as outer city, edge city, metroplex, technourb, post-suburbia, technopolis, heteropolis, exopolis, cities in between, etc.

Global urban population figures by region reflect the scale of urbanisation. In 1950, the urban population represented 30% of the total and was concentrated in North America, Europe, and East and Southeast Asia. During the 1960s, the increase in urban population accelerated, particularly in larger cities, including those designated megacities. In 1950, there were two megacities (New York and Tokyo); by 1980, there were five (with the addition of Osaka, Mexico City and São Paulo); and by 1990, there were nine (with the addition of Mumbai, Kolkata, Los Angeles and Buenos Aires). (Fig. 26.4).

This increase in the number of the largest cities, with populations exceeding 10 million inhabitants, corresponds with the growth of urban populations across all major regions and subregions in the world. In 2022, approximately 65% of the global population lived in cities. By 2050, it is estimated that the urban population

Table 26.1 Planned city in history: a synthesis

Stages	Key concepts	Cities (Examples)
Origins: near east, Egypt, and Harappan culture	• Orthogonal plan • Use of grid (square, rectangle, etc.)	Dur-Sharrukin, Zernaki Tepe, Akhetaten, Mohenjo-Daro
Greek world	• *Polis* • Grid adapted to the topography • Streets at right angles • Public services (*astynomos*, *agoranomoa*)	Miletus, Prieto, Olbia, Syracuse, Piraeus, Rhodes
Roman empire	• Geometric shapes, orthogonal grid • Square or rectangular perimeter • *Decumanus maximus* and *cardo maximus*, *insulae* • Urbanisation is equated with Romanisation • Urban hierarchies (*coloniae*, *municipio*, *civitates*)	Cologne, Timgad, London, Paris, Vienna
Planned mediaeval city	• Without formal models, but with characteristic urban elements • Creation of new cities • Bastides • Foundation of cities by the Dukes of Zähringen	Aïgues-Mortes, Monpazier, Sauvaterre-de-Guyenne, Bern, Freiburg im Breisgau
Renaissance utopia	• Rationalism, Urban Theory and Ideal City Models • Few New Cities in Europe; Geometry and Grid Patterns • Rectilinear Main Streets (Via Nuova in Genoa) • Major Squares (e.g. San Pietro in Rome, Place de la Concorde in Paris, Covent Garden in London, Potsdamer Platz in Berlin, San Marco in Venice) • New Spanish Cities in America; Orthogonal Plan (*Leyes de Indias*)	Palma Nova, Nancy, Neuf-Brisach, Le Havre, Copenhagen Quito, Lima, Mexico City, Buenos Aires, Santiago de Chile

(continued)

Table 26.1 (continued)

Stages	Key concepts	Cities (Examples)
Baroque monumentalism and perspective	• Urban renewal and beautification take precedence over new constructions • Planned implementation takes precedence over theoretical proposals • Gardens (Versailles, Belvedere, Buen Retiro) • Major monumental avenues (Versailles, Karlsruhe)	New: Saint Petersburg, Washington Internal Urban Reforms: Paris, Vienna, London, Amsterdam, Naples, Rome
Urban planning and the industrial city	• Scientific urbanism (legislation) • Urban reform and city expansion (planned) • Creation of new cities (decongestion) • The city within the territory (regional planning) • Movements (City Garden, New Towns, Satellite Towns, Modern Urbanism, etc.)	Internal Urban Reforms: Paris, Vienna, Budapest, Amsterdam, Bologna Urban expansion: Barcelona, Moscow, Berlin New: Letchworth, Milton Keynes, Vällingby, Brasília, Chandigarh

Source By the author

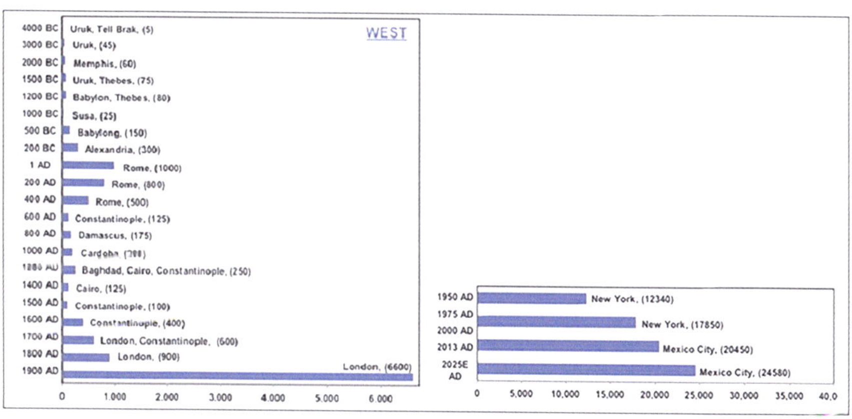

Fig. 26.4 Most populous cities in the West (in thousands of inhabitants) by era. *Source* Morris (2011)

will approach 7 billion, with half residing in Asia. However, Africa is the continent with the highest relative increase (2000–2050), where three of its five subregions are expected to see growth exceeding 500% during this period. Given the issues of inequality and poverty affecting the African continent, this presents an extraordinary challenge for urban planning. In contrast, Europe and North America have mature and highly urbanised models, and therefore have less capacity for growth. Eastern Europe is the only subregion with negative relative values and experiencing a decline in urban population. (Table 26.2).

Table 26.2 Urban population at mid-year by country, 2000–2030-2050 (thousands)

Region, subregion	2000	2030	2050	2000–2050 (%)
WORLD	**2,868,308**	**5,167,258**	**6,679,756**	**132.88**
AFRICA	**285,998**	**824,014**	**1,488,920**	**420.61**
Eastern Africa	54,951	202,579	418,217	661.08
Middle Africa	38,184	133,728	257,254	573.73
Northern Africa	83,357	157,849	230,584	176.62
Southern Africa	28,156	51,909	66,231	135.23
Western Africa	81,350	277,949	516,635	535.07
ASIA	**1,399,722**	**2,802,262**	**3,479,059**	**148.55**
Eastern Asia	635,753	1,222,479	1,290,701	103.02
South-Central Asia	447,089	931,437	1,338,659	199.42
Central Asia	25,366	41,414	57,086	125.05
Southern Asia	421,723	890,024	1,281,572	203.89
South-Eastern Asia	198,781	404,497	526,780	165.01
Western Asia	118,099	243,848	322,919	173.43
EUROPE	**516,725**	**572,890**	**598,857**	**15.89**
Eastern Europe	207,280	203,271	205,314	– 0.95
Northern Europe	73,633	94,053	104,902	42.47
Southern Europe	96,759	112,280	115,021	18.87
Western Europe	139,053	163,286	173,621	24.86
LATIN AMERICA AND THE CARIBBEAN	**397,062**	**600,480**	**685,070**	**72.54**
Caribbean	24,099	35,729	39,824	65.25
Central America	94,801	160,493	194,627	105.30
South America	278,162	404,258	450,619	62.00
NORTHERN AMERICA	**247,471**	**334,780**	**386,690**	**56.25**
OCEANIA	**21,329**	**32,831**	**41,160**	**92.97**
Australia/New Zealand	19,380	29,319	35,386	82.60
Melanesia	1362	2757	4878	258.26
Micronesia	326	421	495	51.75
Polynesia	262	334	400	53.05

Source By the author from United Nations (2018) Revision https://population.un.org/wup/Download/

The extraordinary quality of this volume includes a total of 26 chapters, covering 12 cities across four continents and with contributions from around thirty renowned authors. At first glance, this may seem to be a book on urban history based around case studies. However, it is much more than that, combining theoretical research with highly relevant case studies in urban planning. This ambitious volume manages to interlink deep reflections on the city as a project with analyses of urban history, studies on models, theories and planning across all continents. In addition, emphasis is placed on the major stages in European urbanism due to its global influence, from classical city theory to modern and socialist urban planning.

Acknowledgements Grant project: "Cities in transition. Urban fragmentation and new socio-spatial patterns of inequality in the post-pandemic context. The case of the urban area of Palma (Mallorca)" (PID2021-122410OB-C31), funded by MCIU/AEI/https://doi.org/10.13039/501100011033 and by ERDF, EU.

References

Araújo de Oliveira VM (2022) Urban morphology. An introduction to the study of the physical form of cities. Springer, Cham

Capel H (2002) La morfología de las ciudades. I. Sociedad, cultura y paisaje urbano. Serbal, Barcelona

Costa J (2003) Visiones de la ciudad funcional europea y la ciudad blindada norteamericana en el imaginario del celuloide. Scripta Nova, Revista Electrónica de Geografía y Ciencias Sociales VII:146(037)

Choay F (1983) El Urbanismo. Utopías y Realidades. Lumen, Barcelona

Chueca F (1998) La ciudad del Renacimiento. Breve Historia del Urbanismo. Alianza, Madrid

Galantay EY (1977) Nuevas ciudades. De la Antigüedad a nuestros días. Gustavo Gili, Barcelona

Gravagnuolo B (1998) Historia del Urbanismo en Europa, 1750–1960. Akal, Madrid

Lois-González RC, González-Pérez JM, Escudero-Gómez LA (2012) Los espacios urbanos. El estudio geográfico de la ciudad y la urbanización. Biblioteca Nueva, Madrid

Morris I (2011) Why the west rules—for now. Profile Books, London

Niney F (1994) La ciudad ocupa el 7º Arte. Metrópolis de Fritz Lang. In: Dusser P et al (eds) Visiones Urbanas. La ciudad del artista. La ciudad del arquitecto. Centre de Cultura Contemporània de Barcelona and Sociedad Editorial Electa España. Madrid

Panerai P, Castex J, Depaule JC, Samuels I (2004) Urbans forms—the death and life of the urban block. Architectural Press, London

Rosenau H (1986) La ciudad ideal. Su evolución arquitectónica en Europa. Alianza Editorial, Madrid

Rossi A (1982) La arquitectura de la ciudad. Gustavo Gili, Barcelona

Satoh S (2003) Tokyo. In: Kim KJ (ed) Development pattern and density of selected world cities. Seoul Development Institute, Seoul

Tellier LN (2009) Urban world history. An economic and geographical perspective. Springer, Cham

United Nations, Department of Economic and Social Affairs, Population Division (2018) World urbanization prospects: the 2018 revision. https://population.un.org/wup/Download/

White R (2012) Canadian regional science association/Association canadienne des sciences regionales 35(1/3):51–52

Zoido F, de la Vega S, Piñeiro Á, Morales G, Mas R, Lois RC, González-Pérez JM (2013) Diccionario de urbanismo, geografía urbana y ordenación del territorio. Cátedra, Madrid

MIX
Papier aus verantwortungsvollen Quellen
Paper from responsible sources
FSC® C105338

If you have any concerns about our products,
you can contact us on
ProductSafety@springernature.com

In case Publisher is established outside the EU,
the EU authorized representative is:
Springer Nature Customer Service Center GmbH
Europaplatz 3, 69115 Heidelberg, Germany

Printed by Libri Plureos GmbH
in Hamburg, Germany